Anthropomorphism,
Anecdotes,
and
Animals

SUNY Series in Philosophy and Biology
David Edward Shaner, editor, Furman University

Anthropomorphism,
Anecdotes,
and
Animals

Edited by

Robert W. Mitchell,
Nicholas S. Thompson,
and
H. Lyn Miles

State University of New York Press

Published by
State University of New York Press, Albany

For information, address State University of New York Press,
State University Plaza, Albany, N.Y., 12246

Production by Marilyn P. Semerad
Marketing by Bernadette LaManna

Library of Congress Cataloging-in-Publication Data

Anthropomorphism, anecdotes, and animals / edited by Robert W.
  Mitchell, Nicholas S. Thompson, and H. Lyn Miles.
      p.   cm. — (SUNY series in philosophy and biology)
   Includes bibliographical references and index.
      ISBN 0-7914-3125-8 (hardcover : alk. paper). — ISBN 0-7914-3126-6
(pbk. : alk. paper)
      1. Animal psychology. 2. Animal behavior—Anecdotes.
   3. Anthropomorphism. I. Mitchell, Robert W., 1958–   .
   II. Thompson, Nicholas S. III. Miles, H. Lyn, 1944–   .
   IV. Series.
   QL785.A74 1996
   591.51—dc20   97-26611
                                                                          CIP

10 9 8 7 6 5 4 3 2 1

We dedicate this book to
Judith A. Breuggeman,
Ernest Thompson Seton,
and Robert C. Miles

Contents

List of Illustrations   xi

Foreword   xiii
Frans B. M. de Waal

Acknowledgments   xix

PART I. ATTITUDES, HISTORY, AND CULTURE

1. Taking Anthropomorphism and Anecdotes Seriously   3
Robert W. Mitchell, Nicholas S. Thompson, and H. Lyn Miles

2. Dogs, Darwinism, and English Sensibilities   12
Elizabeth Knoll

3. Why Anthropomorphism Is *Not* Metaphor:
Crossing Concepts and Cultures in Animal Behavior Studies   22
Pamela J. Asquith

PART II. THE NATURE OF ANTHROPOMORPHISM

4. Amorphism, Mechanomorphism, and Anthropomorphism   37
Emanuela Cenami Spada

5. Anthropomorphism: A Definition and a Theory   50
Stewart Elliott Guthrie

6. Why Anthropomorphize?
Folk Psychology and Other Stories   59
Linnda R. Caporael and Cecilia M. Heyes

PART III. ANTHROPOMORPHISM AND
        MENTAL STATE ATTRIBUTION

7. Anthropomorphism and the Evolution of Social Intelligence:
   A Comparative Approach   77
   Gordon G. Gallup, Jr., Lori Marino, and Timothy J. Eddy

8. *Panmorphism*   92
   Daniel J. Povinelli

9. Anthropomorphism and Scientific Evidence
   for Animal Mental States   104
   Hugh Lehman

10. Anthropomorphism in Mother-Infant Interaction:
    Cultural Imperative or Scientific Acumen?   116
    Robert L. Russell

PART IV. ANECDOTES AND ANTHROPOMORPHISM

11. Anecdote, Anthropomorphism, and Animal Behavior   125
    Bernard E. Rollin

12. What's the Use of Anecdotes?
    Distinguishing Psychological Mechanisms
    in Primate Tactical Deception   134
    Richard W. Byrne

13. Anthropomorphic Anecdotalism As Method   151
    Robert W. Mitchell

14. A Pragmatic Approach to the Inference of Animal Mind   170
    Paul S. Silverman

PART V. INTENTIONALITY

15. Varieties of Purposive Behavior   189
    Ruth Garrett Millikan

16. Expressions of Mind in Animal Behavior   198
    Colin Beer

PART VI. CONSCIOUSNESS AND SELF-CONSCIOUSNESS

17. Self-Awareness, with Specific References
to Coleoid Cephalopods   213
Martin H. Moynihan

18. Silent Partners? Observations on Some Systematic Relations
among Observer Perspective, Theory, and Behavior   220
Duane Quiatt

19. Common Sense and the Mental Lives of Animals:
An Empirical Approach   237
Harold A. Herzog and Shelley Galvin

20. Amending Tinbergen: A Fifth Aim for Ethology   254
Gordon M. Burghardt

21. A Phenomenological Approach to the Study
of Nonhuman Animals   277
Kenneth J. Shapiro

22. Anthropomorphism, Anecdotes, and Mirrors   296
Karyl B. Swartz and Siân Evans

PART VII. COGNITION

23. Cognitive Ethology: Slayers, Skeptics, and Proponents   313
Marc Bekoff and Colin Allen

24. Animal Cognition Versus Animal Thinking:
The Anthropomorphic Error   335
Hank Davis

25. Anthropomorphism Is the Null Hypothesis
and Recapitulationism Is the Bogeyman in
Comparative Developmental Evolutionary Studies   348
Sue Taylor Parker

PART VIII. LANGUAGE

26. Anthropocentrism and the Study of Animal Language   365
Judith Kiriazis and Con N. Slobodchikoff

27. Pinnipeds, Porpoises, and Parsimony: Animal Language
     Research Viewed from a Bottom-up Perspective   370
     Ronald J. Schusterman and Robert C. Gisiner

28. Anthropomorphism, Apes, and Language   383
     H. Lyn Miles

PART IX. COMPARING PERSPECTIVES

29. Anthropomorphism and Anecdotes:
     A Guide for the Perplexed   407
     Robert W. Mitchell

List of Contributors   429

References   435

Indexes   499

List of Illustrations

FIGURES

19-1. Mental states of ants, mice, and dolphins 243

19-2. Different species' possession of moderate
or greater levels of consciousness 245

19-3. Capacity of 18 species to experience mental states 246

19-4. Mean factor scores on the cognition factor 248

19-5. Mean factor scores on the affect factor 249

19-6. Mean factor scores on the sentience factor 249

24-1. The error of affirming the consequent 337

24-2. Rat "noses" her way into tunnel 344

25-1. Family tree of primates 358

25-2. Fifth and sixth stages of sensorimotor intelligence
series in primate phylogeny 360

27-1. Path diagram for development of repertoire
of standard combinatorial forms 374

TABLES

18-1. Research strategies, applications, perspectives 227

19-1. Attributions questionnaire 241

19-2. Item loadings from principal components analysis   246

19-3. Contributions of gender and anthropomorphism
       to attitudes toward animal welfare   247

20-1. An amended five aims of ethology   257

25-1. Taxonomic distribution of mental abilities among primates   357

27-1. Training stages in the artificial language   372

27-2. Sequential pairings of signs   373

Foreword

Frans B. M. de Waal

Often, when human visitors walk up to the chimpanzees at the Yerkes Field Station, an adult female named Georgia hurries to the spigot to collect a mouthful of water before they arrive. She then casually mingles with the rest of the colony behind the mesh fence of their outdoor compound, and not even the best observer will notice anything unusual about her. If necessary, Georgia will wait minutes with closed lips until the visitors come near. Then there will be shrieks, laughs, jumps, and sometimes falls, when she suddenly sprays them.

This is not a mere "anecdote" as Georgia does this sort of thing predictably, and I have known quite a few other apes good at surprising naive people. And not only them. Hediger (1955), the great Swiss zoo biologist, recounts how even when he was fully prepared to meet the challenge, paying attention to the ape's every move, he nevertheless got drenched by an old chimpanzee with a lifetime of experience with this game.

Once, finding myself in a similar situation with Georgia (i.e., aware that she had gone to the spigot and was sneaking up on me), I looked her straight into the eyes and pointed my finger at her, warning, in Dutch, "I have seen you!" She immediately stepped away and let part of the water drop, swallowing the rest. I certainly do not wish to claim that she understands Dutch, but she must have sensed that I knew what she was up to, and that I was not going to be an easy target.

The curious situation in which scientists who work with these fascinating creatures find themselves is that they cannot help but

interpret many of their actions in human terms, which then automatically provokes the wrath of philosophers and other scientists
who work with domestic rats, or pigeons, or not with animals at all.
Unable to speak from firsthand experience, these critics must feel
confident indeed when they discard accounts by primatologists as
anthropomorphic, and explain how anthropomorphism is to be
avoided.

Although no reports of spontaneous ambush tactics in rats have
come to my attention, these animals could conceivably be trained with
patient reinforcement to retain water in their mouth and stand amongst
other rats. And if rats can learn to do so, what is the big deal? The message of the critics of anthropomorphism is something along the lines of
"Georgia has no plan; Georgia does not know that she is tricking people; Georgia just learns things faster than a rat." Thus, instead of seeking
the origin of Georgia's actions within her, and attributing intentions to
her, they propose to seek the origin in the environment and the way it
shapes behavior. Rather than being the designer of her own disagree-
able greeting ceremony, this ape fell victim to the irresistible rewards of
human surprise and annoyance. Georgia is innocent!

But why let her off the hook that easily? Why would any human
being who acts this way be scolded, arrested, or held accountable,
whereas any animal, even a species that resembles us so closely, is considered a mere passive instrument of stimulus-response contingencies?
Inasmuch as the absence of intentionality is as difficult to prove as its
presence, and inasmuch as no one has yet demonstrated that animals
differ fundamentally from people in this regard, it is hard to see the
scientific basis for such contrasting assumptions. Surely, the origin of
this dualism is to be found partly outside science.

The dilemma faced by behavioral science today, and the grand
theme underlying the present volume, can be summarized as a choice
between cognitive and evolutionary parsimony (de Waal, 1991). *Cognitive parsimony* is the traditional canon of American behaviorism that
tells us not to invoke higher mental capacities if we can explain a phenomenon with lower ones. This favors a simple explanation, such as
conditioning of a response, over a more complex one, such as intentional deception. *Evolutionary parsimony,* on the other hand, considers
shared phylogeny. It posits that if closely related species act the same,
the underlying mental processes are probably the same, too. The alternative would be to assume the evolution of divergent processes for
similar behavior; a wildly uneconomic assumption for organisms with
only a few million years of separate evolution. If we normally do not
propose different causes for the same behavior in, say, dogs and wolves,

why should we do so for humans and chimpanzees, which are genetically as close, or closer?

In short, the cherished principle of parsimony has taken on two
faces. At the same time that we are supposed to favor low-level over
high-level cognitive explanations, we also should not create a double
standard according to which shared human and ape behavior is
explained differently. If accounts of human behavior commonly invoke
complex cognitive abilities—and they most definitely do (Michel,
1991)—we must carefully consider whether these abilities are perhaps
also present in apes. We do not need to jump to conclusions, but the
possibility should at least be allowed on the table.

Even if the need for this intellectual breathing room is most
urgently felt in relation to our primate relatives, it is limited neither to
this taxonomic group nor to apparent instances of complex cognition.
Students of animal behavior are faced with a choice between classifying
animals as automatons or granting them volition and information-pro-
cessing capacities. Whereas one school warns against assuming things
we cannot prove, another school warns against leaving out what may be
there: even insects and fish come across to the human observer as internally driven, seeking, wanting systems with an awareness of their surroundings. Inasmuch as descriptions from the latter perspective place
animals closer to us than to machines, they adopt a language we cus-
tomarily use for human action. Inevitably, these descriptions sound
anthropomorphic.

The authors in this volume represent different points on a continuum between acceptance and rejection of anthropomorphism. Of
course, if anthropomorphism is defined as the *misattribution* of human
qualities to animals, no one wishes to be associated with it. But much of
the time, a broader definition is employed, namely the description of
animal behavior in human hence intentionalistic terms. No proponent
in his or her right mind would propose to do so uncritically, and even
the staunchest opponent will generally recognize the value of anthropomorphism as a heuristic tool. It is this use of anthropomorphism as a
means to get at the truth, rather than an end by itself, that distinguishes
its use in science from that by the layperson. The ultimate goal of the
anthropomorphizing scientist is emphatically *not* to arrive at the most
satisfactory projection of human feelings onto the animal, but rather at
testable ideas and replicable observations.

This requires great familiarity with the natural history and special
traits of the species under investigation, and an effort to suppress the
questionable assumption that animals feel and think like us. Someone
who cannot imagine that ants taste good cannot successfully anthro-

pomorphize the anteater. So, in order to have any heuristic value at all, anthropomorphism must respect the peculiarities of a species while framing them in a language that strikes a chord in the human experience. Again, this is easier to achieve with animals close to us than with animals, such as dolphins or bats, that move through a different medium or perceive the world through entirely different sensory systems. Appreciation of the diversity of *Umwelten* (von Uexküll, 1909) in the animal kingdom remains one of the major challenges to the student of animal behavior.

The debate about the use and abuse of anthropomorphism, which for years was confined to a small academic circle, has recently been thrust into the spotlight by two books: Kennedy's (1992) *The New Anthropomorphism*, and Marshall Thomas's (1993) *The Hidden Life of Dogs*. Kennedy reiterates the dangers and pitfalls of assuming higher cognitive capacities than can be proven, thus defending cognitive parsimony. Marshall Thomas, on the other hand, does not make any bones about the anthropomorphic bias of her informal study of canid behavior. In her best-seller, the anthropologist lets virgin bitches "save" themselves for future "husbands" (i.e. ignore sexual attentions prior to meeting a favorite male, pp. 56–57), watches a wolf set out for the hunt without "pitying herself" (p. 39), and looks into her dogs' eyes during a vicious gang-attack seeing "no anger, no fear, no threat, no show of aggression, just clarity and overwhelming determination" (p. 68).

There is quite a difference between the use of anthropomorphism for communicatory reasons, or as a tool to generate hypotheses, and anthropomorphism that does little else than project human emotions and intentions onto animals without an attempt at justification, explication, or serious investigation. The uncritical anthropomorphism of Marshall Thomas is precisely what has given the practice a bad name, and has led critics to oppose it in all its forms and disguises. Rather than let them throw out the baby with the bath water, however, the only question that needs to be answered is whether a certain dose of anthropomorphism, used in a critical fashion, helps or hurts the study of animal behavior. Is it something that, as Hebb (1946) already noted, allows us to make sense of animal behavior, and as Cheney and Seyfarth (1990, p. 303) declare, "works" in that it increases the predictability of behavior? Or is it something that, as Kennedy (1992) and others argue, needs to be brought under control, almost like a disease, because it makes animals into humans, something they evidently are not?

I am grateful to Robert Mitchell and Emanuela Cenami Spada for comments. Writing of this paper was supported by a grant of the National Institutes of Health (RR-00165) to the Yerkes Regional Primate Research Center.

Acknowledgments

This book was developed from an invited paper session at the 1989 National Animal Behavior Society meeting held at the State University of New York at Binghampton (although currently less than half of the chapters in this book derive from papers presented at that session). We are grateful to the Animal Behavior Society for (again!) serving as the vehicle for discussion of an intriguing phenomenon, which eventually turned into a book. As the invited paper session was organized to provide a forum for a thoughtful discussion of controversial ideas about psychological interpretations of animal behavior presented in Donald R. Griffin's (1976) *The Question of Animal Awareness* and Andy Whiten and Dick Byrne's (1988) "Tactical deception in primates," we give these authors our thanks for their partisan presentation of controversial methods of interpretation.

We are also grateful for the support, both economical and psychological, provided by the Department of Psychology at Eastern Kentucky University, the Department of Psychology at Clark University, and the Department of Sociology, Anthropology and Geography at the University of Tennessee at Chattanooga. Daniel J. J. Ross offered counsel and support when no one was interested, several anonymous reviewers offered helpful suggestions for improving the book, and Christine Worden and Marilyn P. Semerad offered thoughtful and congenial assistance, for which we are grateful. For their quick action in a crunch, we thank Pam Asquith, Gordon Burghardt, Janet Kazimir, Cathy Clement, Carol Banks, and Pat New. And for their continued interest in the book, editorial assistance, affection, and creation of a congenial work environment, we thank Sue Parker, Randy Huff, Stephen Harper, and Penny Thompson.

We also thank the following publishers and copyright holders for permission to quote from the particular works cited (works are fully represented in the references at the end of the book): C. S. Chihara and J. A. Fodor (1965/1981), from J. A. Fodor (Ed.), *Representations*, with permission from The MIT Press; Gordon G. Gallup, Jr. (1970), Excerpted with permission from Gordon G. Gallup, Jr., (1970). "Chimpanzees: Self-recognition." *Science, 167*, 86–87. Copyright © 1970 American Association for the Advancement of Science; J. Bazar (1980), "Catching Up with the Ape Language Debate," reprinted by Permission of the American Psychological Association; W. Lance Bennett and Martha S. Feldman, *Reconstructing Reality in the Courtroom*, copyright © 1981 by Rutgers, The State University. Reprinted by permission of Rutgers University Press; Thomas A. Sebeok, "Apespeak," reprinted from *The Sciences, 22*, p. 4, copyright © 1981 by Thomas A. Sebeok; P. Carruthers (1989), "Brute Experience," with permission from *The Journal of Philosophy*; K. B. Swartz and S. Evans (1991), "Not All Chimpanzees (*Pan troglodytes*) Show Self-Recognition," copyright © 1991 by Japan Monkey Center, Aichi, Japan, reprinted from *Primates* with permission of Masao Kawai; J. Alcock (1992), "Consciousness Raising III," with permission from *Natural History* (September, 1992), Copyright © 1992 The American Museum of Natural History (1992); H. Cronin (1992), "What Do Animals Want?," Copyright © 1992 by The New York Times Company. Reprinted by permission; D. Jamieson and M. Bekoff (1992b), "Some Problems and Prospects for Cognitive Ethology," with permission from *Between the Species*, where it was first published; J. S. Kennedy (1992), *The New Anthropomorphism*, copyright © Cambridge University Press 1992. Reprinted with the permission of Cambridge University Press; Gordon G. Gallup, Jr. (1994), "Self-Recognition: Research Strategies and Experimental Design," copyright © Cambridge University Press 1994. Reprinted with the permission of Cambridge University Press. In addition, we thank several journal editors for permission to quote from written comments made about a manuscript under review: H. Jane Brockman (editor of *Ethology*), Masao Kawai (editor of *Primates*), Charles T. Snowdon and the American Psychological Association (editor and publisher, respectively, of the *Journal of Comparative Psychology*), and several anonymous editors and reviewers.

Finally, we thank Paddington Press for their 1977 reprinting of the illustrations of J. J. Grandville, which we have exhibited throughout this book. Grandville's delightfully anthropomorphic depictions of animals, now in the public domain, originally complemented the 1842 French novel by Pierre Jules Hetzel translated as *The Public and Private Life of Animals*.

PART I

Attitudes, History, and Culture

1

Taking Anthropmorphism and Anecdotes Seriously

Robert W. Mitchell, Nicholas S. Thompson,
and H. Lyn Miles

This book results from a desire to discuss what are often depicted as twin demons: anthropomorphism and anecdotalism. Although positions against the use of anthropomorphism and anecdotes have seemed an institutionalized scientific doctrine, attitudes toward these approaches to understanding animals are changing within the scientific community. We, as editors, wish to provide a forum for the discussion of scientists' often divergent, yet arguably scientific, views toward anthropomorphism and anecdotes, and to that end we asked researchers studying animals, as well as those studying anthropomorphism and philosophy of science, to present their views. The result is a cornucopia of ideas about the nature of humans, other animals, and the relation between the two.

Because the scientific attitude toward anthropomorphism and anecdotes in the interpretation of animals is profoundly ambivalent, we believe that the topic should be discussed as openly as possible. We ourselves are divided in our attitudes toward anthropomorphism and anecdotal evidence, ranging among mentalism, pragmatic anthropomorphism, and philosophical behaviorism (cf., e.g., Mitchell, 1992, 1993c; Mitchell & Hamm, in press; Miles, this volume; Thompson, 1994), with the result that, as editors, we have tried not to interfere with authors' viewpoints. Each author presents his or her own thoughtful, well-argued personal vision. Not surprisingly, beliefs that one author takes for granted are criticized by another; claims elaborated by one author may be briefly critiqued by another. This approach does not lead to uniformity in perspective; quite the contrary. We ourselves were astounded at the range and complexity of the issues when examined both within and between chapters. This book, then, should not be looked upon as the final answer to an old problem. Rather, it should be used as a means of thinking about how we interpret animals, including ourselves, by examining arguments, concepts, history, and science. To that end, the book takes the reader on a journey of ideas.

Although anthropomorphism and anecdotes have been used throughout the history of science (Agassi, 1973; Guthrie, 1993; Mitchell, 1996), the methodically anthropomorphic analysis of anecdotes of animals had its heyday with Darwin's and Romanes's attempts to explain human psychology through evolution from nonhuman psychology. Knoll examines this fascinating period by relating scientific ideas to cultural conventions, suggesting that evolutionary theory was succored by anthropomorphic attitudes toward animals. Although Darwin's theory predicts psychological continuity between humans and great apes (and perhaps other primates), Darwin and Romanes, in their attempts to provide grand support for the theory, sought and found evidence of psychological continuity between humans and all animals, most particularly dogs! Whereas English sensibilities revolted at the idea of being evolved from a creature such as an ape, they felt solace in kinship with the devoted and faithful family dog. Taking Knoll's ideas one step further by depicting the different paths taken by Western and Japanese comparative psychologists and ethologists, Asquith elaborates the argument that what is labelled "anthropomorphism" depends upon cultural and historical conventions about the place of humans in nature. In writing so, she argues against her former position (Asquith, 1984) that anthropomorphic language about animals is metaphorically related to language distinctly about humans. We use "anthropomorphic" language to characterize animal behavior that is like human behavior in certain salient ways. Because we do not know whether the implications of "anthropomorphic" language—e.g., the intentional agency implied in saying that A "threatens" B—are true of animals or even of all humans so depicted, the relationship between the language and the literal meaning is ambiguous rather than metaphorical.

Although some researchers (e.g., Kennedy, 1992) believe that one should strive for more neutral descriptions of animal behavior so as to avoid anthropomorphism's often uncertain implications, Cenami Spada argues that the notion that one can distinguish anthropomorphic from neutral descriptions a priori is fallacious because it presumes an "amorphic" perspective—paradoxically, a perspective from no particular point of view—by which to determine which description is more accurate. Rather, recognition of some description as anthropomorphic occurs only *after* one has made presuppositions of what is the correct description, and these presuppositions can themselves be anthropomorphic. For example, behaviorists have presumed that human language is subject to the same laws of learning as other behaviors, and therefore that animals can learn language. But if we presume that language is not subject to the same laws of learning as other behaviors, we see that the behaviorists have made an anthropomorphic extrapolation from humans to nonhumans. Cenami Spada argues that the presumption that animals are machines—mechanomorphism (Caporael, 1986)—is itself just as much an unverified assumption as those designated anthropomorphic.

Cenami Spada characterizes anthropomorphism as nothing more than the residue of assumptions about humans applied to animals. Guthrie believes that anthropomorphism is a more directed expectation, an "involuntary perceptual strategy" by which humans guess or expect (unconsciously) that ambiguous or significant stimuli have a humanlike or human cause or form. Although this expectation can be supported or changed by empirical investigation, it appears spontaneously in all humans because other humans and their activities are so influential in our daily experience. Similarly, animism is present in humans and other animals because other living beings are so important in their lives. Guthrie argues against the ideas that anthropomorphism derives from the comfort it gives us in seeing ourselves everywhere, or by extrapolation from the familiarity we have of ourselves, by pointing out that (among other things) anthropomorphic demons provide little comfort and our self-knowledge is not very deep. Caporael and Heyes offer three theories of anthropomorphism—as a cognitive default (similar to Guthrie's "perceptual strategy"), as a system for coordinating interaction which overlaps across species (similar to Guthrie's "animism"), and as a means of making prevalent certain values toward people and other animals. However, the first two theories seem inadequate—we do not know the parameters of the proposed default mechanism, and anthropomorphic interpretations are often notoriously bad predictors of nonhuman *and* human behavior and therefore unlikely to be useful for coordination between species. In their view, anthropomorphism is used because it transforms our relation to other organisms by talking about them as we do about "other" people.

Gallup, Marino, and Eddy presume, contrary to Guthrie and Caporael and Heyes, that we *do* usefully model the mental states of other organisms based on knowledge of our own mental states. Mental state attribution is, therefore, a special case of anthropomorphism based on familiarity with one's mind. In their view, people constantly transform their behavior to take into account psychological deficiencies in other people and in animals, and other animals do the same—more specifically, great apes do. Gallup et al. interpret the evidence to indicate that chimpanzees understand humans and other chimps by attributing mental states to them. Povinelli reflects on the difficulties posed when one assumes that apes and humans use their own mental states to understand those of others. Not only are different understandings of one's own mental states accessible to humans during their development, but members of diverse species may have understandings of their own mental states which are widely divergent from those of humans.

For many researchers, any psychological characterization of nonhumans is simply unverifiable anthropomorphism. In contrast, Lehman posits that the *failure* to characterize an animal as having the same sort of psychological characterization as a human when the evidence warrants it is a form of anthropomorphism. Lehman argues against the private episode view of mental states, proposing that this view represents an inaccurate view of the nature of perception, and proposes that we can observe mental states in animals just as we observe color or hairiness in them. Russell disagrees, arguing that observational definitions which link psychological states to observable behavior should not be mistaken for the name of the behavior (Cenami Spada presents a similar argument). For example, the term "pain" used in the description of an animal's behavior is not used as a simple referent, but instead organizes the animal's behavior to make it explicable from a particular point of view, a point of view dependent upon the value we place on the animal (as Caporael and Heyes also argue). Russell contends that any description of animal or human behavior depends upon some system of description or frame of reference which is inherently anthropomorphic in that it is a system or frame posited by humans.

Purposiveness, intentionality, cognition, and consciousness are all terms which, when applied to animals, are sometimes depicted as unscientific because they are reputed to be unobservable. Yet, as Rollin points out, the idea that psychological attributions are unscientific because scientists cannot accept unobservables is belied by the fact that physicists, those premier scientists, commonly postulate unobservables and seem to do just fine. Indeed, Millikan notes that some form of purposiveness is necessary if we are to have behavioral description at all. However, we need to distinguish such biological purposiveness from intentional purposes, which are a special form of biological purpose.

Millikan distinguishes necessary conditions for intentional purposes, and further proposes that there can be intentional, cognized purposes which are unlike the kind that humans have, so that attributing intentionality to animals need not be viewed as anthropomorphism at all. Beer elaborates by examining the problems encountered when the different interpretations we apply in our understanding of human intentionality are applied to nonhumans, either metaphorically (as in sociobiology) or literally (as in depicting animal behavior). In contrast to Millikan, Beer believes that our understanding of the place of intentions in animal behavior is uncomfortably vague at present.

The use of anecdotes to interpret animal psychology is commonly viewed as a part of any anthropomorphic method, although the two are not inherently tied. In the context of comparative psychology, the term "anecdote" usually refers to a description of a unique (or infrequent) behavior in a narrative, although at times it also refers to any narrative description of behavior. Rollin argues that the method of anthropomorphic analysis of anecdotes is a reasonable source of knowledge about animal psychology if the interpretation has plausibility, which itself depends upon common sense and background knowledge. Whereas Rollin trusts the psychologically rich accounts of animal behavior by people who work closely with animals, such as farmers and zoo keepers, Byrne puts the accounts of primate deception by primatologists to the test. He combines the use of plausibility, common sense, and background knowledge in his analysis and emphasizes that, although it is plausible that many primate deceptions arise from reinforcement contingencies, for other deceptions this interpretation seems *im*plausible and cognitive interpretations more plausible and therefore more accurate. By contrast, Mitchell argues, using the analogy from courtroom testimony, that plausibility presupposes coherence with a particular story about evidence, which results in at least three problems: two different stories can be equally plausible given the evidence, coherence of evidence within a story rather than evidence per se is taken as indicating truth, and stories often make presumptions which are simply accepted without being recognized as assumptions. Although implausibility may allow one to argue that behavior does not cohere with a given story, as Byrne and others (e.g., Morris, 1986) suggest, plausibility does not guarantee accuracy.

When evidence is found to be implausible within a theoretical story, theorists frequently defend their ideas by positing problems in the methodology or observational techniques by which the evidence was obtained, as Swartz and Evans show (see also Collins & Pinch, 1993). Adequate methodology is clearly important, but most researchers know that no methodology is perfect or produces data which are free from all possible alternative interpretations. Silverman argues that the methods employed to make inferences about animal psychology may be as varied as the concerns of the observer. For example, if one is concerned about avoiding being bitten by a dog, one would more likely choose to use any evidence that would indicate an impending bite than if one is concerned about the sorts of circumstances and behaviors that usually indicate biting by dogs. Independent of de Waal's (1991) similar analysis, Silverman depicts the different research methods—anecdotes, quantitative analysis, ecological experiment, and formal experiment—that are used to make inferences to animal mind, and argues that these different methods serve different ends and demand different criteria. Consequently, the anecdotes which one may find useful in detecting patterns in behavior are not appropriate as evidence in supporting formal scientific theories about psychological mechanisms or processes. Similarly, the anthropomorphism we use to understand a pet or caged animal may not be appropriate for controlling that animal's behavior.

Indeed, increased knowledge of animal behavior per se may effect complex psychological interpretation. As a researcher who has studied primates as well as cephalopods, Moynihan finds anthropomorphism, anecdotes, and storytelling inevitable if one wants to study *any* animal's behavior, regardless of how similar these animals are to humans. Perhaps some self-selection is going on here, as Herzog and Galvin suggest that an interest in the psychological experience of nonhuman animals may have in fact led many ethologists to study animal behavior in the first place. Moynihan describes his own observations and conjectures concerning the remarkable behavior patterns of octopus and squid, and concludes that it is most reasonable to believe that these animals know what they are doing and consciously manipulate their performances to influence others. Similar assumptions are made concerning primate behavior based on theoretical beliefs and methodological practices, and, as with cephalopods, their behavior certainly coheres with this view. Quiatt depicts the influences of evolutionary theorizing about individual selection which depicts animals as agents acting to maximize fitness. This theorizing demanded observation of the behavioral "strategies" of individual organisms, leading to the intensive study of individual animals in focal animal sampling. But the understanding of individual animal's behaviors led researchers to experience the world *as if* from that animal's perspective. Quiatt depicts this experiencing, this perspectival shift, as *managed* by researchers, rather than literally experienced, and believes that its *being* managed is important if observers are to gain the kind of information they need to evaluate individual selection theories. Whether or not the assumption of conscious deliberation and reflective thought presumed by primatologists is accurate awaits empirical investigation.

Although primatologists may be accused of presuming that primate behavior indicates a superior psychology when compared with that of other animals even when evidence is in short supply, undergraduates provide a more ambiguous understanding of nonhuman psychology. When undergraduates are asked to compare the psychological capacities of animals for pain, intelligence, self-awareness, etc., directly to those of humans, Herzog and Galvin, as well as Gallup, Marino and Eddy, found that primates, dogs, cats, and dolphins tend to be viewed as far more similar to humans than are other species such as pigs, elephants, birds, and insects, but all animals are viewed as having psychological capacities to an extent less than that found in humans (see also Nakajima, 1992). On the other hand, Mitchell found that, when provided with a story in which an animal exhibits behaviors indicative of jealousy or hiding feelings, undergraduates make few distinctions in their psychological characterizations of humans, children, chimps, monkeys, dogs, bears, elephants, and otters. Apparently the psychological characterizations students have of mammals change when they are presented with behavior indicative of a complex psychology.

The initiation of renewed interest in the empirical investigation of animal consciousness is rightfully credited to Griffin (1976), but as Burghardt recognizes there exists an earlier tradition of interest starting with Romanes and continuing with von Uexküll and phenomenologists such as Merleau-Ponty. Even within this tradition, it is unclear how one is to investigate consciousness, let alone conscious deliberation, in animals. Burghardt calls for just such an investigation, proposing that understanding the private experience of others be added to the traditional four aims of ethological endeavors—proximate causation, function, development, and evolution. He urges that we begin the study of animal consciousness using any methods available—including anthropomorphism, anecdotes, and empathy—but that we be critical in our interpretations, emphasizing plausibility and empirical investigation of predictions. Burghardt's research method has in part been used by others, with contradictory results as to the adequacy of the methodology. Shapiro follows phenomenology in his reliance in part on a kinesthetic empathy to understand his dog, a method clearly anthropomorphic in its reliance on reflection about an individual human's bodily response to dog behavior. But Shapiro also uses knowledge of dogs and other canids, of American attitudes toward dogs, and of the dog's individual development along with an acute and observant eye to produce a picture of this dog's experience of his world which most dog owners and researchers would find plausible (although see Caporael & Heyes, this volume). By contrast, in the study of self-consciousness in animals using self-directed responses to mirrors as evidence, Swartz and Evans show that, rather than leading to a coherent understanding of animal self-consciousness, anthropomorphism and anecdotalism have been used contradictorily both to support theoretical formulations independent of any direct evidence and to deny evidence against these theoretical formulations. Their account of their difficulties defending evidence and a point of view different from that of the mainstream suggests that the *critical* anthropomorphism and accumulation of anecdotes espoused by Burghardt may be difficult for scientists to achieve.

Given Griffin's (1976, 1984) strong expectation that conscious thought of some sort causes complex behavior in most animals, it is not surprising that his view seems often to be singled out for parody by opponents (such as Davis). Bekoff and Allen examine the criticism and support that the views and research of Griffin and his associates have received, and propose that much of the criticism directed toward them seems based on untested presuppositions rather than on unbiased, scientific evaluation of evidence. Although Bekoff and Allen support the research program of cognitive ethology, they are less sanguine than Burghardt about success in the study of animal consciousness. Davis portrays an even gloomier picture of Griffin's concerns about consciousness, and objects to the presumption that conscious thought influences animal behavior and even human behavior. He posits that behavior is, in most if not all cases, amenable to a description which precludes thought as a cause. Although not completely averse to the idea that thought can cause behavior, Davis suggests that we make sure that our anthropomorphic folk psychological beliefs (e.g., that thought causes behavior) are accurate for humans before we extrapolate them to other species. He proposes that, until then, we employ the methods used to study human cognition. He believes these anthropomorphic methods produce replicable evidence and, in his view, reasonable understandings of the mechanisms which control both human and animal behavior. Parker takes an even bolder stand: that the methods used to research cognitive development in humans be directly applied to study psychological development in nonhumans, and that the developmental sequence observed in human ontogeny be applied to primate evolution. Parker supports these applications by the fact that both evolutionary and ontogenetic advances are epigenetic, deriving from previous stages or abilities, and argues that by this technique we can look for evidence of differences between species and chart the evolution of ontogeny.

Language is often viewed as a bugbear for any cross-species comparisons, in that language is believed to be a species-specific behavioral propensity which bears no comparison with the communication systems of other species. Programs orchestrated to teach nonhumans language are objected to as fraught with unverified anthropomorphic inference and description. Miles argues that researchers studying sign-use in apes are in a paradoxical situation—the ape language research paradigm started from an interest in finding out what would happen if apes were raised in a human environment and treated as much as possible as human children are treated, yet researchers are required to abjure anthropomorphic interpretation of their ape subjects! Miles details the different perspectives various ape language researchers use to understand their subjects, and the special problems these researchers experience as they try (or fail to try) to walk the line between anthropomorphism (real and imagined) and the "objectivity" their critics crave. In contrast, research on language in non-ape species does not start by immersing animals as children in a human environment, but rather uses discrimination learning techniques to teach animals behaviors functionally similar to those of humans. Schusterman and Gisiner object to the assumption that results of such studies with dolphins and sea lions should employ terms based on human language, such as syntax, reference and meaning, and demonstrate that language-like behaviors of these animals are less consistent with a linguistic interpretation than with a behaviorist analysis derived from equivalence relations. Paradoxically, their redescription suggests that many aspects of human language learning are also consistent with this analysis! By contrast with either animal language approach, Kiriazis and Slobodchikoff suggest that scientists should avoid anthropocentric and behaviorist assumptions and instead hypothesize that animal communication systems are remarkably complex in relation to their ecological needs, perhaps comparable to human language, in order to discover how complex they might be.

These various perspectives indicate that the issues of anthropomorphism and anecdotalism are not self-contained, but have ramifications for our understanding of psychology, the nature and methods of scientific enquiry and theory building, ethics, and human nature, as well as the nature of nonhuman organisms. The reader is invited to think through his or her own perspective by reflecting on the often contradictory points of view these observers of science portray.

2

Dogs, Darwinism, and English Sensibilities

Elizabeth Knoll

Darwinism is necessarily grandiose, like a few other world-changing theories: if it could not explain everything, it would collapse quickly into explaining nothing. If Darwin's (1859/1968) argument for evolution by natural selection was to be successful as an encompassing explanatory theory of the development and organization of the varieties of life, it had to cover the most difficult case of all—the human species, and the human species at its best. Darwin and his colleagues had to make a convincing argument that every human moral virtue, religious insight, and intellectual capacity arose from characteristics recognizable in animals, and that those human qualities had developed to their present ascendancy by natural processes alone, without any little boosts from a higher power. From his earliest speculations on evolutionary processes, Darwin recognized the human mind as the key citadel to be captured by a successful theory, and he and his younger lieutenant, George John Romanes, set out to capture it.

Although Darwin barely mentioned humans in *The Origin of Species*, it did not take an especially subtle reader to regard his silence as extremely pointed in this intricately argued, copiously illustrated presentation of a universal and entirely naturalistic mechanism for the development of life. Darwin gave no suggestion—quite the contrary—that anything or anyone could stand outside this process. The one extensive analysis of a psychological question—the chapter on instinct—maintains that animal behavioral patterns could arise from the interplay of random variations and environmental pressures. The implication was that human mental characteristics too might have originated in shaped twitches.

And this is exactly what Darwin believed virtually from the beginning, as excavations in his notebooks show, until the end, as the explicit arguments in *The Descent of Man* (1871/1981) and *The Expression of the Emotions in Man and Animals* (1872) reveal. In both the *Descent* and the *Expression of the Emotions*, Darwin began with the evolutionary premise of continuity: "there is no fundamental difference between man and the higher mammals in their mental faculties" (Darwin, 1871/1981, p. 35). His method of argument was to show proto-human psychological characteristics in animals and to describe them in highly anthropomorphic language. Nothing if not consistent, Darwin found mind a long way down, partly by extending the meaning of mentalistic terms. At one point in one of his early notebooks, for instance, he speculated that plants might be said to have some idea of cause and effect since they showed some habitual action (Gruber, 1974, p. 332). Throughout his writings, Darwin's terminological flexibility helped support his general argument, for he tended to use the fundamental (and many would say, fundamentally different) terms of *instinct* and *reason* in wide and overlapping senses.

Much could be (and has been) said about Darwin's theory of instinct; here I want to focus on his anthropomorphism—as an argument, as a persuasive rhetoric, and as a way of seeing and making sense of enormous behavioral variety. Anthropomorphism was used by virtually all the comparative psychology of the nineteenth century, and well before (Thomas, 1983, pp. 124–142), and was one of the key argumentative strategies of Darwinism. After all, Darwin might have attempted to demonstrate continuity by pointing out the bestial qualities in human beings; there is certainly plenty of evidence for their existence. However, that tactic would not only have been off-putting, it would not have answered any triumphant pointing to unique human abilities and virtues. So Darwin argued for continuity by providing examples of the human qualities in beasts. Thus he could simultaneously face the hard theoretical problem and give what he would call a truer and more cheerful view (see Darwin, 1871/1981, p. 184). I want to suggest that the Darwinian account of the mind—"mental evolution"—was as necessarily English and upper middle class in its emotional culture as the Royal Society for the Protection of Children and Animals and *Black Beauty*, and that certain English emotional responses helped shape the presentations of Darwin's theory and helped determine its plausibility to his audience. It may be, in fact, that the anthropomorphizing encouraged by this culture is necessary to evolutionary theorizing.

Anthropomorphism thus makes *The Descent of Man*, which takes up the intensely controversial implication of his theory that Darwin barely acknowledged in *The Origin of Species*, surprisingly appealing reading. One is lulled and lured into accepting evolution by anthropomorphic descriptions of animal behavior. For example:

> Animals . . . love approbation or praise; and a dog carrying a basket for his master exhibits in a high degree self-complacency or pride. There can, I think, be no doubt that a dog feels shame . . . and something very like modesty when begging too often for food. A great dog scorns the snarling of a little dog, and this may be called magnanimity. Several observers have stated that monkeys certainly dislike being laughed at; and they sometimes invent imaginary offences. In the Zoological Gardens I saw a baboon who always got into a furious rage when his keeper took out a letter or book and read it aloud to him . . . (Darwin, 1871/1981, p. 42)

The early chapters of *The Descent of Man* are thick with descriptions of mutually loyal bands of baboons, dreaming cats, and even "ants chasing and pretending to bite each other, like so many puppies" (Darwin, 1871/1981, p. 39). Most of the illustrative anecdotes star monkeys, apes, and dogs—an orphan-adopting female baboon, a small monkey attempting to protect a human from a larger ape's attack, problem-solving retrievers—until Darwin can say, "It has, I think, now been shewn [*sic*] that man and the higher animals, especially the Primates, have some few instincts in common" (1871/1981, p. 48).

Linking man to the primates makes sense: Darwin hypothesized that they shared a common ancestor and were close relatives. But why all the descriptions of dogs? Any evolutionary relationship between dogs and human beings must be far more distant. A careful reading suggests that Darwin gives dogs a slightly but significantly different characterization from the apes. The primates Darwin describes are clever, but comic, or clever, but sneaky—"too clever by half," as the disapproving English cliché has it.

The dogs, on the other hand, are noble. In most of the anecdotes and adjectives, they have the moral edge. As Darwin says, "Our domestic dogs are descended from wolves and jackals, and though they may not have gained in cunning, and may have lost in waryness [*sic*] and suspicion, yet they have progressed in certain moral qualities, such as in affection, trust-worthiness, temper, and probably in general intelligence" (1871/1981, p. 50). When Darwin seeks for embryonic self-consciousness or philosophizing, he finds it in dogs. "No one supposes that one of the lower animals reflects whence he comes or whither he goes,—what is death or what is life, and so forth. But can we feel sure than an old dog with an excellent memory and some power of imagination, as shewn by his dreams, never reflects on his past pleasures in the chase? and this would be a form of self-consciousness" (p. 62). As if aware of how tenuous the connection sounds, Darwin adds a leveling comparison with the lowest form of humanity: "On the other hand . . . how little can the hard-worked wife of a degraded Australian savage, who uses hardly any abstract words and cannot count above four, exert her self-consciousness, or reflect on the nature of her own existence."

It is dogs, too, and not primates, who are supposed to have some flicker of what we would take to be a distinctively human characteristic: religious feeling—and again, the dogs are not much lower, if at all, than savages.

> The tendency in savages to imagine that natural objects and agencies are animated by spiritual or living essences, is perhaps illustrated by a little fact which I once noticed; my dog . . . was lying on the lawn during a hot and still day; but at a little distance a slight breeze occasionally moved an open parasol . . . every time that the parasol slightly moved, the dog growled fiercely and barked. He must, I think, have reasoned to himself in a rapid and unconscious manner, that movement without any apparent cause indicated the presence of some strange living agent . . . (Darwin, 1871/1981, p. 67)

And a dog who loves his master, even more than a monkey who loves his keeper (judging by Darwin's emphasis), shows the closest equivalent in the animal world to that most complex of emotions, religious devotion (Darwin, 1871/1981, p. 68). Many animals show loyalty to their fellows and even help the helpless with something like altruistic satisfaction—"who can say what cows feel, when they surround and stare intently on a dying or dead companion?" (Darwin, 1871/1981, p. 76). "What a strong feeling of inward satisfaction must impel a bird, so full of activity, to brood day after day over her eggs" (p. 79). Again, however, dogs seem to be especially singled out for something a shade superior:

> Besides love and sympathy, animals exhibit other qualities which in us would be called moral; and I agree with Agassiz that dogs possess something very like a conscience. They certainly possess some power of self-command . . . Dogs have long been accepted as the very type of fidelity and obedience. (p. 78)

Descriptions like this make the continuity thesis intuitively plausible and almost attractive to any gentleman whose dog lies at his feet by his fire. Many of Darwin's readers, even those who followed and intellectually accepted his argument, felt some unease and dislocation at the thought of being descended from, in the famous words of the conclusion, "a hairy quadruped, furnished with a tail and pointed ears, probably arboreal in its habits, and an inhabitant of the Old World. . . . classed amongst the Quadrumana, as surely as . . . the common and still more ancient progenitor of the Old and New World monkeys" (p. 389). All the canine comparisons could tame some of the strangeness of this new idea. If Father Adam and Mother Eve had irretrievably vanished from a Darwinian universe, and we were brought face to face with our grandfather the baboon, at least we could still cling to Cousin Rover.

In the late 1870s, a few years after the publication of *The Descent of Man* and as Darwin neared the end of his life, the crucial topic of mental evolution was taken up by his much younger colleague, G. J. Romanes, to whom Darwin gave all his own notes on the subject (E. Romanes, 1908, pp. 73–74, 133). Romanes discussed the problems and conceivable processes of mental evolution more deeply, extensively, and methodically than Darwin and Darwin tended to defer to his opinions on the subject in their extensive correspondence on many aspects of natural history and evolutionary theory.

Among the subjects much discussed in their correspondence was animal intelligence and its relation to human intelligence. Romanes began lecturing on the subject in the late 1870s and Darwin applauded his successes. The Darwins, father and son, suggested research projects; "Frank says you ought to keep an idiot, a deaf mute, a monkey, and a baby in your house!" Note that in this half-joking (but only half) hypothesizing about mental evolution, both Darwin and Romanes assumed that normal human children, intellectually impaired human children, and monkeys were comparable. Romanes's later books, and in particular a remarkable chart at the frontispiece of one of them, shows that when it came to the evolution of mind, he made explicit what was only implicit in Darwin's anthropomorphic language: the idea that mental evolution proceeded along a single continuum, from lowest to highest, in a Chain of Being which included any given species, and human beings at different stages of historical and individual development. Romanes consciously rejected a view of the animal kingdom as a ladder (Romanes, 1896, p. 7), and described the Darwinian model of species relationships as a thickly branching tree. Nonetheless the assumption that all the qualities that make up mind—from memory and primitive attachment to reasoning, social perception, and altruism—could be ranked on a single continuous line pervaded his theoretical writing. Through the late 1870s and 1880s Romanes discussed mental evolution in public lectures, articles, correspondence, and books: *Animal Intelligence* (1882), *Mental Evolution in Animals* (1883), and *Mental Evolution in Man* (1888).

In making his arguments for smooth gradations in capacity across the animal world and up to man, Romanes shared Darwin's method of piling up anthropomorphically described anecdotes. Darwin's definitions of key terms like instinct and mind and thought were somewhat elastic; Romanes was more explicit and systematic about the meanings of his key terms, several of which he invented. In his theory of mind, Romanes intended to show how simple perceptual associations—those which most animals used—could lead smoothly to new and higher forms of thought, until at last the capacity for complex abstraction was achieved. This process took place in parallel ways, Romanes argued, in the development of the individual human being, in the progress of civilization, and in the evolution of the human species from a phylogenetically lower form.

Romanes shared Darwin's most fundamental premise that all human emotions and intellectual abilities have simpler animal counterparts. He defending the premise of continuity with arguments based on two analogies. The first was an anology between human and animal mental processes. The presence of mind was demonstrated by some adaptive and nonreflexive action, which in being nonreflexive shows that some choice had been made. Mental states of other mental beings could be known only as "ejects"—that is, as "ideal projections of our own mental states" (Romanes, 1883, p. 22). Anthropomorphizing, in short, was a basic and necessary methodological rule.

Unsurprisingly, the descriptions resulting from this rule make the mental life of animals sound human, familiar, and engaging. Romanes described a caterpillar receiving ambiguous stimuli as "puzzled"; he ascribed parental affection to earwigs, and jealousy and anger to fish. He distinguished the degrees of vividness and complexity in the "mental images" formed by limpets, spiders, and salmon to show the various stages in the developing imagination. At one point he observed that "the higher Crustacea . . . are able to imagine in a high degree" (Romanes, 1883, p. 153). An animal capable of dreaming or pining for a missing owner would be, he estimated, "about one-third of the distance between the first dawn of the imaginative faculty and its maximum development in a Shakespeare or a Faraday" (pp. 153–154). (Two *English* geniuses, please note.)

If this basic analogy between one's own mental states and those of other beings is accepted, the anecdotes make Romanes's conclusions about mental continuity convincing, even if the anthropomorphic descriptions sometimes seem to have employed an overly developed imaginative faculty. In any case the books make for delightful reading. As in Darwin's work, but at greater length, dogs have a special moral superiority in the vast animal cast; but other domestic animals may show Victorian virtues of self-sacrifice or philanthrophy. A couple of accounts about the disinterested charity shown by generous middle-class cats to poor stranger cats outside the kitchen door (Romanes, 1883, pp. 344–345) are as affecting (although considerably briefer) as Dickens's (1850/1980, 1853/1964) tales of Agnes Wickfield's or Esther Summerson's good works. By the end of his long life, Darwin had given up on serious literature and preferred popular novels with lovable heroines and happy endings; his own and Romanes's illustrations of mental evolution could provide animal counterparts for some of them.

The second key analogy that Romanes used to defend the premise of continuity was a developmental version of the first one. As animal minds could be assumed to work in ways resembling those of the human mind, so "the development of an individual human mind follows the order of mental evolution in the animal kingdom" (Romanes, 1888, p. 5). A tree-like chart that served as the frontispiece for *Mental Evolution in Man* summarized Romanes's hypotheses about the development of emotion, will, and intellect from the roots of simple nervous excitability, and showed the stages of equivalence between human children and animals.

So, according to the chart, a newborn baby, capable of feeling pleasure and pain, falls somewhere between the Coelenterata, which have the lowest form of consciousness, and the Echinodermata. The Mollusca are capable of association by contiguity and fish of association by similarity: so are babies at the ages of seven weeks and twelve weeks. Reason first appears in the baby at the age of fourteen weeks, making him the intellectual equal of the higher Crustacea; at five months he and the Hymenoptera are capable of communicating ideas and feeling sympathy. The last branch of the tree attributes deceitfulness, a sense of the ludicrous, and "indefinite morality" to a fifteen-month-old child, apes, and (of course) dogs.

Some critics of a Darwinian account of mental evolution had pointed out the enormous gap that existed between the sensuous thought characteristic of animals and the conceptual thought of human beings (Müller, 1873a, 1873b, 1873c, 1887; Mivart, 1889). To overcome this difficulty, Romanes proposed a new classificatory scheme, which added a bridge between perceptual and conceptual thought. He postulated what he called "receptual thought." A *recept* was a compound idea of previous similar perceptions, "an idea which differs from a general or abstract idea only in not being consciously fixed and signed as an idea by means of an abstract name" (Romanes, 1888, p. 36). A recept is, so to speak, an idea you are not aware that you have. In receptual thought, compound ideas joining a number of remembered perceptions could be held by very small children and animals, neither of whom had the articulate language necessary for abstract general ideas. Ideas of this kind could form by association, automatically, without any of the reflection required for higher level abstractions. For instance, Romanes suggested, "all the higher animals have general ideas of 'Good-for-Eating' and 'Not-Good-for-Eating,' quite apart from any particular objects of which either of these qualities happens to be characteristic" (p. 27). The weakness of this notion was pointed out by the caustic critic St. George Jackson Mivart, who observed that since Venus flytraps release indigestible objects or very tiny insects, "we might say that plants have abstract ideas of 'Suitable-for-nutrition' and 'Not-suitable-for-nutrition,' and the still more abstract ideas of 'Big-enough-to-be-worth-a-prolonged-effort' and 'Not-big-enough-to-be-worth-a-prolonged effort'" (Mivart, 1889, p. 49).

Romanes adopted the same tactic in analyzing language as in analyzing thought. He postulated an intermediate step between two extremes, a category between purely instinctual cries and fully articulate language. Outward signs, like inner ideas, could be unintentional in the strict philosophical sense. Yet they could still be classified as signs and not simple reflex responses. Such signs might indicate inferences that were based simply on past associations of action with gratification, as when a dog habitually begs, or by inference from a present circumstance, as when one dog sees another dog beg and imitates him to get a similar reward. Thus Romanes linked human language and other sign systems with the signs that ants, bees, and dogs make to communicate with one another.

On Romanes's interpretation, the lowest form of language showed only associative thinking; talk at this stage was "a verbal gesture" (Romanes, 1888, p. 131). Like very young children, animals understood simple words; their failure to use them to convey simple ideas was due only to the accident that their vocal organs were not formed for speech. Romanes noted reports that dogs had been taught to articulate some words. One accomplished English terrier had supposedly been trained by Alexander Graham Bell to say, "How are you, grandmama" (Romanes, 1888, p. 128).

In all developmental movement from associative to abstract thought, from recepts to concepts, from verbal gestures to reflective language, the crucial differentiating point is the awareness that one is performing certain mental acts. But this form of self-consciousness also developed gradually, in the individual, in the human race, and in the world's species. Although the earliest forms of language might be merely associative, language marked a sort of cognitive turning point; the use of words improved communication with others, and could sharpen awareness of the self, particularly awareness of the self as a thinking being (Romanes, 1888, p. 202).

Romanes had planned to write several books on aspects of mental evolution, all dedicated to "bridging the psychological distance which separates the gorilla from the gentleman" (Romanes, 1888, p. 439). However, other philosophical and scientific writings took his time, and death from a brain tumor cut his life short at age 46. His seminal works of comparative psychology presented man as a sort of Senior Fellow of the animal world and at the same time presented animals as comprehensible, and often reasonable and kindly beings, cognitively and emotionally cruder than man but not radically foreign. If the gentleman reading Romanes's books saw that he was there linked to the gorilla, he could take some comfort in the suggestion that the gorilla was after all a relatively gentlemanly fellow himself.

In Darwin and Romanes's writings there clearly is no sign of the anti-anthropomorphic methodological rule laid down in 1894 by the comparative psychologist C. Lloyd Morgan, a protégé, colleague, and critic of Romanes: "In no case may we interpret an action as the outcome of the exercise of a high psychical faculty, if it can be interpreted as the outcome of the exercise of one which stands lower in the psychological scale" (Morgan, 1904, p. 53). Those with a fondness for neatly organized historical eras might say that Morgan's Canon, as it is called, marks the end of the anthropomorphic strategy in psychology and the beginning of twentieth-century behaviorism.

However, Morgan's Canon is a double-edged sword for any defender of human mental distinctiveness. It can cut up as well as down, to have a reductionistic effect on descriptions of human mentality, emotions, and consciousness. The twentieth-century behavioristic rejection of all mentalistic terminology and concepts is a logical extension of Morgan's Canon: if we cannot anthropomorphize the animals, we cannot anthropomorphize ourselves either.

What general conclusions can we draw from all of this?

First, a point useful in a time of much popularized sociobiology: that the psychological and social arrangements that we perceive in animals are more likely to be projections of our own social conventions and assumptions than timeless and universally recognizable accounts of the way animals really are.

More important is the truth that as theorists we tend to see the social structure and interpersonal strategies we are prepared to see. As Bertrand Russell commented in the late 1920s, observing the rise of different schools of would-be scientific psychology,

> One may say broadly that all the animals that have been carefully observed have behaved so as to confirm the philosophy in which the observer believed before his observations began. Nay, more, they have all displayed the national characteristics of the observer. Animals studied by Americans rush about frantically, with an incredible display of hustle and pep, and at last achieved the desired result by chance. Animals observed by Germans sit still and think and at last evolve the solution out of their inner consciousness. (1927, pp. 32–33)

If this is true, then our second and more important conclusion is that anthropomorphizing may be methodologically dangerous and epistemologically unsound, but psychologically almost irresistible. Cognitively, we need to find patterns in ambiguous situations, most pressingly in those that have some emotional importance for us—as animal behavior has for most people who have any dealings with animals, be we whale trainers at Sea World, pet owners, farmers, or comparative psychologists.

Our best solution, then, may be to take our anthropomorphic descriptions, not as final and established truths, but as inescapable and possibly useful preliminary hypotheses and heuristics. We must then attempt to falsify those hypotheses and push those heuristics till they fail—which they almost certainly will, given the remarkable range of animals' sensory capacities, which enable them to live and thrive in social and physical ecologies remote from human experience. Those falsified hypotheses and failed heuristics will remind us, for all our intuitions and wishes to the contrary, that evolution does not produce a linear Chain of Being, that a gorilla is not a potential Victorian gentleman, that a monkey is not a furry baby, and that even our steadfast, intelligent Rover almost certainly has, unknown and unknowable to us, a secret mental life.

3

Why Anthropomorphism
Is Not Metaphor:
Crossing Concepts and Cultures
in Animal Behavior Studies

Pamela J. Asquith

INTRODUCTION

Anthropomorphic description of animals in the behavioral sciences
has had proponents and opponents for well over a century, with the latter comprising by far the majority (see Fisher, 1991, and Mitchell, 1996,
for recent reviews of many of these arguments). One may be forgiven
for perceiving these debates to go round and round, revisiting the same
points and returning to similar conclusions, though the fact that these
discussions keep reappearing is of some interest in itself. Additionally,
the accumulation of new knowledge about animal behavior, of other
cultures' views and studies of animals, and of language use itself, has
rekindled interest in anthropomorphism. At least two factors have
become clear: first, discussions about anthropomorphism are not themselves objective discussions, but are culturally dependent; second, to
maintain that we can use anthropomorphic terminology metaphorically of all animal species, is misleading. Leaving aside, temporarily,
debates about the existence and nature of mental processes in nonhu-

I thank Bob Mitchell for organizing the meetings that prompted these reflections, and for helpful comments on the chapter.

man species, it is contended here that the appellation *anthropomorphism* rests largely on cultural biases and certain linguistic properties of terms, rather than on correspondence to an agreed-upon standard terminology. The term *anthropomorphism* presupposes reference to uniquely human qualities. However, notions about wherein the differences between humans and other species lie, or what are considered to be the most important differences, vary both historically and among cultures. Thus, we cannot assume that anthropomorphism carries exactly the same connotations at all times or for all scientists. Further, to what, exactly, anthropomorphism refers is unclear. Fisher (1991) has pointed out the logical fallacies in our attributions of anthropomorphism as both a term and a concept. He notes, especially, that we do not have empirical data to support the supposition that intentional vocabulary is *in principle* wrongly applied to animals. That is, we lack evidence that nonhuman animals do not have mental and emotional lives. To this I will add that what we consider to be uniquely human is affected as much by culture and historical fashion as by our ignorance of the animals themselves.

To that end, the first part of this chapter notes historical changes in European and American ideas of what is allowable anthropomorphic terminology in animal behavior studies. It also notes cross-cultural differences between Japanese and Western scientists in establishing the defining characteristics of humanness. These differences have resulted in different conceptions of what is serious or unscientific anthropomorphism between the cultures.

The second half of the chapter raises a linguistic problem. The rather common statement that controlled or careful anthropomorphic description of animals is a metaphorical way of speaking, is not, it will be argued, true. To begin with, as Harris (1984) pointed out, while a statement such as "The ship of state foundered" is a metaphor, the statement "Animal A threatened animal B" is not. Why that is so will be addressed below. Second, although it has been maintained that everything we say is in fact metaphorical (e.g., Booth, 1978; Crider & Cirillo, 1992), or that our conceptual system is fundamentally metaphorical (e.g., Lakoff & Johnson, 1980), the metaphorical use of terms or concepts must presuppose the literal use or meaning. Decisions are made, therefore, consciously or unconsciously by the scientist, about the *appropriateness* of terms that are normally used to describe human behavior (the literal meaning) when applied to animal behavior (and claimed to be metaphorical). Thus, however pervasive metaphor is held to be in either ordinary or scientific discourse, the speaker's beliefs about wherein lies the difference between human and nonhuman will affect what is apprehended as anthropomorphism.

It should come as no surprise that true anthropomorphic metaphors that describe things or processes far removed from the grey areas of human uniqueness, such as in the physical sciences, are considered to be creative heuristic devices (e.g., Hoffman, 1985; Miller, 1992). By contrast, where there is any likelihood of attributing "human-like" intentions to the behavior so described, as in the biological sciences, a much closer scrutiny of terms is called for (e.g., Tamir & Zohar, 1991; Kennedy, 1992). In a rather extreme castigation of anthropomorphism in animal behavior studies, Kennedy (1992) has gone so far as to issue a strong warning about this, so invidious a problem does he perceive it to be:

> Since it has taken many centuries to achieve the present measure of emancipation from vitalism and anthropomorphism, we may . . . see this achievement as something to celebrate. . . . Yet there is no room for complacency. . . . The scientific study of animal behaviour was inevitably marked from birth by its anthropomorphic parentage and to a significant extent it still is. It has had to struggle to free itself from this incubus and the struggle is not over. Anthropomorphism remains much more of a problem than most of today's neobehaviourists believe. (pp. 3–4)

The second half of the chapter, then, follows from the first. Our perceptions of "allowable" anthropomorphisms have changed historically and differ cross-culturally. Insofar as metaphor is grounded in the literal meaning of terms, our beliefs about what can be said literally of animals will matter. Further, the kinds of description of animal behavior that are claimed to be metaphors are simply begging the underlying problem of why, or if, they are anthropomorphic.

ANTHROPOMORPHISM AND HISTORICAL ETHOLOGY

The question of mental continuity as addressed by early psychologists and ethologists in the nineteenth century is relevant to understanding evolving ideas about "anthropomorphic license" in description of animal behavior. Modern psychology is referred to in much the same way as it was in the last century, whereas ethology, though based on the nineteenth-century definition, has come to be applied differently. Thus, John Stuart Mill (1843/1948) defined psychology as the science of the "elementary laws of mind," or a philosophy of human nature. He distinguished this from ethology, which he defined as the "science of character formation" which must be approached deductively. Importantly, both were, he felt, susceptible of scientific inquiry. Charles Darwin's theory of evolution was vital to the development of both a comparative psychology and animal ethology, inspiring the "anecdotal school" or "anthropomorphists" to provide evidence for the mental evolution or continuity between humans and animals (Darwin, 1872; Klopfer & Hailman, 1967; Beloff, 1973; Knoll, this volume).

Following Darwin, George Romanes (1882, 1883) collected both scientific and popular accounts of animal behavior (known as the "anecdotal method" [Thorpe, 1979]). This and other such accounts prompted a call for more careful and critical use of anecdotes, coupled with experimental data. Without such, Lloyd Morgan (1894), for instance, questioned whether it was possible to have a true science of comparative psychology, challenging the definition of instinct and concepts of the nature of consciousness (Gray, 1963; Knoll, this volume). Others were completely opposed to the use of anecdotes, and a school of "objectivists" or "scepticists" developed in reaction to the anthropomorphists at the end of the nineteenth century (Bierens de Haan, 1947a). The physiologists Beer, Bethe, and von Uexküll (1899) in Germany and Whitman (1899) in America represented the first objectivist movement. They did not deny subjective experiences in animals, but did not think these could be scientifically studied. An evaluation of anthropomorphism and science by O. Wheeler (1916) argued that we cannot study anything—other humans, animals, or things—objectively, as objects, but only as *ejects*, or in terms of a thinker's own consciousness of himself. Ejective cognition referred to interpreting someone (or something) else's subjective experiences in terms of one's own subjective experiences. Wheeler noted percipiently that this fact has great repercussions for science. J. B. Watson's development of behaviorism in 1913 gained many supporters in the 1920s and was influential through to the 1950s. Among behaviorists, any explanation of behavior that remotely suggested conscious activity was unacceptable. In psychology and biology there were few exceptions to this tradition and to the related stress on parsimonious explanations in biology (see Griffin [1976, pp. 47–50] for some of these exceptions).

Yet naturalistic animal behavior studies were set against this objectivistic trend. The first ethologists studied set patterns of behavior in animals both to gain a better understanding of the function of stereotyped movements and to trace evolutionary affinities among species. Whitman (1898/1986, 1899), the Heinroths (1911, 1924–1933), and W. M. Wheeler (1902, 1905, 1908) showed than an organism's "ethos" or character is an innate attribute of the species and as diagnostic of a species as its physical form. The ethos or character of animals consisted of their life habits and manners or the sort of behavior which in humans was at that time described as gestural, symbolic, or expressive. These characters could be understood often by direct analogy with human character only after a prolonged and intimate acquaintance with the animals. In these studies both the focus and the approach were subjective.

Early ethologists were not, however, thoroughgoing anthropomorphists. Klopfer and Hailman (1967) noted that European ethology was permeated with an empathetic, albeit not necessarily anthropomorphic, view of animal behavior. Perhaps it was the wealth of objective detail contained therein that prompted Thomas (1979) to write:

> I well remember in the early 1950s the sense of welcome escape from anthropomorphism that accompanied my first reading of Konrad Lorenz's *King Solomon's Ring*, a new orientation reinforced by my perusal of Niko Tinbergen's *A Study of Instinct* about the same time. (p. 3)

Two examples written forty-five years apart illustrate changing attitudes to a broad class of what is acceptable anthropomorphic description of animal behavior. One is by Julian Huxley (1923) who played an important part in the early stages of ethology, and the other by Donald Griffin (1976) who founded cognitive ethology.

Huxley studied the behavior of birds. His observations led him to conclude that birds experience emotion, and do so strongly. In comparing birds with other animals he wrote:

> The variety of their emotions is greater, their intensity more striking, than in four-footed beasts, while their power of modifying behaviour by experience is less, the subjection to instinct more complete. (1923, p. 108)

Hence, the implication of the full quotation from which the title of his article (*"Ils n'ont que de l'âme"*) was taken: *"Ils n'ont pas de cerveau—ils n'ont que de l'âme"* [They have no brains, only soul]. Moreover, Huxley felt that an evolutionary explanation could be found for birds to have acquired a capacity for affection, jealousy, joy, fear, curiosity, and other emotions.

Over four decades later, Griffin (1976, 1978) proposed cognitive ethology, or the study of mental experiences in animals. His stated rationale was that it is likely that mental experiences occur in other animals and that these have important effects on their behavior. His definitions of mental experiences and related terms were problematically vague, but he recognized that mental experiences also include feelings, desires, fears, and "sensations" such as pain, rage, and affection. However, as these are private data, directly observable only to the person who experiences them, he suggested that ethologists should concentrate on mental experiences such as intention (or mental images of future events) and consciousness, and not emotionality. Thus, whereas Huxley maintained that feelings and emotions were important to include in study of animal behavior, Griffin and his school felt that their emotional lives are inaccessible and what we might call thinking was more important.

Ethology and comparative psychology had different roots, but they were brought together by comparative psychologist Robert Mearns Yerkes's advocacy of field studies of primates during the 1920s. Yerkes was interested in especially the great apes as a comparative model for humans. He realized that study of the animals in their natural habitat was essential to understanding their behavior in captivity. Yerkes thus sent his students to pursue naturalistic studies of primates in Africa and South America. The most successful of his students was Clarence Ray Carpenter, and his studies established primate ethology in America.

In discussions of attitudes to anthropomorphism in animal behavior studies, primatology provides a rich source of data. The nonhuman primates also lend themselves to richly textured description due to their sociability, intelligence, and morphological similarities to humans, as well as the concentrated efforts of researchers around the world to record their behavior over the past four decades. Primate studies in Western countries have shown a significant increase in the literal application of previously anthropomorphic terms (so designated by quotation marks in earlier studies, or simply avoided and more neutral synonyms used), such as "reconciliation" (de Waal & Roosmalen, 1979), "friendship" (Smuts, 1985), "social strategies" (Strum, 1987, 1988), "deception" (e.g., de Waal, 1982, 1986; Byrne & Whiten, 1988b; Whiten & Byrne, 1988), and "coalitions" and "alliances" (Harcourt & de Waal, 1992), among others. Such terms would earlier have been called merely heuristically useful metaphors, variously labeled "generic anthropomorphism" (Asquith, 1984), "critical anthropomorphism" (Burghardt, 1985a), or "mock anthropomorphism" (Kennedy, 1992), or a form of catachresis, filling a gap in our vocabulary (Boyd, 1979). Although these terms are carefully defined in scientific reports, this change to a literal use of mentalistic terms appears to reflect an increasing acceptance or a phase of acceptance of animal cognition (Asquith, 1984).

CULTURAL DIFFERENCES

Recent studies of comparative attitudes to animals across cultures (e.g., Kellert & Berry, 1981; Callicott, 1982; Kellert, 1991, 1993) have addressed such topics as assessing peoples' primary goals, interactions, and interests with regard to animals and the environment generally. For instance, based upon measures of attitudinal traits labeled as moralistic (primary concern for the right and wrong treatment of animals), utilitarian (primary interest in the practical value of animals), and naturalistic (primary focus on an interest and affection for wildlife and the outdoors), among others, Kellert (1993) compared national views of the ordinary citizen in the United States, Germany, and Japan toward the environment and wildlife. Others have assessed peoples' perception of animal intelligence, which may also reflect cultural biases (e.g., Nakajima, 1992).

Missing in such assessments is what is thought to be the unique defining characteristic(s) of humanness across cultures. For instance, is a moral sense or religiosity considered peculiar to humans; or is rationality or emotionality thought to be the hallmark of humankind? Each culture's view of such things will have obvious effects on the kind of terms which may be felt to be more "explicitly" anthropomorphic. More than anything else, European and derivative American intellectual history from the beginnings of modern science in the 16th century has been concerned with the centrality of the mind, often discussed interchangeably with the soul, as the defining characteristic of humans.[1] By contrast, the possibility of entering into social relations is central to the concept of personhood among the Ojibwa Indians of North America, and one can do this with "nonhuman persons" (Callicott, 1982, p. 302). For the Japanese, emotionality rather than rationality is suggested as central to the human/animal distinction (Ohnuki-Tierney, 1987, 1995).

As the Japanese have contributed to the international scientific discussion of animal behavior, it is interesting to consider the Japanese viewpoint here. A most distinctive characteristic of the human/animal divide for Japanese is the inability of animals to cry (or laugh). A number of Japanese cultural institutions testify to the importance of emotionality in the culture. From medieval literature discussing whether humans alone among animals can cry; to the popularity of a contemporary genre of folk songs known as *enka*, which are characterized by pathos and sadness; to the classification of movies according to the number of handkerchiefs needed to soak up the tears, with the "three-handkerchief movie" being the best (Ohnuki-Tierney, 1995); to a general aesthetic appreciation of art or natural beauty rendered in terms of one's emotional, not intellectual response, can be found evidence of a different set of assumptions regarding humanness. Similarly, the difference in emphasis on rationality between Japan and much of Europe and North America is seen in the definition of "death" for the purposes of organ transplants. Whereas brain death signifies the death of the rational individual in many Western countries, Japan has resisted that criterion and a person is not considered dead until there is no pulse, regardless of the state of the brain (e.g., Norbeck & Lock, 1987).

One can see immediately where this leads, with persisting debates about the degree, nature, or even existence of mind in animals. Whereas "soul" was used interchangeably with "mind" in Western intellectual writings about the human place in nature, one sense of the soul in Japan has been identified with a distinctly Japanese character throughout the centuries. When the Japanese were concerned with retaining their own culture and outlook while importing advanced technologies from the highly developed civilizations of China during the 5th and 6th centuries and of the West at the end of the 19th century, this concept was expressed in such slogans as *wakon kansai* (Japanese soul, Chinese brilliance) or *wakon yôsai* (Japanese soul, Western brilliance). Thus, while adopting foreign techniques, they guarded their own identity, identifying themselves with the Japanese soul, while crediting the Chinese and Westerners with *sai*, that is, brilliance, genius, or talent. This context for *soul* represents a concept closer to emotionality than to rationality in Japan (Ohnuki-Tierney, 1995).

Some of the effects of different conceptions of the human/animal relationship on Japanese and Western studies of primates have been noted (Asquith, 1986). Japanese reports about animals' motives, personalities, and lives were, in their Western colleagues' eyes, highly anthropomorphic. As rationality is so central to the Western debate about human uniqueness, it is not surprising that the strongest invectives against anthropomorphism are about attributing rationality to other animals. Emotionality for the Westerner comprises a subset of arguments about rationality and, as mentioned, there is not universal agreement about it, even among scientists. To the Japanese researchers, questions about the rational uniqueness of humans did not arise and their reports were filled with mentalistic language. Western response to such reportage as unscientific, and hence dismissable, resulted in more than two decades' lag behind the Japanese in certain theoretical developments in primatology (Asquith, 1996).

LANGUAGE, METAPHOR, AND ANTHROPOMORPHISM

With the emphasis this century to include effects of language use itself in analyses of ways of knowing, metaphor is held to be one of the most powerful aspects of language in description, theory formation, and knowing itself. Anthropomorphism in description of animal behavior is often construed to be a metaphorical way of speaking (e.g., Purton, 1970; Stebbins, 1993), sometimes employed for heuristic purposes, or because there is no alternative: it is simply the most appropriate language to use in describing animals' behavior. That is, overly neutral language, such as a purely physical description of behavior, ceases to be meaningful. This construal implies that purposive terms are not being used literally of animals' behavior but serve rather as a convenient shorthand. Yet these apologies for mentalistic terms simply beg the question of why we consider them to be anthropomorphic.

It is argued here that anthropomorphism in animal behavior reports is not always or even usually an instance of metaphor. This point will entail discussion both of what it is we are trying to say about animals when we describe them in so-called anthropomorphic terms, as well as the contention that (*pace* Booth, 1978 and Crider & Cirillo, 1992) we cannot consider *all* language to be metaphorical.

In more than 2,000 years of study, metaphor has withstood all attempts to develop a stringent theory. Today, metaphor is no longer deemed illicit and a violation of scientific discourse requirements of clarity, precision, and verifiability. Rather, it is recognized as one of the deepest and most persistent phenomena of theory building and thinking. The classical distinction between literal and figurative use of language which the Vienna Circle and logical empiricists used was never repudiated, but the apparently obvious merits of literal speech shrivelled upon closer study. With Max Black's 1962 publication of *Models and Metaphors*, the awareness of metaphor as deserving linguistic and philosophical attention grew. Although Black himself did not think that metaphorical language was appropriate in science, he laid the foundation for later work that showed how pervasive and indeed essential metaphor is in science (e.g. Boyd, 1979; Paprotte & Dirven, 1985).

To propose that science itself depends on metaphor may seem odd. For instance, how can something like an engineer's equations for the behavior of stressed concrete relate in any way to metaphor? But they do: the equations come from a metaphor which likens concrete (of all things) to springs (Hoffman, 1985). The use of metaphor in science is of immense practical relevance. The importance of language itself to "knowing" has made the traditional Cartesian dualism of mind vs. matter, or later, subjective vs. objective, yield to a threefold structure: subject, object, and linguistic medium, which play intercausative roles in the formation of what we call reality. Perhaps partly due to the ubiquity of metaphor, anthropomorphic description in ethology is too hastily quoted as an example of it. The distinction between literal and metaphorical meaning is fundamental to any discussion of metaphor. This is necessary due to a tendency in some recent publications to relegate literal meanings to limited cases rather than the other way around when speaking of metaphor (e.g. Crider & Cirillo, 1992; Hesse, 1988).

Two distinctions can be drawn between literal and metaphorical meanings. First, literal meaning is that which is agreed upon by speakers with a common language. The meaning can only be judged right or wrong with reference to common or accepted usage. Literal meanings can, therefore, change. By contrast, a metaphorical word cannot be corrected by referring to proper usage—it can only be criticized as inappropriate or inept. A metaphor will only be appropriate if the meaning of the word used metaphorically can somehow be associated with at least some of the literal meanings of the word used in a metaphorical way, or with other words in the sentence.

Second, metaphorical meaning is parasitic on the literal—that is, the force of the metaphor is derived partly from the literal meaning of the word, but no literal meaning is derived from the metaphorical (except in the case of catachresis). For instance, in the metaphor "cotton-wool clouds," looking white and fluffy are striking characteristics of cotton wool, which, when applying the literal word cotton wool, we take as signs for all the rest of the characteristics of cotton wool. But when applied metaphorically to clouds, we no longer use them as signs. The things called by the literal word "cotton wool" have the striking characteristics and a lot more, whereas the things called by the metaphor have only the striking characteristics. This also holds when no words used literally appear in the metaphor, for example, "a mantle of sorrow" (Crider & Cirillo, 1992, p. 181). A metaphor is a word for one sort of thing applied to a different sort of thing. A metaphorical word has a certain semantic relation to the literal meaning of that word and to the other literal words in a sentence. That association, or between what the association is thought to hold, has formed the basis of different views of metaphor (such as the comparison view or interaction view of metaphor).

Many refinements on the concept and function of metaphor are usefully being produced. Lakoff and Johnson (1980) convincingly argued that our ordinary conceptual system in which we couch our thoughts is fundamentally metaphorical. For instance, the metaphor "Time is money" pervades our way of speaking about time in ways of which we are probably quite unconscious. They provide many examples of the ubiquity of this metaphor and how it structures our thought, as in:

> You're *wasting* my time.
> This gadget will *save* you hours.
> I don't *have* the time to *give* you.
> How do you *spend* your time these days?
> That flat tire *cost* me an hour . . .
> (Lakoff & Johnson, 1980, pp. 7–8)

Nevertheless, it is important to keep these concepts of metaphorical and literal language distinct. The essence of metaphor is the understanding and expression of one kind of thing in terms of another. Metaphorical meaning cannot travel in all the directions that the literal meaning does. For instance, as Lakoff and Johnson (1980) note, in the "Time is money" example, we can spend our time trying to do something and it doesn't work, but we cannot get that same time reimbursed. Metaphorical meaning is different from literal meaning and, as mentioned, is at some level dependent upon the literal meaning, even if it allows expression of something for which we have no literal terms. For example, in Ortony's (1975) metaphor, "the thought slipped my mind," we have no way of characterizing what is a thought, nor how it may move, nor what is mind, yet it describes an experience.

What, then, is inferred when an author maintains that mentalistic terms (such as "threaten," "reconcile," "fear," etc.) to describe animal behavior are being used metaphorically and not literally? First, it is generally agreed that a metaphor or a word used metaphorically has a certain semantic relation to the literal meaning of that word and to the other literal words in a sentence. In this instance, the literal meaning of the term is presumably the way we would use it of human behavior, implying a conscious agent (though a person may not "consciously" threaten someone, the complexities of the human case are beyond the scope of the present argument). However, in applying the term to animal behavior, is it not intended to evoke precisely the literal inference of an agent or of a recipient of a signal acting in some sense in the way we understand the same behavior in ourselves? Obviously, we are not prepared to say that the animal has the complex motives or forethought about possible future repercussions of its behavior that we can attribute generally to humans. And humans often unthinkingly, spontaneously, give and react to signals. However, in using the term "threaten" we are not trying to expand our meaning of animal behavior, nor think in a novel way about animals. We use the term "threaten" because the *circumstances* (rather than the *manifestation*—a fish may signal a threat in an entirely different way than a dog) of the behavior seem to the non-naive observer to be most appropriately described as a threat. The behavior may remind us of comparable behavior in humans, but there is anyway ample evidence that human observers can predict accurately very complex sequences of other species' behavior.

What about the argument that our language is imbued with purposefulness and hence that anthropomorphism is unavoidable? That depends upon what we know of animal cognition. Purposive terms may refer to something that is literally true of animals, but we know very little about that yet. We can try to operationalize our definitions, but that sometimes leads to utter sterility and loss of meaning. However, with care (such as that displayed by Harcourt and de Waal [1992] in their definitions of "alliance" and "coalition" among primates and other animal species), we can in fact mean precisely what we say in the context of our scientific reports, and at the same time test and retest hypotheses about the nature and existence of the behaviors in each new species that we observe.

CONCLUSIONS

What is considered acceptable mentalistic terminology to describe animals differs according to the describers' viewpoints about animals and about what it is to be human. Anthropomorphism has been an ambiguous concept. Fisher's (1991) lucid disambiguation of anthropomorphism draws a useful distinction between situational and categorical anthropomorphism. The former refers to getting it wrong, or misinterpreting the motives or meaning of the behavior of an animal in a particular situation. Thus, for example, we might misinterpret a fear grimace for a threat on seeing the bared teeth of a chimpanzee. That is not to say that a chimpanzee is incapable of feeling or expressing either fear or threat on any occasion (which would make it an instance of categorical anthropomorphism), but only that the behavior was misinterpreted in a particular instance (situational anthropomorphism). However, while we may become consistent in our categorizations of anthropomorphism, a discussion of whether or how far it is acceptable for use in animal behavior description is merely a reflection of our assumptions about the animals' capabilities. We are not talking about anthropomorphism per se but about our perceived place in nature. The flip side of this is what mental capacity is attributed to animals. Ultimately, we do not in most cases *know* that animals lack mental and emotional states, or how they compare with ours, as this knowledge has not been tested empirically (Fisher, 1991). Hence, it may be a misnomer to label terms that imply particular kinds of mental processes as anthropomorphic (as in a chimpanzee forming strategies). While many maintain that we are applying terms to other animals that refer to uniquely human characteristics (or that we are making a logical category mistake, applying a predicate to species to which it cannot apply in principle, under any circumstances), until this is empirically proven, it remains only a matter of opinion and conjecture.

Our presuppositions about the differences between animals and humans, rather than concern about unprofessional reporting, have been at the root of most modern arguments about anthropomorphism. Even with the proliferation of studies of animal cognition (which will hopefully contribute some of the empirical evidence needed to dispassionately determine what is and what is not categorical anthropomorphism), for many reasons we continue to use what we are calling anthropomorphic language to describe animal behavior. The point is, we cannot cite a metaphorical use of this language as a way to avoid the unproven assertions it implies about other animals' mental lives. Insofar as we are incapable of providing evidence for the nature of animal cognition, our discussions about anthropomorphism in fact are discussions about our perceived place in nature. This perception differs with time and place, with data and with cultural viewpoints. Fisher (1991) rightly, I think, remarked:

> The charge of anthropomorphism oversimplifies a complex issue—animal consciousness—and it tries to inhibit consideration of positions that ought to be evaluated in a more open-minded and empirical manner. (p. 51)

With a better understanding of underlying assumptions about other species, our endeavors could more profitably be directed toward gathering the evidence and building up rich, verifiable descriptions of animal lives, rather than debating about the merits and demerits of anthropomorphism per se.

NOTE

1. A large and influential portion of this literature was written by French scholars and the French language itself probably contributed to this interchangeability. The term *l'âme* in French can mean either "mind" or "soul." The restriction of a second historically important characteristic—morality or *humane* behavior—to humans has recently been challenged by primatologist Frans de Waal (1996).

PART II

The Nature of Anthropomorphism

4

Amorphism, Mechanomorphism, and Anthropomorphism

Emanuela Cenami Spada

Our world is a human world, and what is conscious and not
conscious, what has sensations and what doesn't, what is
qualitatively similar to what and what is dissimilar, are all
dependent ultimately on our human judgments of likeness
and difference.

—Putnam, *Reason, Truth and History*

THE PROBLEM

The term "anthropomorphism" comes from the Greek ανθρωπος, man,
and μορφη, form (Oxford English Dictionary, 1971/1987, p. 91). It was
originally described as the inclination to ascribe human appearances
and feelings to any animate or inanimate being, particularly gods. In
Greek religion it could assume a physical or a psychical aspect.

The origin of the negative sense that we attribute to the term
anthropomorphism can be found in Xenophanes' (570 B.C.), who

I thank Bob Mitchell and Debra Forthman for help in editing, and Stefano
Velotti for discussing the topic with me. Preparation of this chapter was sup-
ported by Ministero della Pubblica Istruzione, and by Consiglio Nazionale delle
Ricerche, grant no. 121.05517.

detailed the difficulties that arose from the idea of conceiving deities according to a human form and measure: "But mortals believe the gods to be created by birth, and to have their own (*mortals'*) raiment, voice and body" (Freeman, 1948, p. 22). If gods are born, they must also die. Yet, because gods were thought to be immortal beings, they could not be born. Furthermore, he noticed: "Aethiopians have gods with snub noses and black hair, Thracians have gods with grey eyes and red hair" (p. 22).

> But if oxen (and horses) and lions had hands or could draw with hands and create works of art like those made by men, horses would draw pictures of gods like horses, and oxen of gods like oxen, and they would make the bodies (of their gods) in accordance with the form that each species itself possesses. (p. 22)

In these particular fragments Xenophanes essentially criticized what has been called "physical anthropomorphism." However, the denunciation of the fallacy of conceiving gods as possessing bodily form is tightly connected with the denunciation of spiritual or psychological anthropomorphism. Ascribing human feelings and virtues to deities degrades the figure of the Divine in general. Gods were supposed to be perfect beings, so it is fallacious to ascribe to them human qualities and defects. Xenophanes detected what today can be called a categorical mistake. To anthropomorphize the concept of gods established a contradiction in terms of which characteristics a god should have. It is in this specific sense that the anthropomorphic tendency was considered as a negative or fallacious inclination, as a systematic categorical mistake.

In the study of animal behavior, the use of mental or psychical predicates can be considered a fallacy, that is, a categorical mistake, only if we already know with certainty that animals are completely different from humans. In other words, anthropomorphism corresponds to the same categorical error detected by Xenophanes only if we agree with Descartes (1649/1927) that animals are merely *automata* and humans are not. If we already know that animals are machines, we would commit a categorical mistake every time we ascribe to them a psychological state or a cognitive capacity (a subjective predicate). However, if the problem of anthropomorphism were so simple, it would be an easy task to detect its presence and hence to avoid the mistake. If we are sure that animals are nothing more than machines, we can know in advance that whenever we use subjective terminology for describing animals' behavior, we are doing it only in a metaphorical sense. It is the same metaphorical sense that allows me to say: "This morning my car will not start because it is mad at me."

However, a century ago, evolutionary theory allowed us to be less certain that the attribution of human traits to animals was a systematic categorical mistake. After Darwin, it became clearer not only that animals are not machines, but also that humans are animals. And it is exactly when we are ready to recognize that animals are neither machines nor humans that anthropomorphism becomes a much more complex problem.

Today, anthropomorphism is still considered a threat to the study of animal behavior. What is threatening is the ascription of "subjective traits," while the general process of "ascribing" properties seems safe when based upon "objective observation" and "neutral terminology." Behaviorism proposed a remedy against anthropomorphism which is still accepted and emphasized by many in the scientific community (e.g., Kennedy, 1992): banishing the study of animal mental processes and thereby the use of a mentalistic terminology. Nowadays this appears a mere palliative, not a solution. If anthropomorphism is a disease, this remedy cures its symptoms without affecting its causes.

Consider a concrete example: Let us suppose that each time I drive my car I get a terrible headache. My first guess is that what gives me the headache is driving. Unrelated to the accuracy of my guess, the best treatment for my problem is to avoid driving as much as possible, and, whenever I really need to drive, to take an aspirin. The cause of my headache and its relation to driving still remain unknown.

The remedy proposed by behaviorists to avoid anthropomorphism is very similar: avoid describing observed behaviors in mentalistic terms as much as possible. When I really need to use those terms, I have to retranslate them carefully into an "objective" terminology (as I take an aspirin to avoid the headache). "Mental processes" and "subjective terminology" are the symptoms of the disease known as "anthropomorphism." For some years the prescription seemed to work. Certainly, the use of replicable and rigorous method which behaviorism stressed was important in this success. Yet, more and more often, the complexity of the behavior to be described requires the use of "everyday language." The "objective" and "descriptive" language no longer seems sufficient. Because the use of "non-objective everyday language" seems to be what allows the anthropomorphic disease to re-emerge, two more treatments are required. When using a subjective terminology for describing animal behavior we should accurately translate it into an "objective one," or at least make very clear that we mean it in a "metaphorical sense." Thus, the problem seems to be just the difficulty of translating from a metaphorical language into a literal one (Asquith,

1984; cf. Asquith, this volume). Still, I am not sure that one can solve a problem by simply finding ways through which one can avoid the problem itself.

> The urge is to try to find a description of the phenomena which doesn't use the mentalistic vocabulary. But what is the point in doing that? The facts remain the same. The fact is that mental phenomena have mentalistic properties, just as what goes on in my stomach has digestive properties. We don't get rid of those properties simply by finding an alternative vocabulary. (Searle, 1992, p. 29)

The question is, of course, what are "the facts" that "remain the same." Facts must be described as facts of a certain kind. To describe facts as facts of a certain kind (mentalistic or not) we need a vocabulary of a certain kind. Facts and vocabularies about facts are not clearly separable.

Scientists who think they can get rid of anthropomorphism conclusively think that "facts" are what results from an "objective description" made through a "neutral language" (*amorphism*). By contrast a "subjective description" made through a "metaphorical language" would inevitably lead the research towards a nonscientific and anthropomorphic conclusion. But how can we know what is a subjective vs. objective description? How can we distinguish what is metaphoric from what is literal?

SUBJECTIVE VERSUS OBJECTIVE DESCRIPTIONS; METAPHORICAL VERSUS NEUTRAL LANGUAGE

Let us consider the possibilities that are available if one decides that it is possible to divide the study of animal behavior into two clear-cut and opposed ways: objective/nonanthropomorphic vs. subjective/anthropomorphic. Suppose that, after some years of research, we notice that the complexity of an animal's behavior strongly suggests a description considered exclusively appropriate for the human domain (Hebb, 1946; de Waal, 1982, 1989). Thus, to avoid the trap of anthropomorphism we decide that whenever we use this vocabulary we have to mean it in a metaphorical sense and we need to translate the metaphorical description into objective and descriptive terms. In other words, any correspondence between human and animal behavior is to be considered just a metaphorical (virtually anthropomorphic) analogy, not an objective observation. Kennedy (1992), for example, recognizes the importance of the use of analogies and metaphors in understanding physical phenomena but he warns us against doing the same in the description of animals: "Anthropomorphic analogies for animal behavior are the exception: they readily generate misunderstanding" (p. 159). Here, according to Kennedy, the misunderstandings that metaphors and analogies can generate in the study of animal behavior depend upon the fact that animals are animate systems.

However, the difficulties posed by the use of analogies between humans and animals is the core of the entire puzzle of anthropomorphism. Anthropomorphism cannot simply be avoided; it needs to be understood. As Kant (1790/1991) put it two centuries ago:

> . . . when we compare the formative operations of the lower animals with those of man, we regard the unknown source of such effects in the former case, as compared with the known source of similar effects produced by man, that is by reason, as the analogon of reason. (§ 90, note 22)

Thus, to understand what we cannot perceive directly we must establish an analogy with what we experience.

> But from the similar mode of operation on the part of the lower animals, the source of which we are unable directly to perceive, compared with that of man, of which we are immediately conscious, we may quite correctly infer, *on the strength of the analogy*, that the lower animals, like man, act according to *representations*, and are not machines, as Descartes contends, and that, despite their specific difference, they are living beings and as such generally kindred to man. (§ 90, note 22)

In this sense, the analogical inference is inescapable. When we try to interpret animal behavior we must unavoidably face what Burghardt (1985a, p. 909) called the "subjective analogical inference"; an inference that Romanes (1883/1900) described as follows:

> . . . if I contemplate my own mind, I have an immediate cognizance of a certain flow of thoughts or feelings, which are the more ultimate things—and indeed the only things—of what I am cognizant. But if I contemplate Mind in other persons or organisms, I can have no such immediate cognizance of their thoughts and feelings; I can only *infer* the existence of such thoughts and feelings from the activities of the persons or organisms which appear to exhibit them. (p. 15)

The fact that we have interpretive access to other minds through an analogical subjective inference is certainly not new and is widely acknowledged. What is not easily recognized is that the process of reasoning through analogies does not pertain exclusively to the subjective domain, as if the objective description of animal behavior would be otherwise free from the knowing subject. Such an idea supposes that the subject could step outside him or herself and accurately determine what is subjective and what is objective (from an "amorphic" point of view, i.e., from the paradoxical absence of any point of view). The idea that there are two kinds of vocabulary, two kinds of categories—subjective and objective, with no reference to any already conceptualized field of experience—is ill-conceived. If it were true, we should conceive of our process of knowledge in this way: a subject has an idea (a description, a word, a picture) of an object; this idea or description is true (objective) if it corresponds to the object. The problem with this view is the following: Who can say whether or not that idea or description is true or objective, i.e., corresponds to the object? Only somebody who can stand between our description and the object described. But how can he know how the object really is, if not through another description of the object? We would need a third subject between his description and the object, and so on ad infinitum. As von Glasersfeld (1984) put it, "if knowledge is to be a description or image of the world as such, we need a criterion that might enable us to judge when our descriptions or images are 'right' or 'true'" (p. 26). Obviously, there is no such criterion (we would need another criterion to say that the first criterion is true and so on ad infinitum). This is the persistent problem of the skeptic. But this is not to say that the skeptic is right. The skeptic would be right only if that picture of our knowledge were true. But fortunately it is not so. There are not three distinct entities, the subject, the description, and the object, because there are no objects per se, but only objects "such and such." This means that our subjective categories are the only means to perceive a differentiated reality, and that this appears only through our descriptions. Of course, reality can force us to change our descriptions, because, for instance, we get inconsistent answers.

To explain those considerations with reference to anthropomorphism, an example should be helpful. Let us examine what Kennedy (1992) suggested as a prescription to avoid anthropomorphism in one case: "translating the subjective term 'searching' into the objective one 'scanning' brings us straight down to earth into the real world of behaviour" (p. 164). Kennedy's suggestion is awkward at best.

Certainly, to say that an animal is "scanning" its environment is quite different from saying that an animal is "searching" its environment. "Scanning," in general, is referred to something more mechanistic, which can be easily applied to machines, or physiological processes, as opposed to "searching," which generally implies an active intention of the subject. Now, in order to decide which of the two descriptions is the most adequate for the behavior observed, which one "brings us straight down to earth into the real world of behaviour," we would need to believe that we knew what reality "really is" *prior* to any description of it. We would need to know there is a real and ultimate description of what the animal is doing (amorphism). We would need to be sure that necessarily only one of the two terms (either "scanning" or "searching") corresponds to the behavior we are describing, *before* using one of the two terms (or descriptions) by which we were trying to categorize what we were seeing. This position will inevitably take us to some sort of "x-morphism," either "mechanomorphism" (Caporael, 1986) or "anthropomorphism."

Let us suppose, for instance, that I am observing the behavior of a chimpanzee. It is sleeping surrounded by the other members of its social group and suddenly stands up and "looks around" before it starts moving. At this point I have an unlimited number of different ways of observing and describing the "looking around" behavior, depending on what I am interested in. I can give an accurate description of each of the movements of the animal's head, and, if possible, of its eyes, and call this description "scanning." (How accurate should the description be? At what level of simplicity should I stop to be sure that I have reached the "objective level"?) However, if I was interested in what the animal was "looking for," for example, food or social contact, the "scanning" description would not tell me anything about this. If the focus of my observation is what the animal is "looking for," I need to concentrate on the relation between the more general "searching" behavior and the surrounding objects or individuals. Of course, if we use the term "searching" to describe the relationship between the behavior of the animal and its environment when we are interested only in the behavior of "scanning," that is, the movements of the head, *we need to redescribe, which is different from "translate,"* the same behavior in a different way, and vice versa. At this first level of analysis, both terms can be right or wrong depending upon what the questions are.

The request for a translation from a subjective to an objective terminology comes from something different. It follows from the belief that one term is more adequate than another, more real than another. But the claim that one term is more adequate than another implicitly means that one is supposed to better correspond to the "real" behavior. Yet, in a similar view, as we have seen, we need someone who would stand between the term that we choose for describing the behavior and the behavior itself; someone who would be able to tell us how "the real behavior is" without using any description (amorphism)! Since this person does not exist, ultimately the decision about which of the two terms fits better depends upon how I have categorized the field of experience I am observing (e.g. , animals). If animals are categorized as machines, only the terminology used to describe machines will sound adequate. Any reference to the terminology adopted to describe a psychical world will thereby sound maladroit and perhaps anthropomorphic. If by convenience we were tempted to use such a vocabulary, it would be put between inverted commas and considered just metaphorical. We would then need to translate the description of that behavior according to the vocabulary that better describes machines. However, as far as we know, the opposite position is equally true. If I do not consider an animal a machine, and, therefore, I want to avoid the danger of mechanomorphism, each time I use a term that usually belongs to the vocabulary used to describe machines it would be considered just metaphorical. We would then need to translate the description of the behavior into a mentalistic terminology.

The real problem with the description of animal behavior emerges exactly here: We have classified animals within the same domain as humans, namely the domain of "animate beings." However, there is something different as well as something very similar between animal and human behaviors, and that gap is precisely what we are currently trying to understand and categorize. Thus, anthropomorphism is not a disease, an enemy that stands by itself and that we need to fight as we fight a virus. Anthropomorphism becomes always a possible problem when we try to determine how far we can go in tracing similarities and differences between animal and human behaviors.

Kennedy (1992), for example, concludes his book as follows:

> If scientists, at least, finally cease to make the conscious or unconscious assumption that animals have minds, then the consequences can be expected to go beyond the boundaries of the study of animal behaviour. If the age-old mind-body problem comes to be considered as an exclusively human one, instead of indefinitely extended through the animal kingdom, then that problem too [anthropomorphism] will have been brought nearer to a solution. (p. 168)

Yet, this is a very partial solution of the problem of anthropomorphism. It is to say: If we eliminate the problem we solve it. If we classify "animals" as machines instead of beings within the embarrassing and inconvenient category of "animals," we solve the problem of anthropomorphism. Furthermore, it remains unexplained why what Kennedy calls "everyday language" (as opposed to "scientific language") is sometimes able to explain better the behavior that we want to understand. In those specific cases, i.e., when the "everyday language" works better, Kennedy is asking the scientific community to consider the descriptions to be simply metaphorical and to translate those metaphorical descriptions each time into the real objective terminology. But as we have argued, the real objective terminology exists only within our conceptual schemes of reality. If we categorize animals as machines, it naturally follows that the use of mentalistic and subjective words is inadequate and supposedly anthropomorphic. Yet, the same is true if we do the opposite. If we categorize animals by admitting that humans are animals too, and we describe their behavior using the vocabulary that we use to describe machines, the latter description would be metaphorical. Therefore, we would need to translate that description into a more subjective vocabulary which would be more literal. In this case, the danger is that we would regress into Descartes's mechanomorphism.

In summary, if we agree that our experience is neither a subjective nor an objective affair, but that subject and object are inseparably linked in the understanding of our world, it readily follows that the antithesis "objective/non-anthropomorphic/neutral language" as opposed to "subjective/anthropomorphic/metaphorical language" collapses (Putnam, 1981, 1988; Varela, 1984). What clearly remains is the fact that if we classify animals as machines, anthropomorphism can be considered as the same unavoidable categorical mistake detected by Xenophanes in ancient times. Here, however, "unavoidable" does not mean, as Kennedy claimed, "built into us" or "innate" (1992, p. 5). If anthropomorphism were innately the only way we could perceive animals, we would never be able to detect its presence. We would see our world "anthropomorphically" without being aware of it.

The persistence of anthropomorphism throughout the history of human knowledge should be seen as a sign of the presence of a real problem. If animals are not machines, we need to study to what extent similarities and differences with human behaviors can be drawn. In doing this we unavoidably refer to our experience: what else could we refer to when studying animals! Anthropomorphism is the name we give to a mistake: when we are able to reveal one of our answers to the problem of animal behavior as "too human," that is, once we have understood that we were stretching the analogies between animals and humans too far. Nevertheless, we must run the risk, and we cannot know in advance when we are making that mistake. We will be able to determine the presence of anthropomorphism only a posteriori, that is to say, when we can be sure that the human characteristic we were searching for definitely was not there. And that conclusion, again, does not depend upon an absolute truth, established outside our research (outside science itself). The detection of an anthropomorphic mistake is linked with the dominant paradigm, the unquestioned assumptions that are considered valid at the time of the research and that belong to our conceptual scheme (Candland, 1993).

WHERE THE DANGER OF ANTHROPOMORPHISM HIDES ITSELF

Anthropomorphism is the name of a prejudice that can be detected only after we have been victims of it. Currently, the field of "cognitive ethology" and the research on animal language are often cited as fields at high risk (Burghardt, 1985a; Kennedy, 1992) for repeating the anthropomorphic errors that characterized the study of animals at the beginning of this century.

To show why we can establish *only* a posteriori whether a research is anthropomorphic or not, we can use as an example the controversial area of studies known as Animal Language Research (ALR). As Kennedy (1992) put it:

> Probably the most spectacular examples of professional scientists—animal psychologists in this case—being misled by their subconscious anthropomorphism come from the last series of American attempts to teach animals our language. In this case the mistake was not due to teleological thinking but to anthropomorphism in the different form of assuming that higher animals share our learning abilities. (p. 40)

What needs to be discussed here is when and how we are able to judge if the attempt to teach animals human language is indeed anthropomorphic or not. In order to give an answer, we have to consider the conceptual paradigm which was dominant at the time that research started. The first attempts to teach language to animals were by Kellogg and Kellogg (1933) and Hayes (1951). In both of the studies the researchers tried to verify whether a chimpanzee raised in a human environment (in their home, with their own child) could acquire the ability to express itself vocally. Although the actual learning of language was a failure, these initial "linguistic" attempts with primates are currently easily understood if the epistemological assumptions on the basis of which they were undertaken (i.e., the behavioristic paradigm), are taken into account. In fact, if human language (as well as all other kinds of behavior) is considered just the result of a chain of stimulus-response events, it is reasonable to hypothesize that a more stimulating environment could have been sufficient to elicit a behavior not previously seen in animals. Similarly, when it became evident that for chimpanzees vocal means were inadequate, the attempt to teach animals human language by nonvocal means became plausible (Fouts, 1973; Gardner & Gardner, 1969; Herman, Richards, & Wolz, 1984; Miles, 1983; Patterson, 1978; Pepperberg, 1981; Premack, 1971; Rumbaugh, Gill, & von Glasersfeld, 1973; Savage-Rumbaugh, Rumbaugh, & McDonald, 1985; Schusterman & Krieger, 1984; Terrace, 1979). Now, what ultimately allowed the experimenters to consider their endeavor plausible derived from the behaviorist conception of the *nature of language*. If language, as suggested by Skinner (1957), is viewed as a learned behavior, the attempt to determine how much language can be taught to animals would be a legitimate one. In this vein, since animals did not learn as much language as humans, we could positively conclude, as Kennedy has, that our assumption that higher animals share our learning abilities is wrong. If language is only a learned behavior resting on nothing more neurologically specific than a higher ability to form associations, Animal Language Research (ALR) showed that any organism possessing sufficient intelligence could acquire some language, but not as much as humans. In this sense, no anthropomorphic assumption is involved. It would be anthropomorphic only to say that animals can acquire the same amount of language as we can.

Now let us suppose that we no longer believe in the general behavioristic paradigm and thereby consider language somehow different from a stimulus-response process, and that we look back to the linguistic research done with animals. Almost immediately we would claim that the linguistic question posed to animals was not the legitimate general question, "Do animals have language?," but the anthropomorphic question, "How much language can an animal learn?" (By "legitimate" or "genuine" question I mean a question that does not implicitly presuppose its answer.) In fact, the question was anthropomorphic because of its implied unquestioned answer: Since human language is considered a process learned through stimulus-response associations and animals respond to stimuli, animals must have language; and what the researchers were actually measuring was just "how much" of it they could learn. Here is an extrapolation from an interpretation presumed of human language to other forms. Asking "How much language can an animal learn" implies an anthropomorphic stance, not because it assumes that higher animals share our learning abilities, as suggested by Kennedy (1992), but because it assumes that human language equals learned behavior and therefore any language can be learned by animals, as maintained by the behavioristic paradigm. As a consequence of considering language as just a learned behavior, the analogies between animal and human language have been stretched "too far" (Cenami Spada, 1994; Wallman, 1992).

Anthropomorphism is a risk that we must run, because the questions we pose regarding animals are inevitably drawn from our experience. From what other experience can we draw the contents of our questions? We must be conscious of this risk. But we must also be conscious that we cannot avoid it simply by pretending to look at "how things really are." We can only try to conceptualize better our experience and our scientific categories. We must be ready to abandon our hypothesis concerning animal mind, for instance, when somebody has shown us that the question we were posing contains already an anthropomorphic answer. This danger is always there and we must be ready to criticize our conclusions. But no question, per se, is in principle anthropomorphic. It is not anthropomorphic to ask if animals exhibit minds (Griffin, 1984, 1992; Premack, 1988a; Gallup, 1985), self-awareness (Gallup, 1970, 1975; Povinelli, 1987; Mitchell, 1993b), social cognition (Smuts, 1985; Byrne & Whiten, 1988b; Cheney & Seyfarth, 1990), deception (Mitchell & Thompson, 1986a; de Waal, 1986; Whiten & Byrne, 1988), sympathy or fantasy (Mitchell, 1994a).

When we pose these questions, we certainly need a working definition which should be as adequate as possible to our knowledge in order to identify the phenomena that we want to study while observing animals or conducting experiments. If the working definition is shown to be inadequate (e.g., language equals a learned behavior) then we have begged the question: The answer we believe we have given to the original question is actually not answering it, and we have probably reached a naive anthropomorphic conclusion. Yet, as long as nobody has been able to show that our working definition of the phenomenon we want to study in animals is inadequate, the presence of that phenomenon in animals cannot be considered anthropomorphic.

Let us give a last schematic example:

Question: Do animals feel pain?
Working definition: We can say that animals feel pain if at least the following conditions are satisfied: when an animal has a leg injury, it limps or runs on three legs.
Observation/Experiment: The previous conditions are satisfied.
Answer: Animals feel pain.

If someone wants to criticize this answer as anthropomorphic, he must address his criticism to the working definition, which can be argued to be inadequate or reducible to other simpler causes. It is in the working definition that anthropomorphism can hide itself, not in the question of pain per se, which is a legitimate question and as such cannot be a priori rejected as anthropomorphic. Naturally, if the conclusion has been shown to be anthropomorphic (because of the inadequacy of the working definition adopted), one can always restate the question and reformulate the working definition. This reformulation is probably the only way we can cope with anthropomorphism and hope to make progress in studying animals.

CONCLUDING REMARKS

Anthropomorphism is neither an incurable disease nor a categorical error, but is the name that describes a prejudice detectable only a posteriori. If anthropomorphism were "incurable" in the sense of "unavoidable" (Asquith, 1984) or "innate" (Kennedy, 1992), we would never have been able to detect its presence as a problem. If anthropomorphism were a systematic categorical mistake (e.g., if animals are machines) it would be very easy to detect its presence and thereby avoid it.

Anthropomorphism is a persistent problem, a risk we must run, because we must refer to our human experience in order to formulate questions about animal experience. Since questions per se are not anthropomorphic, neither the avoidance of questions regarding animal mental states nor an agreement upon which language to use while describing animal behavior could help us remove the problematic aspect of anthropomorphism. We experience mental and physical phenomena that we describe with a certain language and we ask ourselves if animals experience those same phenomena too. A neutral language (amorphism) as opposed to a metaphorical language does not exist. Any behavior (either human or animal) is always already described as a behavior "such and such." That is why anthropomorphism cannot be considered just a logical or an empirical problem. We do not have access to an unprejudiced experience that can determine independently the answers to our questions.

Any question regarding animals can be stated. Only the working definitions which we use to conduct experiments or observations, and to come to an answer, can be illegitimate or inadequate. The only available "cure" to anthropomorphism is the continuous critique of our working definitions in order to provide more adequate answers to our questions, and to that embarrassing problem that animals present to us.

5

Anthropomorphism:
A Definition and a Theory

Stewart Elliott Guthrie

There is an universal tendency among mankind to conceive
all beings like themselves. . . . We find human faces in the
moon, armies in the clouds.

—David Hume, *The Natural History of Religion*

In consequence of a well-known though inexplicable instinctive tendency, man attributes purposes, will and causality
similar to his own to all that acts and reacts around him.

—Théodule Ribot, *L'évolution des ideés générales*

Anthropomorphism has been noted for centuries throughout human
thought, yet to many it remains "inexplicable." Indeed, the most basic
questions persist: What do we mean by it? Should it be shunned or
embraced? Why does it occur? Despite some 400 years of general agreement that avoiding anthropomorphism is desirable—and, in science,

This chapter is one product of a Fordham University Faculty Fellowship in
1993–94. I am grateful for editorial help to Phyllis Ann Guthrie and Walter
Guthrie.

The theory presented in this chapter is given in much greater detail in
Faces in the Clouds: A New Theory of Religion (Guthrie, 1993).

central—its definition, its causes, and even its status as an error remain in doubt. In the face of this uncertainty, I suggest that anthropomorphism is best defined as in *Webster's New Collegiate Dictionary* (1977): the "attribution of human characteristics to nonhuman things or events." I further suggest that by definition it should be minimized although by nature it cannot be eliminated; and that it occurs as one result of a perceptual strategy that is both involuntary and necessary.

The earliest known criticism of anthropomorphism appeared some two and a half millennia ago in Xenophanes's often-cited observation (Freeman, 1948) that people model their gods after themselves. Although philosophers and scholars of religion have mulled this observation ever since, the modern concern with anthropomorphism in secular thought as well as in religion began around 1600 with Francis Bacon (1620/1960). Bacon overturned Aristotelian science (and, in a standard view, founded modern science) largely by his critique of Aristotle's teleology, which held that everything behaves as it does in order to achieve some end—in general, to "fulfill" itself. But achieving ends, Bacon maintained, is a specifically human activity, and attributing ends to nature misconstrues nature as humanlike. Indeed, he wrote, most of our perception of the world is colored and misled by a strong tendency to see the world as like ourselves. Our understanding typically relies on causes related to the "nature of man rather than to the nature of the universe" (1620/1960, p. 52).

Since Bacon's time, numerous philosophers, natural scientists, and others have pointed to anthropomorphism throughout human thought, and the great majority has warned against it. Spinoza (1670/1951), for example, wrote that our world view is little more than an extension of our view of ourselves. We see the world as purposeful, for instance, because we ourselves are purposeful, whereas in fact nature has no purposes. Hume (1757/1957) similarly wrote of a "universal tendency" among humans to "conceive all beings like themselves" (p. 29). Goethe wrote that humans "never know how anthropomorphizing they are" (Liebert, 1909, pp. 1–22), and Nietzsche found all human knowledge, including such basic concepts of physics as force, deeply and inevitably anthropomorphic (Stack, 1980). For these writers (Nietzsche, in his late work, is a partial exception) and most others, anthropomorphism clearly is a mistake, to be rooted out wherever possible.

Despite this long scrutiny, no good account of the causes of anthropomorphism has been given. Instead, two standard but flawed explanations—that it comforts us and that it explains the unfamiliar by the familiar—are, singly or together, taken for granted. Because these two are taken for granted, many observers of anthropomorphism find the issue of cause unproblematic. Some think the only question is how we are able to anthropomorphize. Yet disagreement and uncertainty about the nature and prevalence of anthropomorphism recur. Agassi (1973) and Ridley (1992), for example, write that it may sometimes be valid, and Agassi thinks it has, in any case, virtually disappeared from science. Others (e.g., Rollin, this volume) see it as mere common sense, while still others (e.g., Kennedy, 1992) see it as a continuing and even resurgent menace.

The question, whether anthropomorphism necessarily is mistaken, hinges largely on the writer's definition and explanation. As do evaluations, definitions vary widely, ranging from the ascription of human mentality to animals, to the ascription of mentality to the world at large, to the ascription of human features in general—including physical and behavioral ones—to the world. Those who mean by it the positing of similarity between humans and other animals frequently find it justified, pointing out that evolutionary linkages cause real similarities among species. Those who, in the mainstream of opinion established by Bacon, mean by anthropomorphism the attribution of distinctly human traits to what is nonhuman mostly find it a categorical mistake.

The divergences of opinion about the peril, or lack of it, in anthropomorphism thus stem from definitional disagreements bound up in the issue of cause. If one means by anthropomorphism the positing of analogies and homologies between ourselves and other phenomena, then the chief cause is the employment—often apt—of the human mind's powers of pattern-discovery. If one means the misattribution of properly human traits, then the cause is a failure in that pattern-discovery. The failure, in turn, may be caused simply by disparity between the infinite complexity of the world and the limited powers of the mind, or by wishful thinking, or by both.

Confronted by these disagreements, I propose that we keep to traditional usage and mean by anthropomorphism the misattribution of properly human traits. This has the virtue of historical consistency: the first meaning of the term was improperly giving human traits to God, and dominant usage since then, though broadened to giving human traits to anything nonhuman, also assumes impropriety. More important, this usage has the virtue of separating the question of the origin of an idea from that of its validity. Whether we base a model of the world on ourselves or on something else, the most pressing question is not its source but its power to describe or explain.

The separation of origin from validity with regard to finding parallels between ourselves and nonhuman nature again follows usage: although analogies and homologies between ourselves and other aspects of the world abound, we do not call the identification of all these anthropomorphic. For medical research, for example, many animals—pigs, guinea pigs, rats, and mice, among others—resemble us enough physiologically that tests of medicines for humans are made on them; yet this practice is not labeled anthropomorphism. It would be anthropomorphism only if we overestimated the likeness of the physiologies of these animals to our own, as some folklore and religious traditions (Thompson, 1955) and much commercial art (Guthrie, 1993) do with the behavior of animals, portraying them walking upright or speaking a language.

What is best meant by anthropomorphism, then, is what is traditionally meant: not the attribution of likeness, but its overestimation. Once this usage is granted, we can agree we should avoid anthropomorphism simply by definition, as we should avoid repeating any mistake. The question then becomes how to decide which attributions anthropomorphize and which do not. Here no quick answer exists. Instead, there is only a perpetual process in which we scrutinize and compare ourselves and other animals. The differences and similarities among species are empirical issues that must be resolved by ongoing examination.

The next question is for me the most interesting: why do we make the mistake in question and do it so pervasively? For anthropomorphism is not limited to our views of other animals but is broad and deep everywhere in our thoughts and actions (Guthrie, 1993). It colors perception and responses to perception throughout life, as when we speak to our plants, cars, or computers, see a natural disaster as punishment for human misdeeds, or feel that nature shows design.

Indeed, the impulse to find faces in clouds, knotty wood, and other nonhuman forms and to hear a human presence in unidentified sounds in the night appears universal. Literary critics and art historians find it throughout the arts, where they usually call it personification, and ethnographers and folklorists report it in every culture. It arises spontaneously and is no respecter of training or rationality. Even the scientist, Nietzsche (1966) observed, "wrestles for an understanding of the world as a human-like thing and . . . regards the whole world as connected to man, as an infinitely broken echo of an original sound, that of man; as the manifold copy of an original picture, that of man" (p. 316).

Although long recognized as a tendency of thought and as a categorical error, anthropomorphism has elicited little systematic analysis. That such an important and oft-noted tendency should bring so little close scrutiny is a curiosity with several apparent causes. One is simply that it appears as an embarrassment, an irrational aberration of thought of dubious parentage, that is better chastened and closeted than publicly scrutinized. For religionists it is a worrisome limitation on our conceptions of divinity, and for humanists it is a limitation on our conceptions of the natural world and on the power of rationality.

More importantly, however, two superficially adequate explanations of anthropomorphism, the familiarity and the comfort theses, have long been available. Separately and together, they have forestalled more penetrating analyses. According to the familiarity thesis, we use ourselves as models of the world because we have good knowledge of ourselves but not of the nonhuman world and, looking for an explanation of the world, resort to the knowledge that is easiest and most reliable. This view goes back at least to the first systematic critic of anthropomorphism in general, Bacon. The human understanding, he wrote, "struggling toward that which is further off . . . falls back upon that which is nearer at hand" (1620/1960, pp. 51–52). Spinoza later gave a similar account, as did Hume. Hume (1757/1957) wrote that humans, faced with an inscrutable universe, "transfer to every object, those qualities, with which they are familiarly acquainted, and of which they are intimately conscious . . . [Hence] trees, mountains and streams are personified, and the inanimate parts of nature acquire sentiment and passion" (p. 29). According to the familiarity argument, then, our motivation for anthropomorphism is primarily cognitive. We wish to understand the world, and the first criterion for an understanding is that the model be one on which we already can rely.

The comfort thesis, in contrast, refers us not to cognitive but to emotional motives, holding that we are mistrustful of what is nonhuman but reassured by what is human. In some degree, this view, too, goes back at least to Hume, where it mingles as a motive with the desire to know. Hume notes that humans are ignorant or uncertain of major factors affecting their fate and find this unsettling. "These unknown causes, then, become the constant object of our hope and fear; and while the passions are kept in perpetual alarm by an anxious expectation of the events, the imagination is equally employed in forming ideas of those powers, on which we have so entire a dependance" (1757/1957, p. 29). Motivated not only by intellectual but also by emotional needs, we form humanlike models to account for and mitigate events.

A more recent version of the comfort theory, that of Freud, assigns wishful thinking the major role and cognition only a negative one. We anthropomorphize the world (and thus establish religions), Freud claimed, in an irrational attempt to feel we can influence it. "Impersonal forces and destinies cannot be approached; they remain eternally remote. But if the elements have passions that rage as they do in our own souls, if death itself is not something spontaneous but the violent act of an evil Will, if everywhere in nature there are Beings around us of a kind that we know in our own society, then we can breathe freely, can feel at home" (Freud, 1927/1964, p. 22). Here we are motivated not by a need to know but by a need not to know: to deny knowledge and escape into fantasy. As in a popular view, people believe what they want to believe.

Both the familiarity and the comfort explanations of anthropomorphism have wide appeal, numerous variants, and a small modicum of truth. Neither, however, wholly stands up to scrutiny. (For an extended examination, see Guthrie, 1993.) The familiarity thesis, holding that anthropomorphism consists in relying upon what is known to explain what is not, reflects the fact that we do know something of ourselves, without which we could not use ourselves as models at all. To account for our great reliance on ourselves as models of the world, however, our self-knowledge would have to be proportionately great. Yet numerous observers—Shakespeare, Nietzsche, and Freud are only among the most prominent within European culture—affirm that we know ourselves but slenderly. Physiologically as well as psychologically, the proportion of what we know of ourselves to what we do not is no greater than what we know of dogs and cats or stones and streams. Even so gross a fact as the circulation of blood, for example, became known only rather recently. Although we sometimes think we have privileged access to ourselves, in fact the unknown within is as deep as that without. Close or reliable knowledge, then, cannot be the reason we so heavily use ourselves as models.

The comfort thesis, that we feel better to see the world as human-like than not and hence engage in wishful thinking, also seems plausible at first. Certainly humans are gregarious, and solitude may be lonely. But this explanation founders on the fact that while other humans are among our greatest joys, they also are among our greatest perils. Correspondingly, while much anthropomorphism is comforting, much is frightening. When the householder hears the night wind slam a door and imagines an intruder, or the soldier on patrol sees the silhouette of a bush as that of an enemy, their mistakes are not comforting. The grain of truth in the comfort thesis, perhaps, is that any interpretation is better than none. But this is no more helpful than the familiarity thesis in showing why the interpretation we choose so often is humanlike.

Instead, my argument—a cognitive and game-theoretical one— is that in the face of chronic uncertainty about the nature of the world, guessing that some thing or event is humanlike or has a human cause constitutes a good bet. It is a bet because, in a complex and ambiguous world, our knowledge always is uncertain. It is a good bet because if we are right, we gain much by the correct identification, while if we are wrong, we usually lose little. To call it a bet, however, is not to say that it is a deliberate or conscious choice. Instead, like most of the perceptual process, it remains out of our awareness (Kahneman & Tversky, 1982). As a strategy, it is the product of natural selection, not of reason.

This involuntary and mostly unconscious strategy nonetheless resembles a conscious one known as Pascal's Wager. While we cannot ever know whether God exists or not, Pascal held, we should try to believe that He does. If we do and are right, we may be rewarded by eternal joy, whereas if we are mistaken, we lose only the minor pleasure of indulging in a few sins. Conversely, if we disbelieve and are wrong, we risk eternal damnation. The two outcomes are so disproportionate that even if it seems unlikely that God exists, we are well advised to bet He does.

Belief in humanlike presences in general resembles Pascal's belief in God: it follows the principle better safe than sorry. Although we frequently are mistaken in seeing phenomena as humanlike or as caused by humans, these mistakes—collectively and retrospectively known as anthropomorphism—are relatively cheap. Conversely, when we identify something as having human characteristics or causes and it proves to have them, the reward is significant: we can take appropriate measures for fight, flight, or social relationship.

Put another way, our perceptual strategy consists in seeing the world first, neither as what we want to see (as folk wisdom has it) nor as what is most likely, but as what matters most. That is, we scan an ambiguous world, first, with models determined by our most pressing interests. Although our interests vary, humans, because they are highly organized and powerful, figure in them frequently if not constantly. By virtue of their high organization, real humans also appear in a variety of guises and generate a diverse array of phenomena. Scanning an uncertain field with models whose importance and diversity correspond to those of actual human appearance and behavior, we often suppose we have found the humanity for which we are so alert when in fact we have not. When we later decide we were mistaken, we assign these misattributions to the residual category we call anthropomorphism.

A key question remains: what are the characteristics of humanity and how does one judge their presence or absence? Again, no uniform or certain answer is possible. As products of evolution, humans are in many ways continuous with other animals, most obviously with such close relatives as chimpanzees. Because of these continuities, any "characteristics" can be only relative and cannot supply a clear or distinct line between us and others. Tool-making, language, and religion, for example, all have been said at one time or another to set us off from other animals, yet we now know that captive chimpanzees can be taught rudimentary symbolism if not language itself and that wild ones make tools. They may even, in directing threats against thunderstorms (Goodall, 1975), show something like a religious attitude. Hence our characteristics look more like matters of degree than of kind.

On the other hand, much anthropomorphism stands out unencumbered by such complex issues as the degree of our likeness to other animals. Typical features of human behavior include language and culture, with their attendant senses of meaning, purpose, and sociality; and typical physical features include bipedalism and neoteny. When we talk to our cars or find purpose in natural order, or see still lifes as social groupings or see figures in clouds, we attribute—willingly or not—some of our salient features to phenomena that do not have them. We do so because we constantly scan our environments, as we must, for the patterns that most affect us. Whether they affect us for better or for worse, we are highly motivated to find them and do so at every turn. When we find them, or think we do, we naturally look again. Looking again, we have the chance and the need to scrutinize what we have found. Once aware of our perpetual anthropomorphism, we are better able to sort out what is like us and what is not, and in what ways. The task is both empirical and continual.

The account given here of anthropomorphism, as an unavoidable product of a necessary perceptual strategy, also applies to a related tendency, animism. As defined by Piaget (1933) and most subsequent psychologists, animism is the attribution of life to the nonliving. This definition appears broader than, and inclusive of, animism as meant by religionists and anthropologists (i.e., as the attribution of souls or spirits to things that do not have them). It also appears to include an error occurring in nonhuman animals as well. That is, many animals seem occasionally to mistake nonliving things for living ones: birds peck at twigs resembling caterpillars, horses shy at blowing papers, and dogs howl in concert with sirens. Here it might also be called zoomorphism, or the attribution of animal characteristics (we need not assume that a concept of "life" is present) to nonanimal things and events. In any case, animism or zoomorphism as a residual category of mistakes seems to spring from the same strategy as does anthropomorphism: in the face of chronic uncertainty, we all look first for what matters most. For animals, that is other living things. For humans and a few other animals, it is living things and especially humans.

Thus anthropomorphism, usually considered an aberration of thought, a form of wishful thinking, a use of what we know to explain what we do not, or all these, basically is none of them. Instead, it is a reasonable, though in hindsight mistaken, attribution of aspects of what is most important to us, to parts of the world that do not have them. Far from an aberration, it is a class of perceptual errors that is natural, universal, and inevitable. Like the animism or zoomorphism to which it is related by similar causes, this categorical error is a cost of our necessary vigilance for the presence of what most concerns us. Far from fulfilling our wishes, it often alarms us. And far from stemming from conscious intellection, it is rooted in a strategy that usually is out of our awareness and always is out of our control.

6

Why Anthropomorphize?
Folk Psychology and Other Stories

Linnda R. Caporael and Cecilia M. Heyes

INTRODUCTION

In the early sections of *Lectures on Conditioned Reflexes,* Pavlov (1927/1960) tells how he and his assistants puzzled over the behavior of a dog in an experiment. Every time the assistants tried to tie it into its harness, the dog put up a struggle. What caused this behavior? After weeks of thought and discussion, Pavlov and his assistants finally stumbled on the answer: the freedom reflex. Our interests are less in whether dogs have a freedom reflex than in why people—including an early saint of behaviorism—might conclude that they do.

We sketch three accounts of why humans anthropomorphize:[1] anthropomorphism may result from a cognitive default; the perception of overlapping species coordination systems; or a human, species-typical coordination system. The stories illustrate a central point: whether or not humans can know about the mental states of animals is connected to what we believe about human mental states. Beliefs about human mental states influence, and are influenced by, beliefs about the possible attributes of other entities. (At one time, for example, playing chess at a grand master level would have indicated intelligence in computers; today intelligence is the ability to understand a story at the same level as a toddler.) The appropriate question for readers to ask themselves then is not which account is true, or whether one account says animals have mental states while another does not. It is, rather, how useful is one or another account given readers' multiple agendas, be they philosophical, behavioral, psychological, political, or any combination thereof. The three accounts are not mutually exclusive. In fact, elements of each might someday be combined into a theory of context-based mental state attribution. What we do doubt, however, is that any such theory could profitably be grounded in a belief that human cognition is ahistorical, neutral, disembodied, and unbiased by nature or training.

Attributing human characteristics—specifically mental states—to nonhuman entities is pervasive among humans. All cultures have metaphors relating humans, animals, and other entities. Anthropomorphic thinking is characteristic of children, who also show developmental shifts in its use (Inagaki, 1989; Inagaki & Sugiyama, 1988). Anthropomorphism has properties of scale, persisting for the few seconds that a car fails to start or the dog "looks guilty," or for centuries, in various beliefs about a diffuse causal agency in the universe and the human attributes of trees, rivers, and animals. People implore cars to start, dance to cause the rain, and threaten computers. Some authors allow that anthropomorphism may be "built-in" to the human repertoire, but they disagree about the consequences. Kennedy (1992) views an anthropomorphic tendency as a liability, hopelessly distorting our understanding of animal behavior and requiring prophylactic action. Fisher (1990) suggests that humans may have innate conceptual frameworks for understanding humans and other animals and that their *rejection* leads to distortion. Interestingly, neither of these positions denies that humans "naturally" anthropomorphize.

Other authors assume that the attribution of human characteristics to nonhumans is not anthropomorphism at all. Instead it is either veridical (Griffin, 1984; Povinelli, 1987) or at the very least "conceptually innocent" (Dennett, 1978)—a way station enroute to a better understanding of how biological systems are designed:

> Once we have tentatively identified the perils and succors of the environment (relating the constitution of the inhabitants, not ours), we shall be able to estimate which goals and which weighting of goals will be optimal relative to the creature's *needs* (for survival and propagation), which sorts of information about the environment will be *useful* in guiding goal-directed activity, and which activities will be appropriate given the environmental circumstances. Having doped out these conditions (which will always be subject to revision) we can proceed at once to ascribe beliefs and desires to the creatures. . . . It is a sort of anthropomorphizing, to be sure, but it is conceptually innocent anthropomorphizing. (Dennett, 1978, pp. 8–9; cf. Dunbar, 1984)

Just as assumptions of natural anthropomorphism presuppose a cognitive apparatus with certain features (a tendency to anthropomorphize), conceptually innocent anthropomorphism presupposes a cognitive apparatus that can be neutral and unbiased. Scientists might make errors, of course, but errors would be móre or less randomly distributed and certainly not biased toward attributing human characteristics to nonhuman entities. Conceptual innocence and neutral cognition also presuppose that science is generally unaffected by its social context. In particular, the values we place on animals would not influence the beliefs and desires we attribute to them. These are familiar, everyday beliefs about science. However, a mass of research by psychologists shows that humans deviate from ideals of neutral rationality (Kahneman, Slovic, & Tversky, 1982; Faust, 1984; Dawes, 1988), and work by historians and sociologists of knowledge shows that science divorced from its social context is a chimera (Latour, 1987; Pickering, 1992). Although human cognition and the social context of science potentially interact with any scientific inquiry, anthropomorphism appears to be a special case: its elimination was a critical condition for the emergence of modern science (Hansen, 1986).

Still, even if conceptual innocence is improbable, that would not mean anthropomorphism is automatically dangerous or merely unproblematic. Assessing its impact demands more than a better understanding of human cognition—it requires establishing the kind of relation that exists between human minds and animal minds. Currently that relation supposes that folk psychological concepts can be applied to other systems (be they animals, machines, infants, or other humans), and existing debates ensue from that starting point. We suggest that multiple diverse starting points can break the repeated recycling of anthropomorphism's problems and prospects described by Mitchell (1996).

In the following sections, we present three accounts of anthropomorphism. Our first two accounts are intuitively plausible and could be inserted into ongoing debates about animal mental states without severely distorting folk psychological intuitions; the third is plausible, but not intuitive. It arises from the challenge scientific social psychology poses to folk psychology. All three accounts assume "selection for sociality," a scenario for the evolution of human mental systems (Caporael, Dawes, Orbell, & van de Kragt, 1989). The scenario proposes that human mental systems are specialized for face-to-face group living, and that the interface between individual and habitat is a group process. Consequently,

> the human mind/brain evolved for being social (and for learning what that means in our cultures) and not for doing science, philosophy, or

> other sorts of critical reasoning and discourse . . . We expect and find
> cognitive limitations especially under conditions of uncertainty . . .
> These limitations contribute to and interact with various sociocultural
> constructions including folk psychological notions of "human
> nature." . . . Cognitive limitations and the ruses of culture may be
> overcome to some extent by education, environmental feedback or
> "collective rationality" . . . (Caporael et al., 1989, p. 730)

Drawing on empirical research in psychology, selection for sociality
differs from the "evolution of social intelligence" program in important respects. Specifically, it challenges assumptions that behavior can
be explained in terms of genetic or individual self-interest, that group
behavior is the aggregate of individual exchanges for mutual benefits,
that humans engage in complex calculations about their own and
other's self-interests, that humans are "natural psychologists," and that
"cold cognition" or "technical intelligence" is independent of or qualitatively different from social cognition (Alexander, 1989; Byrne &
Whiten, 1988b; Humphrey, 1986; Dunbar, 1984). Selection for sociality is
a minimalist scenario—it does not assume folk psychology provides
an adequate description of behavior; it merely sets a stage for asking
what are the minimal cognitive requirements to negotiate life in social
groups. Hence, there is no single "just-so" story about how anthropomorphism was "really" advantageous in the past. Instead, the scenario
is used to generate three speculations about the psychological foundations of anthropomorphism.

THREE ACCOUNTS OF ANTHROPOMORPHISM

Anthropomorphism as a Cognitive Default

The idea that anthropomorphism is connected to a peculiarity of human
thinking—a type of cognitive default rather than a veridical perception—underwrites much of the controversy about animal minds,
although the peculiarity is seldom specified in much detail ("projection" being a common explanation). C. Lloyd Morgan, in his criticisms
of George Romanes's comparative psychology, explained anthropomorphism as "ejective psychology"—the conscious and superficial bits
of an observer's personal psychological states flung onto other humans
or animals (Richards, 1987). The observer's various "subconscious-
nesses" were the product of past experiences, bodily sensations, and
others' perceptions of the observer.

　　Positing an evolutionary scenario similar to selection for sociality,
Humphrey (1976) proposed humans possessed a distinctive "creative

intelligence," which originated in "the social function of intellect." Individual self-interest was tempered by a sympathetic identification whereby an actor would take another's goals as his own. Social cognition was a "'constraint'" on reasoning, "such as might result if there is a predisposition among men to try to fit nonsocial material into a social mould [*sic*]" (p. 312). Thus, sacrifices and rituals were attempts to bargain with nature, but "nature will not transact with men; she goes her own way regardless—while her would-be interlocutors feel grateful or feel slighted as the case befits" (p. 313). Backing off from this claim a bit, Humphrey asserted that, in fact, nature does sometimes respond. The "relationship of a potter to his clay, a smelter to his ore or a cook to his soup, are all relationships of fluid mutual exchange, again proto-social in character" (p. 314). Technical skill was built upon the foundation of social activity, not distinct from it. In its "descent with modification," Humphrey's novel and distinctive hypothesis was transformed into a familiar folk psychology: the social function of intellect, which was originally to hold society together, became a "Machiavellian intelligence" for manipulating conspecifics in one's own self-interests. Technical and social intelligence were recast as distinct and separate, with an overt warning not to use social intelligence to explain all primate social behavior when comparing the two intelligences (Whiten & Byrne, 1988, p. 50).[2]

A decade after Humphrey's (1976) seminal paper, the social function of intellect was revisited in another context—attributing human characteristics to machine intelligence. Caporael (1986) argued that anomalies such as anthropomorphism or cognitive limitations could be used to make a plausible case that human cognition was fundamentally social cognition. She proposed that anthropomorphism was a "cognitive default" engaged when explaining or predicting the behavior of an entity was important, but no handy explanation for its behavior was immediately available. Anthropomorphism would have a variable time scale, from the desperate moments one spends begging a stalled car to start, to the centuries of custom soliciting supernatural aid. From an evolutionary perspective, attributing human characteristics may be part of a psychological *Bauplan* (process) originating in selection for sociality, and development involves the process of learning when not to default. Piaget (1929/1967), for example, concluded from his work on childhood animism that universal life is the primary assumption of child thought. The attributes of inert matter are gradually detached by thought and experience from the primitive continuum along which all things are living. The argument for a cognitive default is more clear-cut in cases where an entity is inert or "non-life," such as natural systems or computers. Anthropomorphic attributions drop out when an alternative language for describing phenomena becomes available—for example, auto mechanics, computer programs, or weather forecasting.

Sellars (1963) has argued that scientific discovery begins with anthropomorphism. As continued attribution of human characteristics fail to result in effective interactions with nature, other explanations or explanatory metaphors may be sought, giving rise to a more adequate theory. Given that feedback from the environment is frequently irregular (e.g., sometimes it does rain after a rain dance), it should not be surprising that anthropomorphism can persist for long periods of time. A phenomenon may require focused attention as well as new cultural metaphors in order for an alternative explanation to arise. For example, Kepler initially described astronomical phenomena in terms of an intelligent sun that forced planets around violently; it also would languish and grow weaker because of the remoteness of virtue (gravity). Only later did he find another vocabulary that allowed him to envision a celestial machine that works like a clock in which a weight drives the gears (Gordon, 1974). One of the reasons Aristotle's physics is "foreign to the modern mind" is that it is an anthropomorphic one where bodies fall because they seek a specific natural resting place (Wiser & Carey, 1983).

Whatever the specific details of anthropomorphism as a cognitive default, we are left with a conundrum. The hypothesis suggests we are subject to being hopelessly confused about the existence of non-human minds because we will default to attributing human characteristics whenever the going gets rough. The characteristics attributed to animals may or may not exist, but it would take clever experimentation with strict controls to distinguish the operation of their minds from that of our minds.

Overlapping Interspecies Coordination Systems

Like humans, the members of many other species must coordinate their own behavior with that of their conspecifics. They must know when to feed and protect their young, when to avoid a conspecific, and when to approach one. Presumably, appropriate coordination systems would evolve, and at least some systems would overlap, allowing for a sort of "trans-specific recognition" of mental states (Gallup, 1985; Povinelli, 1987). An analogy with human behavior would be the cross-cultural interpretation of smiles indicating approachability and frowns indicating avoidance. Some years ago, one of us (L.R.C.) observed her son, then five years old, interact with a one-year-old orangutan. The pair, who had never interacted with juveniles of each other's species before, executed an "invitation-to-play-by-imitation," initiated by the orangutan and complete with turn-taking, escalation, and variation of imitation. The behavior was identical to the ritual performed by four- and five-year-olds on playgrounds every day. The exchange suggested the possibility that baby humans and orangutans might possess procedural rules appropriate to their own species, but still recognizable and functional across species lines.

Another possibility is suggested by a study undertaken by Berry, Misovich, Kean, and Baron (1992). They used the well-known Heider and Simmel (1944) animated film to elucidate the stimulus properties that give rise to anthropomorphic distinctions. The film shows a big triangle chasing a little triangle and circle around and inside a house (a rectangle with a moving-flap "door"); the little triangle locks the big triangle in the house, which explodes. (Among the many possible reasons for anthropomorphism, economy of description must be one.) Subjects watched an unaltered version of the film as well as a version that disrupted the motion and one that disrupted the shapes. The results for the unaltered film replicated Heider and Simmel's findings: not only do people anthropomorphize the geometric shapes moving on the video screen, they agree on the gender of the moving forms and on a story of aggression, rescue, and escape. Disruption of shapes reduced, but did not eliminate, anthropomorphisms; however, strobe-like stuttering disruption of motion patterns did. Clearly, dynamic transformations are used as evidence of intentionality, but cannot be used to distinguish "real states" from supposed ones.

Evolutionarily, there could be two accounts for overlapping coordination systems. The overlap may be a result of common descent (hence, we could imagine human and orangutan juveniles possessing the same or similar procedural rules). But overlapping coordination systems could also arise from convergent evolution, which would have to be the case for dogs and dolphins as well as other animals (which would make the dynamic transformation hypothesis especially interesting). Positing overlapping coordination systems implies that comparative analysis would be a useful enterprise for determining what common environmental features in the past, given current developmental trajectories, might account for a convergence of coordination. An immediate consequence of this scenario is that the case for mental states where neither homology nor analogy could be invoked (e.g., birds, spiders, etc.) imposes a special burden of explanation to demonstrate that the attribution is not garden-variety anthropomorphism.

Two related complications seem noteworthy. In the boy-orangutan illustration, attributing the desire to play (which the surrounding adults did) was the outcome of observing an ordered sequence of responses. That is, the adults attributed the intentionality after the sequence; the juveniles were simply responding to each other directly. Neither may have attributed to the other a desire to play as the basis for the interaction. Hence, coordination systems may not require attributions of intention at all but simply be a meshing of two systems in an on-line exchange. This would leave the intentionality attributions of the adult observers "dangling" so to speak; at best a linguistic summary of a past event lying outside the coordination system itself. To bring attributions of intention inside the system would require they be involved in the production of behavior, not be merely the linguistic detritus left over from an interaction sequence. But this would leave unexplained the reasons why the adults spontaneously agreed the juveniles wanted to play. In other words, if attributing intentions, beliefs, and desires does not function in an interaction, how do we explain the attributing activity, which itself is fairly complex and seems like it should be useful for something?

Dog trainers—ordinary people who train other ordinary people to interact with their dogs—are real-life, practical demonstrations of cross-species coordination that merit research. A good dog trainer is an intermediary between two species. She discourages her human trainees from attributing "high level" human intentionality because it interferes with dog training. Attributing jealousy, fear, or protectiveness to the dog does not help reduce growling when someone comes to the door. For example, the dog owner, believing Rover is jealous, reassures him by patting his head when he growls at the doorbell—an instrumental conditioning procedure that increases the frequency of growling. The expert trainer teaches the novice human two important lessons. The first is how the owner should behave so that the dog will "canine-morphize" her. Lesson one in dog-training school usually concerns a discussion of how the owner behaves so that Rover recognizes that she is his pack leader. Pack leaders go through doors before the pack; they can take food from pack members' plates, but they never, never permit pack members to take food from their plates. Without emotional yelling and screaming, pack leaders make pack followers lay down in one place for a long time for no other reason than that pack leaders are pack leaders because they enforce their will. Now one result of this descriptive exercise is that it is very clear that to describe behavioral coordination from the point of view of an animal is almost impossible without imagining that the dog is also a folk psychologist (with a somewhat different agenda).

But to actually predict and control the behavior of the dog—essential to training—the owner must be trained as an effective behaviorist.

In scientific circles, this is usually conceived as something psychologists do *to*, rather than *with*, animals. In dog-training circles, behaviorism is conceived as a "a loop, a two-way communication in which an event at one end of the loop changes events at the other, exactly like a cybernetic feedback system" (Pryor, 1984, p. 15). It requires acute attention to the "perils and succors" the species has faced in the past. The pack leader has to be taught to run *away* from the dog if it fails to respond to the "come" command (followers follow, and pack leaders never repeat their commands). Children must be watched carefully around adolescent dogs because a child's darting motions resemble the darting motions of prey. In addition to the behavioral propensities of species and breed, the owner is taught to watch carefully for cues indicating the dog's immediate state. A yawning dog is not a sleepy dog, but a dog under stress.

As the last example suggests, there is a fuzzy distinction, but a distinction nonetheless, to be made between anthropomorphism and understanding the behavioral propensities of an animal. However, it occurs at the level of situated action, played out through response and counter-response (Mitchell & Thompson, 1991). The application of human folk psychology (yawning indicates a desire to sleep) fails to lead to behavior appropriate to the coordination system. The term "stress" is arguably anthropomorphic, but in context (a critical qualification) it summarizes in dog-trainer language the experience of past contingencies. It functions to predict a framework for future contingencies because it constrains the trainer's behavior: a stressed dog will not work well and some other action must be taken. This account allows that some apparently anthropomorphic behavior in humans may not be real anthropomorphism at all. It represents, at a behavioral level of response contingencies, an overlapping among coordinative systems, which may have arisen through common descent (e.g., boy-orang) or through each species having common goals arising through interaction with its relevant milieu. The resultant "common features" may be "read" across species lines, and hence, coordination can occur. This understanding of anthropomorphism is sympathetic to the scientific "critical anthropomorphism" advocated by Burghardt (1985a), but it suggests a particular methodological approach frequently abjured in the investigation of mental states—a behaviorist approach. As Pryor (1984) observes, "Knowing nothing about a particular species but knowing how any subject tends to react to various training events, one can learn more about the nature of a species' social signals in a half hour of training than in a month of watching the animal interact with its own kind" (p. 167).

Species-Specific Group-Level Coordination System

The cognitive default story allowed that folk psychology might be predictive, and defaulting was an epiphenomenon of an otherwise useful "theory of mind." The overlapping species coordination story was somewhat less sanguine, suggesting anthropomorphic attributions were dangling outside the system. They worked to some extent under some conditions as summaries of past behavior, and only as a summary constraining the possibilities for future behavior could anthropomorphism be said to predict behavior. Suppose, however, that we exercise our minds like the White Queen by believing impossible things for a half hour each day (Millikan, 1993), and believe that folk psychology is *not* predictive. We are left with a curious situation: millions of humans talking about beliefs, wants, desires, and motives in widely shared ways *as if* such things were descriptive and predictive, at most arguing about *which* motive is salient, or *what* desire was being pursued, but rarely doubting the validity of the enterprise as a whole. Suppose further that the enterprise, whatever its validity, is so complex and well designed that it appears to be something that should have a function. If not succinct description and prediction, what could that function be, and how might anthropomorphism be related to it?

Our "curious situation" might not be so impossible after all. If the value of intentional description of behavior is that it predicts behavior, it would imply that people's "thoughts and feelings"—needs, wants, and desires—would be informative about behavior, and conversely, behavior would illuminate thoughts and feelings. But this very commonsensical assumption is undermined by behavioral research. Not only do people believe thoughts and feelings are *more* informative than behavior, they are significantly better at predicting thoughts and feelings *given thought and feeling information*, than they are at predicting behavior on the same basis (Andersen & Ross, 1984; Andersen, 1984). Moreover, given behavioral information, people are biased toward attributing dispositional factors as its cause and overlooking situational factors, a widely studied phenomenon known as the "fundamental attribution error" (Jones, 1990; Ross, 1977). If the accurate perception of causes in the immediate past is difficult and easily compromised (e.g., by changes in lighting or the color of clothing—Taylor, 1982), the prospects for accurate predictions of future behavior are not promising. If attributions of intentionality in the conventional folk psychological sense fails to predict human behavior, what else could folk psychology be *for*?

Our minimalist scenario asserts that human mental systems are partly an outcome of selection in groups. If humans are obligately interdependent—obligately social—they require some system for the coordination of behavior. (Pheromones, for example, help coordinate behavior for obligately interdependent social insects.) Talk of intentionality—folk psychological talk—might be a modern instantiation of such a coordination system, a group-level adaptation that is at best weakly connected to individual behavior. Whatever accounts for the behavior of individuals is not motives, goals, intentions, and desires, but some other property or set of properties emerging from ongoing interaction with the environment. Folk psychological talk would organize all the bits of ongoing organism-environment interaction so that behavior would be coordinated. What could folk psychological talk be doing that it could result in behavioral coordination? The clue is in the talk more than the folk psychology.

We deliberately used the word "talk" rather than "language" in the previous paragraph to draw attention to everyday discourse. Everyday talk by ordinary people is "about-talk"—largely concerned about who did what to whom and about what things (shoes, sports teams, dog breeds, weather, ideas, etc.) are good and bad. (Science, philosophy and other sorts of "cold cognitive" critical discourse is derivative [Caporael et al., 1989; Zajonc, 1980].) These are matters of value, not merely of description. Talk has connotative functions in addition to denotative functions and provides an orientation along a good-bad, approach-avoidance dimension for interacting with components of the material and social environment. Connotative functions are privileged: affect frames information content and lingers after content vanishes. People have "first impressions" of liking or disliking other people before an interaction even occurs; they remember disliking or liking a book long after its story or argument has been forgotten. In fact, humans are capable of judging whether or not they like a stimulus even if they cannot identify it, evidence of the primacy of value in perception (Zajonc, 1980). Perception itself is a value-realizing, value-organizing activity.

Even as simple an artifact as a toddler's spouted cup, weighted on the bottom to discourage spilling, connotes a value on independence and self-sufficiency consistent with a culture with early weaning and training for independence (Hodges & Baron, 1992). Indeed, the mere existence of such a device (one of many possible designs for transporting liquids to the mouth) contributes to the direction of the parent's child-rearing activity, influences the child's development, and reverberates through the culture, constraining other activities. Just imagine the clucking that would occur if a woman in an industrialized country were to breastfeed a three-year-old. Humans live in a value-saturated environment. Talk is an ongoing, nonstop negotiation of value that is translated into interaction in a world where "the social" actively inter-acts with "the material" (as in the toddler cup example). Talk about beliefs, intentions, and desires (folk psychology) develops a context for behavior that limits and entrains what actions are conceivable, possible, undesirable, and essential.

In a nutshell, the way folk psychological talk enhances coordina-tion is that it prescribes values, which are realized in the activity of ongo-ing organism-environment interactions; and values coordinate behavior because they limit the degrees of freedom for potential interaction. Folk psychology orients actors toward a widely shared version of common sense, but one that is always subject to negotiation and revision.

From this perspective, attributing human characteristics to ani-mals is a way of changing the values we place on them and how we can behave toward them. Anthropomorphism is part of changing social values; specifically, we suggest, values related to the environment and animal rights, a connection made by several chapters in this volume. Two decades ago, agnostic defenses of anthropomorphism, much less spirited ones, were largely inconceivable for scientists (see Mitchell, 1996). There was little talk about animals in the larger culture, and con-straints on interactions with them were largely associated with animals in the pet category. From primates to rodents, animals were simply one of many components in a mechanistic "Newtonian ecology" that ori-ented human interaction with nature in terms of utility (Boucher, 1985). Changing this perception requires changing values and perceptions in both science and its social context. Before wading into deeper contro-versial waters, we want to make perfectly clear that our comments are speculative, meant to engender both research and dialogue on the com-plex relation between science and society.

Based on a random survey of over 400 animal rights activists, Jamison and Lunch (1992) showed that not only are environmentalism and animal rights functionally related, but that they are also part of a larger liberal and egalitarian social agenda. Additionally, animal rights activists had extremely negative views of scientists (in marked contrast to their views of environmentalists or feminists), ranking scientists with businessmen and politicians, perceiving them as symbols of traditional authority. Over 50% of respondents believed scientists did more harm than good to society. Although the top leadership in the animal rights movements disapprove of pet ownership, and often seem careful to emphasize that their positions are not based on mere sentiment, Jamison and Lunch (1992) found that "intensely emotional experiences with pets" were a significant mobilizing force for the rank and file. Although animal rights movements have occurred in the past, associated with reactions against technological change and exploitation of nature, we suggest anthropomorphism links scientists and like-minded scholars with environmental/animal rights issues in the larger society. An enterprise for "the scientific study of animal minds" creates a floating island between two apparently opposed and hostile communities as well as a refuge and recruitment center for pro-activist scientists and pro-science activists. With the express purpose of demonstrating human-animal continuity, explicitly justifying values for preserving the environment, and promoting humane and egalitarian relations, animal minds research, but not anthropomorphism research, may be a vehicle for changing value-making talk and value-realizing perception.

At this point, some readers will be wondering whose side we are on. We see dangers in the traditional historical privileges of the scientific community, predicated as they were on the privileges of gender, race, and class. For precisely the same reasons, we are uneasy in principle with appeals for social and political values that begins with "science shows. . . ." No matter how much we agree with the values on behalf of which such appeals are made, "science shows . . ." has also been engaged, sometimes successfully, for a variety of discriminatory, noxious, and even deadly social agendas because scientists have special authority in scientific culture. We agree that scientists are not value-neutral, and that values have an important role in critical inquiry and scientific criticism (Longino, 1990); but as we argue in the next section, we believe values should be the products of interactions amongst citizens (who might use scientific information), not of interactions among "scientists" and "ordinary people."

Traditional Newtonian ecology is undergoing a crisis, which Boucher (1985) attributes directly to the environmentalist movement. Students entering ecology in the 1960s were *politically* concerned about environmental issues, and found Newtonian ecology insufficient because it "failed to express the *value* of the environment" (Boucher, 1985, p. 22, emphasis added). Similarly, scientific discussions about animal minds and intentional states are associated with values concerning animal rights and environmentalism (Morton, Burghardt, & Smith, 1990; Plous, 1993a). Bouissac (1989) points out in his call for a scientific "neo-animism" that anthropomorphism is crucial to "the most fundamental dilemma of our time: whether the ever-growing exploitation and control of the environment is worth the risk of its continuing alteration and even annihilation" (p. 498). To attribute human characteristics to animals is a negotiation of value among humans. It changes the way humans perceive animals, and limits and entrains what actions are conceivable, possible, undesirable, and essential.

CONCLUSIONS

Allow us to confess right off: we have no conclusion. We cannot say if anthropomorphism is "really" triggered by ignorance (the cognitive default story), a summary of tentative "predictions" reflecting something "real out there" (overlapping species coordination systems), or a value-making activity of obligately social creatures organizing themselves in a complicated world. We do assert that whatever anthropomorphism is, it is too complex, too multiply stranded, to expect it to be "conceptually innocent."

Clearly, research distinguishing among these accounts would not establish which systems, if any, have mental states. Nevertheless, research on anthropomorphism would have implications for the conduct of empirical investigation on mental state attribution. If anthropomorphism is a cognitive default strategy, then by using effort and imagination to formulate and test alternative accounts of behavior, it should be possible to dislodge anthropomorphic accounts where they are inappropriate. If anthropomorphism arises from shared coordinative strategies, it may be more difficult to dislodge through scientific inquiry. If anthropomorphism is value-making, scientific methods may reveal that an anthropomorphic account of behavior lacks descriptive and/or predictive power relative to some alternative account, but the same methods cannot be used to assess the evaluative function of anthropomorphism. The last possibility suggests a dilemma that should not be overlooked. If anthropomorphism is connected to environmentalism, as we among others suggest (albeit for different reasons), and if it does constrain destructive action in our relation with nature, then empirical ambitions for the discovery of mental states may ultimately be irrelevant. Even if technological fixes were developed to reverse environmental destruction, they would require large-scale shifts in world views (read "values") to be translated into collective behavior (Bouissac, 1989). Anthropomorphism may be an important means for connecting values to action for environmental preservation, and too important to discourage, whatever its foundations, whether or not animals really do have mental states.

We do, however, make a plea for preserving the integrity of both science and values. Scientists do not need to demonstrate that animals have minds (a scientific question) in order to assert that the destruction of nature or the abuse of animals is wrong. The danger of not separating these issues is in conflating scientific values and common sense values—with both being the worse off for it. Rollin (1989, this volume), for example, urges science to follow common sense: animals appear to have mental states, and should be treated accordingly. However, if we insist that science conforms to common sense, we risk losing the identity of science, which in our view is a fallible set of heuristics for finding out how things work despite how they appear to our common sense. Taking the opposite tack, Cavalieri and Singer (1993) assert that there is enough scientific information to make the moral boundary between great apes and humans indefensible. There is a fine line, however, between using research results to inform common sense, and invoking scientific authority to justify it or indicate what it should be: even strong proponents of Cavalieri and Singer's (1993) agenda, proponents intent on "slaying" dissenters, recognize the existence of scientific dissent (Bekoff & Allen, this volume). Our concern is this: If scientific authority is allowed to colonize or subordinate common sense, we risk losing uses for common sense. Common sense, which combines value-making talk and value-realizing perception, is for negotiating and adjudicating our collective moral, social, and political lives. What language and what authority will ordinary citizens have to do this work, particularly if their opponents are scientists? If moral boundaries can be eradicated by "scientific information," what keeps them from being erected with the same justification? "Naturalized" by scientific authority, be it psychology or biology, common sense could no longer serve Everyman and Everywoman through the negotiation of action-organizing values, but would be shifted instead into the hands of science.

NOTES

1. We have decided to use the term *anthropomorphism*, rather than other possibilities such as *mental state attribution* or *subjective analogical inference* (Burghardt, 1985a), because we see it as pointing toward human cognition in much the same way that mental state attribution points toward other entities and away from humans.

2. Arguably, Humphrey (1976, 1986) may be partially responsible for this retreat. His 1976 paper also proposes that humans had "remarkable powers of social foresight and understanding" for calculating the consequences of their own behavior, the behavior of others, and the consequent gains and losses. Notably, his paradigmatic example of social interaction is chess, a zero-sum game. If humans have remarkable powers for calculating the costs and benefits of their own and others' behavior (and the research evidence suggests they do not, as we indicate later), we would expect them to be able to detect that nature does not respond to human overtures.

PART III

Anthropomorphism and Mental State Attribution

7

Anthropomorphism and the Evolution of Social Intelligence: A Comparative Approach

Gordon G. Gallup, Jr., Lori Marino, and Timothy J. Eddy

Why do we anthropomorphize? Why is the tendency to attribute human characteristics to animals so widespread and so difficult to resist? Are these attributions accurate? Can dogs really distinguish between being tripped over and kicked? Are there other species which not only make inferences about mental states among themselves, but generalize these attributions, as we do, to other species as well?

We contend that anthropomorphism is a by-product of self-awareness and the corresponding ability to infer the experience of other humans by using one's own experience as a model (see Gallup, 1982). It is probably true that no two people experience the same event in exactly the same way, but since we are members of the same species we all have similar receptor surfaces and underlying neural circuitry in common. As a consequence, there is bound to be considerable overlap between our experiences. Moreover, given a knowledge of our own mental states and their relationship to external events, we now have a means of modeling each other's mental states. Knowledge of self, in other words, is a vehicle which provides a means of achieving an intuitive knowledge of others. By the same token, to the extent that humans

The authors thank Daniel J. Povinelli for comments on an earlier draft of this chapter.

and other species share similar sensory capacities and mental experiences, anthropomorphism could represent a plausible account of why animals do what they do.

Humphrey (1983) has postulated that the complex interdependence of people in human society was essential for accomplishing the intricacies of survival, and that it became necessary for individuals to become "natural psychologists." Humphrey argues that, early in the hominid line, individuals who were capable of functioning as natural psychologists had a clear selective advantage in such a highly social and interdependent species as our own.

That we routinely make inferences and attributions about different intentions and mental states in one another is obvious and well documented and has been the subject of considerable research by social psychologists (Heider, 1958; Jones & Davis, 1965; Kelley, 1971) under the topic of "attribution theory." This work has generally focused on the tendency for people to make attributions about various mental and emotional states in other people. The interesting effect of this capacity, however, is that we frequently and almost automatically generalize these attributions to species other than our own. Statements such as "the dog feels guilty about having misbehaved" or "the baby monkey is sad and lonely" are commonplace and represent familiar examples of anthropomorphism. Indeed, in our view, anthropomorphism is simply a dramatic instantiation of this more basic introspective modeling capacity (Gallup, 1985, 1992). In order to infer that the baby monkey is sad and lonely, it presupposes prior personal experience with what it is like to feel sad and lonely.

There are a number of interesting questions that follow from this analysis. First, are there limits to the extent to which people make these generalized attributions (e.g., are people as likely to make attributions about intentionality in crickets as they are rhesus monkeys)? What is the shape of the generalization gradient when it comes to inferring mental states in other creatures and, if it is anything but flat, what characteristics seem conducive to making these inferences (cute vs. ugly, warm-blooded vs. cold-blooded, predator vs. prey, etc.)? Second, to what extent are these attributions accurate or unfounded (e.g., can dogs really distinguish between being tripped over and being kicked?). Third, are there creatures other than humans that routinely make inferences about mental states not only among themselves but in other animals as well? And, if so, how widespread is this capacity? Although among humans the object of different mental state attributions is typically other humans, different species and even imaginary entities can elicit such attributions. Anthropomorphism, in other words, is but one of several ways in which people generalize their capacity for mental state attribution.

It is tempting to include inanimate objects as another category of anthropomorphism. Many individuals use "anthropomorphic terms" to describe inanimate objects (e.g., computers, cars). Indeed, much of the early work on attributional capacities (as well as some more recent studies) was focused on how people perceive the "behavior" of moving geometric shapes and simple objects (Heider & Simmel, 1944; Dasser, Ulbaek, & Premack, 1989). However, with the exception of animism in young children, most such instances merely represent the use of descriptive terms in order to conveniently convey information about an inanimate object. We may really *believe* that a dog knows that we trust him to guard our property while we are gone, but few of us would suggest that a computer can actually feel insulted when we shout loathsome remarks at it upon discovering that an important file was destroyed.

Is anthropomorphism widespread simply because it provides a familiar and comfortable account of animal behavior? That is, in describing the behavior of animals do we find ourselves using accounts based on mental state attribution simply because these are the same kinds of accounts we use to describe the behavior of people? Is it merely a matter of convenience, or do we believe these accounts? Do some of our attributions map on to real characteristics of other animals? And if so, which attributions are accurate and to which animals are they applicable?

Mental state attributions cut both ways; i.e., it is one thing to ascribe mental states to animals but quite another to assume that they make mental state attributions of their own. Does the dog love its owner? If it does, is it aware of it? If it is aware of loving its owner, can the dog use its experience with love to infer love in other dogs? Could a dog capable of reflecting on its love of its owner make an attribution about love in its owner? In other species? The mere existence of mental states is no guarantee that organisms will infer comparable states in conspecifics, let alone interspecifics. Indeed, many species may not even be aware of their own mental states (Gallup, 1982). Eddy, Gallup, and Povinelli (1993) found that the extent to which people would ascribe the capacity for mental state attribution to animals was correlated with independent estimates of the extent to which the species was rated as being similar to humans (e.g., mammals were judged as being far more likely to distinguish between being tripped over and being kicked than were cockroaches).

At a fairly pragmatic level there are at least three discernible issues that may be involved in making mental state attributions about animals. First, mental states can be used to generate *descriptive accounts* of their behavior ("the dog loves his owner") which are congruent with comparable accounts that we routinely ascribe to other people and ourselves. Second, mental states can also be used to provide an *explanatory framework* ("the reason the dog is barking and jumping up on his owner is because he loves him"). Finally, mental state attributions involving animals may even have *predictive value*. That is, irrespective of whether the dog really loves his master or not, I might be able to develop a predictive model of the dog's reactions to his owner based on the use of a "love" metaphor. In the latter case, it is important to emphasize that in order for anthropomorphism to have predictive value it does not have to be a valid explanation for why the animal is doing what it is doing.

Many animals act as if they are self-aware but upon closer inspection these often turn out to be instances of hard-wired analogs to self-awareness (Gallup, 1982, 1983). Take the case of a duckling's initial reaction to the silhouette of a hawk. It is not necessary for the baby duck to have any insight into the significant danger posed by hawks or their ostensible intentions. The only thing that matters is that they respond appropriately during their first encounter with the hawk. Indeed, since most hawks would be unlikely to give baby ducks a second chance, if this escape response were not hard-wired it would likely obviate any subsequent need to infer mental states in hawks, other ducks, or anything else for that matter. The point is that attributions about the mental states of other animals can have *heuristic* value regardless of whether such explanations are correct.

ATTRIBUTIONS ABOUT OTHER PEOPLE

Our use of mental state attributions are often altered by assumptions we make about other people who may have a different perspective than our own. Prototypical instances could involve interactions with blind or deaf individuals, persons who speak a different language, young children, mentally retarded persons, and people who are mentally ill. For example, when attempting to communicate with a lip-reading deaf person, hearing people almost invariably increase the volume of their speech as if to compensate for the loss of hearing in the receiver, even though it is obviously ineffectual. In these cases we are not anthropomorphizing in the strict sense of the word, but we do alter our behavior towards these individuals based on certain assumptions we make about their level of social maturity, sensory status, and/or intellectual competence. A relatively pure case of mental state attribution would be to offer someone experience with blindfolds, earplugs, etc. How would the person respond to subsequent confrontations with another person wearing either blindfolds or earplugs? Would they use their experience with these obstructions to infer the other person's impaired sensory experience and adjust their behavior accordingly (e.g., raise the volume of their voice if communicating with a person wearing earplugs)?

Normal adult social exchanges are dependent upon the capacity of the participants to take into account the perceptual, informational, and emotional states of each other. The breakdown of this capacity is illustrated by a mindless conversation. Normally, upon initiating a conversation with someone, we take into account the fact that the other person does not have prior knowledge of our intention to communicate. We must also make attributions about the other person's knowledge state when we ask the person something or engage him or her in a conversation. Therefore, we typically offer "reference points" for the other person by way of an introductory statement with some initial information about our intention to communicate about a particular topic. For example, when initiating a discussion with a comment such as, "Excuse me, I'd like to ask you about the paper you published last month on vocal development in vervet monkeys," we first alert the person of our intention to communicate, and then we orient them in the direction of the topic we wish to pursue. However, sometimes people begin a conversation without taking into account the listener's lack of prior knowledge about their intentions (i.e., they act as if the listener had somehow been privileged to their thoughts or prior experience). Therefore, instead of the statement above, the mindless conversation might begin with "What were their ages?" to inquire as to the ages of the vervet monkeys in the other person's recently published study on vocal development. The one on the receiving end of such a non sequitur finds it nearly impossible to infer the intentions and mental state of the inquirer and therefore cannot respond appropriately. The ability to make attributions about another person's knowledge state is normally used to create a context which facilitates the flow of information between individuals.

When we are faced with persons whose mental states and experiences are perceived as very different from our own the behavioral adjustments we need to make are even greater. This is exemplified by the way adults talk to babies and toddlers. Adults from many different cultures tend to communicate with very young children through a form of slow, high-pitched, over-pronounced speech called "motherese"

(Jacobson, Boersman, Fields, & Olson, 1983; Newport, 1976). Benedict (1976) found that these characteristics of speech are most likely to capture and hold an infant's attention. Children of at least four years of age will often simplify their speech in a similar manner when interacting with mentally retarded peers (Guralnick & Paul-Brown, 1977). We believe that this form of communication is based upon attributions about the mental capacity and knowledge state of the young child or mentally retarded individual. For reasons we will detail (Gallup & Suarez, 1986), there is evidence that children beyond the age when they show signs of self-recognition (between 18 and 24 months) are capable of altering their behavior in accordance with the perceived perceptual and mental state of another person (Guralnick & Paul-Brown, 1977; Shatz & Gelman, 1973). An interesting question to emerge from this evidence is whether the onset of anthropomorphic tendencies towards animals and imaginary entities coincides with the onset of attributional capacities toward other persons in young children. In our view it should, since anthropomorphism is simply a by-product of our more general capacity for self-conception and mental state attribution.

ATTRIBUTIONS ABOUT OTHER SPECIES

Anthropomorphism in Humans

We contend that part and parcel of being able to infer the mental experiences of conspecifics is the tendency to generalize these attributions to species other than our own (Gallup, 1985; Eddy et al., 1993). Not only can humans appreciate how painful it must feel for another human being to stub their toe, but they can also imagine what it must be like for a dog, who, in its haste to run outside, catches its tail in the door. In either case, people use their prior experiences with pain to model the ostensible mental experiences of other organisms in similar situations.

How accurate are attributions of human mental states to other species? Kellert's (1980, 1993) "descriptive" approach catalogs public perceptions and knowledge of animals, where responses given on various questionnaires indicate when people have accurate knowledge (high congruency between personal knowledge and scientific knowledge) or are relatively unknowledgeable (low congruency between personal knowledge and scientific knowledge). Such an approach is important for assessing knowledge as it may relate to attitudes toward animals under a variety of circumstances. However, the attribution of mental characteristics to other species by the public has received little systematic attention with regard to the scientific accuracy perspective of

Kellert. This approach would seem to be a logical extension of social psychological attribution theory, and may be important in the formation of attitudes that people have toward other species (Kellert, 1980), just as such attributions are important in the formation of attitudes toward other people (e.g., if another person's failure is attributed to an enduring personality characteristic as opposed to a situational variable, the attitudes held toward that person are likely to be very different).

Since perceptions of similarity are important in the attribution of various characteristics to other humans (Feinberg, Miller, & Ross, 1981), it seems reasonable to suppose that the same process ought to hold for anthropomorphism. Eddy et al. (1993) conducted an empirical assessment of the factors contributing to cognitive anthropomorphism (see also Herzog & Galvin, this volume). A questionnaire was constructed using a diverse list of 30 animals (e.g., cockroaches, frogs, elephants, pigs, dogs, porpoises, monkeys, chimpanzees, humans, etc.) representing the range of the so-called "phylogenetic scale." (According to this scale, vertebrates are seen as phylogenetically more recent, more complex, and more similar to humans than invertebrates. Likewise, mammals are perceived as "newer," more complex, and more similar to humans than birds, reptiles, amphibians, and fish. Hodos and Campbell [1969] argue that this misconstrues evolutionary relations among species, and we agree. For the purposes of the Eddy et al. [1993] study, the phylogenetic scale was simply used as a classification scheme which would map on to common public perceptions of species differences.) Subjects were asked to rate the degree of similarity between the animal and themselves, and the likelihood the animal could perform certain mental tasks grounded in the capacity for self-awareness and mental state attribution. Furthermore, the tasks represented at least the second order of intentionality derived from Dennett's (1983) scheme.

Eddy et al. (1993) found that attributions of similarity and mental capacities generally corresponded with phylogenetic position. The additional variables of pets and primates emerged as important factors as well. Based on these findings, it appears that cognitive anthropomorphism does not proceed in a blanket or a random fashion, but rather shows properties of a classic generalization gradient. People utilize a number of cues when making judgments about the mental life of other species. In this case, the type of animal in question, particularly its position on the so-called phylogenetic scale, had a direct bearing on the responses that subjects made. The perceived similarity between the human and the target animal also influenced the extent to which subjects attributed higher cognitive processing to them. In addition, being a pet (e.g., a cat or dog) appears to afford an animal special status. Cats and dogs were viewed as more similar and correspondingly more cognitively complex than other mammalian species (e.g., cheetahs) that are not kept as pets, even though there is little or no scientific evidence that such species differ in their cognitive capacities.

Familiarity with the animal is a potent variable in cognitive attributions. Familiarity is likely to enhance affectional bonds as well. Therefore, there may be two mechanisms operating on anthropomorphism. People are likely to attribute similar experiences and cognitive abilities to other animals based on (a) the degree of physical similarity between themselves and the species in question (e.g., primates), and (b) the degree to which they have formed an attachment to the animal (e.g., dogs and cats). Moreover, whereas the Eddy et al. (1993) data could be interpreted as a demonstration of the "false consensus effect" (Suls, Wan, & Sanders, 1988) by which people overestimate the tendency of others to share their beliefs, this would not explain the clear phylogenetic pattern in their results.

Whereas the Eddy et al. (1993) study focused on evaluating the attribution of "intellectual" capacities, Burghardt (1985a) has shown that people regard emotional continuity between humans and nonhumans as even more likely than intellectual continuity, and that the acceptance of evolution as a scientific fact favorably disposes people to believe in both emotional and intellectual continuity. Future studies on attribution should focus on the specific factors of emotional, physical, and intellectual similarity.

"Zoomorphism" in Other Species

Do species other than our own use their experience to model the experience of others, and do they generalize these attributions to species other than their own? Phrased another way, how accurate and appropriate are attributions of human mental states to other animals? Do other animals regularly distinguish between intentional and unintentional acts? Between guessing and knowing? Between sincerity and deception? Do they have the capacity for empathy, sympathy, gratitude, intentional deception, grudging, and sorrow? Formal tests for these capacities would be comparable to those discussed previously in the context of attributions about other people. For example, if we were to give an animal experience with a blindfold, how would it respond to another animal wearing such an obstruction (Gallup, 1988)? Would it use its experience with blindfolds to infer another animal's impaired visual perspective? There are two dimensions to these questions. First, do other animals have the capacity to make mental state attributions to conspecifics? Second, do other animals make attributions about other species? That is, if they do make interspecific mental state attributions, what is the shape of their generalization gradient? Based on recent data which we will describe, there appear to be other species that use their own experience to model the mental states of others, and generalize these attributions to different species. However, the phenomenon appears to be much more restricted than many people realize. To date, documented instances of attributional capacity have only been demonstrated in humans and chimpanzees (e.g., Povinelli, 1993), along with one report of some suggestive evidence for mental state attribution by an orangutan (Miles, 1986).

Povinelli (this volume) has argued that we may need a new set of terminology to describe the phenomenon analogous to anthropomorphism in other species. For instance, it is possible (and perhaps likely) that the attributional capacities of humans and chimpanzees are not completely overlapping, and that each species may attribute its own unique subset of particular mental experience to other organisms. If true, then to be consistent across species, the prefix should identify the taxonomic group to which each species belongs. For example, human beings have the capacity to attribute human cognitive and emotional states to other organisms and this process is called *"anthropo*morphism." To the extent that chimpanzees (*Pan troglodytes*) have the capacity to attribute chimpanzee mental states to others, Povinelli suggests that the term "*Pan*morphism" would be an appropriate designation for their attributional abilities. If studies of orangutans show similar results (as their capacity for self-recognition would suggest), "*pongo*morphism" will be the appropriate terminology, and so forth. (We will also use the general term *zoomorphism* to describe attributional capacities in *any* taxonomic group.) Of course, it is highly probable that a substantial degree of overlap exists among great apes given that all hominoids may possess shared, derived developmental pathways implicated in the capacity for mental state attribution (Povinelli, this volume). Conversely, however, species differences in interspecific mental state attribution could be used to map onto the presence of underlying differences in cognition.

One comparative-evolutionary model of social intelligence bears directly on the issue of anthropomorphism as well as zoomorphism in other species (Gallup, 1982, 1985, 1991). According to this model there should be some heretofore unacknowledged but striking cognitive differences between species as a function of whether or not they are capable of recognizing themselves in mirrors. This prediction is based on the assumption that organisms that show self-recognition can conceive of themselves, and therefore have the capacity to think about themselves in relation to past, present, and future events, and can reflect on their experience and their own mental states. One outcome of this capacity ought to be the ability to develop a variety of introspectively based social strategies such as gratitude, grudging, sympathy, empathy, intentional deception, and sorrow. The model stipulates that organisms that can conceive of themselves are in a unique psychological position of being able to use their own experience as a means of modeling the experience and mental states of others. Can chimpanzees, who recognize themselves in mirrors, use their own experience to attribute mental states to other chimpanzees, and even to other species?

Premack and colleagues conducted the first set of studies designed to answer the question of whether chimpanzees attribute mental states to others (reviewed in Premack & Woodruff, 1978; Premack, 1988a). In one of these tests, four juvenile apes were put into a situation in which they needed the assistance of a human trainer to obtain food from a locked cabinet. The trainer (carrying the key around his neck) willingly followed the animals across a half-acre field to the cabinet if the apes took care to "lead" him by making sure that he did not fall behind. The chimpanzees were able to do this quite naturally and when the trainer arrived at the cabinet he unlocked it and gave the animals a banana. The test came when, one day, the trainer arrived blindfolded and therefore could not follow the animals to the cabinet. Three of the four animals more or less led the trainer across the compound by pulling on the chain around his neck. However, one of the apes, a four year old, responded first by removing the blindfold from the trainer's eyes. Moreover, she did not remove a blindfold placed around other parts of the trainer's face. This animal appeared to understand that the behavior she wanted from the trainer depended upon his unobstructed sight. Whether she understood more complex dimensions of the situation such as what it means to "see," et cetera, is not clear.

In another study, four six- to seven-year-old chimpanzees were given a test that required them to understand the relation between "seeing" and "knowing." Each animal was shown that one of two containers was being baited with food. The animal could not tell which of the two was baited, although it could see that one of two trainers standing by could. The knowledgeable trainer could see because she had an unobstructed view of the containers, whereas the other trainer (like the animal itself) could not see because his view of the containers was blocked by an opaque screen. The task was for the animal to choose one of the two containers. However, before doing so, the animal was allowed to seek the advice of one of the trainers by pulling either of two strings.

The chosen trainer would then tap on the baited cup and the animal would receive the food. Therefore, in order to obtain the food the animal had to choose the trainer who "knew" where the food was. Of the four, two of the animals both chose the knowledgeable trainer and followed her advice consistently. One of the animals did choose the knowledgeable trainer but would not consistently follow her advice. The other performed at chance level. These results suggest that two of the animals understood the relationship between seeing and knowing and were able to correctly attribute a state of "knowing" to the appropriate trainer.

In another paradigm, the female chimpanzee Sarah was shown videotapes of a human actor in a cage experiencing "problems" of one kind or another. For example, in one sequence the actor struggled to reach a bunch of bananas hanging from the ceiling. As the actor was depicted in the middle of attempting to solve the problem the videotape was frozen and Sarah had to choose from a set of pictures depicting the actor in different situations, one of which represented the "solution" to the problem. Sarah performed well above chance on nearly all the problems depicted. Her performance suggests that she was able to attribute states of mind to the human actor.

More recently, Povinelli and his colleagues have undertaken a broad series of comparative studies with chimpanzees, rhesus monkeys, and human children, which have been explicitly designed to test this model about species differences in attributional capacities and self-recognition (Povinelli & deBlois, 1992; Povinelli, Nelson, & Boysen, 1990, 1992; Povinelli, Parks, & Novak, 1991, 1992). The first of these studies corroborated the results of Premack (1988a) by demonstrating in another "guessing" and "knowing" paradigm which incorporated an important transfer test, that chimpanzees can make differential attributions about knowledge states in humans as a function of whether or not the person has witnessed an event (Povinelli et al., 1990). In a similar test, rhesus monkeys (who consistently fail to correctly decipher mirrored information about themselves) appeared incapable of taking into account the knowledge states of other organisms and failed to use their experience to model the experience of others (Povinelli, Parks, & Novak, 1992).

From the standpoint of this chapter, an intriguing aspect of the chimpanzees' performance on the above tasks was that they were required to make inferences about knowledge states in *human* experimenters rather than in other chimpanzees. Thus, these studies can be viewed as amounting to experimental demonstrations of *Panmorphism* by chimpanzees. Since there is likely to be some overlap in the attributional capacities of humans and chimpanzees, it is not clear from these studies whether the chimpanzees were generalizing from chimpanzee abilities or whether they appreciate that humans may have attributional capacities which may differ from or even exceed their own. However, another way to approach this question is to devise a situation (similar to the one where chimpanzees were required to attribute knowledge states to humans) where they could be tested for attributions to other species (e.g., rhesus monkeys). Would a chimpanzee make attributions about rhesus monkey mental abilities which are comparable to those it might make about other chimpanzees and humans? More broadly, what is the shape of the chimpanzee's generalization gradient for attributing mental capacities to other species? In the context of the findings of Eddy et al. (1993), chimpanzees should be able to attribute knowledge states to a variety of organisms, and like their human counterparts these attributions may be related to the perceived similarity between themselves and the animal in question, and/or whether or not the animal is (or was) a companion.

On the heels of an attempt to plot different generalization gradients for mental state attribution between different species capable of using their experience to model the experience of others, an intriguing follow-up question might be whether such species (e.g., the chimpanzee) would come to modify the kinds of attributions they make about other animals after becoming more familiar with the species in question. For example, although chimpanzees might begin by making widespread mental state attributions about rhesus monkeys, would they modify these over time to make them more consistent with various outcomes? As we have already noted, there is growing evidence that rhesus monkeys do not in fact use what they know to model what other rhesus monkeys may or may not know. Would chimpanzees detect these differences? We doubt it. Indeed, most humans have been oblivious to these differences. If anything, the data collected by Eddy et al. (1993) show that humans become more prone to make mental state attributions about species they have become familiar with (such as pets) in spite of the absence of any hard evidence that these attributions are accurate.

The realization that there may be differences among species in their ability to access their own mental states has been a very long time in coming, and has only surfaced as a point of contention in the last decade (Gallup, 1982). Why has it taken humans, who have evolved a highly sophisticated capacity for mental state attribution, so long to begin to seriously grapple with the question of whether these same capacities might be present in other species? One reason relates back to our predilection to anthropomorphize in the first place (see Gallup, 1985). Because we routinely treat other creatures as if they had access to their own mental states (i.e., as if they had minds), the absence of this capacity has been obscured. Because we are so immersed in a social/psychological milieu of mental state attribution we automatically take this capacity for granted when dealing with other species; that is, mental states in ourselves and other people are so obvious that we have been oblivious to their absence in other creatures.

Besides making attributions about humans, there are some experimental data that directly bear on the question of whether chimpanzees make attributions about other chimpanzees. Menzel (1974) conducted a series of elegant studies with captive chimpanzees in a large outdoor enclosure where only one or two chimpanzee "leaders" in a group knew the location of hidden food sources. Menzel's observations of how the leaders' actions varied as a function of the consequences they produced in the actions of the others suggest that both informed and uninformed chimpanzees were able to attribute knowledge states and intentionality to each other. He observed a series of strategies and counterstrategies emerge between rival chimpanzees that suggests they were each able to model and anticipate the intentions of each other as they tried to "mislead" and/or "outwit" the other about the food source and thus secure the food for themselves. Goodall (1971, 1986) reports similar accounts of "misleading" behavior in wild chimpanzees under less systematic conditions. De Waal (1982) provides a detailed and compelling account of chimpanzee interactions in a semi–free-ranging context suggesting that these kind of attributional abilities are routinely played out in their natural social interaction.

Taken together, experimental and observational studies of chimpanzees can be used to suggest that chimpanzees are capable of attributing various mental states to both humans and other chimpanzees. The question of whether chimpanzees make attributions about other animals has yet to receive any systematic attention.

ATTRIBUTIONS ABOUT ABSTRACT AND IMAGINARY ENTITIES

One of the more intriguing ways in which the human propensity for mental state attribution is expressed is through *theism*, the belief in a God or gods. By what mechanism did theism appear as an attribute of some forms of life? In which of the early hominid species did this ability first arise? Why do gods so often possess human forms, personalities, and behaviors? Perhaps much of the nature of our gods is humanlike because our conceptions of God are a by-product of our tendency to anthropomorphize. Maser and Gallup (1990) maintain:

> From the vanity and vengeance of Homeric gods to the jealous God of the Old Testament, to Lucifer the god that lost in the struggle for supremacy, to the modern views of Jesus Christ as the manlike God who offers salvation from the limitations of human life and history, human gods have emotions, hold grudges, and they think, act and sound very much like the humans whose destinies they are said to control. (p. 518)

How might this come to be? According to Maser and Gallup, theism is a by-product of self-awareness and the need to emotionally deliver ourselves from the fear of the inevitability of our own death, which is itself an outcome of our ability to conceive of ourselves. Maser and Gallup propose that conceiving of a God with human characteristics allows us to imagine that God will provide for some type of continued existence for us after death. Within this context, God has become the ultimate attribution—people ascribe all kinds of mental states to God (e.g., God is omnipotent, all knowing, forgiving; God wants us to do certain things and refrain from doing others; God imposes a set of rules on human conduct and applies specific outcomes for compliance or noncompliance). Moreover, a God that shares humanlike characteristics is one that is easier to teach other people about, describe, comprehend, and communicate with, than one that is purely metaphysical. By anthropomorphizing God we make the abstract and enigmatic more concrete and therefore more readily understood, in a manner similar to the way anthropomorphizing animals appears to make them easier to understand.

According to the theory of self-awareness advocated in this chapter, chimpanzees show at least some of the rudiments required for theism (see also Guthrie, 1993, this volume). There is no evidence, however, that they engage in theistic thought. Moreover, within the context of Maser and Gallup's suggestion that the motivation to assuage fears of personal death is the ontological basis for God and religion, it is not clear that chimpanzees have a conception of their existential limitations. Nor is it known at which point in the evolution of our own genus *Homo* the notion of God arose. We might view death rituals (e.g., burial) as evidence of a belief in life after death. The earliest evidence of deliberate and ceremonial burial comes not much more than 100,000 years ago with the Neanderthals. What of *Homo erectus* and *Homo habilis*? What of their australopithecine precursors?

There are other categories of imaginary entities in which anthropomorphizing by humans is commonly instantiated. Notable among these are ghosts and evil spirits, Santa Claus, the Easter Bunny, the tooth fairy, and putative extraterrestrials. Popular interpretations of the recent rash of claims of "abductions" and "medical experimentation" by extraterrestrials who are frequently described as having humanlike eyes, faces, and bodies are obvious examples.

CONCLUSION

We contend that anthropomorphism is a natural by-product of selection for self-conception and the corresponding use of social strategies based on mental state attribution. Not only do people routinely make the same kinds of attributions about mental states in animals that they do in and among themselves, but they also generalize these attributions about various states of mind to inanimate objects and imaginary and abstract entities (e.g., ghosts, evil spirits, religious deities).

Although many people use anthropomorphism as an explanation of animal behavior, there is growing evidence that most species are unable to use what they know to make inferences about knowledge states in others. Indeed, many species appear to not even know what they know. In addition to the use of anthropomorphism as a means of explaining what animals do, it also provides a convenient descriptive account which is parallel to comparable accounts that we give of our own behavior and that of other people. And even though it may not constitute an adequate explanation, it is also possible that anthropomorphism may provide a useful predictive model for anticipating what animals may do under different conditions.

In practice the use of anthropomorphism appears to be influenced by the perceived similarity between humans and animals (e.g., people are more likely to make attributions about complex mental states in mammals and other primates than they are in insects and fish), and the extent to which people have developed an affectional bond with members of the species in questions (e.g., dogs and cats). People also appear more likely to make widespread, indiscriminate attributions about emotional states in diverse species than about complex cognitive states.

So far, chimpanzees are the only species other than humans that have been shown to make mental state attributions about other creatures (i.e., humans). However, at this point virtually nothing is known about the shape of the chimpanzee generalization gradient for mental state attribution. Whether chimpanzees extend their capacity for mental state attribution to the inanimate and/or imaginary realm is also unknown.

8

Pan*morphism*

Daniel J. Povinelli

Anthropomorphism is defined as the attribution of human psychological characteristics to animals. Upon first encountering the term, a naive observer might find it ironic that scientists have dedicated a special term for this tendency, since it is all but inescapable. Who doesn't wonder what the dog is "looking for" as it sniffs around the yard? Who doesn't confidently presume the shorthand of announcing that squirrels "know" where they have hidden their acorns? Pamela Asquith (1984) has even gone as far as to claim that humans cannot avoid describing animals in anthropomorphic ways. We may not always attribute specific humanlike qualities to them such as sentimentality or mean-spiritedness, but, Asquith argues, we at least talk about them in general as if they had intentions, goals, plans, and knowledge states akin to our own. Thus, she concludes that short of extraordinary efforts of verbal self-control, almost all of our descriptions of animals are likely to be anthropomorphisms. This conclusion may seem to reinforce the irony present in the scientist's need to create a special term to describe the obvious. However, part of the purpose of this chapter is to show that having access to the concept of anthropomorphism inadvertently provides us with a powerful tool for rethinking the psychology of other living organisms.

ANTHROPOMORPHISM AND MENTAL STATE ATTRIBUTION

By definition, anthropomorphism requires the capacity (and motivation) to attribute mental states such as intention, desire, knowledge, happiness, envy, joy, and the like, in the first place. Let me explain by way of an example. Suppose, for a moment, that one day while walking through a park you encountered a robot that was searching for a lost three-year-old child. That is, suppose that it could "see" (perceive) the external world through a camera lens, digitally record what it saw, and adjust its future actions on the basis of such information. Thus, upon encountering a tree directly in its path, the robot could easily update its locomotion program, sidestep the obstacle, and avoid all such similar objects in the future. Watching the robot, you might be curious to know how it was achieving such great success. Inevitably (and probably immediately), you would begin to wonder about what it was doing there in the park in the first place, what it wanted, whether it was lost, et cetera. In short, you would engage in a variety of anthropomorphic thoughts about the robot. Furthermore, I suspect that the amount of anthropomorphism that you would engage in would be directly proportional to the degree to which the robot's makers had been concerned with crafting a human-like physical appearance for the machine, or the distance from which you were observing the robot, or both.*

However, suppose the same robot encountered you. From its perspective, you would simply be another obstacle, albeit a complex and moving one. The robot, if equipped only as described above, would have no capacity—or need—to attribute mental characteristics to you. Because it is merely searching for a young child who is lost in the park, and because you are not that child, you represent just another object in its path. In short, because the robot was constructed without the capacity to attribute psychological characteristics to living organisms, it is incapable of entertaining its equivalent of even a fleeting "anthropomorphic" thought. The robot, given its design and mission, has no need to explain your behavior at all. It merely needs to respond to you in ways that will not jeopardize the attainment of its goal.

I do not detail the above example to suggest that machines will never be able to attribute mental states to other living (or nonliving) organisms. Quite to the contrary, I suspect that advances in the field of artificial intelligence will eventually produce machines that do reason about unobservable mental experiences. Indeed, some researchers have already begun to outline the programming procedures necessary to create such systems (Maida, Wainer, & Cho, 1991). The example is intended to demonstrate that humans engage in anthropomorphism only because at some point in the past evolution produced brains suffi-

---

*Editors' note. See Eddy, Gallup and Povinelli (1993) and Herzog and Galvin (this volume) for empirical data on the factors influencing anthropomorphism.

ciently sophisticated to allow us to attribute mental experiences to ourselves and others. Indeed, as an evolutionary biologist, my interest in robot psychology is quite secondary. My primary concern is with the evolutionary processes that created the psychology of biological organisms—processes that organized the first life forms nearly a billion years ago, and which ultimately resulted in the entire scope of both the past and present diversity of life. Viewed in this light, the issue of anthropomorphism becomes one of understanding how and when the capacity to conceive of mental states evolved. In short, understanding the evolutionary history of anthropomorphism largely boils down to understanding the evolutionary history of mental state attribution.

PSYCHOLOGICAL DIVERSITY?

Understanding the evolutionary history of mental state attribution will not be an easy task. To begin with, psychologists have traditionally had a difficult time finding bona fide instances of phylogenetic psychological differences. To be sure, many psychological traits have been offered as potential candidates for phyletic differences in psychological processes. There are even one or two that have undergone fairly rigorous scrutiny, and may ultimately hold up as valid instances of psychological species differences (Bitterman, 1975; Rumbaugh & Pate, 1984). But by and large, comparative psychologists have had such a difficult time convincing themselves that species differences in intelligence exist in the first place, that at least one commentator has repeatedly advocated accepting the null hypothesis—that there are no phylogenetic differences in psychology (McPhail, 1987). Indeed, from one perspective the history of comparative psychology has been one of demonstrating that purported species differences do not exist (see Goldman-Rakic & Preuss, 1987).

In the face of such historical trends, how are we to go about characterizing psychological diversity in mental state attribution? Difficulties abound, and they are not merely methodological and empirical problems. Characterizing psychological diversity of the type described here faces theoretical challenges from three directions, effectively pinning researchers who wish to make progress in the field. From one direction there is a theoretical tradition that views questions related to animal metacognition as unanswerable, and hence unscientific (e.g., Warden, 1927). From another direction, theorists believe that even humans themselves cannot profitably study their own metacognitive processes (e.g., Skinner, 1987). Finally, a third intellectual tradition considers either some, most, or all animals as obviously possessing these capacities (e.g., Griffin, 1976). In the face of such competing agendas, it may seem that the history of comparative psychology is a good guide to its future. This raises the question of whether psychology, somehow, has escaped the trend of continuous, branching diversity.

However, evolutionary biologists interested in the evolution of mental state attribution are likely to approach the issue as they would a given suite of morphological characteristics. Many facets of biological diversity have been well characterized by researchers who study the evolution and function of morphology. The earliest systematic approaches to biology were by naturalists predating Darwin such as Linneaus, Buffon, Lamarck and Cuvier, who organized their thinking about species around similarities and differences in their anatomical structures. At one level, characterizing this diversity was easy, even if tedious and time consuming. Bats have wings, chimpanzees do not. And even beyond anatomy, some of the functional implications of anatomy often escaped controversy. Bats and birds fly because they have wings; on the other hand, chimpanzees lack such wings, and hence they do not fly. Of course this trivializes the tremendous difficulties that functional morphologists face in interpreting form-function complexes (e.g., Oxnard, 1986). But such problems pale in comparison to the status of the field of "functional psychology," which remains in its infancy. Perhaps the surest testament to this difference between the fields of biology and psychology is that the phrase "biological diversity" makes instant sense to biologists, but the comparable phrase—"psychological diversity"—gives most psychologists reason for pause.

It may turn out that there is a rather simple explanation for the apparent lack of diversity in structures of intelligence. It may be that researchers have focused for too long on areas that represent extremely primitive features of animal psychology, such as associative learning. Indeed, as McPhail (1987) has pointed out, "causality is a constraint common to all ecological niches" (p. 645). Hence, we might expect that even the most primitive organisms evolved the capacity to associate one environmental event with another, ultimately producing both instrumental and operant learning. Some researchers have recently emphasized that there may be adaptations in species favoring specific learning strategies (for example, Kamil, 1984). But as for novel domains of intelligence, there may truly be relatively few. But the fact that there are few does not mean that there are none. And I suspect that it is in those domains where the evidence for a psychological process in humans, but not in animals, is obvious, that we are most likely to find psychological characteristics uniquely derived at some as-of-yet-uncertain point in the primate lineage.

The extent to which empirical investigations ultimately support the uniqueness of humans (or even nonhuman primates) in these various domains is irrelevant. What is relevant is that we constrain our searches to areas in which there is reason to suspect that psychological diversity exists. Mental state attribution is one such area. Premack & Woodruff's (1978) research into the capacity of chimpanzees to attribute intentions to others provided the first explicit attempt to provide a controlled demonstration of mental state attribution outside the human species, thus paving the way for future investigations into nonhuman theories of mind. Although the empirical evidence one way or the other remains weak, there are theoretical reasons related to species differences in the capacity for mirror self-recognition for predicting that very few species are aware of the mental states that govern both their own behavior, and that of other organisms (see Gallup, 1982, 1985). Recent comparative investigations have provided some initial (albeit tentative) support for this position by providing evidence that species differences in mental state attribution may mirror species differences in the capacity for self-recognition (Povinelli, Nelson, & Boysen, 1990, 1992; Povinelli, Parks, & Novak, 1991, 1992; but see Povinelli & Eddy, 1996). This research suggests that at some point within the past 25 million years or so, the capacity to conceive of intentions and knowledge evolved within the primate order, resulting in a detectable pattern of psychological diversity in extant representatives of the clade in question. If these differences hold up under the weight of advancing scientific scrutiny, it would mean that for some subset of processes related to mental state attribution, the human case is neither completely unique nor completely trivial. It would mean that humans share certain cognitive capacities in common with several remaining species of a once diverse hominoid radiation. And it would also mean that species that descended from earlier ancestors lack these cognitive features (regardless of whether they were specifically selected or not).

However, it is an open question whether mental state attribution evolved as a complete package as it is found in adult humans. There are many indices along which chimpanzees, orangutans, gorillas, and humans might differ with respect to mental state attribution. Let me take as a case in point knowledge attribution. First, it is possible—and the data from young children suggests that it is even likely—that knowledge attribution is dissociable into several ontogenetic steps. For example, in terms of human development, children may make several transitions in their understanding of knowledge as a mental state. Longitudinal data on the development of the child's use of the word "to know" suggests that its first appearance occurs at around 28 months of age (Shatz, Wellman, & Silber, 1983). Interestingly, preliminary data from our laboratory, using nonverbal tasks, suggests that at around 28 or 29 months of age children may come to conceive of knowledge as an unobservable mental state, even though they have little or no appreciation of exactly how those mental states are formed. Later, perhaps as early as 36 months (possibly earlier), children may come to understand that knowledge governs behavior. That is, they may begin to understand that knowledge acts as a causative agent in directing the behavior of themselves and others. Knowledge may remain "miraculous," however, in the sense that children may still not understand how it is formed either in others or in themselves (Leslie, 1987; Wimmer, Hogrefe, & Perner, 1988; Gopnik & Graf, 1988; O'Neill & Gopnik, 1991; Povinelli & deBlois, 1992). It is only later that children may come to understand how perception (vision, olfaction, audition, etc.) gathers the information from which knowledge is formed. On the heels of this discovery, or perhaps coincident with it, children may also come to realize that knowledge may in many cases really be mere belief, and beliefs, because they are only representations of reality, can be false (Wimmer & Perner, 1983; Moses & Flavell, 1990). Later still, perhaps by about 5 years of age, children come to grasp the idea that individuals can be less than certain about their beliefs, and hence they may come to understand the true meaning of the concept "to guess" or the distinction implied in the contrast between words such as "probably," "possibly," and "maybe" (Johnson & Wellman, 1980; Moore, Pure, & Furrow, 1990).

The broader implication of the dissociable (developmental) nature of the concept of knowledge is that anthropomorphism is not a unitary phenomenon. The situations in which a four-year-old will engage in knowledge attribution to nonhumans (as well as the possible scope of those attributions) must be fundamentally different from those situations when a two-year-old will make such attributions. More generally, at each developmental transition in an individual's appreciation of the mental world, a new form of anthropomorphism emerges. The type and scope of mental states attributed to others will be distinct, and hence the form of anthropomorphisms possible will be unique. A three-year-old child who attributes humanlike characteristics to nonhumans will, by necessity, attribute her own understanding of what it is like to be human, not her mother's or father's or older sibling's. Her anthropomorphism will be sharply delimited by the scope of mental states and activities she is capable of conceiving of in the first place. Lest we think that such transformations of anthropomorphisms are exclusively restricted to infancy and early childhood, it is sobering to reflect upon (and would be even more fascinating to investigate) the ways in which dramatic political, interpersonal, and educational experiences in adulthood transform our appreciation of what it means to be human. Indeed, I suspect that the common quip, "But I am not that naive anymore," usually indexes profound developmental turning points in an adult's psychological life, forever altering the exact form of their anthropomorphisms.

Finally, although I have focused almost exclusively on the cognitive underpinnings of anthropomorphism, the cultural influences that shape our psychological attributions to animals must be considered if we are ever to develop a complete understanding of the phenomenon. The necessity of examining the role of culture is apparent when one considers strong cross-cultural differences in the willingness to attribute mental experiences to animals (for an example see Asquith, 1986).* It might be that purported cross-cultural differences in the tendency to consider mental states such as private knowledge or intention may somehow be correlated with cross-cultural differences in anthropomorphism. Laurie Godfrey and I examined in some detail the arguments advanced by some cultural anthropologists who argue for strong psychological differences between cultures (see Povinelli & Godfrey, 1993). We concluded that although such surface distinctions exist, and are very important from a cultural perspective, there are probably no fundamental underlying differences among human societies in their ascriptions to a fundamental belief in the unobservable mental universe of themselves, and of others. The weight given to various aspects of these subjective experiences and beliefs may differ, but they appear to be present everywhere. In addition, recent cross-cultural experiments suggest that important psychological transitions in theory of mind may be present in cultures that differ radically from our own (Avis & Harris, 1991). This is not to say that cultural differences in conceptions of the mind are nonexistent. Rather, it is simply to say that such differences may be just that, cultural ones imposed over an underlying similar psychology. Thus, although these important cultural distinctions in conceptions of the place of humans in nature may determine the exact form and timing of conscious anthropomorphisms, the underlying subset of psychological states more or less automatically attributed to other species (such as desire, intention, knowledge, belief, jealousy, envy, joy) is probably more or less constant across our species.[1]

Editors' note. See Asquith (this volume).

PANMORPHISM AND BEYOND:
TOWARD AN INFORMED ANTHROPOMORPHISM

What has all of this to do with scientific inquiry into the minds of animals? It should perhaps be obvious by now that our adult position vis-à-vis immature representatives of our species is fundamentally no different than our position vis-à-vis both mature and immature representatives of other species. The same difficulties that hold true in understanding the minds of chimpanzees apply with equal force in the case of understanding the minds of children. For any given psychological process related to comprehension of mental states, developmental psychologists who work with children must seek methods that can rule out simpler accounts of their behavior, including their linguistic behavior. Some researchers in the field of child development have missed this irony, and have lamented the fact that they work with a species which speaks (Chandler, Fritz, & Hala, 1989). From their point of view, linguistic responses cloud the issue by confounding the development of language and the development of metacognition. However, at some point in the child's development at least the linguistic and behavioral evidence match each other in confirming the presence of the child's awareness of a given mental state or activity (Povinelli & deBlois, 1992). This will literally never be true in the case of apes, who do not develop language. Admittedly, with respect to children this is the trivial case, because by the time that both linguistic responses and spontaneous and elicited behavior converge, there is little reason to debate the existence of the capacity in question. But before dismissing this instance simply because it is "trivial," we should at least draw from it the clear lesson that in human development the typical question about metacognition is when, but with respect to apes and other animals it is whether it develops at all.[2]

It should also be obvious by now that the evolutionary history of mental state attribution (reflected in reconstructed phylogenies) may have created an analog to the developmental anthropomorphisms described above. In other words, beginning with the evolutionary advent of an awareness of the mental world, each new evolutionary innovation in mental state attribution created a novel "anthropomorphism." However, in reality they were not (and are not) *anthropomorphisms* at all. If species other than humans engage in mental state attribution, they must (by definition) be engaged in an analog of anthropomorphism. However, because the exact scope and type of attributions they might make will depend on their own evolutionary histories, it is clear that there is a need to distinguish among the "morphisms" of which they may be guilty. For example, to the extent that there are chimpanzees (*Pan troglodytes*) that engage in mental state attribution, I propose that their unique constellation of attributions to members of other species be referred to as *Pan*morphism. Likewise, orangutans (*Pongo pygmaeus*), who are also likely candidates for mental state attribution (Gallup, 1982), should see members of other species through their own eyes and hence be guilty of *Pongo*morphism. In principle, this would be true for representatives of genera ranging from *Gorilla* to *Gallus*. Fortunately, we may be spared a terminological nightmare by the very real possibility that the evolution of metacognition may have been a relatively recent occurrence (Povinelli, 1993; Povinelli & Eddy, 1996). At any rate, before genera-specific (or, if need be, species-specific) morphisms are invoked, it will first be necessary to demonstrate the existence of mental state attribution using rigorous experimental or observational procedures. In the context of *Pan*morphism, this need is highlighted by the still relatively poor evidence that even chimpanzees attribute mental states to themselves and others (Premack, 1988b; Povinelli, 1993; Povinelli & Eddy, 1996). Not only are we nowhere near defining the scope of *Pan*morphism, we cannot as of yet determine with much certainty whether it even exists.

One implication of accepting the concept of species-specific morphisms is that it turns the question of the utility of anthropomorphism around. Instead of endlessly debating whether anthropomorphism clouds our judgment, produces poor science, or is a valuable heuristic tool, attempts to understand *Pan*morphism, for example (that is, attempts to characterize the theories of mind that chimpanzees possess) will ultimately inform us as to the accuracy of our inevitable anthropomorphic outlook on the world. Asquith (1984) may have been correct in arguing that we cannot escape anthropomorphism, but we may not need to escape it in order to make scientific progress in understanding the minds that reside in other species. Studies of mental state attribution in a diverse array of animal species may reveal when our anthropomorphisms reflect reality, and when they merely reflect the biases inherent in our mental machinery. After all, simply because as young infants we believe that the moon is a person, this does not necessarily make it so.

There are two final caveats that are worth considering, both of which involve the issue of diversity. The first is that just as humans may have evolved unique specializations or elaborations of ancestral psychological traits, so too may have other species. Thus, in the case of chimpanzees, *Pan*morphism may not merely be a restricted subset of human mental attributions, but may include other attributional capacities not shared by humans. For example, Povinelli and Eddy (1996) have investigated the possibility that chimpanzees appreciate underlying mental states such as attention in others, but base their attributions on the presence or absence of behaviors specific to their species. Thus, individual species may have evolved autapomorphic (uniquely derived) features of cognition related to mental state attribution. In short, *Pan*morphism and anthropomorphism may be conceptualized as Venn diagrams of either completely overlapping or partially overlapping mental state attributions.[3] The final point about diversity I wish to raise concerns within-species variation. Not all individuals within a species will possess all (or even any) psychological capacities for metacognition. Thus, we must always keep in mind that our species-specific morphisms may not incorporate all members of a species. We are accustomed to dealing with this fact in our own species by defining certain human populations as "exceptional," and hence restricting our generalizations to the majority of human cases. But when it comes to other species, individual variation in psychological characteristics related to mental state attribution may be far greater than in our own (for example, potential individual differences in self-recognition, Swartz & Evans, 1991; Povinelli, Rulf, Landau, & Bierschwale, 1993).

CONCLUSION

Notions about the internal awareness of animals have a way of cycling through time. What is common sense today will be passé tommorrow, only to be resurrected the following week as the obvious truth. The only method available to us for breaking this pattern is hypothesis testing—theory construction and falsification. Questions about metacognition in chimpanzees, for instance, cannot be answered by the mere force of poetry or prose. To be sure, such artistic literary endeavors can forever play a vital role at the cutting edge of science by organizing our ignorance and highlighting our humility. And as humans who practice science at least part-time this is a fact well worth remembering. Even Charles Darwin, logical positivist par excellence, was not above turning to metaphor when ignorance dictated the need:

> An anthropomorphous ape, if he could take a dispassionate view of his own case, would admit that though he could form an artful plan to plunder a garden—though he could use stones for fighting or for breaking open nuts, yet that the thought of fashioning a stone into a tool was quite beyond his scope. Still less, as he would admit, could he follow out a train of metaphysical reasoning, or solve a mathematical

problem, or reflect on God, or admire a grand natural scene. Some apes, however, would probably declare that they could and did admire the beauty of the coloured skin and fur of their partners in marriage. They would admit, that though they could make other apes understand by cries some of their perceptions and simpler wants, the notion of expressing definite ideas by definite sounds had never crossed their minds. They might insist that they were ready to aid their fellow-apes of the same troop in many ways, to risk their lives for them, and to take charge of their orphans; but they would be forced to acknowledge that disinterested love for all living creatures, the most noble attribute of man, was quite beyond their comprehension (Darwin, 1871/1981, pp. 104–105).

If we assume that Darwin's ape was a chimpanzee, then he had, in granting it the ability to move between its own psychology and that of humans, created a creature able to detail its *Pan*morphic world view through conversation alone. But no such anthropomorphous ape is likely to rescue evolutionary biologists who follow in Darwin's footsteps. Anthropomorphic apes are part of the fictional world, and are thus only valuable at the edge of our knowledge. Our legacy, as David Premack (1986) has pointed out, will be written with the behavioral results of questions we ask of apes through behavioral means. Such explorations remain in their primal infancy, but they threaten to reach adolecence soon. What this adolescent will have to say, and how much of it we will be able (or willing) to understand, may surprise us all. Darwin may have misjudged the chimpanzee's *Pan*morphic outlook on the problem of mind, but it remains to be seen by how much.

NOTES

1. Some limited evidence on the emotional end of this point comes from investigations by Ekman (1973) and colleagues who have gathered intriguing evidence that humans are capable of accurately judging the emotions behind facial expressions from individuals of cultures very different from their own. Other indirect evidence has been gathered by cultural anthropologists in the 1940s and 1950s who used a number of projective psychological tests such as the Thematic Apperception Test in a number of different cultures. The important point does not concern the validity of the interpretation of such tests, but rather the fact that in all cases the subjects' responses are laden with psychological attributions related to knowledge and intention (for examples see Wallace, [1952] on the Iroquois of New York and Gladwin & Sarason [1953] on the Truk of Micronesia).

2. The study of autistic children represents an excellent exception to this rule. The typical question about an autistic child's theory of mind for a given

metacognitive capacity (for example, their ability to respresent false beliefs) is whether they develop it at all. (See, for example, Perner, Frith, Leslie, & Leekman, 1989.)

3. The extent to which a situation could exist in which the relationship between two species' capacities for mental state attribution would need to be represented as a non-overlapping Venn diagram is an intriguing philosophical problem, although I admit that such a situation seems unlikely. In terms of evolutionary biology this would translate into a situation in which two species had evolved completely separate theories about the mental world. But even at the lowest order such species would seem to overlap at least insofar as they believed in the *existence* of the mental world.

9

Anthropomorphism and Scientific Evidence for Animal Mental States

Hugh Lehman

Because they experience feelings, emotions, and perceptions, animals are considered sentient beings—beings which experience mental states. Animals experience pain, contentment, anger, or boredom; they see, smell, and touch objects, themselves, and other animals. However, sometimes when animals are characterized as having a particular mental state, there is a considerable degree of unease. Quotation marks are used to indicate that an author has some doubts about the appropriateness of the term. Sometimes controversies arise and some scientists claim that the characterization is anthropomorphic and therefore dubious. Yet, given that animals experience some mental states, it is unclear where anthropomorphism ends and accurate description begins. In this chapter I define the meaning of the term *anthropomorphism* so as to be clear what is included under this rubric, discuss the nature of evidence used to apply mental characterizations to nonhumans, and counter arguments that attribution of mental characteristics to animals is unreasonable.

ANTHROPOMORPHISM

We are concerned with the question of attributing states of consciousness (mental characteristics), which we agree that humans experience, to animals. When such attributions are called "anthropomorphism" the implication is that it is a kind of error or fallacy. Now, clearly it is not an error to attribute a human mental characteristic to an animal of some other species if that animal possesses the characteristic. For example, both an animal and a human being can be correctly and unequivocally described as submissive. If submissiveness is a characteristic that some human beings possess, then it may be called a human characteristic. (Calling a characteristic "human" does not imply that only humans have this characteristic.) Since submissiveness appears to be a trait of mind or personality we may call it a human mental characteristic. It hardly seems to be an error to describe some dogs as submissive even though, if we so describe a dog, we are attributing a human mental characteristic to that dog. In light of this example we shall restrict the sense of the term *anthropomorphism* to erroneous or unwarranted attributions of human mental characteristics to animals. Given that a dog is submissive, we are not guilty of anthropomorphism if we so describe it.

If we define anthropomorphism as the attribution of a human mental characteristic to an animal that does not have that characteristic, then anthropomorphism is clearly a kind of mistake. Furthermore, it seems reasonable to use the same term to refer to the opposite kind of mistake also, namely to the denial that an animal possesses a human mental characteristic when, in fact, the animal does possess that characteristic. So, let us understand anthropomorphism in this broadened sense. An advantage of this way of speaking is that it denies a rhetorical advantage to those who are prone to deny that nonhumans have any significant mental characteristics or to deny such attributions in particular cases. If those who are prone to make such charges recklessly know that they may be accused of committing such a serious error as is suggested by the term anthropomorphism, they may become less reckless.

There appears to be another possible error to which those who use the term anthropomorphism may refer. As we have defined the term above, anthropomorphism refers either to false attributions or to incorrect denials, that is, to a kind of factual error. But, users of this term may also mean to refer to a kind of inferential error, that is, to an inference that something is or is not the case when there is insufficient evidence to draw such a conclusion. For example, if someone concluded that a cat was happy because the position of the cat's mouth resembled a smiling human mouth, we might say that such a person was guilty of anthropomorphism. In saying that in this case, we would not mean to imply that the cat was not happy or that a cat could not be happy. What we would mean to imply is that the observation of the shape of the cat's mouth is not sufficient evidence to justify concluding that the cat is happy. Let us include an error of this sort under the scope of the term anthropomorphism also. Again, there is an opposite kind of mistake, namely, failure to conclude that an animal has a human mental characteristic when the evidence available clearly warrants drawing such a conclusion. Such a failure may also be called anthropomorphism. Hence, anthropomorphism as conceived here consists in any one of four possible kinds of error: A case of anthropomorphism is either

1. an affirmation that an animal has a human mental characteristic when it lacks that characteristic; or
2. a denial that an animal has a human mental characteristic when it possesses that characteristic; or
3. an inference that an animal has a human mental characteristic on insufficient evidence; or
4. a failure to draw a conclusion that an animal has a human mental characteristic when the evidence clearly warrants drawing such a conclusion.

EVIDENCE, OBSERVATION, AND HYPOTHESIS

Tough-minded scientists, especially those with materialist or behaviorist sympathies, have often expressed the view that we should not attribute mental states to animals without proof. For example, in a recent survey of the life sciences, the following remark appeared in the chapter called "The Biology of Behavior": "Humanlike motives, feelings, or insights are not attributed to animals without the strongest possible evidence" (Handler, 1970, p. 379). Other tough-minded individuals are, of course, willing to make such attributions, presumably without the strongest scientific evidence. For example, it is accepted among broiler producers that chickens may be bored and that boredom produces behavior that is undesirable from the producers' point of view (Harrison, 1964, p. 18). What is the nature of the evidence available for such inferences?

Scientific evidence consists in observations. Statements that purport to describe observations normally may be called observation reports. Consider, as an example of an observation report, the statement "The chicken clucked loudly when the man entered the yard." Such a statement may be an observation report for a suitably located observer. Let us say that a statement is an observation report for an observer, providing that the statement would be affirmed by any person in the same circumstances as the given observer who understood the language used, was paying proper attention, was honest, et cetera. In consequence of this definition we might expect all suitably situated, et cetera, observers to agree to any given observation report. This is, I think, the way most scientists would understand the term *observation report*. According to a more stringent notion an observation report is a statement that can be learned ostensively (see Quine & Ullian, 1970)

The statement concerning the chicken is not an observation report because of any feature intrinsic to the statement itself. It is an observation report to an observer. To someone else the same statement may be a hypothesis. Hypotheses may be proved and may express facts without thereby ceasing to be hypotheses. A hypothesis is a statement that a person either does not know to be true or that he has discovered to be true by some process other than observing the state of affairs reported in the statement. (This is *not* to say that we can come to know hypotheses in ways that don't depend on observation at all.) Just as a hypothesis is not necessarily proved, so an observation report is not necessarily true.

EVIDENCE FOR ANIMAL FEELINGS

On the basis of observations a scientist may be justified in accepting either (a) observation reports or (b) certain hypotheses that enter essentially into the explanation of those observations. We wish to maintain here that some scientific observation reports and hypotheses about animal feelings are warranted on the basis of evidence. We may observe that an animal is in pain or is contented. A statement such as "That animal's paw hurts" may be an observation report. Further, statements that attribute mental states to animals may help to explain our observations and so may be corroborated or confirmed by such observations.

Consider the following example in which a scientist implies that a hypothesis is warranted:

> Another posture which occurs in hostile encounters is the Upright Threat Posture. . . . In interpreting the motivation underlying this posture, therefore, we use an additional criterion—the actual form of the posture. . . . It is just on such occasions, when the attacker is obviously held back by fear, that one sees the most intensive posturing. (Tinbergen, 1968, p. 225)

In this paragraph, Tinbergen speaks of "interpreting the motivation underlying this posture." It appears that what he is doing is inferring the bird's internal mental state, namely, its fear. The basis for this inference is that the assumption that the bird is afraid contributes to an explanation of the occurrence of the posture on the particular occasion.

Many people may be surprised at our contention that we can have knowledge of the mental states of animals directly in the form of observation reports. We shall try to support this contention by providing a critique of the major reason people may have for rejecting it: what we will call the *private episode view* of mental states such as feelings. Other people object to the contention that such beliefs (that animals have mental states) are scientifically justified because they provide explanations of observations, because they subscribe to a view we will call *materialism*. We shall try to raise serious questions for both the private episode view and materialism. The objections show that these positions may be rejected and thereby clear the way for accepting our contention concerning knowledge of animal mental states.

The suggestion, that a statement attributing feelings or sensations to an animal could be an observation report, will be dismissed by some people as absurd. Pain or contentment, they might say, are feelings and, as such, are private episodes. By calling such feelings private, such people mean to imply that the feelings are not observable by any creature other than the creature whose alleged feelings are reported. In other words, such people might affirm that reports of feelings are not observation reports unless such reports are first-person reports. Since animals cannot express such first-person reports, there can, in this view, be no observation reports of animal feelings. Tinbergen expressed the private episode view in the following remark:

> We feel there are, in principle, two types of observables when dealing with behavior. One type, which includes the movements of animals, can be shared by different observers . . . The other observables are the subjective phenomena that coincide with behavior, which we observe, each of us, in ourselves; and these are, by definition observable only to the subject. (1960, p. 185)

Let us critically evaluate the private episode view. We maintained above that if a statement is an observation report then we may expect there to be a consensus among all suitably located observers as to the acceptability of the statement. Such a consensus presumably is dependent on the observations of the phenomena described in the report since many people who are not in a position to observe the phenomena in question might doubt or deny the report. But clearly many statements concerning animal feelings may satisfy this condition, namely they are accepted by observers of the animal and not by others. While we cannot conclude from this that such statements are necessarily observation reports, we clearly have strong reason for the claim that such statements are observation reports.

We noted above that some writers defend a more restrictive notion of observation reports in accord with which a statement is an observation report if and only if it can be learned ostensively. A statement can be learned ostensively if it can be learned by a person who does not know the language well enough to understand verbal explanations of the meaning of the statement. For example, the statement "There is a dog" might be learned ostensively from a person who points at a dog while saying "There is a dog." However, reports of pain or contentment in animals are statements that may be learned ostensively. For example, if a dog is visibly limping one may teach a toddler the expression "Doggy's paw hurts" through getting the child to notice the limping dog (perhaps by pointing him or her at it) and saying "Doggy's paw hurts." So, a statement which refers to a dog's mental state can be learned ostensively and so satisfies an even more restrictive concept of an observation report.

Whatever plausibility the private episode view of feelings may have, may derive from an unacceptable analysis of perception. Common sense theories about perception include three positions concerning the nature of perception. One of these is acceptable and the other two are not. Often all three positions are intermingled in a confusing jumble. To explain these three positions let us introduce the term *sensory percept* to refer to mental states of which we become conscious as a consequence of our sensory organs being suitably stimulated by physical objects or energy. For simplicity, we may consider seeing a green leaf as an example. We may have a visual percept of the surface of the leaf when light reflected from the leaf stimulates our retinas. The three views of perception may be explained as follows: Naive realism is the view that the sensory percept one has on seeing the leaf is identical with the surface of the leaf. Representative realism is the view that the sensory percept is a near duplicate of the surface of the leaf. Causal realism is the view that the sensory percept is caused by a certain set of circumstances in which the leaf plays an essential role and light reflected from the leaf strikes our retina. In this work we shall simply assume that of the three views, causal realism is correct. Our definition of sensory percept reflects this assumption. The other views lead to conceptual difficulties that can be overcome only by accepting inherently implausible theories about reality and perception. (In the naive realist view the surface of the leaf is a state of consciousness—a sensory percept. In the representative realist view questions arise as to how the surface of the leaf can exactly resemble all the varied sensory percepts caused by observing the leaf.)

Now let us consider two reports concerning an animal, let us say, "This animal is in pain" and "This animal is gray." There are, no doubt, differences between these two reports but it is not necessarily the case that one of the differences is that the report concerning color is an observation report while the other is a hypothesis concerning an unobservable state (unobservable to the observer of the animal). Both these reports might be learned ostensively and both might be such that any honest, suitably positioned observer, who understands English, would assent to them. Of course, it is clear that when we observe that the animal is in pain, none of our sensory percepts need be either identical with nor closely similar to the animal's pain. However, when we observe that the animal is gray the same is true. Our sensory percepts needn't be either identical with or closely similar to the animal's color. Someone who allows that there may be observation reports concerning an animal's color, but not concerning an animal's pain, may be disposed toward this view in consequence of accepting a naive or representative realist view of color perception even though he would reject such views in regard to perception of animal feelings. Once we learn that naive and representative realist analyses of color perception are as unsatisfactory as such analyses are for perception of pain, then we may be less inclined to object to the contention that reports of pain in animals may be observation reports.

Still others may claim that the report of pain is not an observation report on the grounds that we don't really observe the pain. We see the animal writhing or limping or hear it crying or whimpering. They may say that when we judge that the animal is in pain it is in consequence of an "interpretation" of what we really observe. But, we may ask, what is implied by saying that this judgment is the result of an interpretation? The reference to interpretation here may refer to the fact that the judgment depends on our brain's being able to "integrate" the pattern of stimuli it receives in light of neural patterns acquired genetically or through past experience. But, if this is what the reference to "interpretation" implies, then the judgment that the animal is gray is also the result of an interpretation. Should we therefore say that we don't really see the animal limping or don't really hear it crying? Clearly, we don't have to agree to that and equally we don't have to accept the contention that we don't really observe the pain.

Some people may say that such reports as "This animal feels contented" or "This animal is in pain" cannot be observation reports because these reports may be mistaken. Someone on seeing an animal may think that it is not in pain when it is. This could be a result of the person's not interpreting the animal's behavior correctly. Similarly, a person could think that an animal is in pain when it is only feigning some injury. However, while it is correct that such reports as these may be mistaken, that does not show that they are not observation reports. Observation reports are not necessarily true. Even such reports as "The color of this animal's coat is gray" may be mistaken. Misinterpretation of sensory stimuli is always possible.

Let us move on to a consideration of materialism. Materialism, as here conceived, is the view that hypotheses concerning mental states have no explanatory power. That is, materialism is the view that we cannot appeal to such hypotheses if we want to explain any observable phenomena. A more traditional term for materialism is *epiphenomenalism*.

The materialist view is not self-evidently correct. Indeed, our natural inclination would be to reject it as false. What could be more natural than to try to explain a dog's behavior by saying, for example, that the dog is wagging its tail because it perceives its master and is pleased thereby? In this example, we have an explanation of the dog's behavior that refers to two of the dog's mental states: the dog's perception of its master and the dog's pleasure in this perception. Yet, pointing at such commonplace examples is not enough to refute the materialist position. The materialist may contend that in the example in question we do not have an explanation of the dog's behavior. He or she may call it a pseudo-explanation and disparage it as "folk psychology."

Let us divide the materialist's reasons for objecting to positing mental states in animals into two categories. On the one hand there are reasons having to do with the evidence for the explanatory hypotheses. On the other hand there are reasons having to do with the possibility of causal relations between mental states and the observable behavior. Briefly, let us say that there are evidential objections and there are causal objections.

The materialist may express his contention that there is no evidence for the explanatory hypothesis in a number of different ways. He may contend, for example, that to explain the dog's behavior by reference to its mental states is circular since, he may allege, there is no evidence for the existence of the mental states other than the observations of the behavior. Basically the same objection may be expressed by challenging us to provide the evidence that supports our contention that the dog perceives its master or that the dog feels pleasure in this perception without referring to the tail wagging or other behavior which the hypothesis purportedly explains.

But, the materialist's objection is unfounded. To see this let us consider two types of error that may be present in explanations that are allegedly circular. First, a circular explanation may be fallacious in that the explanans (the statements that do the explaining) are merely a redescription of the same data or facts reported in the explanandum (the statements reporting occurrences or facts to be explained). Second, a circular explanation may be fallacious in that there is no evidence for the explanans other than the observed behavior which the explanans purportedly explain. However, with regard to our example concerning the dog's behavior, we can see that our explanation is guilty of neither of these fallacies—or at least that it need not be so. Saying that the dog feels pleasure in the perception of its master is not simply another way of saying that the dog is wagging its tail. The dog might feel such pleasure and not be able to wag its tail. (This might be the case if the dog's tail wagging apparatus was rendered temporarily nonfunctional.) Again, there may be much evidence that the dog is pleased at the perception of its master other than the fact that it is wagging its tail. There are other aspects of the dog's behavior that indicate its pleasure.

Let us turn to the materialist's other reasons for objecting—the causal reasons. It is natural to hold that in order to explain an animal's behavior one must cite a causal pattern which culminates in that behavior. Now, the materialist may maintain that mental states or events can play no essential role in the explanation of an animal's observable behavior. But what reasons can be given for saying that mental events cannot be part of a causal pattern which culminates in some sort of animal behavior? Again there appear to be two sorts of reasons. First, it may be alleged that if an event or state of type A is a cause of an event or state of type B then it must be true that whenever an A occurs then a B occurs. However, it may be claimed that if A is a mental event or state such as a dog's perception and feeling of pleasure, it may not be true that whenever an A occurs, a B occurs. For example, it may not be true that whenever the dog takes pleasure in the perception of its master it wags its tail. Second, it may be alleged that if an event or state of type A is a cause of an event or state of type B then there must be a physical mechanism of some sort which connects cases of A to cases of B. However, it may be alleged that there can be no such mechanism in cases in which A is a mental state and B is a physical state.

Again, the reasons offered by the materialist are unfounded. With regard to his first reason we can grant that if A is the cause of B then whenever an A occurs, a B occurs. But it is not necessary to maintain that the animal's feeling is the cause of the animal's behavior. The behavior is, no doubt, the result of a complex of factors of which the animal's feeling and perception is an essential component. In saying that the feeling and perception are essential components of the cause, we are not saying that these factors are, in themselves, sufficient to bring about the effect. What we are saying is that they are a necessary part of a set of conditions which is sufficient to bring about the effect. Thus, pointing out that it may not be true that whenever the dog takes pleasure in seeing its master it wags its tail does not show that the perception and feeling do not play an essential role in the explanation of the behavior.

Needless to say, there are many other cases where, in explaining some event or condition, we cite some other condition that is not sufficient but is only a necessary part of a sufficient condition. In many of these cases the cause and effect are both physical events. For example, if one claims that the existence at present of a valley is the result of an ancient river, one does not mean that the river is itself sufficient for the present existence of the valley. After all, the valley might have eroded before the present time had it consisted of softer materials or had the climate been significantly different in the intervening period.

The materialist's other reason for saying that mental events cannot play an essential role in explaining behavior is unwarranted also. The contention that a mental event or state cannot be connected to the animal's behavior by a physical mechanism implies that the mental event or state is not identical to some physical event or condition. But for all anyone (even the materialist) knows at the present time, mental events or states may indeed be identical with physical events or states. Since we do not know that mental events or states are not identical with physical events or states, we do not know that mental events cannot be connected to behavior by physical mechanisms.

ARGUMENT FROM ANALOGY AND ANTHROPOMORPHISM

We have alleged that attributions of feelings of pain or contentment may be warranted observation reports or hypotheses citing factors essential to explaining observations—particularly in explanations of behavior. Many people use argument by analogy as a basis for claiming that animals feel pain or contentment. However, I have not argued, for example, that since at certain times animal behavior is similar to the behavior that human beings manifest when they are in pain, that the animal is in pain at those times. I have not appealed to argument by analogy because I do not believe that our knowledge of animal feelings rests on observable similarities between human and animal behavior. Argument by analogy is incapable of providing sufficient evidential support to warrant claims to knowledge.

In an argument by analogy one is arguing that because an object (event, state of affairs, etc.) A is like an object B in some respects, that A is like B in some other respects. The reason why this sort of argument is weak is that any two things, no matter how dissimilar, are similar in a great many respects. To see this it may help to consider two things that seem to be quite dissimilar, let us say a sneeze and an elephant. There are many ways in which a sneeze is similar to an elephant. They both exist at a certain place and for a certain length of time. They both can inspire a poem. They both may be frightening to some creature. They are both beneficial to man. They can both spread germs, et cetera. Clearly, arguing that since sneezes and elephants are similar in these respects and that since the elephant is also heavy, the sneeze is heavy too, is logically and methodologically weak. Or, again consider the analogical argument: Horses snort, people snort; people have a sense of humor, therefore horses have a sense of humor. It may be that horses have a sense of humor. But this argument does nothing to establish that claim.

There are some respects in which pain-manifesting behavior of some animals is similar to normal human pain-manifesting behavior. For example, when in pain both a human being and a horse may groan (Fraser, 1974, p. 170). Still, we may take the groan of a horse as indicating that the horse is in pain not because the horse's groan is similar to a human groan but rather because we take the horse's pain as explaining the groan. There is a causal explanation of groaning in which pain plays an essential role.

Sometimes the inferences concerning animal mental states based on arguments from analogy are examples of one of the types of anthropomorphism distinguished above, in particular the types of anthropomorphism involving an unwarranted inference. For example, to infer that a dog is worried because it has an expression on its face that resembles the expression of a worried human, is a case of drawing a conclusion about an animal mental state without sufficient evidence (anthropomorphism of type 3). Again, to infer that an animal is suffering because in similar circumstances a human would be suffering is an argument from analogy and is also a case of the same fallacious inference. If a human had to live in a yard with chickens all the time he or she would be distressed. But this need not distress a chicken.

It is sometimes argued that an animal is *not* suffering because its behavior is similar to behavior of humans who are not suffering. For example, if an animal is not crying out or groaning, it may be inferred that it is not suffering. This is an argument from analogy. But it is an instance of the fourth type of anthropomorphism. Some animals do not cry out or groan even if they are in pain. That behavior may be an evolutionary adaptation to circumstances in which groaning would attract predators.

Our consideration of the evidence available for the attribution of mental states to animals suggests that there may be rational ways of resolving controversies concerning animal welfare. In recent years, the greatest controversies in animal agriculture have concerned calves raised for veal and chickens raised in battery cages. The calves may be raised in pens that allow them no room to move about. Further the calves may be kept in darkness except for brief intervals when they are to be fed. Chickens in battery cages may be unable to extend their wings and may be subject to vicious pecks from their cage-mates. It will not do to say that clearly such animals are suffering because humans would suffer in such close confinement or if kept in darkness for almost the entire day. But there are ways of determining whether the animal is suffering. Experiments can be done to determine the animal's preferences. A condition that causes suffering will often serve as an aversive stimulus for an animal. Further, careful observation of an animal's behavior following release from close confinement may indicate whether it is behaving in a way that serves to relieve discomfort caused by the close confinement. Further, there is other evidence regarding distress such as boredom in animals. Animals suffering from boredom develop anomalous behavior patterns (Hafez, 1969, p. 352; Harrison, 1964, p. 18). Of course, undertaking scientific experiments to determine animal mental states is not easy. Great care must be taken in the design of experimental research in order to determine correct explanations.

SUMMARY

In this chapter I have touched briefly on a number of topics relevant to the assessment of attributions of mental states in animals. The unjustified or false attribution or denial of a pain or other mental state to an animal is anthropomorphism. Such attributions may be justified in some cases by observation but in many cases by an inference to a hypothesis which explains the observations. In the course of our discussion of this matter we have been led to criticize both the private episode view (that pain and other mental states are observable only by the creature whose mental states they are) and the materialist view or epiphenomenalism (that mental states cannot enter into explanations of behavior). Finally, we have argued that arguments by analogy do not provide sufficient evidence for attributions of mental states to animals.

10

Anthropomorphism in Mother-Infant Interaction: Cultural Imperative or Scientific Acumen?

Robert L. Russell

In a series of micro-analytic studies of filmed human mother-infant interaction in quasi-naturalistic settings, the developmental psychologist Colwyn Trevarthen attempts to document and explain the development of subjectivity and various levels of intersubjectivity (e.g., Trevarthen, 1979a, 1979b, 1980; Trevarthen & Hubley, 1978). As a nativistic psychobiologist by training, Trevarthen explains the emergence of subjectivity and the various levels of intersubjectivity on the basis of maturational processes of the infants' brain, even claiming that human communication develops out of "a fundamentally coherent field of intentionality which is . . . already anatomically partitioned [and localizable in the brain] at birth" (Trevarthen & Hubley, 1978, p. 213). Of more interest and concern, however, is the strategy and language Trevarthen uses in attempting to describe and document the behavioral instances of subjectivity and intersubjectivity that he wants to explain on the basis of the developing brain. Unless and until such brain structures are actually localized, it is from the field of behavior that "evidence" for the development of subjectivity and intersubjectivity must be gleaned.

Focusing on the human mother's and infant's field of behavior, Trevarthen's strategy to defend the use of such terms as subjectivity and intersubjectivity is to analyze them conceptually via ordinary language usage and to link such analyses with hard-boiled ostensive definition. He contends that such words as subjectivity and intersubjectivity "may be defined clearly and defended scientifically by reference to how infants behave" (1979b, p. 530). For example, Trevarthen defines subjectivity as "the ability to *show* by coordinated acts that purposes are being consciously regulated. Subjectivity implies that infants master the difficulty of relating objects and situations to themselves and predict consequences, not merely in hidden cognitive processes but in manifest, intelligible actions" (1979a, p. 322). The list of infant behaviors that demonstrate subjectivity include focusing attention on things, handling and exploring objects with interest in the consequence, orienting or avoiding while anticipating the course of events and meeting or evading them. In short, subjectivity is considered to be present when behavior is coordinated and goal directed. Intentionality is similarly defined, and instances are putatively recorded by film footage that shows the infant adjusting his/her perceptual receptors in preparation for locating and tracking changes in stimuli. Such adjustments show "rudimentary intentions because they are directed to the obtaining of particular kinds of consequence, or particular directions of experience" (Trevarthen, 1977, p. 232). It is the ability of the infant to show his/her intentions in actions that makes the infant's subjectivity manifest (Vedeler, 1987). Thus, the newborn is described both as a subject and as engaging in rudimentary intentional acts on the basis of what appears to be simply rudimentary goal-directed behavior. Such behavior manifests the infant's subjectivity.

Trevarthen describes primary intersubjectivity, which emerges at two or three months of age, as "mutual intentionality and sharing of mental state" (Trevarthen, 1977, p. 241); an "[i]nterlacing of subjective perspectives" (1980, p. 334); or "the linking of subjects who are active in transmitting their understanding to each other" (1979a, p. 347). Instances of primary intersubjectivity are putatively recorded in film footage that shows such sequences as: the mother catches the infant's attention, the infant smiles and makes certain gestures and/or vocalizations toward the mother, the mother reacts with pleasure, smiling and admiring, and perhaps mimics or imitates the infant's actions, to which the infant responds with pleasure and further gestures and so on. Thus, the infant's behavior can literally be *seen* as an instance of primary intersubjectivity or not according to Trevarthen.

Finally, secondary intersubjectivity, which develops at 9 or 10 months of age, is defined as a deliberately sought sharing of experience about events and things, that is, the combining of the infant's interests in the physical, privately known reality near him/her and acts of communication addressed to persons (Trevarthen & Hubley, 1978, p. 184). Instances of secondary intersubjectivity are putatively recorded in film footage that show, for example, an infant repeatedly looking up at the mother's face when receiving an object, pausing as if to acknowledge receipt, or an infant gently moving the mother's hand aside so she could get to beads beneath it.

Everyone will note that Trevarthen uses mentalistic terminology in his behavioral descriptions. Moreover, the types of interchanges that are singled out as exemplars of subjectivity, primary and even secondary intersubjectivity, must be admitted to be evident in other species as well. Thus, unless there is something peculiar about the topography of the human infant's behavior that Trevarthen has not sufficiently distinguished from other infant's behavior, we are left with the plausible conclusion that subjectivity and these two levels of intersubjectivity are attributable to any infant/mother pair that exhibits coordinated, goal-directed behavior. In other words, we can conclude that it would be appropriate to use such descriptions whatever species is being investigated and "anthropomorphisms" would be considered legitimate. Alternatively, we might conclude that Trevarthen's descriptions of human mother-infant interaction are themselves misguided, that is, "anthropomorphic" in just the same ways that descriptions of nonhuman behavior are labeled as "anthropomorphic" when a mentalistic idiom is employed for their description. At this level, then, we would be left with two possible conclusions: that Trevarthen's mentalistic terminology does away with anthropomorphism altogether, since subjectivity and intersubjectivity, as defined and cinematically recorded, are distributed *across* species; or that some human infant behavior is itself described in what should be considered anthropomorphic terms.

One alternative to this either/or predicament would be to bicker about the way in which Trevarthen has employed his terms. This is the tack that I want to pursue. First, I will raise questions about Trevarthen's use of an indicative or referential theory of description. Such theories presume a clear demarcation between the world as it is given independent of our descriptions and the language in which we do our describing. I'll marshal two quick arguments against his theory of description. They will go a long way to dispel Trevarthen's strategy for establishing the emergence of subjectivity and intersubjectivity through the use of "exemplars" captured on film. I will then develop a somewhat quixotic position about language, not unrelated to the first two arguments. This position will indicate to what extent bits of behavior are infused with inextricable symbolic significance. This fact will be used to explain the tremendous pressure on human infant researchers and others to see their infants as demonstrating subjectivity and inter-subjectivity.

To set the stage for the first argument, consider the dilemma that confronts the philosopher trying to understand two simple statements: "The sun always moves" and "The sun never moves." Both statements are equally true but at odds with each other. Following Nelson Goodman (1978), we can then ask:

> Shall we say, then, that they describe different worlds, and indeed there are as many different worlds as there are such mutually exclusive truths? Probably not. We can, instead, say that the strings of words are not complete statements with truth values of their own but rather very elliptical statements for such statements as "Under frame of reference A, the sun always moves.", and "Under frame of reference B, the sun never moves."—statements that may both be true of the same world. . . . With these less elliptical statements, we can come to see that frames of reference do not belong to the thing described as one of its material attributes but rather to systems of description, terminological frameworks. Each of the two statements relates whatever is described to a particular system. (pp. 2–3)

Goodman thus concludes:

> If you ask about the world, you can offer to tell me how it is under one or more frames of reference; but if I insist that you tell me how it is apart from all frames, what can you say? We are confined to ways of describing what is described. Our universe so to speak consists of these ways rather than of a world or of worlds.

Thus, we should say that terms such as subjectivity and intersubjectivity belong to a particular frame of reference, and belong more to the system of description than to the things described. It follows, then, that Trevarthen's pointing to particular cinematically recorded behaviors as instances of subjectivity and intersubjectivity is wrong-headed. What he seems to be doing is this: he pushes the ideality of his mentalistic frame of reference into the material reality of the behaviors recorded. This is accomplished by a kind of slight of hand, in that we only receive the elliptical descriptions and not the frame of reference in terms of which they gain warrant. Having completed this slight of hand, it is then possible for him to point at instances of cinematically recorded infant behavior as if they were things amongst other things in an always already given objective world. The only difference between one type of infant behavior and another is that the behavior of interest has an additional attribute, that of subjectivity, or primary or secondary intersubjectivity. Far from endorsing Trevarthen's strategy, Goodman's argument can be extended with very little ingenuity to show that in some fundamental sense any and all of our descriptions are anthropomorphic, since they are given in terms of frames of reference that are posited explicitly by us or implicitly by the language we choose to use. In other words, Trevarthen cannot tell us about the infant's behavior as it is in itself. He must use some system of description.

The second point about Trevarthen's strategy follows from the first. If the terms subjectivity and intersubjectivity are part of an orienting frame of reference, they then would be expected to function more like abstract terms. Take for example the term *university*. It organizes concrete observables, such as the chemistry building, the psychology building, the administration building, et cetera, but university would not be "found" among the instances of observables that it organizes. You could not take a cinematic picture of the buildings and grounds on a piece of state property, and then point to one of them and say, "See, the university is here!" Similarly, it can be argued that as abstract terms subjectivity and primary and secondary intersubjectivity would not be something observable that could be pointed to on the films.

The terms, of course, could have other functions. They could be used in causal statements. Even so, just because these terms serve in the role of explanatory concepts or in explanatory sentences, that would still not warrant taking them as referential terms pointing to substantial realities. Instead, they would then simply serve the role of inference tickets, as Gilbert Ryle (1949) called them, allowing narratives about predictions or regularities concerning particular matters of fact. For example, sentences like "Bacteria x causes disease y" have a special function. It is to causally connect statements of fact such as "Here is bacteria x" and "Here is disease y." Such causal sentences, however, do not attain the same status as the sentences they attempt to connect. They do not then take on a referential function pointing to a thing as do the sentences that they organize causally. Thus, even if subjectivity and intersubjectivity were to serve in causal sentences or lawlike statements, that in itself would not be justification to take them to have simple referents that could be observed on film.

The final point to be made extends the first two in a somewhat quixotic way, following Kenneth Burke. It is obvious that Trevarthen subscribes to the referential or indicative theory of language, whereby the primary function of words is to serve as the signs of things. A word is ostensively pointing to a thing. However, Burke makes an interesting case for reversing this formula and taking things as the signs of words.

I will here outline Burke's arguments for this reversal. I will try to get the gist across through the use of brief examples. First, we must come to consider sentences and words to be in an "entitling" relation to nonverbal situations rather than simply pointing to an unambiguous thing. For example, everyone can understand the sentence "The man walks down the street." Note, however, that as stated it could not be illustrated. To do that one would need far more detail—how tall of a man, what kind of a street, in the night or in the day, et cetera. Thus, the sentence serves as a way of entitling or summing up what could be quite varied and complex nonverbal circumstances. This can also be shown for single terms—like strolling, or street, or, most importantly, human being. Single words, like sentences, are abbreviations; in fact, they are verbal abbreviations of nonverbal abbreviations. If we can come to see this relation, then we can consider how it might be extended, reduplicated, on other realms. For example, just as this sort of entitling can occur in the verbal realm so too can it occur in the nonverbal realm. Everyone has had the experience of a favorite tune or ambience at sunset functioning as a sign or an abbreviation or as an entitling of a whole scene or circumstance that comes streaming to mind. We can now imagine the summarizing nonverbal object or scene being paired with the summarizing words or network of narratives. But note—and this is the crucial point—we have now the circumstance where the particular nonverbal object, the thing, can be taken as the visual manifestation of all of what has been entitled or abbreviated in the corresponding word or network of words made into narratives. Thus, to use Burke's example, if you consider the individual human being as abstracted from the kinds of contexts in which alone a living human being is possible, s/he has become, from the linguistic point of view, a sign or manifestation or imperfect exemplar of all that the entitling word "human being" has abbreviated. And since entitling and abbreviations follow lines etched by social values, custom, myth, et cetera, it can be said that "the things of the world become material exemplars of the values which the tribal idiom has placed upon them" (Burke, 1966, p. 361).

Construing the relationship between things and words in this way provides our third vantage point to assess Trevarthen's research, and in fact, any research endeavoring to understand human infant behavior or action. For we can now begin to understand the infant as the material exemplar of the values and attributes that have come to be entitled in such words as human being, humanity, or humankind. The infant is the material exemplar of that whole linguistic network of concepts and narratives in terms of which we have come to understand ourselves, our heritage, and our progeny. That is, insofar as we have learned a philo-

sophical anthropology from our everyday linguistic commerce, insofar as that anthropology is itself embedded in our very language, and
insofar as it is impossible to entirely escape from that linguistic determination, just so will the infant be "seen" to be exemplifying (in proto-
forms, or in rudimentary forms) all of that which is entitled in its name
as a human being.

That terms like subjectivity and intersubjectivity serve as abstract
terms in frames of reference argues against construing them as ostensive
labels for concrete observable behaviors. The choice to use them is no
more or less than a choice of frameworks—for the description of human
or nonhuman behavior. In the case of human infant behavior there is a
strong bias to choose a framework in which they are included—the
infant as a material exemplar of all of that which is entitled in its name
as a human being strongly suggests a terminology comprised of mentalistic terms. That, however, neither bars such a terminology for the
description of other species' behavior nor secures its use for human
behavior as the most warrantable. To decide on what is most useful or
warrantable requires specification of "usefulness" criteria. Formulation
of such criteria would not be a strictly empirical task. Consequently, it
seems evident from what I have said that the legitimacy of the use of
mentalistic terms for the description of any behavior has less to do with
the behavior itself than a choice between frameworks and their commensurate criteria of usefulness.

To conclude, then, there seem to be no good grounds of the sort
given by Trevarthen to see subjectivity and intersubjectivity inscribed in
the films of the mother and infant interaction. Instead, such terms seem
on three counts to be part of a descriptive framework that is abstract in
character, social in origin, and constitutive of who we and others are.
Such terminology may be considered right, nevertheless, on philosophical and pragmatic grounds, not on the grounds that instances
have been cinematically recorded. The same must be said for their use
in the description of nonhuman behavior.

PART IV

Anecdotes and Anthropomorphism

11

Anecdote, Anthropomorphism, and Animal Behavior

Bernard E. Rollin

Although scientists studying animal behavior commonly make use of anecdotes and anthropomorphism in their everyday life and in books for the general public, they often explicitly disavow anecdotal evidence or anthropomorphic interpretation in published scientific discourse (Crocker, 1984). This ambivalence among scientists suggests that anecdotes and anthropomorphism are important for thinking about or understanding animal behavior, yet, if research is to be deemed acceptable as science, they must be weeded out. In this chapter I argue that such weeding out is unnecessary. Anecdotalism and anthropomorphism, despite their bad reputations, are not significantly different from or worse than many practices of demonstration and argument that scientists readily accept.

THE COMMON SENSE OF SCIENCE

Scientists in general share a common sense of how to do their work based on philosophical assumptions taken for granted and empirical material taught as fact (Rollin, 1989). For the most part, working scientists believe that science can deal only with what can be directly observed or what is subject to experimental verification. Failure to mark these precepts has historically led sciences such as psychology, biology, and physics to understandings unnecessarily fraught with speculation, metaphysics, and theology. Science is also believed to be value free, and thus can allegedly make no moral claims since moral judgments are unverifiable.

When this common sense of science is applied to the study of animal behavior, it leads to the belief that the thoughts, feelings, concepts, desires, and intentions of animals are not the sorts of things that can be either perceived or explored experimentally; thus, the psychology of animals in a mentalistic sense is not a legitimate object of study, and attributions of psychological states to animals—commonly labeled as anthropomorphism—are unacceptable. Individual instances of behavior in context—commonly called anecdotes—are similarly unacceptable because they are too often subject to overinterpretation and anthropomorphism.

With such a formidable arsenal of arguments arrayed against talking anthropomorphically of mental states in animals and buttressing such claims anecdotally, scientists' reluctance to countenance such talk is understandable. This common sense of science, implicit in some versions of positivism and explicit in others, found clear expression in Watson's behaviorism and exerted major influence even on thinkers otherwise inimical to behaviorism, such as Lorenz and Tinbergen, who also wished to avoid talk of the mental states of animals (Rollin, 1989). It is a staple of most training programs in biology and biomedicine, and even of many in comparative psychology, and has resulted in a skepticism toward anything that cannot be directly perceived or experimentally verified.

Although initially such a skepticism seems quite reasonable, if such skepticism is systematically adhered to, doing science would be impossible. Presuppositional to scientific activity are assumptions that flagrantly violate the claim that everything in science must be observable or subject to direct experimental confirmation. For example, scientists assume that a real, public, intersubjectively accessible world exists, independent of their perception; that others perceive this world and think more or less as they do; and that accurate and inaccurate reports about that world can be distinguished. And scientists assume that the past exists even though they cannot experience it directly, and that the universe was not created three seconds ago with fossils and memories created on the spot. Yet none of these beliefs can be confirmed by observation or directly tested by experiment, and few scientists are disposed to reject them even though they conflict with what is entailed by scientific common sense.

As soon as one abandons a hard-line verificationism which only admits direct observables into science and accepts that particular non-verifiable beliefs are admissible on the grounds of plausibility (e.g., that an external world exists independent of observers and that it is commonly accessible), one has replaced a rigid logical criterion for scientific admissibility with a pragmatic one, in which one needs to *argue for* exclusion of certain notions from science rather than simply apply a mechanical test. Such a replacement is evident in the history of physics. Contemporary physics, traditionally the hardest of hard sciences, has proliferated notions that violate the common sense of science (see papers in Kitchener, 1988). Physicists talk of all sorts of entities and processes, from gravitation to black holes, which are not directly verifiable or directly tied to experiments. Such theoretical notions are accepted because they help us to understand the physical world far better than we would without them.

THE REAPPROPRIATION OF ORDINARY COMMON SENSE

The scientific study of animal behavior has itself experienced a dramatic change in accepting more pragmatic criteria for psychological states in animals (e.g., Menzel, 1974; Griffin, 1976; Breuggeman, 1978; Dennett, 1978; Mitchell & Thompson, 1986b; Whiten & Byrne, 1988; Cheney & Seyfarth, 1990). This decline in skepticism about animal psychology is the result of many factors (Rollin, 1989), but perhaps of greatest significance is the fact that people have become increasingly conscious of moral obligations to animals and increasingly unwilling to have scientists work unimpeded by moral concerns. In particular, public attention has riveted on the issue of pain in animals used in research and has resulted in federal laws mandating control of pain and suffering in laboratory animals (Rollin, 1987). Traditional scientific common sense described research about animal pain as research into pain mechanisms and behavior, and ignored any talk of the subjective experiential dimension (see, for example, the majority of papers in Kitchell & Erickson, 1983). However, more recently (Panel Report, 1987), the report of a symposium by the American Veterinary Medical Association acknowledged that animals feel pain, and pointed out that animal research is used to model human beings and is therefore based on a tacit assumption of anthropomorphism. A similarity between human and animal pain is presuppositional to doing pain research and analgesia screening in animals and extrapolating the results to people—a *feeling* common to both is assumed, not merely similarity in plumbing and groaning. Thus, at least in the case of pain, scientific attribution of mental states to animals is based on anthropomorphism as presuppositional to its intelligibility. If one can in principle extrapolate from animals to humans (Rollin, 1982/1992), one can do the reverse as well.

In the face of federal laws prohibiting animal pain (reflective of the new social ethic about animals), it is inappropriate for scientists to express skepticism about their ability to know that animals feel. The concern for moral obligation to animals has forced upon science a reappropriation of ordinary common sense (Forguson, 1989) about animal thought and feeling, and a recognition that science is not value free and independent of moral concerns (Rollin, 1989). In ordinary common sense, animal behavior is described anthropomorphically, in that mentalistic attributions are made to animals. Ordinary common sense assumes that animals, like people, feel pain, fear, curiosity, and other mental states. Indeed, as Hume (1739/1960, p. 272) points out, few things are more repugnant to people using ordinary common sense than skepticism about animal mind.

In ordinary common sense, organisms are viewed as sentient beings with perceptions, emotions, ideas, intentions, desires, and beliefs which influence their behavior; mentalistic attribution to animals provides a plausible theoretical structure for explaining and predicting animal behavior (Dennett, 1978). Descriptions of animal behavior have a mentalistic component as well as a behavioral component, and the mentalistic component explains the behavioral component (Hebb, 1946). Saying that a dog is in pain, for example, means that the dog is exhibiting a particular range of behaviors or responses, such as cringing, avoidance and loss of appetite, which are most plausibly explained by the dog's *feeling* something—hurt—which is experientially similar to what we feel when we are hurt.

Much of the plausibility of any particular psychological attribution to animals derives not from any conscious reasoning on the part of an observer, but from experience interacting with these animals. Although criteria for assessing pain and its degrees in animals can be set, usually the best source of information about animal pain are farmers, ranchers, animal caretakers, trainers—in short those whose lives are spent in the company of animals and who make their living through animals (Morton & Griffiths, 1985). Similarly, although some scientists could get along perfectly well in laboratory situations expressing agnosticism about animal pain, emotion, and other psychological states, the caregivers of these animals work most effectively with the animals using anthropomorphic mentalistic attributions (Hebb, 1946). If people work directly with animals and don't recognize pain, fear, anger, and other mentation, they are highly vulnerable to injury, unable to control or train their charges, and likely to lose their livelihood. Of course these people may have no explicit criteria for recognizing psychological states in animals, and in that sense their reports may have anecdotal qualities—that is, they are not obtained in laboratories under controlled conditions. But given that scientists specifically disavowed the reality of nonhuman psychology and made no attempt to study it, it is perfectly proper to look to those who have been *compelled* to understand nonhuman psychology for millennia (Morton & Griffiths, 1985). To preclude data on (and interpretations of) animal behavior simply because the data were not observed by "accredited scientists" or garnered in laboratories (which are highly unnatural conditions for animals) is against the spirit of what science should be. Some philosophies of science (see, e.g., Feyerabend, 1975) suggest that science should be democratic in its admission of data sources, but stricter in the theories of explanations it graduates. To be sure, ordinary common sense is "theory laden" with often problematic categories and interpretations, but so too is the common sense of science. For example, how intelligent educated scientists bought whole-hog into behaviorism for most of this century is still hard to understand.

The case supporting anecdotal and anthropomorphic attribution to animals may seem most plausible in relation to pain. This plausibility is based not only on the fact that humans and nonhumans have similar behavioral responses to stimuli that are painful in humans, but also on similar neurochemistry and physiological mechanisms for pain, phylogenetic continuity, and even on the fact that humans who do not feel pain do not fare well at all (Singer, 1975; Rollin, 1989). One could argue, of course, that as long as simple, fundamental, primitive mental experiences like pain are looked at, anecdotes and anthropomorphism seem reasonable; but that they are not reasonable once higher mental processes are invoked. Ordinary common sense is far too disposed to provoke exaggerated intelligence, planning, reasoning, emotional complexity, and other unwarranted conclusions. Precisely such romantic, unbounded anthropomorphism and exaggerated anecdotes led to the behaviorist reaction against animal mentality, and a return to it now can only lead to a similar reaction.

One might respond with the idea that the ability to feel and respond appropriately to pain bespeaks mental sophistication beyond mere sensation (Buytendijk, 1943/1961). Pain in and of itself would be of little value uncoupled with some ability to choose among alternative strategies of response, that is, fighting or fleeing, hiding, et cetera. Indeed, the evolutionary utility of pain likely consists in the ability of the noxious stimuli to evoke not only motivation to alleviate it, but strategies to deal with it as well. But even if this is not the case, the argument and strategy constructed for using anecdotal and anthropomorphic information to identify pain is in principle no different from using the same approach to understand "higher" (or other) mental processes. The relevant distinction is not pain or sensation versus thought or higher mental processes—it is rather good versus bad anthropomorphism, reasonable versus unreasonable anecdote.

PLAUSIBILITY

Once again the key notion is *plausibility*, the same sort of measure used to attribute thoughts, plans, feelings, and motives to other humans, be it in daily life or when serving on a jury. We do not experience other people's mental states, and language and behavior can be used to conceal and deceive. How, then, do we judge other humans' mental states? I suggest that what we do is a combination of weighing of evidence and "me-morphisms"—extrapolations from one's own mental life to that of others. For example, if a friend of mine with a normal propensity for jealousy suddenly finds his wife, whom he loves deeply, running around flagrantly with another man and tells me he bears her no ill will, I am skeptical. I can in principle be convinced, but this would require very extensive observation and interaction with him to trump my more plausible interpretation that he is jealous. On the other hand, if he tells me that his *is* jealous and angry, or behaves that way, it is certainly reasonable to assume that he is, for that fits with what I know of myself and others.

Although we would be hard pressed to articulate them, we all have canons for judging the plausibility of anecdotes about other people, and of explanations of their behavior. A man who tells me that a woman obviously has a crush on him and cites as evidence that she ran into him twice in one week at the grocery and said "Hi" may reasonably not be taken seriously. In fact, the vast majority of our knowledge of human behavior does not come from scientific research, but from life experiences and would be dismissed by the common sense of science as anecdotal.

Like Romanes (1882), I claim that anecdote is, in principle, as plausible a source of knowledge about animal behavior as it is about human behavior, provided it is tested by common sense and background knowledge (Rollin, 1989). When someone interprets his dog's restlessness as evidence that the dog knows that it is her birthday, we can dismiss that since we have not reason to believe that the dog has—or even can have—a concept of birthday. On the other hand, if the person telling the anecdote explains the dog's excitement by saying that he has learned that when his masters cook and clean all day and frequently look out of the window, guests are coming, that is consonant with what we know of dogs' abilities.

More difficult cases occur when the anthropomorphic anecdote concerns a species of animal with which we do not enjoy the familiarity we do with dogs, though here the problem is in principle no different than when we deal with people who come from cultures significantly different from ours. When they belch loudly after a meal, we may label them as rude people out of ignorance of their culture, wherein such an act is a polite compliment. We can make the same mistake looking at unfamiliar animals, as when the child or urban adult reports equine sex-play as fighting.

Once one has in principle allowed the possibility of anthropomorphic, anecdotal information about animal mentation, one must proceed to distinguish between plausible and implausible anecdotes, plausible and implausible anthropomorphic attributions, remembering that, even if we are right to be skeptical, implausible accounts may turn out to be true. Many people tell outrageous anecdotes, or interpret them in highly fanciful or unlikely ways, and even publish such nonsense. This should no more blind us to the plethora of plausible anecdotes and reasonable interpretations forthcoming from people with significant experience of the animals in question, than should the presence of outlandish stories about or outrageous interpretations of human behavior cause us to doubt all accounts of human behavior.

Anecdotes and their interpretation may be judged by many of the sort of principles Romanes (1882) relates in his classic introduction to *Animal Intelligence*. Does the anecdote cohere with other knowledge we have of animals of that sort? Have similar accounts been given by other disinterested observers at other times and in other places? Does the interpretation of the anecdote rely upon problematic theoretical notions (such as imputing a grasp of "birthday" to a dog)? How well does the data license the interpretation? Does the person relating the anecdote have a vested interest in either the tale or its interpretation?

One point which has escaped notice is that anecdotes are logically no further off base than reports of scientific experiments and their interpretation. In some ways the latter may be more suspect. Reports of data falsification, fraud, and dishonesty in scientific publications indicate that scientists are as human as anyone else. Given a "publish or perish" system for science, scientists must produce or else they must effectively give up their careers. If this is so, then researchers have a strong vested interest in obtaining results, which in turn should excite our natural suspicion. Of course scientific reports are replicable in principle, but little money exists for such replication as long as the result coheres with data in the field. Anecdotes are also replicable in principle, either by experiment or observation. In the final analysis, any report of an experiment is by definition an anecdote, not a confirmed hypothesis. Are the multiplicity of theoretical biases which scientists carry by virtue of their training any less or more or equally pernicious to their observational capacity than the theoretical biases built into a nonscientifically trained but intelligent observer?

The methods suggested for analysis of anecdotes and animal behavior in general can be applied to a singular incident that occurred at the Denver Zoo and was videotaped by Denver television. An African elephant was down and refused to get up, a condition known to lead to death if not corrected. All efforts to get the elephant to stand up—including bringing in a crane to lift the elephant—had failed. By chance, several Asian elephants were herded by the afflicted elephant. The Asian elephants broke ranks, approached the fallen elephant, and nudged and poked him until he stood up. They then supported him until he stood on his own. Thus far an anecdotal narrative is presented with little or no theoretical bias intruding and no interpretation offered. The story is surely relevant to the study of elephant behavior. Although the TV station had a vested interest in dramatic stories, the events are on video and other observers presented the sequence of events in the same way.

The common sense interpretation of the data offered by the station and other observers is that the elephants were altruistically helping another elephant, albeit of a different species. Such an interpretation is more problematic than the simple reportage of the events, since "help" is ambiguous and speculative and the events are certainly open to other interpretations. When, however, one juxtaposes that story with the many other stories of elephants showing helpful behavior to other elephants, together with the extensive data we have on the problem-solving ability and social nature of elephants (Moss, 1987), the interpretation gains in plausibility.

CONCLUSIONS

Mentalistic attribution to animals provides a very plausible theoretical structure for explaining and predicting animal behavior. Anthropomorphism, if tested against reasonable canons of evidence, is a plausible theoretical approach to assessing animal behavior. Since far more ordinary people than scientists observe animals, it would be a pity to rule out anecdote, critically assessed, as a potentially valuable source of information and interpretation of animal behavior.

12

What's the Use of Anecdotes? Distinguishing Psychological Mechanisms in Primate Tactical Deception

Richard W. Byrne

In the behavioral sciences, unanticipated events often make for fascinating and suggestive observations. Since they are unanticipated, the best that can be said is that they "were collected on an *ad lib* basis" (Altmann, 1974). Often, they are dismissed as "anecdotes," although there is undeniably a subjective element in what is dismissed: "my data, your observations, their anecdotes." Rightly, wherever it is feasible, the descriptive record of one of these unanticipated events is used to inspire ideas to test with systematic, controlled observations or experiments.

Problems arise, however, when these normal methods are particularly difficult to apply, yet the issue cannot be ignored. Perhaps the most obvious example with which to defend this case is that of mammalian infanticide: only in lions and Hanuman langurs is infanticide so regular that a systematic study of the behavior can be executed successfully (e.g., Hrdy, 1977; Packer & Pusey, 1983), but infanticide occurs far more widely (see Hausfater & Hrdy, 1984). Because of the obvious genetic importance of infanticide, cases are no longer dismissed as anecdotes (as they used to be: see Hrdy, 1977), but recorded with care and in detail. The same is now true of extra-pair copulations in pair-forming species of birds, though in the past occasional sightings of acts at variance with monogamy were forgotten as mere oddities. The lesson is this: careful and unbiased recording of unanticipated or rare events, followed by collation and an attempt at systematic analysis, cannot be harmful. At worst, the exercise will be superseded and made redundant by methods that give greater control; at best, the collated data may become important to theory.

This chapter will concentrate on observations of primate deception. The behavioral tactics used by primates are considered by many researchers to be an important matter for a proper understanding of primate mentality and the evolution of human cognition (e.g., Trivers, 1985; Mitchell & Thompson, 1986b; Jolly, 1991; Byrne & Whiten, 1988a). Nearly all the available data on primate deception are anecdotal, in the sense that they were collected during the course of other studies, ad lib, with no easy means of evaluating frequency against any proper control. However, they are not anecdotal in the pejorative sense; they are not the casual observations of inexperienced observers, embellished by multiple retelling and rife with implicit interpretation. On the contrary, deception in primates has repeatedly been watched and carefully recorded by highly experienced primatologists, each with years of familiarity with the species concerned, and training in behavioral recording (e.g., Goodall, 1971; Menzel, 1974; de Waal, 1982, 1986; Byrne & Whiten, 1985; and the records of many observers collated in Whiten & Byrne, 1986, and in Byrne & Whiten, 1990). These observed acts are very varied in character, but many are goal-directed acts, whose success depends upon another animal being deceived. In the early 1980s, confronted by the dearth of published data on primate deception, yet finding frequent informal admissions of interesting observations, Whiten and I attacked the problem by surveying primatologists for any accounts that matched what we called "tactical deception," and collating their records to see if any patterns would emerge. The definition of *tactical deception* that we used[1] was: "acts from the normal repertoire of the agent, deployed such that another individual is likely to misinterpret what the acts signify, to the advantage of the agent" (Byrne & Whiten, 1988b, p. 271). Independently of our work, Mitchell (1986) proposed a hierarchy of animal deception that made a similar distinction between the inflexible sort of deception shown by animal mimicry or injury feigning (most likely a result of differential replication of genes), and behavioral deception that is modifiable by learning or planning, and so not shown by every individual under similar circumstances. Mitchell's Level 1 and Level 2 together correspond to what we called "strategic deception"; his Level 3 and Level 4 map onto our "tactical deception." A wide range of haplorhine primate species (monkeys and apes) are now known to perform tactical deception (Byrne & Whiten, 1990, 1992), and in some cases the evidence shows the same pattern being independently reported by different researchers in different populations.

The definition of tactical deception explicitly does *not* include a phrase such as "with the intention of misleading," which would be an essential feature of any definition of *lying*, in the sense used by philosophers of human behavior. Our intention was to capture *functional* deception, not to require evidence of intent to create false belief. Partly, this was because such evidence is difficult to get, and study of the whole topic of deceit in animals liable to founder on this difficulty. Partly, it was because we believed that most primates had no intentions to create false beliefs, yet their deceptive behavior was nevertheless biologically important and psychologically interesting (Byrne & Whiten, 1985, 1988b). However, whether in fact records of primate tactical deception include some cases in which causing false belief—as well as gaining profit—was intended by the agent is naturally of special importance for theories of the evolution of mind. Although we explored the implications for primate mental representation (in each category of primate tactical deception), of the possibility that we might one day gain adequate evidence of a primate's intent to cause false belief (Whiten & Byrne, 1988), little evidence was available at the time (but see Menzel, 1974; Mitchell, 1986; de Waal, 1986). Sufficient evidence to claim that deception was *intended* existed in only the merest handful of records, and far more commonly records lacked the essential historical and contextual data which would have allowed the determination of their status, intended or otherwise (Byrne & Whiten, 1988b; Mitchell, 1986, came to a similar conclusion about Level 4 deception, which corresponds exactly to our "intentional tactical deception").

This indeterminacy was perhaps unsurprising in a retrospective survey. Now, a much larger corpus of data is available (see Byrne & Whiten, 1990, for a phylogenetically organized catalogue), and many records include good contextual information. Conversely, the rarity, spontaneity, and subtlety of tactical deception have hindered experimental analysis of the phenomenon, and little advance has been made in this direction since the pioneering work of Menzel (1974) and papers in Mitchell and Thompson (1986b). Important psychological issues are raised by the possible mechanisms and origins of tactical deception, and behaviors like this which naturally occur at very low frequencies may nevertheless be biologically important. We ought therefore to assess properly what evidence there is, not ignore it and wait for a time when we have better evidence—a time that may be slow in coming.

LIMITED USE OF ANECDOTAL EVIDENCE

The argument of this chapter is that progress can be made now, provided anecdotal data are handled carefully. The type of conclusion that can be drawn from their analysis will naturally be different from that obtained by systematic observations. The situation is no different from many other cases in science: for instance, in order to state life expectancy at birth one requires much systematically collected data, whereas to state that humans can live to 100 years one merely requires reliability of one record, and the claim is not strengthened by many unauthenticated records. The aim of the current chapter, then, is to explore the possible interpretations of some records of tactical deception, in order to suggest how this kind of rather precious observational data might best be handled, and to illustrate just what extra information it would be desirable for observers to record in the future.

It should be emphasized that information collected, as in this case, by the survey method can never provide precisely reliable *quantitative* data on the relative frequencies with which acts are used in different primate populations or species, for several contributory reasons (see Altmann, 1974). Obviously, there are immense differences in the number of hours for which different species have been studied. Furthermore, the inevitable subtlety of a behavior like tactical deception means that positive evidence of occurrence is only possible in studies which are concerned with intimate social details, whereas many studies of primates are focused instead on ecological recording. If an observer's effort is fully engaged in noting food type and interanimal distances at frequent intervals, subtle interactions will not be noticed.

The ability of observers to detect tactical deception will also vary with their own skills in interpreting the communicative acts of the animals, which in turn is bound to vary with the relative similarity of the animals' communication system to human nonverbal communication. Ape nonverbal communication shares more in common with our own than any monkey's, and strepsirhines (the group of primates including lemurs and bushbabies, characterized by their wet rhinarium) rely extensively on olfaction. Consequently, there will be a bias towards recording more tactical deception in species closely related to humans.

Finally, for two reasons tactical deception is much more likely to be recorded in species that range in a highly dispersed way or whose social organization is of a fission-fusion type. Both reasons relate to the level of "obviousness" needed for tactical deception to be detected. On the one hand, in a dispersed group it will frequently be the case that one individual or subgroup will be in possession of information that others lack, and withholding information is—in various ways—one of the commoner forms of deception observed. (Although some would dispute that withholding information strictly constitutes deception; see Mitchell, 1988, and Byrne & Whiten, 1988b, for discussion of this point.) We should therefore expect deception to pay more often for individuals in dispersed groups. On the other hand, it will often be the case that the human observer of a dispersed group will gain information that some of the animals do not share; thus, human observers of noncohesive or fission-fusion groups are in a better position to detect tactical deception. In very cohesive groups, by contrast, most tactical deception would need to be of a higher order of subtlety. As a result, deception would be both less likely to occur and less likely to be detected.

But there is no reason in principle why high-quality observation should not be sufficient to confirm the *presence* of tactical deception in a particular species, and disproportionately high or low frequencies may be approximately indicated by comparison with the number of studies conducted that *might* have detected deceptive tactics (Byrne & Whiten, 1992). Absence of the ability to deceive others, though logically unprovable, may be suggested by "strong" negative records—a continued absence of behavioral tactics of deception after many hours of intimate observation of circumstances where individuals could clearly profit from their use.

QUALITY OF EVIDENCE

Immediately we are confronted with a problem in the use of the term *anecdote*: is its use still appropriate here? Often, the term anecdote suggests a single, casual observation by an inexperienced observer, unfamiliar with the species observed and with the concepts of explanation in animal behavior. Of course, any attempt to interpret such data is rash, simply because it is untrustworthy. Published cases of tactical deception in primates, by contrast, come from experienced observers of behavior who have many years of experience of their species, and who were often trained in behavioral explanation. Independent, interobserver agreement on a phenomenon in multiple records of its occurrence is far from the everyday, dismissive sense of an "anecdote." To draw attention to the difference, I will here use the term *record* instead. The careful observations of experienced pet owners, unfamiliar with behaviorist theory but knowing their animals well, form a category of evidence intermediate between casual anecdotes and scientific records.

In some cases observers have been lucky enough to be in a position to record data that can take us even further, and can rule out some possible hypotheses of the origins of the behavior. In order to illustrate how this can sometimes be done, I will repeatedly use as illustration one record of deception, a manipulation of a "target" animal by an "agent" using a "social tool," in chacma baboons. This record, and the terms for the animals' roles, were originally described by Byrne & Whiten (1985, p. 670) and discussed further by Whiten & Byrne (1988, p. 240). The observation was as follows:

> Adult female Mel is digging, probably to obtain a deep-growing corm. The young juvenile Paul approaches to 2 metres and looks at her, then scans around; no other baboons are in view. Paul looks back at Mel and screams. Adult female Spats runs into view towards them, then chases Mel over a slight cliff and out of view. Spats, who is Paul's mother, normally defends him from attack. When both females are out of sight, Paul walks forward and continues digging in Mel's hole.

What are the possible interpretations of this observation? I will begin at the most dismissive possibility, and work gradually towards the most mentally complex; this, of course, follows the approach recommended by Lloyd Morgan (1894).

LEVELS OF TACTICAL DECEPTION

Evidence Level-0: Not Tactical Deception

Possibility (a): The Ethological Description of the Animal's Actions Was Inappropriate. If the agent Paul's actions have genuinely caused the actions of Spats (the social tool) and hence those of the target Mel, interpretation of this record as *deception* depends on what the scream communicates. When a young baboon screams and his mother runs aggressively to the scene and attacks the nearest other animal, was the message conveyed by the scream one of fear? In principle it need not have been. Baboon screams have not been studied experimentally in the field and may encode several messages in variants which differ only slightly in acoustic form. If in fact the scream used were a variant, a call for aid in food competition, it might well be that the mother was not deceived. On this interpretation, Paul used a call in a normal, nontactical way, indicating the need for aid in food competition, and his mother (correctly interpreting the call) acted with the goal of improving her juvenile's nutrition.

For this record, such an interpretation is unlikely for two reasons. This population of baboons was living through an ecological "bottleneck" of low nutrition (Whiten, Byrne, & Henzi, 1987), and support from relatives was observed rarely and only then when a relative was threatened in agonism, not in food-directed coalitions (Byrne, Whiten, & Henzi, 1987). Maternal support in food competition is thus highly unlikely. Secondly, we also recorded the same animal, Paul, using the tactic *against* his mother Spats, in this case using the leader male of the group as the social tool. The plausibility of the leader male supporting his offspring in feeding competition against one of his females is even smaller.

In other records, such an interpretation may well be correct. For instance, Stuart Altmann reports that in yellow baboons, "mothers with weanlings will often ignore (avoid visual contact with) their milk-begging offspring. Ignoring as a social strategy used in many situations is widespread" (Byrne & Whiten, 1990). Two interpretations are possible here. In one the avoidance of visual contact is tactical: then, for the tactic to be effective, the offspring must be deceived into believing their mothers have not noticed their screams. In the other, the avoidance of visual contact is a communicative act of "refusing." It is not deception of any kind since it functions effectively when correctly interpreted by the offspring; nor is the act tactical since it subserves its normal function. (Note that it could be held that the normal function of "looking" is by definition to see, not to communicate; on this view, the act of looking away would be classified as Level 0 (c), see below.) Since the young animal's behavior would presumably be rather similar in both cases, it is difficult to be sure that this record should be classified as tactical deception, and we did not do so (Byrne & Whiten, 1990).

The same is true of the "alarm calls" given when no predator is present by sentinel species in mixed flocks of Neotropical birds, often securing food for the caller by causing another bird to hesitate (Munn, 1986). Although Munn interprets this as deception, the fact that the "false" uses are actually commoner than the "true" ones raises the possibility that the original designation of "alarm call" was incorrect. Suppose, for instance, that "threat" was a better description: might a bird not sometimes threaten a predator, and also (and more often) threaten a conspecific in possession of a coveted resource? Of course, a threat display at the sort of predators that are highly likely to kill the bird would not be expected. The most serious predators for Neotropical rainforest passerines are probably forest falcons, a species notoriously difficult for humans to detect. This difficulty has two consequences. Firstly, it makes it likely that assessment of call function has been made when birds were seen with other, less serious predators. Secondly, it makes it difficult to accept that Munn could be perfectly sure no forest falcon was, in fact, in range during some observations. This renders any evaluation of these "alarm calls" particularly difficult. Until we have a very clear understanding of call functions in a species, claims of deceptive uses need to be treated with caution.

Possibility (b): The Conjunction of Events Was a Coincidence. In our baboon example, suppose that Paul's mother Spats, for reasons unconnected with anything described in the record, just happened to attack the other female, Mel, at precisely the moment when the juvenile was watching her feeding. On this assumption, the juvenile's scream was unconnected, and caused by something else which went unnoticed. This must be taken seriously as a possibility. A big danger of any retrospective survey is that a striking chain of events that appears goal directed is considerably more memorable than the (far more common) cases of chains of events that do not appear goal directed. Coincidences do happen, and soliciting a trawl of ad lib records from many observers is a good way of finding some. What is required in general is control data.

In the case of the young baboon's scream, the observation was in fact collected as part of a longer term study, and we can be sure that both the screams of juveniles and aggressive interactions between adult females are rare events; thus this particular coincidence is an inherently unlikely one. The tactic was observed three times in three weeks, and so we can rule out the possibility of simple coincidence. This reasoning emphasizes both the use of control data in interpreting an unusual record, and the difficulty of interpreting a pattern of behavior that is observed only once, both points which have been made before (Whiten & Byrne, 1988).

A more interesting form of coincidence is also possible, in which the "goal" is a coincidental windfall of an action of the agent intended for another purpose. If the juvenile Paul, attracted by the chance of a valuable subterranean corm, had approached the adult female Mel closer than she was prepared to tolerate, she would have threatened him. Then he would have screamed, in unfeigned fear, causing his mother Spats to attack the unrelated adult threatener of her juvenile, Mel. In this scenario, the corm which Paul proceeded to eat was merely a windfall benefit of the interaction between the two females, itself prompted by the *correct* interpretation by Spats of a scream of fear from Paul.

This account requires that the observer had failed to notice the threat by Mel, which is plausible for one observation, but less so for a series of three, independently observed by two different primatologists over a course of weeks. Furthermore, if it is to be argued that the goal

that caused the young baboon to scream was to avoid an attack threatened in a very subtle way, then exactly the same pattern should be
shown whether or not his mother was in visual contact. Indeed it could
be argued that screaming should be *more* likely if his mother were near
at hand. In fact, in this field study it was a commonplace situation for a
young animal to see an adult excavating a deep-growing corm, but the
only cases which then led to the young animals screaming were ones in
which the mother was distant or out of view.

Although any interpretation of coincidence is therefore unlikely to
apply to the record here, the likelihood of windfall rewards has wider
implications. Just such a scenario—of threat, scream and rescue, and
windfall reward—is very likely to happen sometimes. Any food reward
is bound to act as a *reinforcer*. Its (natural, coincidental) contingency
with the antecedent behavior would serve to *condition* the act as a
response to a particular stimulus. In future, the response of screaming
may then occur to gain food reward rather than to avoid threat. This
possibility will be examined further below.

Possibility (c): The Tactic Was Manipulative But Did Not Rely on Deception.
In some cases an action, used in a tactical way to manipulate a social
companion so as to achieve a goal, may work effectively yet not involve
deception. For an example, consider the following record of baboon
behavior:

> Every day, one can see females approach mothers, pretend to be pri-
> marily interested in grooming the mother when what they are really
> after is an opportunity to sniff, touch, or even hold her infant. (As a
> result, mothers with small babies are often very well groomed.) But is
> the mother really deceived? Surely the multiparous ones know exactly
> what's going on! (S. A. Altmann, in Byrne & Whiten, 1990, p. 41)

Here, as Altmann notes, deception is unlikely to be the mechanism of an
evidently effective tactical behavior. Instead, some account in terms of
trading grooming for access, or in terms of the message of peaceful-
ness and lack of threat of a groomer, would make more sense. In the
example of Paul's scream that we have analyzed in detail here, the idea
of a manipulative but not deceptive tactic does not appear to apply; if it
works as a manipulative tactic, the juvenile's behavior must have
deceived the mother.

For this record, then, we can be confident that a classification of it
as tactical deception is warranted: the act was used tactically, the goal
that controlled the agent's behavior was that of obtaining food, and
this goal could only have been achieved if the social tool was deceived
as to the nature of the interaction between the agent and the target.

Given that this record is a case of tactical deception, we must next ask: could it be consistent with our conventional understanding of animal learning, or do we need some cognitively richer interpretation?

Evidence Level 1: Learned Tactical Deception

To show that a tactic of deception had been conditioned by a history of coincidental reinforcements would ideally require focal animal data-recording throughout the subject animal's lifetime. To have much chance of success, the sampling would need to span much of the animal's waking hours if—as in wild baboons—social interactions over food could occur at any time. As such, the labor of data collection would be immense, yet just this has been seriously suggested at times: perhaps first by Mills (1898), and recently by Strum (1988), who had actually done it for a month with a baboon. More practically, we should at least endeavor to use normal developmental data on ontogeny as an aid in determining mechanism of acquisition (as noted by Byrne & Whiten, 1985, and Mitchell, 1986, 1988, who points out that this was Lloyd Morgan's remedy for cavalier acceptance of anecdotes in 1894!). Unfortunately, at present even this is only available on deception in a few captive animals (Miles, 1986; Savage-Rumbaugh & McDonald, 1988).

However, using the extensive knowledge of reinforcement available from laboratory experiments, we can assess the *plausibility* that an explanation in terms of reinforcement is right. Taking Lloyd Morgan's canon seriously, we should accept learning by reinforcement as the explanation unless it is highly implausible to do so.

In the case of the young baboon obtaining a corm by screaming, then, is reinforcement plausible? In other words, is it likely that the scenario of coincidental reinforcement, which we sketched above and rejected as an explanation of the particular record (the tactic at the time when it was observed), had in fact occurred in the past? Might the young animal have at some point (1) approached too close to an adult feeding on a desirable resource, (2) been threatened, (3) screamed with unfeigned fear, (4) been consequently supported by a more powerful relative, and thus (5) gained a food reward as an accidental windfall? The answer is surely yes. This history alone would not serve to condition the behavior pattern as observed, since the young baboon did not scream when he met just the same situation but with his potential helper nearby or in full view. A more prolonged or fortunate history of events would be required to shape the precise eliciting pattern of "food held by animal subordinate to available helper, who is itself out of visual contact." Such a history is not likely to have occurred very often, which suggests that baboons possess an ability to learn rapidly in social situations. This suggestion is consistent with the statistically significant overrepresentation of baboons over other monkeys in showing tactical deception (Byrne & Whiten, 1992), and also with recent experimental work. Coussi-Korbel (1994) hid food tidbits in the extensive cage of a group of mangabeys *Cercocebus torquatus*. Rapide, a nondominant male, was shown the food before the whole group was released. On the first trial, he went straight to the food, but lost it to the dominant male, Boss. Next, Rapide held back and Boss stayed with him; the food was found by another animal, which ate it. In the third trial, Rapide moved in the opposite direction to the food, Boss followed him, and Rapide—apparently seeing his opportunity—then rushed for the food and gained it. On the fourth and subsequent trials, it was clear that Rapide used this "leading away" tactically, and was often successful in diverting Boss and so gaining the food. Mangabeys are closely related to baboons, and evidently learn with sufficient rapidity to acquire a deceptive tactic in four trials.

In some cases of tactical deception, a much less elaborate "possible history" has to be invented to explain the tactics as based on acquisition by conventional learning mechanisms. For instance, if limping after an attack happens to have an effect like that of a submissive display, and so inhibits further aggression, then this release-from-attack could well act as a positive reinforcer and so condition the animal to limp only in the presence of a potential attacker. De Waal (1982) has described such behavior in a captive chimpanzee, and has argued for what he calls the "daring interpretation": "Yeroen [the agent] was playacting. He wanted to make Nikki [the target] believe that he had been badly hurt in their fight" (pp. 47–48). However, given the plausibility of a history of reinforcements, more evidence would be required before we could accept claims that the animal intended to deceive, as de Waal (1986) later acknowledges. Very similar tactical use of limping has been observed several times in domestic dogs (for instance, Goodall, 1986).

To some extent the necessary further evidence could come from control data on the frequencies of those events, as they occur independently in the animals' lives; the events would need to occur together to condition the tactical deception, and we can estimate the probability of this from the independent rates of occurrence. Consider the case of a subadult male baboon, attacked by several adults, who stood on his hindlegs and stared fixedly into the distance (Byrne & Whiten, 1985), as if an alarming stimulus such as a predator were present. His staring served in fact to direct the attackers' attention to a distant hillside, and the subadult thus avoided the attack. It was argued (Whiten & Byrne, 1988) that in principle, by calculating the frequency and dynamics of such aggressive encounters, and the rates of detection of genuine causes for alarm in the particular baboon population, that it would be possible to calculate the probability of a genuinely "alarming" event occurring precisely in the narrow time window between an attack being launched and its conclusion, and the probability that the attacked animal would notice this event before the attackers.

In this sort of case, where all the independent events are of rather low frequency, then without knowing their probability distributions it might be unwise to make the natural assumption that the probabilities were straightforward to calculate. Consider the analogous case of airplane crashes. In such disasters, it is usually the case that several very improbable events are found to have happened simultaneously (R. M. Dawes, personal communication, March 1, 1988). Simple multiplication of the independent probabilities of each event would suggest that airplane crashes should virtually never occur, given the actual number of flights that take place; yet crashes do occur. In our baboon case, perhaps we should be cautious of reading too much into the apparently very low probability that a predator or similarly dramatic stimulus should happen to be sighted by a pursued animal in the few seconds before the pursuit is over and the attack carried out.

Nevertheless, the method is certainly useful in *comparing* the probability that individuals in different populations of animals might have learnt by reinforcement essentially the same tactic of deception. For example, it is frequently reported that domestic dogs will use the same tactic as these baboons, of orientation and gaze towards a (non-existent) cause for alarm, in order to obtain a reward. An example is the greyhound that, to obtain access to a coveted dogbed, ran to a window and stared out fixedly until the other greyhound then occupying the bed also ran to the window to look out, whereupon the first quickly ran to the bed and sat in it (E. Lennon, personal communication, 1993). In the case of pets, causes for alarm (e.g., postmen, visitors, etc.) are undoubtedly more frequent than sightings that alarm baboons in the field. Also, the time window in which a lucky conjunction of events would be effective in causing conditioning (i.e., the time when another dog is occupying a coveted dogbed) is long. Thus we can be sure that coincidental circumstances, having the power to reinforce the contingencies of stimulus and response that amount to this tactic, are commoner for domestic dogs than they are for wild baboons. Either dogs should show the tactic more often than baboons, or baboons learn much quicker than dogs.

Just because tactics of deception can be explained most parsimoniously by an explanation consistent with normal learning theory, their interest to psychologists and biologists should not be diminished. The conditional stimuli which trigger the behaviors can clearly be highly specific ones, as they not only include the presence or the absence of particular individual animals in the visual field, but even in some cases what those animals are able to see from their own viewpoints (Whiten & Byrne, 1988). The tactics are often highly effective in gaining resources or safety for the agents who deploy them. And finally, simple conditional linkages between circumstances and behaviors are potentially a powerful way of problem solving. When a series of such linkages can be elicited in turn by the same problem description, we have what is called a *production system*, and psychologists have found these to be effective in simulating human problem solving in chess, formal logic, and other areas (e. g., Newell & Simon, 1972). Much human problem solving can thus be understood without any need to attribute to the solver any knowledge of the beliefs and thoughts of other individuals (and see Byrne & Whiten, 1991, for use of the production system formalism in discussing cases of primate deception which appear intentional).

Evidence Level 2: Intentional Tactical Deception

All tactical deception (including all records coded Level 1 and Level 2) is necessarily "intentional" just in the sense that the agent acts as if it intends to achieve a goal by using the procedure. However, a deeper level of intentionality is possible, in which the agent acts as if it intends another animal to have a certain belief, in order that the agent's goal can be achieved. This deeper level has been termed second-order intentionality by Dennett (1983), and has been suggested to have had a key role in the development of human consciousness (Humphrey, 1983). In fact, because deception often involves X intending Y to believe that X thinks something false, intentional deception is often third-order intentionality, in these terms. However, once given the recursive capability of second-order intentionality, third-order intentions are simply double-embeddings; presumably any individual capable of embedding intentions at all is capable of embedding again, continuing the recursion. The (potentially infinite) recursion this leads to is most likely curtailed by memory limits, just as real humans are limited to about three levels of syntactic embedding in their sentences, unlike the transformational grammars of linguists. (For discussion of "counting levels" see Perner, 1988; Whiten & Perner, 1991.)

Second-order intentionality amounts to a mental simulation of another individual's mind (Byrne, 1995). Craik (1943) has argued that *thinking* is best understood as a process of simulation, in the mind, of events that have not yet or could not happen in the world. In a system with second-order intentionality, the process is taken one stage further: an individual is able to simulate not just events in the physical world but events in the mind of another individual. The implications of this process are obviously of great importance for the evolution of cognition.

How could we be sure of intentional tactical deception in animals? In the absence of complete historical records of potential reinforcement histories, which would be required to make the very difficult proof of the absence of conditioning, what evidence would convince? To judge from our experience as humans there are at least three possibilities. The first is used routinely by cartoonists, and that is smirking or concealed laughter by the agent at the success of the tactic. Obviously such visible amusement is not necessary for the performance of the act, and if detected it would be liable to lead to counterdeception by any animal capable of understanding second-order intentionality. It is, therefore, perhaps not surprising that no such amusement has been detected in primate agents. Even if it were detected it would, for parsimony, be explained away as pleasure at gaining the reward.

More usefully, Humphrey (1988) has suggested that a duped animal could sometimes realize the deceit and show "moral outrage." Byrne and Whiten (1988b) suggest one record that fits this description. In it, a chimpanzee researcher (Frans Plooij) used the tactic of orienting and staring into the distance to get an infant female chimpanzee to leave him, just as in the baboon record quoted on p. 144. The youngster had shown every appearance of enjoying Plooij's presence; she had climbed on him and groomed his hair. If Plooij had pushed her away he would have risked an attack by the mother. When the young chimpanzee returned a few minutes after leaving Plooij, she did not resume her friendly attentions to him. Instead, she hit him and for the rest of the day avoided his presence. This strongly suggests that she understood that he had intended to deceive her.

One single record is insufficient evidence on which to hang such an important claim as second-order intentionality in chimpanzees, and as yet no comparable cases have been forthcoming. That chimpanzees are capable of "moral" judgments based on intention, is, however, supported by the experimental work of Povinelli (1991). He arranged that two humans both failed to deliver a coveted drink to a caged chimpanzee. One experimenter spilled the drink "accidentally," the other did so with clear deliberation; the chimpanzee thereafter showed a preference for the human who had only "accidentally" failed to cooperate.

A third possible indication of an understanding of deception comes when animals show techniques that anticipate and counter deception by others, which are themselves unusual behaviors and thus implausible candidates for reinforcement histories. One example concerns the record (again from Frans Plooij) of a chimpanzee that hid behind a tree and peeped out, looking back at another chimpanzee—one that was, in fact, deceiving him by concealing its knowledge of a food source. Chimpanzees do not normally hide-and-peep at all; my own experience of chimpanzees is limited to four months in which no animal at Mahale did this, but Goodall (1986) writes after 25 years at Gombe and records no such behavior except during rare intergroup interactions. The fact that this hide-and-peep action was effective in countering the deception, combined with its improbability of occurrence, convinces me that this animal understood that it had been deceived. Mitchell (1988) suggests that the second animal might "be thinking that something is odd about the first's actions and be waiting to see what it does" (p. 260); but the second animal concealed itself before it waited, an action more consistent with an understanding that the "oddness" of the other's actions concerned itself, in a way not to its advantage. Similarly, the arms race described by Menzel (1974) between a subordinate female chimpanzee, experimentally given knowledge of a hidden food source, and a more powerful male chimpanzee who regularly robbed her of the food once it was discovered, powerfully suggests that both protagonists understood that the other intended to manipulate them. Apparently with no opportunity for coincidental reinforcement of chance actions, these animals escalated their tactics, finally reaching such subtleties as the male walking away as if unconcerned and then suddenly wheeling round to attend to the female's path of movement and then extrapolating it to the food.

Each record of this nature, as an isolated case, can be questioned, and an elaborate—though often implausible—explanation constructed on reinforcement lines (e.g., Byrne & Whiten, 1991). However, the overall pattern is harder to explain: good evidence of *intentional* tactical deception is found in the four species of great ape but not in monkeys, whereas monkey records contribute the great majority of all cases of primate tactical deception (Byrne & Whiten, 1992).

From time to time it is argued that observational data of this kind cannot in principle distinguish second-order intentionality from lower level explanations; most recently this argument has been put forward by Heyes (1993). There is a sense in which this must be true: radical behaviorists remain convinced that all human behavior can be explained without need of mental state attribution, since every case has an undeniable history of observable cues and rewards that led to it, and the phenomenological inner states remain unobservable. However, the latter is a clumsy way to redescribe behaviors elegantly captured by intentional explanation, and it does not seem to be what Heyes means. Instead, she illustrates her argument with one particular record from Whiten and Byrne's (1988) catalogue, and explains—quite rightly—that associative learning might have led to the tactic being used. But no claim to the contrary had been made in the 1988 report about that particular record (of a female baboon distracting a male by grooming, then snatching his prized food), and when soon afterward it *was* explicitly catalogued, the most parsimonious classification was judged to be "conditioned tactical deception" (Byrne & Whiten, 1990). A better test of Heyes's belief would be to attempt the same exercise for the most compelling observational records of intentional tactical deception, as done by Byrne and Whiten (1991). While they were still able to redescribe each in behaviorist terms, the accounts so produced were clumsy and unlikely, and, as for adult human action, (second-order) intentional accounts began to seem more parsimonious.

Records of tactical deception in monkeys lack any comparable signs of second-order intentionality, and it therefore seems unlikely that monkeys can really understand the deceptions they certainly practice. That is, while monkeys may know that under certain specific circumstances a tactic works reliably to obtain the results they desire, they do not understand the mechanism by which it does so. Monkeys do not realize that their actions cause others to have false beliefs. They would therefore be unlikely to invent novel deceptive tactics without a history of rewards to shape their chance behavior. This lack of comprehension and inventiveness is striking in contrast to the fact that monkeys have a well-deserved reputation for social sophistication. Monkeys rely on social support in tripartite interactions, they know not only their own but others' kin and rank, and they use grooming to build up alliances and friendships rather than simply using force to achieve goals (e.g., see chapters 6, 7, 9, 11, and 19 in Byrne & Whiten, 1988a). The ability to understand another's intentions would so clearly benefit such socially oriented species as the monkeys, that the lack of evidence for this insight outside apes is surely significant.

CONCLUSION

Rich observational data should no longer be ignored. There is, however, a fate worse for anecdotes than being ignored. Anecdotes have often been used to illustrate a particular interpretation of some systematic data, perhaps the results of an experiment. In these cases there is a danger that what is in fact happening is that the systematic data does not entail the conclusions drawn, and the reader's acceptance of these conclusions really owes more to the anecdote. This way of using anecdotal evidence is particularly dangerous if the anecdote is cleaned up by all-too-fallible human memory and nowhere published in its original form.

The approach of this chapter, using a scheme for sorting evidence, is intended to enable a proper scientific treatment of the rather messy but potentially important data that result from collecting good but unsystematic observations of rare behaviors. I would argue that it is always better to apply clear and overt criteria of assessment to records quoted in full from the original source (often a field notebook), than what so often befalls anecdotes at present.

The next stage beyond the careful and scientific evaluation of unsystematic records should obviously involve systematic observations. These may well take the form of decisive experiments, but given Dennett's (1983) point that the frequent trials required for laboratory experimentation are a rich source of potential reinforcement schedules to condition behavior patterns, it would be unwise to pin everything on a single experimental test. In addition, sophisticated and careful observation of natural behavior, backed by sufficient control data on the animals' normal behavior, could prove to be at least as powerful as laboratory experiment in determining—as in this case—which species can truly understand the deceptions they perpetrate.

NOTE

1. Note that the original definition (Byrne & Whiten, 1985, p. 670) differed slightly from the one quoted here by including "used at low frequency" in the specification of tactical deception. Commentators to Whiten and Byrne (1988a) pointed out that, while deceptions might normally be at low frequencies, under certain circumstances deceptive tactics might be stable even at rather high frequencies, so we modified the definition accordingly.

13

Anthropomorphic Anecdotalism
As Method

Robert W. Mitchell

Anthropomorphism—the extrapolation of characteristics from humans to nonhumans—and anecdotalism—the psychological interpretation of an individual's actions presented in a brief story—have been subjects of considerable scientific debate in relation to the psychology of nonhuman animals. When combined, anthropomorphism and anecdotalism create a method of interpretation which derives from a method of interpretation of human behavior utilized in our legal system. In this chapter, I depict this method of anthropomorphism and anecdotalism, a precise (but flawed) set of rules for interpreting stories based on analogy from contemporaneous views of human psychology. Because I believe that the presentation of animal behavior within a story is the factor which most saliently propels us to a humanlike (and potentially accurate) psychological interpretation, I also describe two studies showing that an animal's behavior as depicted in a story, rather than the type of animal which enacts the behavior, is what people (or at least undergraduates) use to determine its psychology (cf. Gallup, Marino, & Eddy, this volume; Herzog & Galvin, this volume).

I thank March Hamm, Roy Lockard, and Shon Goodwin for assistance in data collection and questionnaire creation; Lyn Miles, Dick Byrne, Hal Herzog, Sue Parker, Marc Bekoff, and Linnda Caporael for commentary on the chapter; and Miya Hamai of the Japan Monkey Centre and Chiaki Nakayama for translating some Japanese work for me.

To avoid the bitterness often engendered by "accusations" of anthropomorphism, I begin by stating emphatically that the use of anthropomorphism per se to examine nonhuman psychology is necessarily neither good nor bad for scientific understanding. Scientists both lauded and declaimed have used anthropomorphism in their interpretations of nonhumans; many of those who argue against the anthropomorphism of others use it to advance their own ideas and interpretations; and a studied failure to recognize similarities between humans and other species seems a ludicrous position to entertain (see Mitchell, 1996). Much the same can be said in relation to anecdotes. Having said that, I wish to argue that anthropomorphism can be problematic when it is presupposed to the extent that a scientist fails to examine or be concerned with the evidence needed to support the anthropomorphism. Viewed in this way, anthropomorphism is no more nor less problematic than any other presupposition which is unexamined; but it differs in that it is often difficult to acknowledge what seems so obviously true.

EXPECTATIONS, KNOWLEDGE, AND ANTHROPOMORPHISM

We are all aware that expectations can be brought into awareness when one observes things contrary to these expectations, as when one visits a different culture, and one might expect that the same is true with anthropomorphic expectations about nonhumans. Set up as a general rule, this idea is: when observations do not fit our expectations, we revise our expectations to make them consistent with our observations. Indeed, this view seems essential to empirical inquiry in science. We may believe, developing this viewpoint, that in research with animals hypotheses can be tested with precise predictions which can confirm or disconfirm so-called "anthropomorphic" interpretations, such that anthropomorphism can be viewed as simply a means of generating hypotheses which will be examined empirically (via observation or experiment) to determine their validity.

Unfortunately, the suggested means of thwarting unexamined anthropomorphism does not always work; humans are often remarkably averse to revising their ideas based upon factual observations. For example, "the idea that homosexuality equals a lack of male behavior, which in turn equals a deficiency of male sex hormones, persisted from the 1940s through the late 1970s despite the failure of hormonal treatments to influence sexual orientation and despite the fact that most studies failed to find any association between adult hormone levels and sexual orientation" (Byne & Parsons, 1993, p. 228). In general, evidence against a particular proposition is not consistently taken as indicating that the proposition is false. Researchers whose findings contradict those predicted by theories and hypotheses are often denounced as bad experimentalists (Collins & Pinch, 1993; see, e.g., Gallup, 1994), and fault is found in their inadequate experimental designs, even if these designs are less faulty than those that support hypotheses (Swartz & Evans, this volume). All in all, "The meaning of an experimental result does not . . . depend only upon the care with which it is designed and carried out, it depends upon what people are ready to believe" (Collins & Pinch, 1993, p. 42). And what people are ready to believe in studies of animals often depends upon anthropomorphic assumptions that are without an empirical basis. To have an empirical basis, anthropomorphism depends upon three things: knowledge of human behavior, knowledge of nonhuman behavior, and accurate recognition of similarity between human and nonhuman behavior.

Very often, our knowledge of human behavior is inadequate as a means of understanding not only nonhumans but also other humans (Caporael & Heyes, this volume; Davis, this volume), including prehistoric humans. For example, when I asked Richard Leakey (personal communication, October 14, 1988) if anthropomorphism is a problem in the interpretation of fossil hominids, he claimed it could not be because they are human. But this claim seems naive; recent accounts acknowledge that erroneous presuppositions of what "humans" are like or experience produce inadequate interpretations of prehistoric humans. For example, the extended survival of early humans who were disabled or severely crippled was taken by many scientists to indicate that these humans were part of a social network which exhibited compassion toward them, thereby indicating their similarity to modern-day humans. However, many modern-day humans with the same disabilities are shown little compassion by members of their culture, yet manage to survive to old age (Dettwyler, 1991). Thus, the anthropomorphic (and ethnocentric) assumption that people with disabilities require compassion for survival is empirically unfounded. In a similar example, claims that dead Neanderthals were intentionally buried and provided with grave goods were used to support arguments for Neanderthal belief in "spiritual" concerns, affection for conspecifics, and similarity to modern humans, but when evaluated the evidence was inadequate to support these arguments and claims (Gargett, 1989). People can feel affection for a person who died without burying him or her; elephants bury their dead and return to the site of the burial, but they are not more human as a result. Although some argue that any "disposal of a corpse in a pit (even a natural one) and covering it with dirt or rocks constitutes a mortuary rite" (Frayer & Montet-White, 1989, p. 180), in arguing so the new terms, "disposal of a corpse," cease to have the same connotations as the old terms, "intentional burial," and the question of whether burial is even necessarily present in modern-day humans—the original point of comparison—is ignored. In another example, early human cave paintings are often taken to be depictive and representational. The basis of such an inference is our perception of the resemblance between the paintings and animals; but we know that among humans what an image resembles need not be what it depicts. For example, the Egyptian hieroglyph that looks like an owl to us did not represent an owl to Egyptians (Davis, 1989). If the extrapolation from one set of humans to another is problematic, the extrapolation from one set of humans (or one human alone) to another species seems more markedly so. On the other hand, if we have knowledge about humans, this knowledge suggests ideas to explore about other humans and other organisms. For example, the fact that cones in our retina allow us to see colors suggests that cones in other animals' retinas allow them to see color as well, and that animals (including humans) without cones in their retinas would not see colors, and these explicitly anthropomorphic extrapolations can be explored scientifically.

Having knowledge of nonhuman subjects is the second problem for anthropomorphic analysis. Although perceptual observation of animal behavior may appear to be direct, such that we see what is actually present, there is a problem here in that normal perception is not divorced from anthropomorphic expectations (although see Millikan, this volume). This problem can be exemplified by my own surprise at my own perceptual anthropomorphism. In doing research on dog-human play, I videotaped people playing with their own and an unfamiliar dog. I devised a coding scheme by which I could directly compare the activities of dogs and people. While videotaping these interactions, I became convinced (much as was Darwin [1874/1896, p. 71]) that both dogs and people played a similar game of object-keep-away which was suggestive of complex understanding of the other's psychology. Both dogs and people sometimes dropped the ball their play partner was trying to capture from them, seemingly in an attempt to entice the partner to try to obtain it, and then regained the ball before the other could obtain it. I was intrigued by my observations, because I felt that here was evidence that people and dogs showed a similar psychology. When I examined the videotaped evidence in depth, however, I found that the behaviors of dogs and people were very different. People dropped or placed the ball near the dog and watched the dog for signs of movement toward the ball before regaining it (which actually differs from object-keepaway, and is called "fake-out"—see Mitchell & Thompson, 1991). Dogs, by contrast, dropped the ball in two ways: usually after the person had shown no interest in the ball, a dog dropped the ball far away from the person, looked away from the person, and only picked up the ball when the person neared it; or a dog dropped and regained the ball as it ran, independent of its nearness to the person. In no case did a dog try to engage a person's attention to the ball and play with that attentiveness when it dropped the ball. Thus, my recognition of a surface similarity between the games of dogs and people caused a perception of identity at a level where none existed. Imagine my surprise when, after describing the differences between dogs and people at the International Ethological Congress in 1987, I was informed by numerous audience members, themselves well trained in objective observation, that *their* dogs performed fake-outs just like those of people! This experience made me even more cautious about interpreting my perceptions of behavior psychologically, particularly when knowledge of the development of the behavior is unknown (see Mitchell, 1986; see also Davis, 1989). Real-time observational reports, in and of themselves, can be subject to overinterpretation.

This caution should not stop scientists from making observational reports, only from presuming that one perceives accurately and without bias. Much-needed knowledge is gained by observational methods, especially when combined with a concern to search for the behavioral basis of psychological descriptions (e.g., de Waal, 1986). Indeed, our account of the psychological nature of other human beings is based on our knowledge of the behaviors they enact that are akin to the experiences we have. For example, if my friend and I experience grief, my knowledge of my own grief comes from my experience and expression of grief, but my knowledge of his grief comes from my observation of his expression of his experiences of grief; both sources of knowledge allow appropriate use of the word "grief" with reference to myself and my friend (for further discussion, see Mitchell, this volume, chap. 29).

The third problem for anthropomorphism is the accurate recognition of similarity between human and nonhuman behavior. Accurate recognition of similarity is problematic because "recognition" of similarity depends upon other criteria than actual similarity (and even when similarity is present it need not indicate identity). For example, consider the following anecdote:

> A grizzly cub in Yellowstone Park found a big ham skin—a prized delicacy. Just as the little fellow was lifting it to his mouth a big bear

> appeared. He instantly dropped the ham skin, sat on it, and pretended
> to be greatly interested in watching something in the edge of the
> woods. (Mills, 1919/1976, p. 3)

When this story is presented to scientists, they often discount it as
unlikely (stating that the observations themselves are probably flawed!)
because they believe that a bear is not capable of hiding objects and
feigning indifference. However, imagine that the story had been told
with a chimpanzee as the protagonist:

> A young chimpanzee at Gombe found some bananas—a prized deli-
> cacy. Just as the young chimp was lifting one to his mouth an adult
> chimp appeared. He instantly dropped the bananas, sat on them, and
> pretended to be greatly interested in watching something in the edge
> of the forest.

This story, by contrast, suggests strongly to scientists that the animal
was hiding the bananas, and also suggests that the animal knew that if
he sat on the bananas the older animal would not see them, and that
his apparent interest in something on the horizon might dissuade the
older animal from examining him too closely. For some reason—what I
would call an anthropomorphic assumption—the anecdote seems more
suggestive of psychological knowledge on the part of the young chimp
than on the part of the young bear because scientists instantly think of
reasons to believe it. We expect chimpanzees—phylogenetically close
relatives—to show behaviors similar to us; bears, not being phylogenetically close relatives, we consider less likely to show similar behaviors
(see Mivart, 1898, p. 194). Yet there is no reason why other nonprimate
organisms cannot show behaviors just as similar to humans as those of
chimpanzees. Indeed, according to naturalists, bears exhibit many of
the same behaviors taken as evidence of social and technical intelligence
in chimpanzees (see Burghardt, 1992; Mitchell, 1993d; also compare
Mills, 1919/1976, 1920; and McNamee, 1982, p. 69; with Byrne & Whiten,
1990, pp. 84–86). Of course one should be cautious in interpreting observations, whether of bears or chimps or humans. Perhaps chimpanzees'
behavior provides better evidence of psychological understanding than
does bears'; I'm not so sure. But what is sure is that the chimpanzee has
replaced the bear as the "most human of animals" in American scientific
imagination, with the result that the demotion of the bear in American
thought (its new "symbolic identity") "may or may not accurately reflect
the animal's nature but . . . arises out of individual and cultural perceptions of that species" (Lawrence, 1990, p. 140).

Theory rather than evidence is often used to justify distinctions
between primates and nonprimates even when their behavior is osten-

sively the same. For example, in trying to argue that false alarm calls by dogs are different from false alarm calls by baboons, Byrne (this volume) states that dogs have greater experiences of true alarm calls and therefore more chances to recognize the effect of the call upon its housemates than do baboons. "Thus we can be sure that coincidental circumstances, having the power to reinforce the contingencies of stimulus and response that amount to this tactic, are commoner for domestic dogs than they are for wild baboons" (Byrne, this volume, p. 145). However, no evidence is provided that dogs *must* enact or experience frequent alarm calls before they catch on to their functional value; it may be that dogs catch on right away, even before the presumed extensive conditioning is supposed to occur, whereas baboons require several takes. But the idea that dogs take a long time to learn to deceive makes a good story, even without evidence, because it fits preconceptions about the superiority of primates over canids (though see Thompson, 1976, for different assumptions).

Even when one acknowledges the anthropomorphic basis of one's perceptions, theories, and hypotheses, unsubstantiated assumptions can be maintained. For example, in interpreting the evidence of mirror self-recognition in chimpanzees, researchers (Gallup, McClure, Hill, & Bundy, 1971; Gallup, 1977b, p. 335; 1983) explicitly used Cooley's (1912) ideas about the development of the self-concept in humans. Cooley argued that a self-concept is a result of other's appraisals of oneself, which provide the "opportunity to examine one's self from another's point of view" and therefore a means of self-objectification (Gallup, 1977b, p. 335). These researchers assumed that mirror self-recognition represented a rudimentary self-concept, and argued that if Cooley's arguments were true, then mirror self-recognition in chimpanzees would require social interaction. They showed that group-reared chimpanzees showed mirror self-recognition, whereas isolation-reared chimpanzees did not, which they claimed supported their conclusion that chimpanzee mirror self-recognition indicates a self-concept. Note, however, that the existence of social interaction in chimpanzees was taken to support the anthropomorphic assumption that mirror self-recognition represents a self-concept; the researchers never sought evidence of any evaluative appraisal between chimpanzees which would give rise to a Cooleyesque self-concept, and without this evidence Cooley's argument fails to make sense in relation to chimpanzees. Soon simply the fact of mirror self-recognition in chimpanzees indicated to these researchers that chimpanzees had knowledge of other minds—a conclusion quite opposite to Cooley's. In this inverted view, mirror self-recognition indicates a self-awareness "within" which chimps model the minds of other creatures (Gallup, 1985, p. 633; Gallup & Suarez, 1986, pp. 14–15). By this inversion, the lack of evidence of evaluative appraisals by chimpanzees was no longer problematic. Unfortunately, this anthropomorphic theory thwarted the search for other interpretations and denied any "self" to organisms that failed to recognize themselves in a mirror (see Mitchell, 1993a, 1993c, 1994b; Swartz & Evans, this volume).

Anthropomorphic inference also hindered publication of observations showing that not all normal chimpanzees recognize themselves in mirrors (Swartz & Evans, this volume). Reviewers argued that there must be a flaw in the design which failed to allow for mirror-self-recognition, rather than believe that chimpanzees failed to know their own image as such. I think that the basis for argument here is an implicit anthropomorphic assumption, which can be stated as follows: If mirror-self-recognition is the same in humans and chimpanzees, then chimpanzees should, like humans, always recognize themselves in mirrors; therefore, it is impossible that chimpanzees could fail to recognize themselves in mirrors. But just because chimpanzee behavior looks like human behavior on the surface need not indicate that identical processes explain both (although they may—see Parker, 1991; Mitchell, 1993b).

What I think happens when scientists interpret animal behavior is that they are looking for evidence to support a story (often presented barebones as a theory), but unfortunately sometimes the story is so convincing that evidence is unnecessary to convince.

STORIES, JURIES, AND COGNITIVE ETHOLOGISTS

Given the problems inherent in interpreting anecdotes anthropomorphically, it is surprising that the same interpretational method is used to understand humans. We use anecdotes of people to evaluate their abilities, understand their personality, and assign blame. These anecdotes are fashioned in the form of stories, much as are anecdotal accounts of animals. This method of interpretation is institutionalized not only in common practice, but in our legal system. In what follows, I compare the method of anthropomorphic anecdotalism vis-à-vis humans and nonhumans so that we can see what standards are applied to humans and then discern their applicability to nonhumans. This comparison is neatly made by looking at how essentially anecdotal evidence is evaluated in criminal trials.

In their book *Reconstructing Reality in the Courtroom*, Bennett and Feldman (1981) come to conclusions highly relevant to the interpretation of nonhumans as well as other humans. As they put it,

Almost any act can be associated with diverse causes, effects, and meanings. . . . In addition to having the potential for multiple significance, social actions are so complex that exhaustive descriptions are impossible. . . . Most important, the meaning of an action has little or no connection to the sheer quantity of detail in accounts about it. Constructing an interpretation for a problematic social action . . . requires the use of some communications device that simplifies the natural event, selects out a set of information about it, symbolizes the information in some way, and organizes it so that [listeners] can make an unambiguous interpretation and judge its validity. Stories are the most elegant and widely used communication devices for these purposes. (pp. 66–67)

[Thus,] in order to understand, take part in, and communicate about criminal trials, people transform . . . evidence . . . into stories about the alleged criminal activities. . . . Stories, as the mechanisms that selectively determine the relevance of various courtroom factors in any given case, provide a basis for transforming the statistical inventory of trial variables into a theoretical framework. . . . Evidence gains coherence through categorical connections to story elements such as the time frames, the characters, the motives, the settings, and the means. (pp. 4, 6, 8)

The interpretation of anecdotes about both animal behavior and criminal activities share the common element of requiring stories within which evidence is incorporated and through which evidence is interpreted (see also Haraway, 1989). Without this incorporation into a story, observations are not evidence because calling something "evidence" presupposes that the something supports a particular story or "theoretical framework," that is, presupposes that it is evidence *for* a particular proposition. (Anecdotes also tend to support more emotionally based than intellectually based beliefs [Sappington, 1990], perhaps because they allow the hearer to identify with the character in the story.) The fact that evidence is incorporated into a story, though necessary, has some surprising consequences. As Bennett and Feldman describe it,

. . . since most audiences for stories . . . have not been directly exposed to the events or actions in them, they have little recourse but to base their judgments about the credibility of stories on assessments about story structure. These assessments involve either applying interpretive rules to story symbols to connect them to a common theme, or fitting the set of connections in the story into some "frame" or model of social action . . . to determine the consistency and completeness of the network of connections as it bears on the central action in the story. To the extent that these assessments yield ambiguities, the story will be both difficult to interpret and regarded as implausible. (p. 68)

> Evidence that cannot be organized within a developing story structure can be held up immediately as a possible sign of . . . deceptions, or as an indication that the emerging story is not adequate. (p. 8)

Thus, the less information storytellers provide that is inconsistent with the main theme, the more believable their story becomes. In the end, "a well-constructed story may sway judgments even when evidence is in short supply" (Bennett & Feldman, 1981, p. 68). The judgment of not only the audience may be swayed by a good story, but of the observer telling the story as well.

So far, the analogy between anecdotes about humans and nonhumans elucidates several similarities: first, both rely on stories to conceptualize evidence; second, such stories take one away from the evidence per se and constrain one to rely on consistency within the story structure for the truth; and third, inconsistencies within the story, or between the story and the evidence, are taken as a sign of either the inadequacy of the story or of deception by the alleged criminal. In the interpretation of animal psychology and behavior, inconsistencies within a story suggest that the observations are inadequate, and inadequacies between story and evidence may suggest that observers are deceiving, as when Humphrey (1988) argued (unbelievably) that Fossey's descriptions of gorilla deception were lies. Also, the fact that an organism's behavior does not fit one's expectations may indeed cause one to think that *the organism* is deceiving (e.g., Morris, 1986, p. 184), just as we do when a person's story is inadequate to understand his or her behavior.

A final way in which anecdotes of humans and nonhumans are similar is the approach taken in the courtroom when "the evidence introduced by a witness cannot be incorporated into the story"; then, "the dominant rhetorical tendency is to place the evidence in the comparative context of what an ordinary person would do in the situation or what the evidence would mean to an ordinary person" (Bennett & Feldman, 1981, p. 130). A similar strategy is employed by cognitive ethologists such as Griffin, as when he (1976, p. 19; 1981, p. 72) claims, in relation to deceptive activities of chimpanzees, that if these same activities had been performed by humans, we would have no doubt of the performer's intention to deceive, and therefore the chimps intended to deceive. But the "ordinary person" axiom should not be utilized so uncritically (see Mitchell, 1986, p. 15). By the same argument, if the deceptive activities of fireflies had been performed by humans, we would have no doubt of the performer's intention to deceive; but that would not lead us to believe that fireflies intended to deceive. Even in interpreting humans there are problems with the ordinary person axiom. Bennett and Feldman (1981) provide an excerpt

> from the cross-examination of the defendant in a larceny trial [in which the prosecution] set up the inference that the defendant must have known that the merchandise he bought was stolen because he didn't take the precautions that an ordinary person would have taken prior to purchasing the goods . . . [The prosecutor asked questions such as,] Did you ask these gentlemen where they had gotten the calculators? . . . Did you ask them how long they had the calculators? . . . Did you ask them how old the calculators were? . . .
>
> The prosecutor went on to ask whether the defendant was aware that he got a particularly good deal on the calculators, whether he was surprised that they were so new, and so forth. The effect of these questions was to support the inference that the defendant should have suspected that the calculators were stolen. In fact, the defendant wrote a check for the merchandise and typed up a bill of sale that both he and the seller signed. Although these do not seem to be the acts of someone who is convinced he is buying stolen goods, these factors were lost in the barrage of prosecution questions about whether he took the precautions that an ordinary person would have taken to make sure the goods weren't stolen. (pp. 130–131)

The significance of this instance for the ordinary person axiom is twofold. First, by ignoring some particulars and emphasizing others, one can build a case that one intention is in progress when in fact the actions are consistent with an entirely different intention. Even if it is true that "Everything can be given at least two equally cogent explanations" (Hirsch, 1985, p. 29), our legal system suggests that the more likely of the two can usually be determined; still, hung juries occur. Second, "People who have different understandings about society and its norms may disagree about the plausibility of a story" (Bennett & Feldman, 1981, p. 171), as was obviously the case in the different perspectives of the prosecutor and the defendant.

> [Asking] what an ordinary person would do in the situation or what the evidence would mean to an ordinary person . . . is often an appealing tactic since it asks jurors to consider what they would have done under the circumstances and what the evidence means to them. Since, however, this tactic involves removing the evidence from a context provided by the witness and placing it in a context tailored to the juror's experience, it may result in inferences based narrowly on the juror's past experiences. In many cases, what is "ordinary" for the typical lower-class or minority defendant may be quite extraordinary from the perspective of the typical white middle-class juror. (1981, p. 130)

Similarly, what is ordinary for the typically white middle-class observer of animals may be quite extraordinary for that observer's nonhuman subjects. Assuming that the perspective of the other is the same as one's own is akin to the understanding displayed by those children in Piagetian tasks who think that a person on the opposite side of a mountain sees the same things as themselves, although starting out with the assumption of perspectival identity may subsequently be quite revealing of differences and similarities. In the case of criminal justice, Bennett and Feldman (1981) write, "even if differences in social outlook could be documented, most of them could not be regarded as sources of bias without acknowledging the inherent subjectivity and relativity of legal judgment[, yet the] overwhelming popular commitment to the myths of fairness and objectivity in the justice process virtually prohibits [this acknowledgment]" (p. 181). In the case of ethology, however, the recognition of differences in social outlook between ethologists and their nonhuman subjects has become a topic of concern, such that both an easy anthropomorphism and a blatant denial of subjectivity in relation to nonhumans are no longer tenable (Nagel, 1974).

The analogy with courtroom testimony supports a profound ambivalence toward anthropomorphic anecdotalism. If we can send people to death or life imprisonment based upon anthropomorphic analysis of anecdotes, surely we can use the same method to understand animals. However, if the anthropomorphic analysis of anecdotes leads to error in the interpretation of humans in the courtroom, surely it will lead to error in the interpretation of animals.

Let's examine a story about chimpanzees to see how contrastive interpretations, themselves stories about a story, make it difficult to choose between them. Just as in the courtroom, once two story structures are provided to account for evidence, which story is to be believed becomes problematic. Take for instance a story proffered by Byrne (this volume) and Whiten and Byrne (1988) as evidence of counterdeception. In this story (described by Plooij),

> One chimp was alone in the feeding area and was going to be fed bananas. A metal box was opened from a distance. Just at the moment when the box was opened, another chimp approached at the border of the clearing. The first chimp quickly closed the metal box and walked away several metres, sat down and looked around as if nothing had happened. The second chimp left the feeding area again, but as soon as he was out of sight, hid behind a tree and peered at the individual in the feeding area. As soon as that individual approached and opened the metal box again, the hiding individual approached, replaced the other and ate the banana. (Whiten & Byrne, 1988, p. 242)

Byrne (this volume) argues that "The fact that this hide-and-peep action was effective in countering the deception, combined with its improbability of occurrence [i.e., chimps rarely enact such actions], convinces me that . . . [the hiding chimp] understood that it had been deceived" (p. 148). By contrast, Mitchell (1988) wondered "why the [hiding] chimp need be thinking anything about the other's psychological processes. Might not the [hiding chimp] be thinking that something is odd about the first's actions and be waiting to see what it does?" (p. 260). Here are two different but equally justifiable interpretations of the hiding chimp's actions. Either the chimp recognizes that the other chimp is feigning disinterest and therefore hides to see what the other is hiding, or it recognizes that there is something fishy about the other chimp and hides to see what it will do when no other chimps are around. Byrne (this volume) states that the acts of hiding and observing are "more consistent with an understanding that the 'oddness' of the other's actions concerned . . . [the hiding chimp] itself, in a way not to its advantage" (p. 148). But even if this understanding is so, it does not mean that the hiding chimp recognized that the other was deceiving it (or, more accurately, concealing from it). The hiding chimp could just as easily have thought that the "oddness" of the other chimp's actions might be worth looking into, without specifying to itself why. Similarly, people unaware they are being deceived by their lover sometimes only get a feeling that something is wrong and even look for evidence to find out what, and do not come to recognize deception until later (Werth & Flaherty, 1986).

Note that this attempt at interpretation follows much of the same strategy as antagonists in the courtroom. An interpretation of evidence (the original description) is provided in the form of a story about the evidence (i.e., it is about counterdeception vs. it is about hiding to see what will happen) which provides a structure within which the evidence makes sense. The evidence is consistent with both stories. No new evidence is introduced to decide between alternatives; the same evidence is used by both. An appeal is made (implicitly by Byrne, explicitly by me) to what an "ordinary person" would experience in the same situation. But either interpretation is one which an ordinary person might experience. Thus, in the story of "counterdeception" we are stuck with (at least) two equally commensurate interpretations, and no way to judge which is correct: a hung jury.

STORIES AND ANIMAL PSYCHOLOGY, AS JUDGED BY A JURY OF UNDERGRADUATES

Whatever their problems for interpretation, stories are probably how we get our categories for initially interpreting the psychology of nonhumans. We interpret animal behavior as capable of being narrated, presuming the who, what, where, when, how, and why structure of newspaper accounts of human activities. Although research suggests that our psychological interpretation of nonhumans follows from their similarity to us, such that animals viewed as similar are also viewed as more "humanlike" psychologically (Eddy, Gallup, & Povinelli, 1993; Plous, 1993a; Gallup et al., this volume; Herzog & Galvin, this volume), these interpretations were discerned independent of any particular behavior the animals exhibit. That is, subjects were asked how much of some psychological attribute they believe that particular animals have, or how uncomfortable they would feel eating a particular animal. However, if my interpretation of the method of anthropomorphic anecdotalism is correct, our interpretation of animals depends more upon what animals do (that is, what story we are told about them) than upon how similar they are to us.

To test this idea, and also to test the significance of anecdotal stories upon our interpretation of animals, I provided undergraduates with stories in which different animal species engaged in the same behaviors (see Mitchell & Hamm, in press, for greater detail). The students were asked to evaluate their agreement or disagreement, on a scale from one (strongly disagree) to six (strongly agree), to a series of statements about the psychological states of one of the animal protagonists.

In the first study, students read the following story:

> Patricia Ekman studies nonverbal behavior of chimpanzees. She observed the following interaction in a nature park:
> B, a male, is with S, a female, comfortably stroking her. G, another male, moves to S, and begins to stroke her. B turns away from S and looks intently at his hand.

This story is taken from a similar depiction of "feigning interest" by chimpanzees in de Waal (1986). Six different animal species were used: people, chimpanzees, elephants, bears, dogs, and otters; the elephant looked at his foot; the bear, the dog, and the otter each looked at his paw; and the dog nuzzled rather than stroked S. Each subject ($N=748$) received only one story. (In addition, two other versions were presented; both emphasized the species of the animal depicted, and one also showed a photograph of a member of the species. None of these additions produced significant differences in interpretation—see Mitchell & Hamm [in press].) After reading the story, subjects were asked to "Please read carefully the following statements about B and circle the number that corresponds to your interpretation of B." Six statements asked about B's emotional state after G stroked S: was B upset, embarrassed, delighted, angry, afraid, or jealous? Nine statements asked about B's emotional state when B looks at his paw: is B very interested in looking at it, pretending to be interested in looking at it, thoughtful, thinking about what to do next, indifferent to what just happened with G and S, feigning indifference, trying to deceive G about his indifference, trying to save face, or trying to hide his feelings?

Other than the animal's behavior in the story, four factors might be important in subjects' interpretations: physical similarity, phylogenetic closeness, familiarity, and cultural stereotype of the animal as like humans. If physical similarity to humans were important for anthropomorphism, then chimpanzees and bears would be viewed as psychologically like humans, but not dogs, elephants, and otters; if phylogenetic closeness were important, then only chimpanzees would be viewed as like humans; if familiarity were important, then only dogs would be viewed as like humans; and if cultural stereotype were important, then all animals but otters would be viewed as like humans. The null hypothesis is that anthropomorphism is dependent upon the behaviors of the animal and independent of which animal is depicted.

Surprisingly, all animals were viewed similarly on almost all psychological depictions. In general, subjects agreed (i.e., gave rankings greater than 3.75) that B was upset, jealous, and angry after G stroked S, and that B is thinking about what to do next when looking at his hand/foot/paw; they disagreed (i.e., gave rankings less than 3.25) that, after G stroked S, B was delighted, afraid, or embarrassed, and that when looking at his hand/foot/paw B is deceiving about his indifference, trying to save face, indifferent to what just happened with G and S, and very interested in looking at his hand/foot/paw. Subjects were uncertain (i.e., gave rankings between 3.25 and 3.75) that when looking at his hand/foot/paw B is trying to hide feelings, feigning indifference, or thoughtful. Subjects were also uncertain as to whether the dog was pretending to be interested in his paw (ranking of 3.55), but disagreed that all other animals (including humans) were pretending to be interested. Note that, in many of their rankings, subjects disagreed with or were uncertain about de Waal's original interpretation of the chimpanzee as "feigning interest."

In another study (also presented in Mitchell & Hamm, in press), subjects read the following story:

> Dr. Gallagher studies nonverbal behavior of people and their dogs. She observed the following interaction.
> M, a male, and F, a female, sit on the ground. M reaches his arm to

touch F, and M's dog H rushes up to them and moves between M
and F, facing and staring at M [or F] while touching him [or her].

This story depicts what many dog owners describe as jealousy in their
dog. The story was also presented in two other forms, as being about
"young children in a schoolyard" and about "rhesus monkeys," where
H is "M's younger sibling." These stories agree with depictions of jealousy in monkeys (Breuggeman, 1978) and young children (Stern, 1924).
In addition, within the story gender was manipulated for which organism (M or F) reaches his or her hand to touch the other (i.e., which ini-
tiates), which organism (M or F) owns (for the dog) or is the sibling of H
(for the monkey and child), and which organism (M or F) H is facing
and staring at while touching. The results of these manipulations allow
one to determine not only if gender of other animals is a variable in
the interpretation of an animal's behavior, but also whether H's facing
its owner/sibling or the other, or H's facing the initiator of reaching or
the other who is reached for, influences the interpretation of H's psychology. Again, each subject ($N=713$) received only one story.

The subjects were again asked to rank on the same 6-point scale
their agreement or disagreement, and in this experiment 18 questions
were asked: nine about how H felt when H moves between M and F,
and nine about how H is feeling when H directs actions specifically
toward M (or F). The same feelings were named in the first and second nine questions: H is feeling jealous, protective, neglected, left out,
angry, or delighted; H is being aggressive or helping M and F to interact; and H wants attention.

Again, the stories attempted to pick up on factors of physical similarity, phylogenetic closeness, familiarity, and cultural stereotype.
However, in this instance physical similarity and phylogenetic closeness
were both instantiated by the monkey, familiarity by the dog, and cultural stereotype as humanlike by both monkey and dog. If physical
similarity and phylogenetic closeness were important for anthropomorphism, then the monkey would be rated as more similar psychologically to the child than would the dog; if familiarity were important,
the dog would be rated more similar to the child than would the monkey. The null hypothesis that there are no species differences supports
two ideas: that the cultural stereotype of the animal as humanlike, as
well as the animal's behavior in the story, influence its psychological
interpretation.

Once again, there were no major differences in how the animals
were ranked. Subjects agreed that all animals felt jealous, left out,
neglected, and protective, and wanted attention; and disagreed that all

animals felt delighted or angry, or were helping M and F to interact. Subjects were undecided as to whether the animals were aggressive, but leaned toward agreement more for the child than for either the dog or the monkey (see Mitchell & Hamm, in press, for further details). In addition, subjects interpreted all animals as showing signs of attention getting and neglect when it faced the owner (for the dog) or the sibling (for the child and monkey), but were less certain how to interpret the animals' behavior when it faced the other organism in the story; and subjects interpreted all animals as showing a need for attention when the owner or sibling directed its actions and attention to the other organism, but as more protective when the other organism directed its actions toward the owner or sibling (perhaps because the latter could be viewed as potentially aggressive). Generally, dogs, young children, and rhesus monkeys were all viewed as exhibiting jealous (and sometimes protective) feelings when they move between a familiar and an unfamiliar person and then stare at and touch one of them.

CONCLUSIONS

Apparently, once mammals are described as actors into a story, what they do in that story determines how anthropomorphically they are characterized (see Haraway, 1989), independent of what species of mammal they are. Our narrative depictions of animals in stories, cartoons, and films powerfully influence us in our anthropomorphizing and empathizing with animals (Cartmill, 1993; Guthrie, 1993). Contrary to other accounts (Nakajima, 1992; Eddy, Gallup & Povinelli, 1993; Gallup, Marino & Eddy, this volume; Herzog & Galvin, this volume), undergraduates fail to apply the "phylogenetic scale" presumed to underlie their attributions of psychological states to animals when animals are depicted as behaving actors. Although it will be interesting to see if other animals are also viewed anthropomorphically in similar stories, depicting animals such as goldfish or cockroaches in stories of jealousy and embarrassment may prove exceedingly difficult, in that their behaviors are not similar in detail to those of humans. Overall, these data suggest that what animals *do* may be the most important determinant for people's perception of their psychology.

This suggestion is in keeping with the idea that an animal's behavior, more than the similarity of the animal or its behavior to humans or their behavior, is what *should* determine our understanding of the animal's psychology (Bennett, 1964/1971, p. 100). "Anthropomorphism is the null hypothesis" for undergraduates as well as for comparative developmental evolutionary studies (Parker, this volume). However, different people may supply different weightings to the factors present in any story. Unlike undergraduates, primatologists may find phylogenetic closeness to humans an important additional factor in discerning a humanlike psychology in the protagonist in a story because of their concerns about human evolution, and people who don't like animals may use only the fact of the protagonist's being an animal to deny it all human-like characteristics. The ways in which our prejudices influence our interpretations as scientists should clearly be a topic of interest and concern.

There is a difference between interpretations by juries and undergraduates and interpretations by ethologists, in that ethologists sometimes stick with their subjects for an extended period of observation. In this way, ethologists' observations of nonhumans can be similar to those of cultural anthropologists (Griffin, 1976, p. 88). Humans take for granted assumptions about other people, and typically recognize these *as* assumptions only when (and only sometimes when) these assumptions are violated, which they are by people from other cultures (and by animals). When first interacting with people from another culture, "We *expect* others to be like us, but they aren't. Thus, a cultural *incident* occurs, causing a *reaction* (anger, fear, etc.), and we *withdraw*" (Stolti, 1990, p. 61). If we are at all reflective about our experience, however, "We become *aware* of our reaction. We *reflect* on *its cause*. And our reaction *subsides*. We *observe* the situation, which results in developing culturally appropriate *expectations*" (pp. 61–62). Just such a method of reflective analysis of nonhumans is elaborated by Burghardt (1985a), Rollin, Byrne, and others in this volume, and is expected of seasoned field observers (Byrne & Whiten, 1988a; Whiten & Byrne, 1988) and others, such as zoo keepers, animal caretakers, and farmers, who have frequent intimate encounters with animals (Hebb, 1946; Rollin, this volume).

It is an old truism that we interpret our observations in terms of our theories and are directed by our assumptions when we observe and interpret. Reflection about our assumptions and examination of scientists' and other people's methods of interpretation may bring greater objectivity to the study of animal behavior and psychology. In examining the evidence of similarity between nonhuman primates and humans, for example, we can no longer ignore the similarity of nonprimates to primates (including humans), or discount evidence of differences between nonhuman primates and humans, and between humans in different (or even the same) cultures. We need to be explicit about all of these similarities and differences if we are to have accurate understandings of continuity and discontinuity between humans and other species, and therefore of the usefulness of anthropomorphism. But cautious arguments are often punished in science when they are against the mainstream view (Collins & Pinch, 1993). As a result, unexamined anthropomorphism and storytelling in science are here to stay, because they allow us to create truths we can believe in, however inaccurate they may be.

14

A Pragmatic Approach to the Inference of Animal Mind

Paul S. Silverman

Over the last decade, the perennial question of whether and when we should attribute human mental states to animals has again been vigorously debated. Participants to this debate have generally assumed that a single, rational set of criteria for mental states must be applied both to our daily interaction with animals and to our attempts to build formal theories of the evolution of mind. In this chapter, I argue that this assumption is unwarranted. Furthermore, a consensus about appropriate criteria is unlikely, given the significant role played by intuition and personal values. Instead, our criteria for attributing mental states should be designed for the situations in which they are going to be used, bearing in mind the consequences that will flow from choosing one set of criteria rather than another.

Many researchers who work closely with animals are familiar with the conflict that occurs between scientific psychological attributions they make based on conservative inferences from behavioral observations, and the intuitive beliefs about mental states which they develop based on personal interactions. Few have explored it. Years ago, D. O. Hebb (1946) reported that in contrast to his own fruitless two-year effort to describe in behavioral terms the temperaments of captive chimpanzees, his animal keepers' intuitive and spontaneous attributions of emotions and attitudes such as fear, nervousness, and shyness, provided the staff with "an intuitive and practical guide to behavior" (p. 88). While Hebb employed logical criteria to identify mental attributes, and failed to find them, his practical-minded assistants did "find" mental phenomena by applying mentalistic terms that worked for them. The present thesis is that it is both futile and unnecessary to resolve this apparent conflict.

THE VARIETY OF CRITERIA FOR ATTRIBUTING MENTAL STATES

In the course of the contemporary debate, authors have proposed a great variety of criteria for deciding whether an other animal experiences the same mental states as a human in a given situation. These criteria have been of two types: *Philosophical criteria* are those which address the epistemological question of how we can know that a particular mental property exists in animals. What would we have to observe and how should we draw conclusions? *Methodological criteria* are those which deal with determining permissible procedures for obtaining such knowledge. How should we observe? At both levels, there are fundamental disagreements about which criteria are acceptable and which are not.

Philosophical Criteria

Implicit in most attributions of mind to animals is a proposal first articulated by Romanes (1882). One observes the behavior of an animal and the context in which it occurs, then imagines oneself in the animal's place and simply asks the question, "If I were behaving in that way, what would be the necessary mental phenomena producing my behavior?" Unfortunately, this process is intrinsically problematic for two reasons. It can be difficult both to identify animal behaviors that are analogous to human ones, and to determine what mental processes are necessary to produce the behaviors. Putting one's self "in the animal's place" is a formidable task.

The problem of identifying analogous behaviors has been characterized by biologists as judging whether behaviors are "homologous"— that is, whether they serve similar biological functions. Criteria for making such judgments have been proposed (Wickler, 1965; Parker, this volume; see articles in Rajecki, 1983; Nitecki & Kitchell, 1986), but the process can be viewed as largely intuitive and aesthetic, determined, at base, by "an ethologist's practiced eye" (Beer, 1980, p. 47).

Homology aside, there is little agreement about the types, range, or number of behaviors that are necessary to infer particular mental phenomena. I offer three examples:

1. It has been proposed that "awareness" can be inferred from the existence of sense organs or avoidance behavior (Rollin, 1981), "versatile communication systems" (Griffin, 1976), "versatile adaptability of behavior" (Griffin, 1984), mirror self-recognition (Gallup, 1985), and symbol use (Piaget, 1976).
2. A similar dilemma occurs in the inference of "intentionality." While evidence for negative feedback control is necessary (Rosenblueth, Wiener, & Bigelow, 1943), it is not sufficient. The degree of complexity of a feedback system and the extent to which its goals are humanlike determine whether it is attributed mental "intentions" (Silverman, 1983). But how complex and how humanlike must such goals be?
3. Finally, claims that some primates understand the mental states of others employ various criteria: mirror self-recognition (Gallup, 1985), symbolic communication and planned deceit (Mitchell, 1986), or identification of others' needs (Premack & Woodruff, 1978).

In addition to Romanes's proposal, some writers (Menzel, 1986; Suomi & Immelmann, 1983; Walker, 1983) have argued that judgments about biological relatedness should also be prerequisite to or strengthen cross-species generalizations pertaining to mind. This proposal raises the issues of how degree of relatedness should be determined and how it should affect decisions about the existence of mental processes.

Commenting on a recent attempt to draw inferences about mental capacities from anecdotal accounts of animal behavior (Whiten & Byrne, 1988), Dennett (1988) suggests that intuitions and values are always at play in such an enterprise:

> we should set aside the illusory hope of finding "conclusive" evidence of creative intelligence, evidence that can withstand all skeptical attempts at "demoting" interpretation. When we recognize that the concept of intelligence is not a neutral one or entirely objective one, this should not surprise us. (p. 253)

In addition to the question of what phenomena constitute adequate evidence for mentalistic attributions is the issue of the role of evidence in the decision-making process. Let us briefly consider two topics: the definition of "parsimony" and the testing of hypotheses. These too present difficulties for consensus.

A contemporary of Romanes, C. Lloyd Morgan (1894), cautioned that nonmental, mechanistic explanations of animal behavior should always take precedence over mentalistic ones. This view has become the standard for "parsimonious" explanation in psychology. However, those whose intuitions and values suggest the likelihood of psychological characteristics shared by man and other animals propose alternatives to Morgan's definition of parsimony. In defending his description of insight in chimpanzees, Wolfgang Köhler proposed that the principle of parsimony should be balanced by a "principle of maximum fertility," by which a phenomenon should be examined from several viewpoints (Boakes, 1984). Contemporary attacks on the singular importance of parsimony support arguments that primates may attribute mental states to one another (Humphrey, 1976; Bennett, 1978; Menzel & Everett, 1978, Premack & Woodruff, 1978; Gallup & Suarez, 1983; Griffin, 1984; Whiten & Byrne, 1988; de Waal, 1991; cf. Kummer, Dasser, & Hoyningen-Huene, 1990). For example, Whiten and Byrne (1988, p. 235) propose that "a psychological representation is an 'economical' one" since a behaviorist account of animal behavior would require a much longer description and more complex inferences.

A similar disparity among approaches to testing hypotheses exists. Experimental and statistical procedures in the behavioral sciences generally presume that the Type II error (mistakenly accepting the null hypothesis) is more acceptable than the alternative Type I error (mistakenly rejecting the null hypothesis). (The "null" is the hypothesis that the treatment has no effect on the measured variable.) However, despite acceptance of this approach and the underlying principle that hypotheses be tested through efforts to disconfirm them (Popper, 1959), in practice both scientists and laypersons share a "confirmation bias": they tend to seek information that supports their hypotheses, rather than information that might disconfirm them (Mahoney & DeMonbreun, 1978; Mynatt, Doherty, & Tweney, 1978). Moreover this bias may be justified under two conditions of uncertainty:

1. The confirmation bias may be scientifically fruitful when either the data or the hypothesis are imprecise, since disconfirmation may be mistaken due to lack of precision in measurement, precision of the hypothesis, or precision in the relation understood to exist between the two (Doherty, Tweney, & Mynatt, 1981). Clearly, such imprecision predominates in research on animal cognition and animal mind. For example, evidence of self-recognition in chimpanzees and orangutans has been interpreted as suggesting self-awareness (Suarez & Gallup, 1981), although the relation between mirror-recognition and self-awareness is unclear, and other evidence of self-awareness is either lacking or ambiguous (Mitchell, 1993b).

2. The confirmation bias may be morally imperative when the consequences of a Type II error are more harmful than those of a Type I

error. Certainly such considerations are made in decisions regarding the testing of new medical interventions, clinical diagnoses, and treatment of medical and psychological disorders, and in the implementation and evaluation of many social and educational programs. The argument can be easily extended to questions of animal welfare.

Methodological Criteria

Not only have scientists debated about behaviors indicative of mental states and criteria for making decisions, they have also argued about methods adequate for producing and observing criterial behaviors. These methods can be ranked by the degree of certainty in hypotheses about mental processes that they potentially generate: anecdotal reports, quantified naturalistic observations, the ecological experiment, and the formal experiment. This ranking also corresponds to some of the philosophical choices outlined above. To the extent that one accepts that generous criteria for behavioral analogy are adequate, that the parsimonious explanation is the mentalistic one, that hypothesis confirmation rather than disconfirmation is acceptable, and/or that certain low-frequency behaviors are representative of mind, then the constraints on acceptable evidence are loosened. Below, I discuss the limitations and advantages of each method, along with views about the kinds of inferences that they make possible.

Anecdotes. An observer describes an animal's behavior, or sequence of behaviors, and the context in which it occurs. The description is then used to formulate a proposal about the psychological processes that generated the behavior. This technique can capture spontaneous low-frequency behaviors which are difficult to trigger artificially and permits the generation of novel hypotheses. When the anecdote results from observations made in the natural setting, there is the additional advantage of ecological validity.

Relatively rich and sensitive interpretations can be stimulated by the observer's familiarity with a particular animal. However, inaccurate interpretations based on the anthropomorphic projections of the observer may also result. Like parents who unintentionally overattribute the intentions and meanings of their children's utterances as they facilitate language acquisition (Snow, 1977), researchers who live with or raise their animals are most susceptible to problems of overinterpretation (Goodall, 1988; Patterson & Linden, 1981).

There are other disadvantages of anecdotal reports. Low-frequency behaviors which appear to reflect complex intellectual capacities may simply represent rare coincidences of behavior combinations or the results of trial and error learning. Reports of positive instances of a behavior generally occur without the context of reports of negative instances—situations in which the behavior, though appropriate, did not occur. The reliability or accuracy of such reports cannot be determined. Hypotheses cannot be tested or theories constructed.

While this approach has a long and controversial history, the most ambitious recent use of this method has been the effort by Whiten and Byrne (1988; Byrne & Whiten, 1990) to identify and interpret the use of deceptive tactics in nonhuman primates. They (Byrne & Whiten, 1990) attempt to categorize anecdotes (generated by many researchers) as *"candidate* records of tactical deception" (p. 4) in a psychological taxonomy involving the intentions of the deceiver to conceal, distract, create an image, use another animal as a tool, and deflect responsibility to a fall guy. However, whether the taxonomy represents the animals' mental capacities or the observers' description of presumed functions remains unclear. Whiten and Byrne (1988) suggest that "no 'single' observation can be regarded as definitive evidence in support of a hypothesis (e.g. that a certain species is capable of intentional deception)." But they argue that with multiple records, "we can be confident that a real pattern of behavior exists" (p. 243). Perhaps. Confidence in the meaning of the pattern, however, is not improved.

Quantified Naturalistic Observations. This approach, frequently employed by ethologists exploring social interactions, involves the collection of detailed descriptions of frequencies of qualitatively different behaviors or other characteristics, contingencies among them (between and within animals), or contingencies between these and various events. The range and representativeness of behaviors and events described can be judged to be ecologically valid. Because of interpretational constraints on descriptive and correlational approaches, the degree of certainty one can place in mentalistic interpretations is limited. An unusual example of this approach, applied to infer mental characteristics, is the application of factor-analysis to infer personality traits from behavioral observations of nonhuman primates (Caine, Earle, & Reite, 1983; Stevenson-Hinde & Zunz, 1978). We must ask whether such descriptions are true explanatory constructs (Does a monkey ignore the whereabouts of other group members *because* he is "confident"?) or useful labels which contribute to predictions. In another example of the power and limits of naturalistic observations, Cheney and Seyfarth (1990) conclude that vervets discriminate among individuals in their own and in neighboring groups, form and assess hierarchical relationships which have the logical property of transitivity (If A>B and B>C, then A>C), and redirect aggression at particular targets by social categories. However, the mental processes underlying such behavior cannot be specified.

The recent invention of structural equation modeling techniques, frequently used in studies of social influences on human development, does allow for certain causal interpretations of correlational longitudinal data (Biddle & Marlin, 1987) using large numbers of cases (Martin, 1987). The method requires rather precise hypotheses of causal networks and measurement of associated variables. Although the technique has potential for facilitating mentalistic interpretations of behavior patterns, it has not yet been applied to this end.

Ecological Experiments. This approach, also known as the "field experiment" and used to brilliant effect by Cheney and Seyfarth (1980) in the area of social cognition among vervet monkeys, has been described by philosopher Daniel Dennett (1983) as "the Sherlock Holmes method." The researcher predicts criterial behavior in a specific circumstance and essentially generates anecdotes by setting up situations which are likely to provoke the "tell-tale move" by the animal. This procedure presents a compromise between anecdotes, naturalistic observation, and true experimentation, in which the researcher manipulates the animal's natural environment. The ideals of experimental control, counterbalancing, and measurement of extraneous variables can be approximated with this approach, although these manipulations and measurements are limited. Describing the results of several studies using this method, Cheney, Seyfarth, and Smuts appropriately qualify and conservatively phrase their conclusions. For example: "Females behaved *as if* they were able to associate particular screams with particular juveniles, and these juveniles with particular adult females, *suggesting that* monkeys . . . recognize the *associations* that exist among other group members" (1986, p. 1363) (emphasis added).

Formal Experiments. The formal experiment, of course, has a long tradition in comparative psychology and tests of animal cognition, and its strengths provide the foundation of science: it permits the testing of hypotheses through the artificial control of variables; it is objective and replicable. However, the fruitfulness of even this method of investigating mental phenomena in animals is debatable. Gordon Gallup (1985) has argued that this approach is particularly suitable to the purpose of investigating animal mind. Because many organisms have evolved genetically controlled behavior which mimics behavior under mental control, evidence of mind is most reliably revealed by behaviors which are "biologically arbitrary." On the other hand, because animal intelligence may have been evolutionarily selected primarily in the social domain, it may be difficult to elicit evidence for mind outside of this domain (Cheney et al., 1986). In addition, when artificiality is introduced, questions about external validity are raised and the probability of the Type II error is increased—it becomes difficult to infer lack of a mental capacity from failure on a task (Silverman, 1986). Finally, the multiple training and testing trials which are commonly used in this paradigm often permit operant conditioning explanations of task mastery.

The two extremes of this range of methodologies highlight the limitations all impose on mentalistic inferences: formal experiments are often ecologically vacuous and therefore may be inadequate to illuminate true mental competence; anecdotes are arbitrary and permit little confidence in interpretation. In practice, theorists can rely on evidence converging from several methodologies, presuming that various weaknesses nullify each other. In principle, the validity of each method for drawing conclusions about animal mind remains problematic.

I propose that what one intends to "do" with the conclusions reached should play a significant role in determining one's philosophical stance, one's need for scientific certainty, and the consequent criteria for adequate evidence. This theme is developed below.

A PRAGMATIC APPROACH TO THE USE OF EVIDENCE

Attributions of mental states to animals occur in two quite different domains: in our daily interactions with animals and in our attempts to build a scientific theory of the distribution of mind in nature. These different domains demand very different criteria for evidence of mind. Two examples illustrate the problem of mixing these applications. Whiten and Byrne (1988) propose that "If the language-learning child, working from observable behavior, finds psychological terms both indispensable and unproblematic, it is surely wise to assess their utility for scientists working under a similar constraint" (p. 235). But the constraints are not similar and the purposes to which the psychological terms are to be put are quite different. The child wishes to achieve practical (action-oriented) ends quickly and completely. The scientist wishes to develop a formal theory which is internally consistent, falsifiable, and ultimately useful. This difference does not mean that the scientist should avoid the use of psychological terms. However, the justification of their use in one domain does not automatically qualify their use in another. To cite another instance, Fox (1985), considering ethical issues of animal treatment, accuses scientists of "'protective-objectification'

—the denial of others' subjectivity—in order to avoid closeness, responsibility, and the burdens of empathy" (p. 66). While any activity, including the practice of science (or the pursuit of animal rights), can function as a psychological defense mechanism, theory building and daily interactions are different enterprises requiring different standards of conduct and decision making.

Summarizing pragmatic philosophy as it applies to scientific knowledge, Thayer (1972) writes: "theories are inference policies, neither true nor false (except pragmatically) but nonetheless critically assessable as to their utility and clarity and the fruitfulness of the consequences that result from adopting them" (p. 435). We can expand this statement to encompass knowledge—neither true nor false in the sense of reflecting an objective Nature—but susceptible to value judgments based on its utility and fruitfulness. Below I discuss the two domains of application—daily interactions with animals and theory building—in an effort to determine "inference policies" which are appropriate to each.

Daily Interactions with Animals

Humans interact with animals in three important ways, each of which raises somewhat different issues. First, there is the area of protection of animals from humans (encompassing the issue of ethics and animal rights); second, the need to predict and control animal behavior; third, the use of animals as companions.

Ethics and Animal Rights. Some arguments for the application of ethical standards to animals are strictly philosophical and require no empirical justification. For instance, Singer (1975) suggests that many of us are "speciesists" in the same sense that some are racists. Zak (1989) argues that membership in a moral community extends beyond membership in the human community: equality is a moral presumption and empirical efforts to compare species are irrelevant to this conclusion.

However, many efforts to justify moral positions do use evidence. Often such evidence involves liberal use of anecdotal material, rich inferences of mental characteristics based on relatively simple behaviors and requiring many assumptions, and preference for the Type I error of mistakenly inferring presence of a mental characteristic. Since the consequences to the animal may be profound, this approach is justifiable and conservative in the same sense that the legal system presumes innocence. Writing in a veterinary research journal, Morton and Griffiths (1985) conclude that "If one can extrapolate from animals to humans [re animal models] then it is equally legitimate to do the reverse and, therefore, painful conditions in humans should be assumed to be painful in animals until evidence can be produced to the contrary" (p. 432). This presumption has been codified in regulations implementing the 1985 amendment to the Animal Welfare Act (U.S. Congress, 1989), with less than 10% of the scientific community objecting that the approach was too anthropomorphic. In general, the regulations take an anthropomorphic stance, proposing that distress can and should be reduced by allowing social contact and interaction with conspecifics and/or humans, exercise, and enrichment of the physical environment, with these factors increasing in emphasis as one moves along a presumed phylogenetic scale.

Adopting an empirically based anthropomorphic position may prove difficult. For example, the phenomenology of human distress (e.g., depression) is probably quite different than that for animals (see discussion below). Even the definition of what constitutes an enriched environment is problematical, if "enrichment" is to be defined as providing psychological benefits (Beaver, 1989).

Despite inherent problems, we have little choice but to adopt the anthropomorphic position with its liberal assumptions. Noting that "there are no reliable biochemical markers for pain" (1985, p. 432), Morton and Griffiths use informal anecdotal reports and impressions of animal technicians, researchers, and veterinarians to determine criteria for judging when an animal is in distress. They present a scheme for assessing the magnitude of suffering involving assessment of decreased body weight, appearance, clinical (physiological) signs, unprovoked (abnormal) behavior, and (abnormal) responses to an appropriate stimulus. Novak and Suomi (1988) propose similar standards for defining the psychological well-being of primates in captivity.

Prediction and Control of Animal Behavior. While working at a primate research laboratory, I noted that the competence of each animal keeper appeared to be related to the degree to which he empathized with the stumptail macaques under his care. The keeper who felt a kinship with the animals was most likely to treat them humanely. This keeper, who intuitively understood the animals well enough to coax them from one area of the group cage to another and who established a "relationship of equals" with the dominant male, was respected by the research staff. While I attempted to determine the complexity of the animals' social cognitions through ponderous experimentation, reaching somewhat indeterminate conclusions (Silverman, 1986), like Hebb's (1946) associates the keeper simply acted intuitively and effectively. For him, the "parsimonious" attribution was one that could be made quickly and which generally worked in handling the animals.

The efficient prediction and control of an animal may require the attribution of mental characteristics or the application of operant learning principles, depending on the particular circumstances and the predilections of the human. Decisions about what attributions to make are based on what works. Anecdotes, experimental results, assumptions of mental abilities or of "mindless" behavioral principles, are all of equal potential usefulness in this intrinsically eclectic process.

In the practical situations which present themselves for prediction and control, it is likely that most people attempt to achieve satisfactory rather than optimal results (Simon, 1979). A suggestion for the making of decisions concerning the diagnosis and treatment of disease by medical students is broadly applicable to cases of practical interactions with animals. Elstein, Shulman, and Sprafka (1978, p. 297–298) propose the following heuristics which incorporate both scientific principles and pragmatic concerns with potential utility and costs:

I. Generate a list of alternative hypotheses.
    A. Generate multiple competing hypotheses.
    B. Consider the most likely explanation first.
    C. Consider the potential utility of the hypothesis.
II. Gather data.
    A. Test the most likely hypothesis first and then the ones most likely to prove useful.
    B. "Screen" explanations by using easily administered (even though less exact) tests.
    C. Consider the costs (to the animal, decision maker, and beneficiary of the conclusions) against the benefits of a test.
    D. Be no more precise in the testing than is required for the application of the results.
III. Combine the data and select a course of action.
    A. Find and evaluate disconfirmatory evidence.
    B. Consider multiple explanations if each may lead to a different application.
    C. Estimate and maximize the expected value of the application.

Using these criteria, our animal keeper, entering the cage of a potentially vicious male stumptail, might through various observations judiciously conclude that the animal is able to infer motives and is capable of complex deception. On the other hand, he might not bother to make such observations or draw a conclusion about a juvenile who presents less of a threat.

Animals As Companions. The use of animals as companions involves the liberal attribution of a variety of human thoughts and feelings to the animal. One might describe this attribution as erroneous anthropomorphism or egocentric projection. In either case the process is an intuitive one requiring no empirical justification. In fact, it is quite likely that a skeptical attitude towards one's own attributions would impede the formation of an emotional bond. Advocating the use of animals in psychotherapy, Levinson (1984) describes the rich inferences that people make about their pets:

> An owner who has a close relationship with his companion animal feels that the animal understands him, accepts him with all his foibles and hang ups, cares about him and will defend him. The master also feels that the companion animal is communicating his regard for his owner by his affectionate behavior and has the same emotions as his master. (p. 138)

Efforts to Build Formal Theories of Animal Mind

The purpose of the comparative study of animal behavior has been stated as "assessing the universality of behavioral phenomena or their variance . . . and of discovering and defining new idiosyncratic phenomena" (Bornstein, 1980, p. 3), establishing the general laws of function of evolutionary structures (Eibl-Eibesfeldt, 1983), and building "an intellectual framework from which to conceptualize and interpret human behavior and human evolution" (Gallup & Suarez, 1983, p. 23). We study animals to identify their relationships to us and our relationship to nature.

Whether we are motivated by a search for differences or similarities, the process of science is the same. The fundamental basis for theory construction is the procedure of hypothesis testing. Theories of animal mind have been generated by the testing of hypotheses concerning the nature of animals and by tests of hypotheses about human nature using animal models. In both cases, one adopts precisely the opposite stance of the ethicist. I have already discussed the characteristics of formal and ecological experiments. Clearly, these approaches, along with a starting assumption of nonmental processes underlying behavior, conservative inferences of mental characteristics (or mindlike characteristics), and preference for Type II errors, are optimal. In practice, however, theories are supported by accumulated evidence from a variety of

sources. Recent cases in point are a review of research of Piagetian studies of animal cognition and cognitive development (Doré & Dumas,
1987), and the use of multiple sources of evidence in speculating about
"how monkeys see the world" (Cheney & Seyfarth, 1990).

Where no single measure or method engenders certainty, the results
of multiple indices and methods may provide convergent support
(Gallup, 1988). However, this eclectic approach creates problems when
mixtures of sources and approaches use incompatible assumptions and,
as noted earlier, when research results are used to draw conclusions
about subjective experiences. Consider the development of an animal
model of depression. An early discussion of this topic determined that
one of the minimum requirements of such a model is that "Independent
observers should agree on objective criteria for drawing conclusions
about the subjective state" (McKinney & Bunney, 1969, p. 246). McKinney
and Bunney then cited behavioral evidence of depression in animals in
controlled studies in which manipulations resulted in huddling,
decreased social interaction, appetite and sleep disturbances, and screaming by infant rhesus monkeys (separated from their mothers). Anecdotal
reports of illness and death, unresponsiveness, loss of appetite, weight
loss, listlessness, self-mutilation, agitation, anorexia, and social with-
drawal following separation from assumed attachment figures in a variety of species were also considered. It is not clear that all observers would
agree that either or both the experimentally induced behaviors and the
anecdotal reports represent behavioral evidence for depression.

McKinney and Bunney also stated that an animal model of a
human disorder can be created through the establishment of identities
in cause, symptoms, and cure. While similar behavioral symptoms of
depression in humans are often described (weight loss or weight gain,
insomnia or hypersomnia, fatigue, psychomotor agitation or retardation), subjective factors such as depressed or irritable mood, apathy,
diminished ability to concentrate, and recurrent thoughts of death are
also criterial (American Psychiatric Association, 1987). Graham and
Rutter (1985) suggest additional subjective factors, all typically com-
municated through language: feelings of self-blame, self-deprecation,
helplessness to alter one's life circumstances, hopelessness about the
future. Since alleviation of such symptoms constitute identification of a
"cure," and cognitive causes of human depression have been identified (Beck, 1974), the criteria for a comprehensive animal model are not
met. At best, animal behavior models only a portion of the indicators of
human depression.

In the domain of modeling, we are left again with the problem of
the insufficiency of even stringent criteria for making inferences from

behavior to mental states. It would appear that we would be prudent to interpret research results as suggesting evidence for "mindlike" (in this case, "depressionlike") characteristics, being careful to clearly state the logic of analogy rather than of identity (Silverman, 1983).

NEGOTIATING THE JUDGMENT OF ANIMAL MIND

The attribution of mental characteristics to animals requires philosophical decisions about the reasonableness of inferring mind from behavior, what constitutes descriptive or explanatory efficiency, the acceptability of different types of judgmental error, and the appropriate methods permitting such attributions. These decisions, in turn, are influenced by the uses to which they will be put. Applications of decision include practical interactions with animals and efforts to build formal theories of animal mind.

But this taxonomy is overly simplistic. Both the decision-making and application categories which I have created overlap: uses of judgments may be multiple and occasionally are unforeseeable; and the decision maker may find that different seemingly incompatible decisions must be applied to a single animal or on a single occasion. The animal-as-research-subject may be a companion animal, and in any case deserves ethical consideration. Sometimes the conclusions reached in one domain of application are applied to another and this application can produce a dilemma. Scientific theories influence practical interactions. For example, to the extent that animal models of depression approximate the human condition, one must ask whether it is ethical to use the model to induce depression in animals.

Speculations, practical interactions, and theory construction require increasingly less open and more logically constrained sets of decisions. This requirement may result from two mediating influences: The cost of error and the constraints of time. Glass and Holyoak (1986) propose that "when the cost of error is not high, applying more effortful . . . judgment procedures may not be worthwhile" (p. 361). Speculations clearly involve little cost of error since the process is a kind of moratorium with minimal consequences. Errors in practical interactions (e.g., prediction and control, ethical treatment, companionship) often do involve costs, to the animal and to the person. Erroneous theories may also be harmful, perhaps more so, to the extent that their results serve the purpose of broad practical application. Where the presence or absence of an act involves possible costs to the animal and the converse creates potential costs for the human (as occurs in research, pet ownership, sports, and farming) the question of the relative cost of error becomes key. The cost to the animal of mistakenly concluding that it does not experience distress under a particular circumstance must be weighed against the cost to the human of mistakenly concluding that it does experience distress. One finds concerns among researchers that the literal cost of compliance with stringent animal welfare regulations would threaten research contributions to human problems. This threat suggests that attributions of mental states to animals should be made conservatively. On the other hand, those focusing primarily on animal welfare consider that the mistaken denial of such attributions would be costly for the animal. Both sides call for additional research to resolve the dispute. While new information will emerge regarding what constitutes natural behavior and what provokes abnormal behavior, it is unlikely that there will be agreement on the interpretation of the conclusions. This lack of agreement results because the "cost of error" has different meanings for each position.

A second influence on the relation between the decision-making process and applications is the time available to form and enact the decision. Where a judgment must be instantaneous, effortful procedures are less expedient than heuristics. As I brush my dog and she startles, whines, or snaps when the brush touches her dysplastic hip, I immediately interpret her response as pain and a desire to escape it. I grip the leash more tightly to ensure her presence and proceed to calm her. Her behavior changes and I infer that she has been comforted and feels no distress. In this example, the costs incurred if these attributions are wrong are minimal and the pressure of time is great. I simply presumed a mental state and acted parsimoniously, that is, with speed.

CONCLUSION

I have argued that the attribution of a characteristic of animal mind rests on choices among fundamental issues such as appropriate philosophical assumptions, theoretical positions, and methodologies. These choices, in turn, are influenced by their anticipated applications. It is likely that the perceived cost of judgmental error and constraints on decision-making time mediate the way in which applications affect the positions we take on such issues.

Arguments about animal mind often are founded on citations of empirical evidence and the "correct" interpretation of such evidence. An interesting conclusion of the preceding analysis is that different and contradictory interpretations may be simultaneously correct, in the sense that each is appropriate to its application. Where disputes occur about what interpretation is the appropriate one, the underlying disagreement may rest on beliefs about potential costs of error in judgment. The costs for the person engaged in speculation, the researcher testing a hypothesis, and the person engaged in a practical interaction are different. At times, this "person" may be the same individual.

The mental attributions we apply to animals may, indeed *must*, be inconsistent. I may talk to, even identify with, the same animal who has failed a formal test for some mental capacity or process. We are legally obliged to assure the "psychological well-being" and avoidance of "distress" in research animals. This may require liberal application of anthropomorphic assumptions, while at the same time we attempt to identify what these terms mean by the use of more conservative assumptions and methodologies. If this is formally inconsistent, so be it. Animal mind cannot be said to exist in any absolute sense. And philosophical, theoretical, and methodological approaches to inferring animal mind cannot be said to be good or bad, adequate or inadequate, without considering their potential applications.

PART V

Intentionality

15

Varieties of Purposive Behavior

Ruth Garrett Millikan

In anthropomorphism, animals are credited with having intentions, plans, or humanlike purposes, that is, cognitions. Concerns about anthropomorphism can focus on whether any one (or all) of three parts of this attribution is accurate: that animals have purposes at all, that animals have cognitions, and that animals have humanlike cognitions. In order to know whether such concerns are appropriate, one needs some sort of theory concerning what purposes, cognitions, and human cognitions *are*—some theory about what exactly it *is* that one is attributing to the animal. In this chapter, I supply such a theory, which differentiates two broad kinds of purposiveness in behavior: biological purposiveness, and "intentional" purposiveness—the kind that involves cognizing purposes, having plans. I argue that there can be no study of behavior that fails to rely, at least implicitly, upon speculation concerning the biological purposiveness of that behavior. Thus, the attribution of purposiveness per se to animals is not something to be avoided, indeed, it is best if the necessary attributions are made as explicit as possible.

Intentional purposiveness is more problematic. Intentional purposiveness is a *form* of biological purposiveness. It involves a particular kind of mechanism for the implementation of biological purposes. Intentional purposiveness undoubtedly comes in numerous forms, some much less complex than those exemplified by human intentions. To attribute cognition to nonhuman animals need not be, and surely in most cases should not be, to attribute the human kind of cognition to them. There are intermediate possibilities.

I begin by describing biological purposes and showing how behavior is recognized with reference to biological purposes. A much fuller treatment of this theme may be found in Millikan (1993, chap. 7). I then describe intentional purposes and cognitions, suggest how human cognitions may differ from those of nonhumans, and describe some forms which nonhuman cognition may take. Fuller treatments may be found in Millikan (1984; 1993, chaps. 3–9).

By a "biological purpose" or a "biological function" (I intend these as synonyms) I mean the kind of purpose that the heart's beating has. The purpose of the heart's beating is to circulate the blood. Similarly, the purpose of your skin's flushing in the heat is to dispel heat from your blood, the purpose of the frog's flicking out its tongue when the right kind of shadow crosses its retina is to effect the ingestion of flies. In general, the biological purpose of a behavior is whatever salutary effects this behavior has had often enough, during the evolutionary history of the species, to help account for the current presence in the species of the mechanisms that produce the behavior. Crudely, biological function is historic survival value.

My first task will be to clarify why it is that one cannot study behavior without making at least implicit reference to the behavior's biological purpose. Indeed, one cannot even pick out behaviors as the objects of one's study antiseptically, in a way that is theory free, free from all speculation about biological function. What observations count as among the data for any particular science is never a matter that can be settled apart from theory—a point that is universally recognized by reflective contemporary scientists and philosophers of science. For example, classical chemistry studied chemical compounds and not solutions or mixtures, but what is a compound and what is a solution or mixture is a matter that is determined by chemical theory, not prior to theory. How is this general principle manifested in the sciences of behavior? Indulge me for a moment as I display some species-specific philosopher's behavior. I propose to point out something so close to our noses that we tend not to notice it. Then, when you have noticed it, and noticed how familiar and obvious it is once pointed out, hence how trivial it is, I will insist that in fact it is deep and profound.

The invisible yet obvious fact to which I direct your attention is that there is a literal infinity of different possible descriptions that might be given of any animal's behavior at any given time, only a tiny few of which descriptions have any relevance for behavioral science. These are the descriptions that connect in a pertinent way with function. Consider, for example, the various motions that an animal makes. A good portion of the behaviors of animals are motions of one kind or another.

But motions can only be described relationally. Relative to what do we describe the animal's motions? Consider Amos, the mouse, there on my kitchen floor. He runs in the opposite direction from the waiting cat. In the same motion, he runs toward the waiting broom. He also runs between a black square on the linoleum and a tomato stain, towards the kitchen clock and towards London, away from magnetic north, thirteen times as fast as the clock's second hand swings, five seconds after the last perceptible motion on my part, 0.05 seconds after the last agitated flick of the cat's tail. Clearly this list of descriptions of Amos's running could be extended indefinitely. And wait; is his behavior even a running? Perhaps Amos's motion should be described merely as a rhythmic beating of the paws, which happens to carry Amos across the floor, just as it happens to carry him toward London.

Nor will it help to relativize Amos's motions to his own body. Notice Amos as he blinks. Are his eyelids momentarily covering his eyes? Or are his eyelashes being disentangled momentarily from his eyebrows? Or do the eyelashes point momentarily to the navel, or is it to the toes? Perhaps all that occurs is a rotation of the eyelids. Do they rotate at an angular velocity of 1000 degrees/sec, or is the rate equal to three revolutions per mouse heartbeat?

Nor are motions in any way peculiar with regard to the infinity of their possible descriptions. Amos can make squeaks, chattering sounds, sneezes, coughs, choking sounds, or he can be silent—silent except that, if you listen closely, he makes breathing sounds, and little thumping sounds with his feet (danger signals, or just foot patter?), and also with his heart. Which of these sounds, and which silences, are subject matter for behavioral science and which are not? How should the sounds be described? By pitch, or inflection, duration, periodicity, harmonic structure, rhythmic structure, amplitude, pattern of repetitions? Consider the sounds that a human makes. Some of these, such as screams and laughs, can be described relatively crudely. Others, the speech sounds, need to be described in great detail, and in accordance with principles of such subtlety that these are not yet fully understood. Still other sounds, such as sounds made while choking or urinating, sounds made by the heart and, normally, those made in breathing, do not need to be described at all. Sometimes silences need to be described and sometimes they do not.

Given the infinity of possible descriptions of behavior, what determines the forms of description relevant to behavioral science? Which behaviors, which describable outputs, are true Behaviors, using a capital B—outputs that are proper subject matter for animal behavior studies?

Does one look, perhaps, for repeated behavioral units, for patterns that recur? That mice run away from cats, for example, is a recurrent phenomenon, that they run toward waiting brooms is not. But that cannot be the answer. For the heart says pit-a-pat with wonderful regularity, every mouse eyeblink is a momentary disentanglement of its eyelashes from its eyebrows, every mouse foot touching the floor makes a minuscule thump, and choking is a distinctive and reliably reproducible sound, under the right stimulus conditions, yet none of these are Behaviors with a capital *B*. Equally, it is a mistake to think that what we are after is whatever behavior falls under laws. All of the above behaviors fall under laws, indeed, unusually reliable laws. Consider the knee jerk reflex, which falls under a wonderfully reliable law. The physiologist looking for clues concerning the engineering of the body is interested in knee jerks. But if knee jerks are ever of interest to students of behavior, this is not because they are Behaviors, but only because they can be used to help diagnose the condition or give clues about the inner structure of mechanisms that *are* responsible for true Behaviors.

No. What makes it that this output of an animal but not that is a proper piece of Behavior, or that an output *described* this way rather than that is a proper piece of Behavior, is always some kind of intimate relevance to biological function. The knee jerk is not a Behavior because it has (so far as we know) no function. It doesn't effect anything that has aided survival. For the same reason, it is unlikely that choking sounds are Behaviors, nor sneezing *sounds* nor coughing *sounds*. On the other hand, sneezing and coughing themselves probably are Behaviors; probably having mechanisms that produce sneezes and coughs has survival value. Blinking is properly described as covering the eyes with the eyelids because that is what effects, say, keeping out the sand, while disentangling the eyelashes from the eyebrows and pointing with the eyelashes toward the toes have no functional effects. Amos's behavior is not just a rhythmical moving of his paws but a full-fledged running across the floor because it is historical runnings and not just historical rhythmic paw motions on the part of Amos's ancestors that, characteristically, contributed to survival. On similar grounds, whatever may or may not be "in his mind," Amos's Behavior is surely a running away from the cat, and just as surely is not a running toward the broom or toward London. That the perceptual-motor systems Amos inherited from his ancestors and that are responsible for his current running are systems that, operating so, sometimes effected removal of those ancestors from the vicinity of predators, certainly helped account for their proliferation; that these mechanisms effected approaches to brooms or, say, to large cities, certainly did not.

What counts as a true Behavior is what is assumed to have, under that description, a biological purpose. Often this point of theory drops into the background because it is so evident which behaviors must be functional, which not. It is evident, for example, that Amos is to be described as running, but surely not as running toward London—so evident, indeed, that it is hard to discern how any deep point of theory could possibly be lurking here. Other times, however, it is not a bit obvious what an animal is doing that constitutes its true Behavior. Colin Beer, for example, tells an involved story about his struggles to discover where the true behaviors lie within the vocalizations and displays of laughing gulls (Beer, 1975, 1976). And it is well to be aware that every description, every classification, of a behavior makes an implicit reference to known or unknown biological purpose. If one is not aware that theory is inevitably involved even at this level, one will surely be more likely to import implicit *bad* theory, for example, to import unexamined anthropomorphism, into one's descriptions of behavior. The unreflective human, asked to describe the behavior of another species, naturally relies on projection: What would I be doing if I were acting like that?

Attributing biological purposes to behaviors is not, of course, the same thing as attributing thoughts or cognitions. It may be a function of the eye-blink to effect keeping out the sand, but neither the eye nor the animal as a whole must cognize this goal in order to blink. Similarly, suppose that you condition my operant eye-blink response by reinforcing it with your smile. My blinking will then have acquired a new biological goal, as it produces your smiles in accordance with the biological design of my learning systems. Compare: the pigment rearranging mechanisms in the skin of an old world chameleon can be said to flip from having the biological purpose of making the chameleon appear brown to having the biological purpose of making it appear green as the chameleon moves from a brown to a green surface. Similarly, the biological purpose of a conditioned response can be said to depend upon the environmental circumstances that produced it. For a fuller treatment of this theme, see Millikan (1984, chapter 2; 1993, chapter 9). But though my blinking may have making you smile as its biological purpose, I will not cognize this purpose, I will not blink because I think, either consciously or unconsciously, of this purpose. S-R learning does not, in itself, produce cognition. Whether learned or unlearned, differential responses are not *thoughts*. What then are cognized purposes, intentional purposes? How do they differ from learned or unlearned biological purposes?

The word "intentional" is used by philosophers to refer to items that are *about* other things, as, for example, the sentence "Paris is pretty" and a map of Paris are both about Paris. The belief that Paris is pretty, the desire to visit Paris and the intention to visit Paris are of course also about Paris. All of these items exhibit "intentionality." External items that exhibit intentionality, such as sentences, graphs and charts, maps, road signs, sheet music, and representational paintings, are called "representations." Similarly, a dominant theory to which I subscribe proposes that inner intentional items such as beliefs, hopes, desires and intentions are similarly representations. More generally, all cognitions are inner representations, that is, inner models, in the most abstract mathematical sense, of what they are about (Millikan 1984; 1993, chapters 3–5). The difference between merely biological purposes and intentional purposes is that in the latter case the animal's biological purposes are implemented via the manufacture and use of inner representations—representations of the environment and/or representations of the animal's goals. Human beliefs, desires, and intentions are then differentiated by being embodied, at least in part, in an especially sophisticated system of inner representation, on which I will comment soon. Without doubt, however, humans also continue to rely on other levels of cognition, upon more primitive forms of representation.

To give you some idea of how abstract the models that are representations can be, English sentences are such models. Significant transformations of English sentences (substitution transformations, for the most part) correspond to transformations in what the sentences are about; the domain of true sentences maps onto, or bears an abstract isomorhism to, the domain that is the real world. Or for those acquainted with this field, neural network modeling is in large part an investigation of certain very abstract forms that mental representation may possibly take. The animal that cognizes is capable of constructing a variety of alternative inner models, presumably states of its nervous system, which accord, by certain abstract rules of correspondence, with what the animal thinks about. These abstract models, abstract mathematical "maps", may function as maps of the environment, modeling facts about the animal's world, or as blueprints showing desirable outcomes of actions, or plans adopted for action.

Let me try to make this more concrete by calling on a close analogy. Consider for a moment the dance of the honeybee. Transformations of the dance (e.g., a rotation of the long axis of the dance by so many degrees) correspond to transformations in what is represented (changes in the represented direction of nectar from the hive). The dance is an abstract mathematical map, a representation, of the location of nectar. The biological purpose, which is to get the worker bees to supplies of nectar that have been spotted by their coworkers, is effected by the cooperation of two kinds of systems. The first produces maps of where the nectar is. These serve as blueprints of concrete biological goals to be achieved, the goals, namely, of getting to this place, or that place, where the nectar is. The second kind of system "reads" these blueprints, that is, reacts to them appropriately, in a manner such as to achieve the projected biological goals. In this case some bees dance while others watch: the representations are not inside the bees but outside. The dances are representations, but not, as such, cognitive representations or cognitions. Suppose, however, that the mechanisms responsible for manufacturing a model of the environment, the model itself, and the mechanisms responsible for interpreting it, were all within the same organism. Then you would have primitive *cognitions*— thoughts (though not necessarily conscious ones) of the layout of the environment, of goals to be achieved.

To attribute *intentional* purposes to an animal is to attribute to it some kind of inner representational system, some way of mapping the world and its goals which serves as its means of achieving those goals. The truly interesting part, however, is to reflect on the variety of ways that mapping principles might be employed by biological systems. For example, there are a number of very fundamental respects in which human beliefs and desires must differ from representations like bee dances (besides, of course, that bee dances are not inside the organism, hence not cognitions). I will mention three such differences. Various other crucial differences are enumerated in Millikan (1993, chap. 4, sect. 5).

First, bee dances are undifferentiated between the indicative or fact-stating mood and the imperative or direction-giving mood. The dance tells the worker bees where the nectar is (facts); equally, it tells them where to go (directions). The step from this kind of primitive representation to human beliefs and intentions is an enormous one, for it involves the separation of indicative from imperative functions of the representational system. Representations that are undifferentiated between indicative and imperative connect states of affairs directly to actions, to specific things to be done in the face of those states of affairs. Human beliefs are not tied directly to actions. Unless combined with appropriate desires, human beliefs produce no action. And human desires are equally impotent unless combined with beliefs about how to fulfill the desires. But there is no reason to suppose that the capacity to retain purely factual knowledge, knowledge disconnected from any specific projected uses, or to harbor explicit desires disconnected from any specific ideas about how to fulfill them, is a feature of every animal that cognizes. Many may harbor only undifferentiated inner representations (Millikan, 1995).

Second, since indicative and imperative functions are separated in the inner representational systems of humans, they need to be reintegrated in order to produce actions. Thus humans engage in practical inference, combining beliefs and desires in novel ways to yield first intentions, and then action. Humans also combine beliefs with beliefs to yield new beliefs. But it may be that other species do not have such inference capacities, or have them to a more limited degree.

Third, the representational system to which the bee dance belongs does not contain negation. Indeed, it does not even contain contrary representations. If two bees dance different dances at the same time, these dances do not conflict, for it may well be that there is indeed nectar in two different locations at once. (On the other hand, the bees can't go two places at once.) But in a representational system without negation no contradictions can occur. If the inner representations of a cognizing animal were to lack negation, hence the potential for contradiction, this would be highly significant. If we follow philosophical tradition, the law of contradiction plays an absolutely crucial role during acquisition and development of all concepts not definitionally tied to a given significance for action. Animals lacking negation in their inner representational systems would be unable to learn new concepts except insofar as these were directly tied to action. All of their concepts would have to be either purely practical, or static, hereditary, built in. Moreover, I have argued that negation is dependent upon subject-predicate, that is, propositional, structure and vice versa. Representations that are simpler do not express propositional content (Millikan, 1984). Even such sophisticated representations as maps, charts and sheet music do not contain subjects and predicates. Animals that thought exclusively in representations without subject-predicate structure would not think propositions.

To attempt to express the contents of the cognitions of animals that were undifferentiated between indicative and imperative mood, or cognitions without subject-predicate structure, or that were not subject to negation, or that did not participate in processes of inference or information transformation, by translating these into or correlating them with English sentences, would not be at all accurate. Taking an extreme example, consider the "fly detector" in a frog's optic nerve. It might be considered to produce an extremely elementary sort of inner representation. The firing of the detector at a certain time and place maps the presence of a fly at a certain time and place—the same time and place. Firing at another time and place represents a fly at a different time and place. Or should we say instead that the firing is really an imperative representation that tells the frog to snap at a particular time and place? Does it say "There's a fly here now" or does it say "Quick, snap here now"? Yet the firing of the detector really tells the frog (or its brain) neither of these things. To say "There's a fly here now" it would have to contrast with a possible representation that said "There's *not* a fly here now," and with possible representations that said, say, "There's a *beebee* here now" or "There's a *cat* here now" or "There was a fish on Tuesday at four."

What is really needed in order to understand nonpropositional animal cognition is not a translation into English, but explicit description of the *kinds* of representational systems such animals in fact use, and their ways of using them. The ultimate goal must be to construct and test models of the cognitive systems of each of the various animal species, much as human psychologists are beginning to construct and test models of human information processing. The ultimate confirmation of such models will lie in the minute physiologies of the various species. At the moment, however, for the most part we can only speculate about what various kinds of representational systems are in principle possible, and which of these might in fact be used by biological systems. But we will surely go astray if we fail to keep in mind that between human propositional thought, and no thought, there are many intermediate possibilities.

16

Expressions of Mind in Animal Behavior

Colin Beer

This chapter mixes memory and desire. The memory is of how science in general and ethology in particular were conceived in the tough-minded, positivistic tradition in which I was brought up as a student. The desire is for the possibility that, with the questioning of this positivistic tradition in general and the emergence of cognitive ethology in particular, issues concerning animal mentality and intentionality, which the older views kept in the dark, might now be looked at in a new light. The philosophical developments critical of the older positivism bear on these issues in ways that are both encouraging and challenging to cognitive ethology. In what follows I first sketch some of the history of ideas relevant to the transition from classical ethology to cognitive ethology, and then discuss some of the philosophical problems with which I think current attempts to be usefully anthropomorphic about animal behavior have to contend.

I.

When I started out in ethology, which was in the mid-1950s, there was still talk of "action-specific energy," "innate releasing mechanisms," "consummatory acts," and the other notions that went into classical ethological motivation theory. Already, however, both within and without ethology, people were questioning the soundness of that theory.

The classical ethologists (e.g., Lorenz, 1950; Tinbergen, 1951), with their speculative constructs, were being confronted by tough-minded young critics who, to use one of William James's expressions popular with positivist philosophers at that time, demanded to know the "cash value" of the language of ethological theory in empirical terms.

This kind of demand for empirical grounding of scientific concepts had earlier been made in physics and psychology. The movement known as operationism (or operationalism) (Bridgeman, 1927, 1938; Stevens, 1939; Feigl, 1945) dictated that all the expressions used in science, aside from the logical connectives, were to be given definition in terms of observational, experimental, or measurement procedures. In behavioristic psychology, for example, "thirst" is pegged to duration of water deprivation, and accordingly purged of the idea that it is a subjective state of feeling parched.

Such attempts by scientists to "purify the language of the tribe" are continuous with a much older preoccupation of philosophers: the grounding of epistemology—theory of knowledge—on absolutely secure foundations. For Descartes self-evident truths and deductive inference gave certainty to understanding. For the British empiricists we could be sure only of what sensory experience provided, everything in the mind having its derivation in the deliveries of the senses. For Kant the bedrock of knowledge was jointly in the a priori structure of the mind and the experienced nature of the world. For philosophy in general, at least in the Western intellectual tradition, what Rorty and others have called "foundationalism" has been a dominating theme (Rorty, 1980). As J. L. Austin (1962) said: "The pursuit of the incorrigible is one of the most venerable bugbears in the history of philosophy" (p. 104).

Operationism sought to give meaning rather than knowledge a foundational grounding, but specifically in science. At about the same time—the 1920s and 1930s—logical positivism took a similar stance towards meaning in general (see Ayer, 1936, 1959). Logical positivism originated with a group of philosophers known as the Vienna Circle, who regarded science as the successor to metaphysics on questions about the structure, contents, and dynamics of reality, specifically because, in their view, scientific statements had meaning and metaphysical statements did not. The theory of meaning implying this judgment was verificationism, which derives from the distinction that Kant drew between analytic truths and synthetic truths.

In analytic statements the predicate is an expression equivalent to the subject, that is, the statement is definitional or tautological, as in "bachelors are unmarried men." In synthetic statements the predicate says something about the subject term which is not contained in the meaning of that term, and the denial of which would still make sense, for example, "Herbert Spencer was a bachelor." The logical positivists maintained that all statements are either analytic or synthetic, and that the meaning of a synthetic statement is equivalent to what would empirically verify the statement. Thus the meaning of the statement about Spencer consists in the availability of biographical evidence that he was never married. Closer to home, for a statement about inclusive fitness to be meaningful there must be ways of assessing the direct and indirect components of fitness and the genetic relatedness of the organisms involved. The statements of metaphysics fail the verification test, since they purport to be about what transcends experienced reality. Consequently, according to logical positivism, they are meaningless.

However, logical positivism in its original form proved to be too crude to sustain its case. For one thing there was the question of the status of the verification principle itself: it appeared to be neither analytic nor synthetic, and so a third category of meaning had to be admitted, that is, what is conveyed by a rule or stipulation. But once that foot was in the door all sorts of other kinds of utterances turned up to crowd the entry into the meaningful, including questions, commands, promises, and counterfactuals.

With Wittgenstein's *Philosophical Investigations* (1953), attention turned to the variety of ways in which language works, and the emphasis shifted from preoccupation with meaning to preoccupation with use. As Urmson (1956) put it:

> In place of the dogmatic "The meaning of a statement is the method of its verification" we were advised "Don't ask for the meaning, ask for the use" and told that "Every statement has its own logic." (p. 179)

Wittgenstein likened the logic of language to the rules of games: just as proper play in the conduct of different kinds of game entails adherence to different sets of rules, so coherent use of language in different kinds of contexts requires following the different rules of the "language games" involved. Following in Wittgenstein's wake, Ryle (1949) diagnosed a number of philosophical puzzles as consequences of failure to follow or recognize differences of conceptual form in statements of the same grammatical form, and hence as "category mistakes." Thus to ask for the telephone number of the average man, or the temperature of a molecule, is to categorize these terms in a manner contrary to the statistical logic of averaging and the kinetic theory of heat. Similarly, Ryle argued, you can weigh a brain but you cannot weigh a mind in the same sense, so to suppose identity between mind and brain is to be caught in a category confusion.

However, according to Chihara and Fodor (1965/1981), both Wittgenstein and Ryle can be viewed as continuing to pursue a foundationalist program like that of operationism by trying to show that statements about mental states or processes or events are really statements about behavior; or at least that there are logical or conceptual relations between statements about the mental and statements about behavior. For example, Ryle's version of this logical or philosophical behaviorism, as the position came to be described, argued that attribution of a mental state can be cashed as a prediction about how someone will act given certain circumstances. Accordingly, to say that Jones intends to go to the opera is to say that, unless obstacles arise to prevent or dissuade him from doing so, Jones will take whatever steps he thinks will get him to the opera. On this analysis the intention of the man is comparable to the solubility of a lump of sugar or the fragility of a piece of crystal: a dispositional property conditional for its manifestation on, respectively, opera-going opportunity, submergence in water, and rough handling. The intention is not, therefore, to be thought of as something you could find by opening up the head, any more than the solubility is to be thought of as something you might find by breaking up the sugar.

Philosophical behaviorism, unlike the psychological behaviorism of Skinner and Co., might have some appeal to people who want to be anthropomorphic about animal behavior. If ascribing intentions, beliefs, feelings, and so forth to an animal is no more than a way of saying what the animal will be likely to do when it is in such and such a situation, the usual objections about invoking unprovable subjective states no longer have any force. However, the move to ground mentalistic talk in behavioral dispositions appears incapable of being carried through. One problem is that the range of circumstances conceivably relevant to how a mental state might be expressed is indefinite, if not boundless. Another and more intractable problem is that further reference to mentalistic or intentionalistic notions inevitably enters into the specification of the disposition or the circumstance. Thus our opera-goer was to take whatever steps he *thought* would get him to the show. A further dispositional analysis, to deal with what he thought, would bring in still more intentionalistic reference, for which, in turn, the process would have to be repeated, and so on ad infinitum (see Churchland, 1988, p. 24). Like trying to turn quickly enough to catch your shadow before it moves, the attempt to find behavioral coin to cash the concepts of mind continually comes up short.

Emerging from critiques of logical behaviorism is the view that our mental concepts, rather than being anchored in outward criteria or behavioral dispositions, constitute a network in which the meanings depend upon relationships between constituents such as terms and propositions. As Chihara and Fodor (1965/1981) expressed it:

> Perhaps what we all learn in learning what such terms as "pain" and "dream" mean are not criterial connections which map these terms severally onto characteristic patterns of behavior. We may instead form complex conceptual connections which interrelate a wide variety of mental states. It is to such a conceptual system that we appeal when we attempt to explain someone's behavior by reference to his motives, intentions, beliefs, desires, or sensations. In other words, in learning the language, we develop a number of intricately related "mental concepts" which we use in dealing with, coming to terms with, understanding, interpreting, etc., the behavior of other human beings (as well as our own). (p. 57)

Chihara and Fodor go on to liken this system to theories in science: "It thus seems *possible* that the correct view of the functioning of ordinary language mental predicates would assimilate applying them to the sorts of processes of theoretical inference operative in scientific psychological explanation" (p. 62).

As this last quotation suggests, ideas about scientific explanation had shifted emphasis from the empirical basis in observation and experiment insisted on by the operationists, to a recognition of how conceptual framework and methodological practice can affect what is to count as a fact and constitute data. Representative of this shift, and important contributors to it, are Popper's (e.g., 1968) opposition of conjecture and refutation to induction and verification, Kuhn's (1962) deployment of his notion of "paradigms," and Hanson's *Patterns of Discovery* (1958) with its insistence that "there is more to seeing than meets the eyeball" (p. 61) (see also Toulmin, 1961, and Feyerabend, 1962).

II.

The view of the language of mind as a theoretical network was anticipated by a number of philosophers, including Quine (1960), who construed the analytic/synthetic distinction and reductionism as unfounded "dogmas of empiricism"; and Sellars (1956), who branded foundationalism as a myth—"the myth of the given." Several prominent philosophers of mind hold this view, and they usually assume it when they talk about the psychology of propositional attitudes or the "folk psychology" of common sense. If they are right it is what we assume when we apply mental predicates to animals.

However, this holistic conception of our understanding of mental life includes much that is exclusive to human beings. Our possession of language makes a profound difference. In using folk psychology to explain or interpret or make sense of what people do, we often impute kinds of knowledge and inferential capacities unthinkable in the absence of linguistic capability, and hence beyond the grasp of animals. Indeed one can think of language as part of the fabric of the network, as Fodor argues, in effect, in *The Language of Thought* (1975). For instance, one of the logical characteristics of intentional mental states is that they are referentially opaque; that is to say substitution of co-referring terms may not preserve truth (cf. the mistaken description of referential opacity given by Colgan, 1989, p. 67). For example, from the fact that I believe that George Eliot was the author of *Middlemarch* it does not necessarily follow that I believe Mary Ann Evans or the author of *Silas Marner* to have been the author of *Middlemarch*, for I may be ignorant of the fact that all three were one and the same. But co-reference of terms and hence referential opacity more or less imply propositional form in the representation of intentional states. Since nonlinguistic animals have no terms for things (except, perhaps, in rare instances such as the signaling of locations by honeybees, and the alarm-calling of vervet monkeys [Cheney and Seyfarth, 1990]), the possibility of referential opacity as we know it presumably does not arise in their case, and so if they can be said to have beliefs, desires, and so forth the manner in which they have them must differ significantly from ours.

Aside from the difference that language makes, the reach and complexity of connections attaching to ideas in the human case will usually far exceed what is conceivable for any animal. Take the case of a cat crouching beside a hole down which it has just chased a mouse (cf. Stich, 1983, pp. 104–105). We should be inclined to say that the cat thought there was a mouse down the hole. But consider what thinking that would mean to us: it would mean that there was a furry mammal down the hole, a tetrapod vertebrate, a whiskered rodent, a warm-blooded cheese-eater, and a whole lot more that could not possibly occur to the cat. Only a small part of the network within which mouse-ness is nested for us extends into the cat's world. Similarly, as Dretske (1988, p. 153) has pointed out, the surprising ability of pigeons to identify trucks in photographs (Herrnstein and Loveland, 1964) does not imply that the birds have any notion of what the word "truck" connotes.

However, holism can be carried to extremes amounting to a reductio ad absurdum. If the meaning of every word in one's vocabulary depends on the meaning of every other word then no two of us could ever think precisely the same thing, for our vocabularies are bound to differ in some respects. And since I am continually acquiring new words and ideas, what I think about something now could not be the same as what I thought about it last week, or last year, or when I was eight. Likewise Shakespeare's sense of "red" must have been completely different from ours, for he knew nothing about traffic lights, tomatoes, or the Communist Party. Because of considerations like these Fodor now thinks that holism applied to meaning is "simply preposterous"; "a really crazy idea" (Fodor & Lapore, 1992, p. xii). But a holistic position does not have to maintain that *all* the concepts and connections in a cognitive or semantic network are affected by *every* change in the network, or *every* difference between one such network and another. Some meanings may be so distant or insulated from the sites of change or difference as to remain impervious, just as, to quote Dretske (1988), "An organization can grow larger (employ more people) without necessarily changing the functions of the other employees" (p. 155). Dretske admits that there may be no principled way to draw the lines around functional dependencies within a network, but he thinks that *what* a term represents—its reference or what it stands for—is likely to stay the same even when what is signified *about* what the term refers to differs or undergoes change. So, to return to our cat and mouse, the cat's mental representation of the mouse will have the same extension as ours—that particular gray object seen disappearing down the hole—even though what the cat can think about that object and what we think about it are vastly different.

The use of intentionalistic language in talk about animals can therefore be problematic because much of what goes with that language when applied to people may have to be left behind when applied to animals. Yet we find such usage virtually impossible to avoid, for we often have no other language available with which to describe what an animal is doing as we perceive it. Take the events leading up to the cat and mouse standoff. We should describe the cat as pursuing and the mouse as fleeing. What else could we say without losing the essence of what went on? But to talk of pursuing and fleeing is to describe the behavior in terms of actions rather than movements: that is to say the behavior is categorized according to its means-end configuration, its "goal-directedness," rather than according to its motor patterning. Now in contrast to motor patterning, which can be accounted for in terms of things going on in nerves and muscles, actions typically call for explanation in terms of intentions, wants, or motives, and awareness of the circumstances in which they are to be satisfied (e.g., Taylor, 1964). Dretske (1988) argues for a concept of behavior that is much like what I (following Taylor) have just referred to as action: "a complex causal process, a structure wherein certain internal conditions or events (C) produce certain external movements or changes (M)" (p. 21). The internal conditions, he argues, can include representation of the circumstances in which the organism is placed, and states of belief and desire providing reasons for acting appropriate to those circumstances. If describing the cat as pursuing and the mouse as fleeing thus imply such intentionality in the animals, we are going beyond what most ethologists would countenance. Nevertheless, ethologists use action terms like pursuing and fleeing all the time, and, as I have said, could hardly do without them. Even Colgan, who, in his recent book on motivation (1989), is dismissive of cognitive ethology, has to resort to talk about the signaling of intentions when he gets to communication.

This use of intentionalistic imbued language in description of animal behavior constitutes a latent and pervasive kind of anthropomorphism. When applied to animals, action terms and other terms involving intentionality are tacitly more or less cut adrift from their places in the network of folk psychology, and either hooked up in another network of conceptual and logical relationships, or used, as it were, with invisible quotation marks—in tacitly understood qualified senses excluding the questionable connotations attaching in the human context. A third possibility is that the words are used metaphorically to explore whether reflection from the human context might illuminate the animal context.

III.

The double network idea has, in effect, been acknowledged in at least one instance. In his reply to Mary Midgley's (1979) attack on *The Selfish Gene*, Richard Dawkins stated that the motivational language that he used in his book was not to be taken as such, and that to take it literally was a "misunderstanding of the definitional conventions of the whole science of 'sociobiology'":

> When biologists talk about "selfishness" or "altruism" we are emphatically not talking about emotional nature, whether of human beings, other animals, or genes. We do not even mean the words in a *metaphorical* sense. We *define* altruism and selfishness in purely behaviouristic ways ... (Dawkins, 1979, pp. 556–557)

To explain what he meant by this he quoted from *The Selfish Gene*:

> An entity ... is said to be altruistic if it behaves in such a way as to increase another entity's welfare at the expense of its own. Selfish

behaviour has exactly the opposite effect. "Welfare" is defined as "chances of survival", even if the effect on actual life and death prospects is . . . small. . . . It is important to realize that the above definitions of altruism and selfishness are *behavioural*, not subjective. I am not concerned here with the psychology of motives. . . . that is not what this book is about. My definition is concerned only with whether the *effect* of an act is to lower or raise the survival prospects of the presumed altruist and the survival prospects of the presumed beneficiary. (Dawkins, 1976, pp. 4–5)

Presumably the same kind of construal is to be given to such terms as "decision," "assessment," "strategy," "investment," "choice," and "deception" when they occur in the writings of sociobiologists. There is an at least implicit systematic remapping of the connotations and interrelationships of these originally intentionalistic terms onto the universe of selectionist discourse. And there is direct historical precedence for this practice: Darwin (1859/1968) adopted and adapted some of the language of laissez faire economics in explicating his theory of natural selection in *The Origin of Species*. Indeed "selection" itself, in this context, has to be understood as implying nothing about conscious choice.

However, it is an open question whether the sociobiological use of intentionalistic language has completely and consistently divested itself of all influence from what the terms mean in their original contexts. Even dead metaphors can be haunted by the ghosts of their original meanings (Empson, 1947, p. 25). This may not necessarily be a bad thing. Indeed there can be heuristic value in keeping the metaphors alive. Max Black (1962), among others, has argued that metaphors have a powerful and central part to play in the history of science. For some of the contexts in which sociobiologists deny that their intentionalistic language has any intentionalistic force it may be worth considering whether the original literal senses of the terms might after all apply in some way. This is particularly so with regard to some aspects of animal communication.

In a paper which sparked a controversy, Dawkins and Krebs (1978) argued that communication is better conceived in terms of manipulation than in terms of information conveyance. The controversy turned on whether these two conceptions exclude or complement one another, but I do not want to be concerned with that here. My interest in the issue is that it was couched in selectionist terms, and the motivational question was explicitly put aside:

If an animal can benefit by "predicting" the behaviour of other animals, he will tend to do so. Needless to say, there is no implication of

conscious prediction. Predicting means, here, behaving as if in anticipation of another animal's future behaviour. (Dawkins and Krebs,
1978, p. 288)

This, of course, is consistent with what Dawkins was earlier quoted as
saying about definitional convention in sociobiology. But the motivational question is there to be asked, and is surely not without interest. Is
it not conceivable that the animal performing a display or giving a signal is *trying* to manipulate or inform its audience, as is assumed to be
the case when these terms are used of people? The question is not idle,
for it can generate predictions for observation and experiment to try to
verify.

IV.

Encouragement for the pursuit of the question of intent in animal
behavior comes from the philosopher Daniel Dennett (1978, 1987a).
Dennett has developed an approach that he calls intentional system
theory, which, he argues, is necessary to deal with machines like computers that are too complex to come to grips with in terms of their physical states or design features, and can also be usefully applied to animal
behavior. In adopting an intentional stance towards a machine or an
animal one regards the thing or the creature as if it had beliefs and
desires, and the rationality to act in the light of the beliefs so as to attain
what would satisfy the desires. This is a pragmatic or heuristic strategy,
which does not include trying to answer the question of whether the
machine or the animal really has beliefs and desires as we experience
them. Adoption of the approach by ethologists has already led to observational and experimental work on animals to test intentional stance
predictions (e.g., Ristau [1986] on injury feigning in plovers; Gyger,
Karakashian, & Marler [1986] on the "audience effect" in chickens). I
think it likely that there is much that is waiting to be explored with this
approach, if only we can find the ingenuity to pose the questions about
animal intentionality in testable terms. In the meantime I offer from
my gull studies an example of the sort of thing that seems to me to
warrant adoption of an intentional stance.

When Laughing Gull chicks are between one and two weeks of
age post-hatching they typically call and approach in immediate
response to crooning or ke-hah calling by their parents (Beer, 1969,
1970). Older chicks call as vociferously as do younger chicks when they
hear their parents call, but they are much less likely to approach immediately, and tend to stand their ground as if they were trying to get the

parents to come to them. The parents show a similar pattern: when chicks are nestlings the parents almost always approach when the chicks call, but when the chicks are older than two weeks or so the parents become less and less inclined to go to calling chicks, and instead call to the chicks as if trying to induce the chicks to come to them (Beer, 1979). Such vocal exchanges can last for several minutes, until one or other of the parties gives way or gives up. The impression is that the parents and chicks are engaged in a contest of wills.

Much of the display behavior occurring between adult gulls is similarly a protracted affair in which each bird seems to be trying to coerce the other(s) into retreating, approaching, or engaging in some more specific form of social activity. The behavior looks as though it were aimed at manipulation in the original literal sense of the term. In light of this, the recalcitrance shown in interactions between older chicks and their parents is conceivably involved in a chick's acquiring competence in social skills required for success in communication as an adult, such as motivational self-control and "mind reading" (Krebs and Dawkins, 1984). I am uncertain about how to proceed further in studying this behavior. Perhaps experimentally manipulating the contingencies experienced by chicks during the "tug-of-war" stage would provide a way of getting at what they are trying to do, and how the experience might bear on later social competence. For the present I am persuaded that currently available ethological ideas about motivation do little to illuminate the behavior as I perceive it.

In general I think that, with the rise of sociobiology, ideas about motivation and social development in animals have been left in a state that is primitive or misses the point of much that seems to be going on in social interaction. Colgan's (1989) book on motivation may be a sign of a more general renewal of ethological interest in motivational questions, but this book takes a predominantly reductionistic and operationistic stand, which virtually ignores issues of the kind that I have been concerned with here. In any case, to return to my starting point, the motivation theory of classical ethology is now generally regarded as defunct, and, apart from the work on imprinting, classical ethology did little to advance understanding of social development.

However, tucked away and little noticed in an early paper of Lorenz's (1937, trans. in Schiller, 1957), and further developed in Lorenz's *On Aggression* (1966), there is an idea that could turn out to be an embryonic anticipation of an animal motivation theory along intentionalistic lines. Lorenz said that "instinctive action is a craved reflex" (in Schiller, p. 172). As I have indicated above, a philosopher, Dretske (1988), has already attempted to work out a theory in which a notion of craving or desire, as well as representation and belief, can be included in explanation of animal behavior. The prospects for such a theory are problematic. Indeed, there are philosophers who question whether belief/desire psychology has a future even for the human case (P. M. Churchland, 1981; Stich, 1983; P. S. Churchland, 1986). Extending such psychology to animals faces formidable obstacles, including the problem I mentioned earlier about how the language of intentionality is designed for humans rather than animals. At times using that language for animal cases seems like using a map of New York to find your way about New Delhi. But if cognitive ethology is to justify itself as something more than anecdote and special pleading, I think it will have to try to develop a theory of animal motivation and cognition in a philosophically principled manner. This may require taking the human case as a starting point. If so, old-fashioned anthropomorphism could be in for a new lease on life.

Such a position is a far cry from the operationally grounded theory envisaged by the critics who attacked the ideas of Lorenz and Tinbergen back in the fifties. However, as I have tried to show, one can argue that at least some of those critics had a mistaken view of the nature of scientific theory. And while the idea that human intentionality might offer a model for animal intentionality is at present only a vague possibility, I find encouragement from William James, who said, in a comment about Herbert Spencer's notion of evolution: "At a certain stage in the development of every science a degree of vagueness is what best consists with fertility" (James 1890, p. 7). I think that ethology is still in such a stage with regard to questions of motivation and cognitive awareness.

PART VI

Consciousness and Self-consciousness

17

Self-awareness, with Specific References to Coleoid Cephalopods

Martin H. Moynihan

Many animals behave as if they "know" what they are doing and/or what they look like. Coleoid cephalopods, the squids, cuttlefishes and octopuses, provide examples of both kinds of apparent consciousness. In these examples, the terms used may appear to be anthropomorphic. The evidence itself may seem anecdotal. I think anthropomorphism and anecdotalism are inevitable in all descriptions of complex social behavior, in fact, if not in theory. They can no more be avoided in descriptions of exotic invertebrates than in observations of "higher" animals.

Anthropomorphism is inevitable because all the actions and reactions that we observe can be interpreted only in terms of our own experiences and conceptions. The 1989 Animal Behavior Society symposium from which this book developed was held after something called the "anthropic principle" had become a focus of a best-selling book (Hawking, 1988) and of articles in fashionable magazines (such as *The Economist*: see Anonymous, 1989). Hawking's account is brief and sharp: We see the universe the way we do because, if it were different, we would not be here to observe it. A behaviorist might use a comparable phrase: We see an animal's behavior the way we do because we can not do otherwise.

Anecdotalism is inevitable because it is practical. Some sequences of social patterns, courtship for instance, are so long and intricate that they are difficult for a human observer to follow closely or continuously. In most cases, sequences also are performed relatively rarely, occasionally, or in special circumstances. Quantification and statistics can be applied only to simpler and/or more repetitive phenomena. Rare and complex interactions have to be described one by one. Storytelling is a science—at least a practical technique—as well as an art. Every story or message has its own internal logic or "grammar"—naturally different for different species and cases. Perceptions of the same stories are not always identical or even consistent.

The social behavior of coleoid cephalopods is remarkable less for the general principles involved than for the methods employed. The signal systems of most of these animals are largely visual, certainly not acoustic and apparently less olfactory or tactile than the corresponding systems of many mammals and arthropods. Coleoids use special postures and movements to encode and convey information. They can emit inks as decoys. More important, they use color changes in almost endless ways. Many species, certainly all the forms of well-lighted waters, have pigment cells in the superficial levels of their thin and fragile skins; chromatophores with one or more kinds of melanins (red, brown, black), iridocytes (silver), and/or leucophores (white). Each of these pigment cells is an elastic sack. Each sack has its own innervation directly from the central nervous system and can be contracted or expanded independently. The pertinent stimuli must come from, be perceived and classified by, elements of the well-developed visual system. (The functional convergence of cephalopod and vertebrate eyes is a classic reference in biology textbooks.) The transitions between full expansion and extreme contraction of the elastic sacks can be almost instantaneous. No other system of color change of other animals is as rapid. Nor as complicated. Given the innervation, there are opportunities for many combinations. Every patch of skin may change colors independently of every other patch. While I would not be prepared to swear on oath that coleoids can produce plaids, still, they can manage approximations. Some of the relevant literature is cited in Hardy (1956), Packard and Hochberg (1977), Hanlon (1982), Packard (1982), Andrews, Messenger, and Tansey (1983), and Moynihan (1985). The species that has been studied most thoroughly or at greatest length in the field is the Caribbean Reef Squid, *Sepioteuthis sepioidea*. But many other species probably pursue similar strategies and use similar tactics (e.g., *Euprymna scolopes* and *Idiosepius pygmaeus*).

All the visual patterns encode information. The effective functions of the social patterns of *Sepioteuthis* are both various and often variable. Sometimes they are overlooked or ignored. When noticed, they can induce different responses. They can be either intraspecific and/or interspecific. They can be "bivalent." Many patterns are shown at least as frequently to potential predators or other species as to colleagues, rivals, and competitors of the same species. This bivalence can be misleading. In fact, it is adapted to be misleading in some but not all circumstances and to some but not all observers.

Particular patterns may occur either exclusively in interspecific situations or exclusively during intraspecific encounters. There are many examples. But there are also examples of patterns that occur in both sorts of encounters. In species such as *Sepioteuthis sepioidea*, living in and around crowded tropical reefs, there usually are both conspecifics and individuals of other species in the vicinity of a performer. Both classes of neighbors can perceive the same performances. They do not always draw the same conclusions. For a potential predator, performances may be cryptic (overlooked or misleading), baffling (unpredictable), or even frightening. On the other hand, individuals of the same species as the performer probably can interpret such performances usefully. At the very least, they should be able to calculate probabilities of advance or retreat with some reasonable degree of confidence.

Some patterns may tell partial lies. A Caribbean Reef Squid may mimic vegetation (see below), black sponges, or corals. It may, therefore, hope to pass unperceived by predators; it may also truthfully suggest the presence of plants, sponges, or corals in the neighborhood.

There are other dimensions or kinds of truth and honesty. A hopefully cryptic animal may or may not pass unperceived; but when it is actually noticed, as by conspecifics, then the information proffered should be of such a nature that it can be interpreted correctly. Even mimic patterns seem to accurately reflect motivation. Thus alert, experienced, and sophisticated observers should trust their own eyes. Only the naive may be misled. And then, again, perhaps only partially.

Coleoid cephalopods do not bluff. A favorite subject of behavioral discussions a few years ago was "game theory" (see, for instance, Maynard Smith, 1974, 1979, 1982; Maynard Smith & Parker, 1976). Much attention was paid to competition and asymmetrical contests. It was suggested that many individuals might benefit, in many contexts, by misrepresenting their intentions. The idea is dubious. This sort of dishonesty has never yet been shown to be widespread (comments in Moynihan, 1982). As far as one can tell, it is almost or completely absent in coleoid cephalopods. These animals may conceal their identities. They do not conceal their intentions.

Common or "ordinary" kinds of mimicry may appear to be easy to explain at first. The easiness disappears on second or third consideration. In theory, turning black beside a black object or turning white against a white background might be supposed to be a direct, almost unedited one-to-one response. But other reactions are not so straightforward. Consider a mixed background that is white in one area and black in another area. A coleoid may adapt by going black on one part of its body and white on another part. The lines between black and white are distinct—no wishy-washy grays. The arrangement is never random. It is precisely the parts of the body that are near white that go white. The same for black. The parts may be left, right, front, rear, top, and/or bottom. All combinations are possible. One part may be black during one performance, white during the next, depending upon circumstances. More complex backgrounds may encourage more complex combinations of color patterns.

What must be happening, in fact, is that the sensory input is being transformed or redistributed in the central nervous system. There must be a switchboard linked to something like a cognitive map (Tolman, 1948; Thorpe, 1974b; Olton, 1979; Griffin, 1981). Incoming data must be sorted before responses can be sent out. A cognitive map is not all that is required. Of course, any performer should know the ins and outs of its surrounding topography. It should also know its interlocutors, the other individuals and organisms with which it may have to interact. It can play with or against them. Interlocutors come and go. They change colors and shapes. What a performer really needs is a cognitive "color snapshot" or "cinefilm" of its surroundings.

Doubtless such aids are available. In the case of coleoid cephalopods, there may be a paradox. The animals are supposed to be color-blind. At least, no one has been able to prove that the Common European Octopus, *Octopus vulgaris*, can distinguish among hues in the laboratory. What should one conclude? Perhaps not very much. The laboratory situation is restrictive and exceptional. It took a long time to demonstrate that cats could discriminate hues when called upon. It is also true that *O. vulgaris* is somewhat exceptional. Some other coleoids, including *Sepioteuthis*, show changes of hues during intraspecific encounters, especially courtship. Certainly, many natural observers, for example most teleost fishes, have excellent color vision.

There has been much experimentation upon the central nervous system of some coleoid cephalopods (references in Young, 1977; Wells, 1978). Their brains may be better known than those of any other animal, with the possible exception of humans. Unfortunately, less attention has been paid to the central controls of communication and color changes, as distinguished from the innervation of single pigment cells, than to the corresponding controls of other activities such as hunting and prey capture. On general grounds, one may suppose that the particular neural tracts known as the optic lobes and the higher integrative centers such as the vertical lobe are crucial. But, to my knowledge, the details remain to be worked out.

Let me give some examples of particularly complex behavioral sequences. They are taken from the Caribbean Reef Squid as described in Moynihan and Rodaniche (1982) and Moynihan (1985) with little change.

Courting parties of the species usually resolve themselves first into trios of two males and one female and then into real bisexual pairs. The dominant or favored male usually assumes a distinctive color pattern. He remains "normally" dark on the side of his body facing (visible to) the female while turning brilliant glittering silver on the side facing away from her. The silver is repellent, probably a threat to discourage approaches by rivals. Courting individuals may change relative positions rapidly and repeatedly. When and if so, the displaying male shifts his silver from side to side, always with appropriate orientation. In my experience, there are no mistakes.

Individuals of the species may perform "protean" display before predators. These are sequences of different-appearing patterns, various color arrangements, upward and downward pointing and curling movements of the arms, etc., all performed in rapid alternations. Each of the patterns involved can be performed by itself or in various combinations. Alone or as a distinct entity, it has its own special significance. The individual significances are not, however, easily detectable in the protean sequences. Here it is the very alternations, the visual contrasts, that seem to be crucial. The effect is kaleidoscopic and baffling. Most outsiders could hardly read the sequences at all.

A singular incident can be quoted directly from Moynihan (1985):

> Two young *Sepioteuthis sepioidea* were discovered in very shallow water over white sand at midday. They were in more or less basic coloration when first seen. Since they could not easily escape from the observer, they first shot out blobs of ink and then turned Dark and floated with twisted arms, in obvious imitation of the blobs which were slowly disintegrating. The animals were, in fact, mimicking their own decoys. (pp. 104–105)

Still another anti-predator reaction is pertinent. When surprised and frightened in more variegated environments than white sand, Caribbean Reef Squid, and especially young individuals, may assume color patterns of transverse Bars or longitudinal Streaks. The two patterns are hostile, produced when a tendency to escape is stronger than a tendency to advance or attack. Both can be cryptic against certain backgrounds, such as, turtle grass, *Thalassia*, or drifting *Sargassum* weed. When actually seen and noticed, however, they can be disruptive or disconcerting. Like many other patterns of coleoids, they serve different functions in different individuals or in the same individual at different times. When a group of Caribbean Reef Squids is mildly disturbed by a potential predator, most members of the group adopt the same patterns, either Bars or Streaks as the case may be. But there is often an "odd man out," an individual that goes Bar while its companions are in Streak or, conversely, goes Streak while its companions are in Bar. The odd man out is often the last individual to display. It seems, to a human observer, to be *trying* to look different from its associates. Perhaps looking different is protective. The difference might confuse or circumvent the visual search image of potential predators. It could be another way of passing unrecognized without necessarily being cryptic.

An interesting aspect of such behavior is that the appropriate color changes are complete. Performing individuals may not, in fact, be able to see their rear ends. Still, they are able to ensure that everything conforms to the right image. The sorting and redistribution systems of the central nervous system do seem to work. Only the transcription must be difficult.

These examples are picturesque. They are not widely known among students of the behavior of other animals. They deserve a better press. Cephalopods can provide useful material for comparison with vertebrates and arthropods. They should be cited more frequently than has been the case in recent years.

The flexibility and opportunism of coleoid behavior is suggestive. These animals have considerable abilities to solve different social problems and to select different solutions to similar problems in different circumstances. Choices are overt and decisive. I cannot believe that they are not deliberate and, in some sense, conscious. Alternative explanations would seem to be even more complicated, with even more loops and feedbacks, a system of Ptolemaic epicycles.

The frequency with which coleoids have to make choices among complicated inputs and outputs may help to explain an(other) interesting aspect of their biology. They are supposed to have evolved relatively larger as well as more complex brains than teleost fishes, their principal competitors (Packard, 1972). It could be, perhaps in part, because they have to make difficult or delicate decisions more frequently than their rivals.

The general point of much of the preceding argument might be phrased concisely as another extension or parody of the anthropic principle. Perhaps "awareness should be assumed to exist whenever its existence is the simplest possible assumption."

The same point can be made less flippantly. All individuals, at least among vertebrates, arthropods and coleoid cephalopods, distinguish between themselves and "others," other individuals, other sexes, other species. On logical grounds, it is difficult to understand how one could distinguish others without having an idea or impression, conscious or not, of what one is oneself. Biologists, sociologists, historians suggest different emphases. The differences are more apparent than real. Biologists are concerned with how one distinguishes others from oneself. Social scientists phrase the problem in reverse: how does one distinguish oneself from the others? A student of human behavior asks: "How can one coexist with another without questioning or considering one's own existence?" (Etienne, 1989, p. 183). A historian says "For people, as for societies, an inspection of the 'other' is a prerequisite of a recognition of oneself" (Peyrefitte, 1989, p. 466). (My translations.) The alternative approaches may be tweedledee and tweedledum, *blanc bonnet* or *bonnet blanc*. Yet a conclusion is obvious and plausible. There is some sort of personal identity.

What this might be among different animals is the next problem to be solved.

18

Silent Partners? Observations on Some Systematic Relations among Observer Perspective, Theory, and Behavior

Duane Quiatt

PRIMATE FIELD STUDIES AND OBSERVATION METHODS

Shirley Strum, in *Almost Human* (1987), says that when she set out in 1972 to test the so-called "baboon model" of early hominid life she had two goals:

> The first was to apply new observation techniques . . . more rigorous and systematic than . . . methods [previously employed. The second was to] identify all the individuals in the group, not just the adult males, as the early baboon watchers had done. By identifying all the individuals, watching a representative sample of them, giving females equal time with males and using the new observation techniques, I hoped to eliminate any selective biases that might unconsciously have been incorporated into the previous picture of baboon society. (p. 13)

This sounds like a reasonable program, but in *Almost Human* Strum never clarifies exactly what those "new observation techniques" may have involved besides leaving the Land Rover to follow her Pumphouse Gang of baboons on foot—evidently a big decision at Gilgil in 1972 but no great shakes for field primatologists elsewhere, surely, including baboon watchers, at that late date.

Cynthia Moss, a student of elephant behavior who worked out of Kekopey when the Pumphouse Gang was in residence there, discusses primate field studies in her book, *Portraits in the Wild: Behavior Studies of East African Mammals* (1975), in a chapter on baboon behavior. Moss notes that "the main cause for conflicting descriptions" of baboon social organization—that is, between older accounts focused mainly on the agonistic interactions and sexual escapades of adult males and later descriptions neither so male centered nor so heavily preoccupied with sex—"seems to lie in observer methods":

> Today [that is, in the early to mid-1970s], far more sophisticated and systematic methods for collecting data on behavior are being used. Researchers have worked out data sheets in which interactions can be recorded systematically and later quantified by computer analysis. In "focal sampling," the observer concentrates on the behavior of one individual for, say, an hour, recording onto special sheets everything that the animal does. After several hundred hours of focal sampling on the various individuals in a troop, one has substantial results with which to work. One can note, for instance, that out of thirty observed hours Male A spent ten hours within 3 feet of Female B and one hour with Female C. It is these quantitative results that are revealing the most telling new information on baboon behavior. (p. 201)

Strum, says Moss,

> learned to recognize all . . . individuals in the sixty-six-member troop. She too used systematic sampling methods . . . logg[ing] 1250 hours of observations [in her 16-month study], including more than 700 hours of focal animal sampling. From the material she collected on all kinds of interactions from avoidance and indifference to fighting, grooming, mating, and playing, a picture began to emerge: certain baboons spent more time with certain other baboons. (p. 204)

So, in Moss's account, a picture begins to emerge of the "new methods." They are quantitative, based on theory-grounded operational definitions; they involve scrupulous attention to sampling procedures, with an emphasis on focal animal sampling; they include a concern for acts of omission ("avoidance and indifference to fighting") as well as commission; and they afford comparisons of preferential relationships in terms of temporal duration and spatial association ("certain baboons spen[d] more time with certain other baboons").

Just how new such methods were, even in these relatively early years of primatological field research, is open to debate. Focal animal studies appear to have been pioneered in the early 1960s by researchers studying macaques in provisioned groups (see, e.g., Fisler, 1967). The power of focal animal sampling as applied to analysis of social behavior depends, first, on identification of individuals with whom a focal animal interacts and, second, on recognition of genealogical relationships that obtain within the group or groups of concern. Be that as may, more important than the introduction of new methods and techniques of observation has been a change in perspective. The adoption of new perspectives in the study of animal and especially nonhuman primate behavior has changed the relationship between theory and method, and it is that change with which I will be concerned in what follows. First, however, let us take a closer look at those new methods of observation referred to by Strum (1987) and also by Smuts (1985).

Focus on the Individual and on Individual Relationships

Irven DeVore (1985), a pioneer in the study of primates in nature, has described what happened as the early, relatively short-term, descriptive studies of primate behavior, in the 1950s and early 1960s, gave way to long-term projects:

> [D]etailed observations by successive teams of scientists made it possible to document the behavior of each individual baboon through most of its lifetime. Correlational analyses of large sets of quantitative data were used to test general hypotheses about social organization and life history variables. (p. xii)

There followed, in the 1970s, a period in which field studies of baboon social behavior were structured by the revolution then under way in evolutionary vertebrate ecology: "newly refined hypotheses were formulated and tested against this new paradigm. Research emphases shifted to questions about the 'strategies' in baboon life" (p. xii)— strategies relating to inclusive fitness, optimal foraging, resource competition, and life history decisions. Throughout this time, as DeVore says:

> Coding and sampling methods were developed to minimize observer bias, . . . to "objectify" the animals and their behavior in order to purge primate studies of the easy anthropomorphism of the turn of the century. . . . By wedding such precise methods to the powerful predictive theories of the new behavioral ecology, scientists believed that the major intellectual questions in baboon studies would soon be answered; what remained was the careful documentation of such principles as kin selection, resource competition, and game theory analyses of aggression. (pp. xii–xiii)

In retrospect, according to DeVore—and this is the foreword writer speaking, partly (these passages are from an introduction to Smuts [1985] *Sex and Friendship in Baboons*):

it is clear that at every stage of research we human observers have underestimated the baboon. . . . Baboon behavior could not be neatly accommodated by [traditional] research methods and theoretical expectations. As this volume so elegantly illustrates, new methods were needed that allowed observers to focus on those behaviors that were most meaningful to the animals themselves, and new theories had to be developed to explain the often surprising findings that emerged. (p. xiii)

Smuts (1985) herself writes that her study "involved two major problems: one related to theory and the other to observation" (p. 7). The theoretical "problem" boiled down to an assumption that friendship is encouraged by natural selection. That assumption, she writes, "allows one to take full advantage of a series of powerful theories for explaining behavior developed by evolutionary biologists over the last 100 years," theories which can be translated in the field into a series of "much more immediate, often compelling questions about phenomena *of immediate concern to the baboon[s] themselves*" (Smuts, 1985, pp. 7–8, my emphasis, with correction). Here are the questions to which Smuts (p. 8) refers:

How does a friendship form? What makes it survive or, instead, falter in the early stages? Why does [one] male have half a dozen female friends, [another] none? . . . What exactly is friendship to a baboon? (p. 8)

The last question, Smuts notes, is more basic than the first three, and I want to consider the problems which that question must raise for the observer. We can begin by asking what baboon friendship is to Smuts. Operationally, not a lot; she defines friendship in two dimensions, evidenced by grooming and proximity. The challenge, as she (p. 8) puts it, was to reformulate, "in a way that would permit both scientific scrutiny and objective communication to others without completely sacrificing the immediacy and vividness of the new observations," her conviction that interactions between friends differed in frequency and in kind from those between other dyads "and that friends did indeed constitute a distinct category of relationships" (p. 8). So here we have the second of her concerns, the one relating to observation and method. "Meeting this methodological challenge," she concludes, "was a second major goal of the study."

Moving into the World of the Baboon

Smuts's methodology, very like Strum's, is one with which field primatologists today are thoroughly familiar. It begins with identification of individual animals and involves a combination of focal animal, ad lib,

and scan sampling. Spatial associations are recorded along with interactions, and, ideally, close attention is paid to variations in the intensity
and duration of bouts and interaction sequences, in order to characterize and assess small differences in the character of behavior and of relationships. Therein lies perhaps our best hope of getting at the essence of
a relationship such as friendship, for "friendship is not just proximity
but a collection of relationship attributes, seeking the other out and
protecting the interests of the other. Proximity may help you locate
friends, but 'friendship' is irreducible" (R. W. Mitchell, pers. comm.,
28 May 1992).

Smuts (1985) says that in deciding what behaviors to record she
adopted "the attitude of an ethnographer confronted with a previously
undescribed society. "I made a determined effort to forget everything I
knew about how baboons are supposed to behave. Instead, I tried to let
the baboons themselves 'tell' me what was important," (p. 30), what
behaviors—to refer back to the passage from DeVore's (1985) foreword—"were most meaningful to the animals themselves" (p. xiii).

Only a spoilsport would suggest that she is unlikely to recognize
what is important in baboon behavior until she puts it in perspective
with all that she has learned. And, anyway, isn't this what we all want,
at some level?—to throw off the blinders of convention and habit and
peer directly into the heart of behavior? Stuart Altmann seems to have
aspired toward some such immediacy as early as the 1950s, watching
rhesus monkeys on Cayo Santiago, for he wrote: "I left it to the monkeys
to tell me what are the basic units of social behavior" (1962, p. 373).
Louis S. B. Leakey must have had it in mind when, arranging for Jane
Goodall to work at Gombe, he insisted that he wanted someone unham-
pered by academic prejudices. The problem for the observer, of course,
is how to proceed, how to get on with the job in the absence of defining
expectations, and then how to manage that direct contact not just across
cultures but across species.

Smuts (1985) refers to the close observation that makes it possible
to record social interaction "in fine detail," which she says enabled her
to see her subjects "from a baboon's perspective" (p. 27). Strum (1987,
pp. 27, 35–37) speaks repeatedly of this translocated perspective—the
subtitle of her book, "a journey into the world of the baboon," is to be
read with that translocation in mind. Once out of the Land Rover, she
moves "gradually . . . closer, and finally, on foot, into the troop itself,"
where, she says, "pink-nosed and squinting, I became the intrepid
baboon watcher, going wherever the troop went . . . observing, taking
notes, thinking." What she thinks about, to begin with, is her determination to preserve her role "as a nonentity . . . tolerated but unobtru-

sive." But preserving her role as a nonentity proves difficult. One day she is approached by an immigrant male, Ray, who, positioning himself between Strum and two resident males, solicits her support in agonistic interaction with them. Strum refuses to cooperate. "Ray won his struggle alone," she writes, "but I shall never forget how honored I felt by the compliment he paid me."

This incident, it turns out, marks Strum's arrival at journey's end. She has relocated herself cognitively in the world of the baboon. "There was nothing different about the day," she reflects. "*I* was different" (p. 37). Just how different becomes evident on an anticlimactic, stock-taking day. She has taken to tasting the foods that the baboons eat (she jokes about it being a convention of primatological field work), and now, "Munching, writing, crunching, gazing, I realized suddenly that I was looking at the Kekopey landscape in an entirely different way—as a baboon, eyeing what was next on the daily menu" (p. 56).

Changes in Observer Perspective and in Underlying Theory

What are we to make of such a shift in perspective? There could hardly be a better description of the cognitive leap that fieldworkers commonly manage. I am choosing my words carefully—it is important that it *be* managed. But is it no more than a leap of the anthropomorphic imagination? Is it wholly metaphorical? As it clearly is in part—indeed, a very literary and romantic way of describing one's professional work. How instructive might it be to give this metaphor a closer if not more literal reading? More to the point, how is the cognitive shift which brings Strum into the world of the baboon facilitated by those "new" and highly touted methods of observation and by the theoretical models with which they are connected?

Three Observer Perspectives

I have already noted that what seems to me new in such accounts as Strum's is not so much the methods described as the perspective adopted, which I think cannot be considered apart from the theory or theories of behavior to which it is related. In asking how observer perspective may relate to theories of animal behavior, and to the behavior of animals observed, it will be useful to consider three different vantage points, which I will call:

1. outside the arena of behavior, looking in;
2. inside the arena of behavior, looking out (and around); and
3. translocative and coincident with that of an animal observed.

I don't like these terms much, but I hope they get across the notion of reduced psychological distance between observer and observed from one vantage point to the next. Distances, of course, cannot be measured in feet and inches; it is cognitive perspective that I have in mind, with no suggestion, in the case of the last, that an observer's visual faculty can be rooted out of its developmental housing and relocated elsewhere.

These perspectives can, however, be described in physical and, to begin with, in reasonably literal terms. From the first standpoint the observer can be said to look in and perhaps slightly down on his subjects, as an extraterrestrial agent, God on His creation, the experimental primatologist on his created habitat. From the second standpoint, the observer can be said to look out from a reserved or isolated position inside the arena of behavior, not necessarily from a constructed blind or a Land Rover but, in any event, from an unexamined state of cultural confinement—and in that sense still at a point of some remove from the subject animals' social environment.

I should emphasize that the confinement just noted is not my invention, it is implicit in Smuts's and especially Strum's discussions of "breaking away" from confining habits. Metaphorical, it nevertheless calls to mind the literal cage in the forest of that turn-of-the-century optimist, R. A. Garner, within which he waited for chimpanzees to amble by, conversing with one another (Garner, 1900). The cage that I have in mind is not dissimilar to Garner's. It is that of our own cognitive limitations, which we hardly know how to take into account but cannot ignore when considering observer perspective.

The third vantage point of course is fictive; adopting this perspective the observer proceeds "as if" viewing the world through the eyes of a focal animal subject. Methodological invention is employed to break down the barriers of cultural confinement. Ideally, the "new methods" to which both Strum and Smuts refer would move the observer out into the world (or, much the same, into the mind) of the baboon.

Fitting Observer Perspective to Research Strategy

I will argue that changes in observer perspective, in theoretical perspective, and in methods of observing and recording behavior are systematically linked. Different theoretical assumptions underlie different observer perspectives. The connections are historical, but with levels of meaning which cannot be accounted for historically, that is, they do not arise simply one from the other. To get an idea of what I mean, consider four different research strategies in relation to appropriate observer perspectives, as in Table 18-1.

TABLE 18-1
Research strategies, applications, perspectives

Research Strategy	Knowledge Application	Observer Perspective
1. Nonhuman primates as stand-ins for human beings	Medicine and industry	Outside looking in
2. Modeling behavior, i.e., constructing/testing conceptual models of behavioral process	Pure research in, e.g., physiology or psychology	Outside looking in
3. Making evolutionary comparisons across species and across groups within species	Maximizing the relevance of conceptual models to understanding human behavior	Inside looking out
4. Making socioecological comparisons of behavior within and across groups	Testing alternative models of competition for local resources	Translocative and moving

The first three strategies were reviewed by William Mason some years ago, in an examination of the scope and potential of primate research (Mason, 1968). Strategies 1 and 2, whether they involve treating primates as whole- or part-system models of human beings (as in Strategy 1) or employing them in the construction of models of general biological process (as in strategy 2), entail methods that are manipulative and experimental, and an observer standpoint that is "objective," positioned at a psychological and, usually, a physical remove from subjects and their behavior.

Scientists who adopt the second strategy, to investigate, say, "concept formation, or the mechanisms of hunger, or the influence of hormones on sexual performance" (Mason, 1968, p. 104) may or may not have as a goal applying their findings to understanding human behavior. According to Mason, we probably can assume that researchers studying nonhuman primate behavior are interested in human behavior as well, first because the economics of research forbids the use of monkeys and apes unless the problem at hand requires it, second because problems which require monkeys or apes as subjects usually "are . . . formulated in a way that betrays a fundamental interest in man" to begin with (p. 105).

Whatever animals are used in modeling behavior, the conceptual models strategy "must be fitted within a broader approach . . . [one that] is evolutionary and comparative" (1968, p. 106). This evolution-ary-comparative approach (Strategy 3 in Table 18-1) Mason identified as a third and final research strategy, which in his view was of particular importance to studies in which monkeys and apes were "used to inves-tigate such complex and peculiarly human issues as the nature of intel-ligence, the sources of neurotic behavior, and the origins of love. How far are we justified," he asked, "in applying conceptual models devel-oped on the nonhuman primate to the behavior of man?" (1968, p. 105). Only comparisons of species-characteristic behavior, observed in mon-keys and apes in their natural habitats (across a variety of such habitats) could inform us.

Such comparisons were central to field studies of primates in the 1950s and 1960s. Central too was the assumption, typological and cir-cular, that in naturally formed groups the behavior repertoires of indi-viduals assumed to be representative of their age and sex could be described as "species-characteristic." Behavioral variability of individ-uals was not exactly ignored, but it tended to be minimized methodi-cally in analyses which, while attending to differences between age/sex classes, pooled individual contributions within class and thus, in effect, treated them as identical. The observer standpoint appropriate to this research, conducted in nature, in the habitat of a subject species, was that of an objective scientist still removed psychologically and often physically from the behavioral situations in which his subjects involved themselves—not looking in from the outside but looking out from inside, from a vantage point physically or psychologically insulated, by intent and design a neutral observer on neutral ground.

In the 1970s, the theoretical bases of animal and perhaps espe-cially nonhuman primate behavior studies underwent reconstruction. It began in fact a bit earlier, with William Hamilton's elegant account (1964) of the circumstances under which the concept of individual reproductive gain could be extended to include collateral as well as direct descendants, followed by George Williams's clarification (1966) of Darwinian theory in terms strictly of individual selection—both at least partly in response to V. C. Wynne-Edwards's provocative argument for group selection (1962). A main virtue of Wynne-Edwards's argu-ment was that it provided description in detail of specific structures presumed to underlie group selection, thus lending itself to constructive debate, and it produced a groundswell of response from evolutionary biologists.

The individual versus group selection debate continued to pre-occupy biologists through the middle and late 1960s, triggering pro-duction of a number of individual selectionist subtheories—by, for example, Robert Trivers (1971, 1972), John Maynard Smith (1965, 1974), George A. Parker (Maynard Smith & Parker, 1971), and of course Hamilton himself (1971)—which together provided a foundation for the development of a behavioral ecology of the individual. This gave rise to a fourth broad research strategy (Table 18-1), one that is evolutionary and comparative, to be sure, like Mason's third strategy, but not at all in the sense of those terms as previously construed. Basic to it is deployment of the "powerful predictive theories of . . . behavioral ecology" (DeVore, 1985, p. xii) against a rich base of information concerning individual differences in social interaction. And that deployment, I suggest, both encouraged and was stimulated by a shift of observer perspective to the third vantage point I have described, a standpoint parallel with that of individual subjects assumed to be exercising one or another alternative strategy of behavior.

On Not Mistaking Changes in Perspective for Advances in Method

Field studies of primate behavior today focus on differences in the behavior of individuals, within and across groups, not just to understand a particular group's adjustment to its habitat—and so to explain differences between groups—but to help fit general models of individual selection to solution by individuals of the diverse problems which they are likely to encounter in the course of their particular lives. Such description necessarily involves specifying how alternative solutions may be dictated (or their ranges of efficacy delimited) by environmental constraints in conjunction with learned, habitual behavior.

Where social behavior is concerned, the focus of field studies since the mid-1970s has been on the interacting individual per se, as an entity whose fitness, and whose customary patterns of behavior on which fitness is contingent, may be subject to dramatic change as a consequence of a few bouts or isolated sequences of behavior. The more we take learning into account, the more likely we are to focus on individuals as strategists and tacticians (leaving out of account, for the moment, questions of motivation and intent).

Because of this connection between developing theory and method, certain habits of data collection to which I think most field primatologists are long accustomed—close observation, identification of individuals, precise definition of operational meaning and observational units, and consistent attention to the temporal-spatial aspects of social interaction—have taken on new kinds of significance, so that it may be tempting to think of them as "new" methods. Individual identification, for instance, is no longer simply a practical means of organizing observations so that pooling of data within age-sex class can be a matter of choice rather than necessity. Focal animal sampling is not viewed as but one way, among many, of breaking down the universe of interactions observed, to ensure against loss of material in narrowly designated spatiotemporal continuums. In combination with sequential analysis, focal animal sampling provides us with a tool that may be indispensable to research into social action, that is, into behavior as the physical execution of intentional strategies by a reflexively conscious animal.

I put it thus tentatively—"may be indispensable"—because it is not clear as yet whether our understanding of primate social behavior is well described in terms so strongly mentalistic and intentional. It is important that we distinguish between intention in that sense of a reflexive consciousness at work as opposed to intention in the much less problematic sense of purposeful, goal-oriented activity. The distinction is important, given our current understanding of behavior, not so that we may recognize things for what they are, but to help us settle on research goals that are achievable and appropriate to questions arising. As yet we know relatively little about nonhuman primate cognition, not to mention our own. Certainly we don't know enough to characterize with anything like precision differences in the nature of intention between, say, chimpanzees and baboons, despite the fact that so far only for the former do we have evidence indicative of self-recognition and self-reflexive behavior (Byrne & Whiten, 1985; Whiten and Byrne, 1988). It is probably sufficient at this point that we appear to be on the way toward integrating selfish gene explanations of cause with models of behavior more appropriate to the individual level of analysis, that whole-organism level at which more than lip service can be paid to proximate constraints, external and social as well as internal and physiological.

However, we should not mistake a shift in observer perspective for methodological advance. Primate field studies today—and I have in mind work like Strum's (1987) and Smuts's (1985) rather different studies of baboon friendship, Robin Dunbar's careful analysis of gelada reproductive decisions (1984), Frans de Waal's studies of chimpanzee politics and primate peace making (1982, 1989), Andrew Whiten and Richard Byrne's compilation of accounts of "tactical deception" (1988), and Dorothy Cheney's and Robert Seyfarth's analyses of vervet monkey vocalizations (1990)—have been most instructive concerning the implementation of behavioral strategies, deception, role playing, social manipulation, and information processing. Some of these primatologists, Strum and Smuts most obviously, have recommended anthropomorphic interpretations of primate behavior on theoretical grounds, sounding rather like advocates of a superior method. Adopting what I have called a translocative perspective, by which (again) I do not mean one that is more than figuratively detached from its visual anatomical base, they do their best to view the landscape, the local habitat with its resources—including, most particularly, social resources in the form of conspecific baboons, vervet monkeys, or chimpanzees—through the eyes and from the reflective standpoint of particular actors, their focal animal subjects.

Justifying the Translocative Perspective

For 20 years and more students of primate behavior, field researchers especially, have been conditioned toward this detachment and displacement of "normal" observer perspective by our method of choice, that is, focal animal sampling, and by theory—by all those ideas that are central to our understanding of behavioral ecology as the study of individuals engaged in competition for and investment in local resources. We are by now habituated to weighing transactions in social behavior from the standpoint of each individual involved—as, to take for example a paradigmatic behavior, infant handling by unrelated "aunts," in which consideration can (and must) be given to effects on reproductive fitness of each participant—aunt, infant, and biological mother.

Of course, to repeat the obvious, we cannot really see the world through eyes other than our own. Most of us will acknowledge having committed abysmal misreadings of minds which, if any, we should have known. In the case of nonhuman primates, with whom in most cases certainly (I won't say all) we cannot even converse about misreadings, our assessments of cognitive intent seem to be less, as well as more, presumptuous than is the case when it is a family member, friend, or some other human being whom we presume to understand. Intentional readings of nonhuman primate behavior arguably are less presumptuous in the sense that a human being watching baboons, chimpanzees, or vervet monkeys ordinarily is not himself a prominent actor in the situations under study (not by intent, at least!). The observer has therefore a kind of rationale for the objectivist assumption that he can squeeze himself metaphorically into the mind of a focal animal, look out at the world through its eyes, and not cramp the "steersman" or gum up the works otherwise just by being there—as we so often do when we try officiously to align with our own the views we presume to be central to the behavior of those human beings with whom we wish to interact harmoniously. How it is more presumptuous for a human observer to try to shoehorn himself into the brain of a nonhuman primate is so obvious that scientific discourse on the translocation seems appropriate only to special symposia like the one for which this chapter originally was written.

I have asserted that primate ethologists today are conditioned by individual selectionist theory and, more specifically, by the several subtheories that concern competition for local resources, in which cost-benefit accounting, logical decision making, and rule-based execution processes are central. Thus conditioned, we find it useful to adopt an observer perspective that is detached and moving, one in which we more or less automatically (and of course metaphorically) seat ourselves next to whatever steersman it is that runs a focal animal's cognitive show. But theoretical conditioning is not theoretical justification. Is there, then, justification for this egregiously anthropomorphic shift?

The Argument for Self-conscious Deliberation in Other Animals

Justification can be looked for in the evolutionist view of the development of reflexive consciousness and human intellect, as opposed to the essentialist assumption that these are the outcome, only remotely connected with cognitive adaptations preceding, of some genetic dawn breaking on *Homo* if not *sapient* behavior. The evolutionist view is that it is primate behavior, and perhaps as Nicholas Humphrey has argued (1976), more specifically primate *social behavior*, with its peculiarly transactional manipulation of individuals and relationships—as opposed to tool behavior, with its unidirectional and straightforwardly instrumental manipulation of physical objects—to which we must look for an explanation of the development and the character of human intellect. Primate society in rudimentary form, beginning, Humphrey suggests, with early prosimians similar in behavior to the social lemurs of today (see also Jolly, 1966), "preceded the growth of primate intelligence, made it possible, and determined its nature" (Jolly, 1966, p. 506).

This is a doctrine of perhaps greater interest to anthropologists and primatologists than to those whose primary concerns lie in the behavior of wolves, bighorn sheep, reptiles, or fish. It has been of particular interest to field primatologists because (1) chimpanzee termiting aside, tool behavior is less likely to be observed in natural habitats than in enriched captive settings and provisioned contexts, and (2) social behavior in nature is, if not more "natural," thicker in texture at least than in all but a few carefully structured captive settings. The argument that a peculiar kind of intelligence is involved in social cognition goes hand in hand with the argument that naturalistic studies of the social behavior of free-ranging primates afford a peculiar insight into the nature of that intelligence.

The Need for (and Problems with) Empirical Demonstration

Both arguments may strike some as dubious; nevertheless, Humphrey's treatment of "creative intellect" has provided the basis for a narrow, justified anthropomorphic attribution to monkeys and apes of something like our own reflexive self-consciousness. And it makes sense, if we deem it fruitful to attribute self-consciousness to other animals, to look first at those species that are closest to us phylogenetically. Theoretical justification of course is not the same as empirical demonstration, and empirical demonstration may be desirable not just for its own sake, as *proof*, but to provide the foundation for a methodology by which we can examine similarities and differences in "social cognition" within and across species instead of just insisting on phylogenetic continuity. Humphrey himself, it should be noted, though he made much of their "creative intellect," did not grant monkeys and apes a "reflexive consciousness," and neither perhaps do most of the primatologists whom I have been citing except by implication in their descriptions of social manipulation and tactical deception.

Mirror recognition of oneself as an individual provides an objective measure of self-consiousness (Gallup, 1977b), though there remain arguments as to exactly what is being measured (Mitchell, 1993a). Whatever it is, chimpanzees and orangutans appear to have it, but monkeys and, surprisingly, gorillas in most instances have not treated their mirrored image in ways that would suggest that they recognize it as self-reflective. The gorilla Koko apparently does not share this deficiency, if deficiency it is. Francine Patterson has maintained for years that self-reflection is an important element of Koko's consciousness. Patterson's claims (see especially Patterson, 1986) do not seem to have carried much weight with students of primate behavior (although this may be changing—see comments on Patterson & Cohn, 1994, in Parker, Mitchell & Boccia, 1994), and I am not sure why. We allow ourselves, perhaps, to think of Koko as a special case, or of Koko and Patterson as a special case—friends, primarily, rather than ape and human scientist, so that whatever passes between them can be treated as friendly transactions and not as matter for science. But Patterson's work deserves careful reading.

Jane Goodall has inured us to anthropomorphic description of chimpanzee life (see especially Goodall, 1967, 1971, 1986) and to depictions of friendship transcending species boundaries. We are not much troubled, I think, by Goodall's references to friendship with, of course, "the wild chimpanzees" en masse, but also with particular individuals among them. The late Diane Fossey taught us to look at gorillas in a similar light (Fossey, 1983), albeit with an uncomfortable feeling that Fossey may have gone a bit too far in preferring gorillas to human beings; and now Smuts, Strum, and others encourage us to look at baboons and, by implication, Old World monkeys in general as potential friends. In the context of recent speculation concerning the evolution of human intellect, speculation that attributes to primates in general social intelligence of a special kind—if not "almost human" then certainly closer to human than is that of other animals (cf., e.g., Cheney, Seyfarth, & Smuts, 1986; Essock-Vitale & Seyfarth, 1987; Quiatt & Reynolds, 1993)—claims to friendship with the subjects of one's primatological research take on an awkward edge of meaning. But such claims should not be dismissed as one-sided expressions of sentimental longing, for they reflect important changes in the relationships between at least some field workers studying the behavior of Old World monkeys and apes and the subjects of their study. If those changes have so far been described primarily in autobiographical how-I-did-it accounts focused on experiential learning by the observer in relation to animals observed, the other way round only incidentally, the accounts nevertheless convey well the transactional character of the research on which they are based.

The Transactional Character of Field Studies in Primate Behavior

Humphrey's (1976) essay on the social function of intellect has influenced more than any other theoretical work current speculation concerning social cognition, deception, and what Byrne and Whiten (1988b) have termed "Machiavellian intelligence." Humphrey suggested that the social gamesman or games animal must, like a chess player, "be capable of a special sort of forward planning." For, as in chess,

> a social interaction is typically a *trans*action between social partners. One animal may . . . wish by his own behavior to change the behavior of another; but since the second animal is himself reactive and intelligent [not necessarily self-conscious, note] the interaction soon becomes a two-way argument where each "player" must be ready to change his tactics—and maybe his goals—as the game proceeds. (Humphrey, 1976, p. 309)

Those influenced by Humphrey have sometimes overemphasized the manipulative aspects of such engagements, and have introduced in varying degree notions of devious intent that are virtually absent from Humphrey's statement. Meanwhile, they have largely ignored the implications for research conduct of what Humphrey had to say about our own transactional thinking. Humphrey (1976) argued that the self-ishness of social animals (including human beings) is tempered by "sympathy . . . a tendency on the part of one social partner to identify himself with the other and so make the other's goals to some extent his own." Sympathy, he suggests, following Waddington (1960), "is a bio-logically adaptive feature of the social behavior of both men and other animals" (p. 313). Humphrey's view is that transactional behavior in conspecific interactions is a fact of animal social life; human beings, however, are so prone to thinking transactionally that they "behave inappropriately in contexts where a transaction cannot in principle take place," treating inanimate entities, plants, and other animals as if they were "people" (p. 313). Such mistakes may be "creative" (p. 314), for example, in domestication of plant and animal species, certain aspects of which must have involved a kind of systematic anthropomorphizing, practical and applied—though I want to be careful to distinguish between, on the one hand, responses of individual plants and animals to being treated as "one of the family" and, on the other hand, that genetic adaptation which is at the definitional heart of domestication. It would be stretching metaphors beyond use to suggest that mutation and inter-generational transformation were in any meaningful sense transactional events.

But I don't think it is stretching metaphors to acknowledge that individual primates who live out their "natural" lives in close, daily interaction with human observers can be expected to adjust to the socially powerful presence of those observers (as their parents before them, in many instances), in ways that would have been impossible to predict at the outset of a 10- or 20- or 30-year study, and that will be dif-ficult to interpret at its close precisely because of the transactional factor. In referring to "close, daily interaction" between human observers and primates in nature, I of course do not mean to suggest that field prima-tologists for the most part do anything but strive to remain detached, passive in external appearance, unresponsive to overtures; however, the fit between intention and realization is loose at best, and minimal interaction, whatever that may be, is not ever equal to no interaction. I would remind readers that, with primates more than with most species, the problem is not how to refrain from intervening in social situations (easy enough if it were left up to the observer); it is how to avoid being included in the transactional games of a group of individuals that are habituated to one's presence. There is no solution, finally; one simply cannot avoid being roped in on occasion. I would remind readers too that my emphasis throughout this chapter has been on the focal animal approach, which, in its nature, is likely to entail a special and continuing relationship with particular individual subject animals.

Consequences for Research

Eugene Linden, writing on ape language studies, called the subjects of those studies "silent partners" in research (Linden, 1986). *Silent Partners* is a terrific book title, with several levels of ironic reference, none too subtle for the common reader. But it turns out that these and other subjects of primate research have been our partners in a very real sense, in transactional associations, and from a transactional standpoint not so silent after all. How many field primatologists have forsaken the study of behavior for conservation work? And who is working harder for the preservation of primate species and their habitats? In some sense, and I think in a very important sense, the activist clamor for animal rights can be interpreted as a highly vocal expression of what primatologists in cooperation with their nonhuman partners have long demanded and to a great extent have achieved. Perhaps, if there is foundation for that assertion, it could be extended to apply to animal behaviorists more generally, but my argument has been that the study of primate social cognition, claimed by some to be a special case, may have to be treated as a special case if only because for going on two decades now it has been an area of practical and increasingly conscious application of theoretically justified anthropomorphizing. The character of research into primate behavior appears to me to have been greatly affected by attributing to monkeys and apes a special kind of intelligence and treating their interactions as an outcome of that intelligence, with very little concern, so far, for experimental verification—a subject which I have left to other contributors to this book to explore. My point has been that change is in progress; it does not appear to depend on, and can only be quickened by, experimental verification of that emotional and intellectual kinship that most of us have come to accept and, indeed, on which most field primatologists probably rely.

19

Common Sense and the Mental Lives of Animals: An Empirical Approach

Harold A. Herzog and Shelley Galvin

Is it possible, *at certain moments we cannot imagine,* a horse can add its sufferings together—the non-stop jerks and jabs that are its daily life—and turn them into grief?

—Peter Shaffer,
Equus: A Play in Two Acts
(emphasis added)

While philosophers and scientists may debate the existence of mental states in animals, most people readily offer opinions as to what is going on in the heads of their pet dogs and cats, the pigeons they feed in city parks, and the monkeys behind the bars at the zoo. Beliefs among laypeople about psychological processes have been called variously "folk psy-

We thank Robbie Pittman for his statistical expertise, and Mary Jean Herzog, Bethany Stillion, D. W. Rajecki, Marc Bekoff, Nick Thompson, Bob Mitchell, Gordon Burghardt, and Mickey Randolph for their comments on the manuscript. Roberta Dewick allowed us access to the University of Tennessee students who served as subjects. This research was conducted with the assistance of a Western Carolina University Hunter Scholar Award to HAH.

chologies" (Beer, 1991, this volume), "everyday understanding" (Semin & Gergen, 1990), "implicit theories" (Sternberg, Conway, Ketron, & Bernstein, 1981), "lay theories" (Furnam, 1988) and simply "common sense" (Rollin, 1989). Such theories are a central feature of contemporary functionalist theories of the human mind (e.g., Stich, 1983), but they are equally relevant to our interactions with animals (Rasmussen, Rajecki, & Craft, 1993). Beliefs concerning other species are often anthropomorphic, and they influence behaviors toward animals, attitudes about how animals should be treated, and even research strategies.

While remaining agnostic about the mental lives of animals, we can collect useful data about the manner in which human beings think about animals' experiential worlds. There are several reasons why it is important that we understand the psychology underlying the attribution of mental states to animals. Among these are the revival of interest in the possibility of animal consciousness as a subject for ethological research and the changing consensus within our society over the morality of our treatment of other species. We will leave it for others to debate the utility of understanding the minds of animals via the "intentional stance" (Dennett, 1983), "critical anthropomorphism" (Burghardt, 1985a, 1991, this volume), "reflective ethology" (Bekoff & Jamieson, 1991), "applied anthropomorphism" (Lockwood, 1985), and phenomenological techniques (Shapiro, 1990; this volume). We will, however, address the relationship between ideas about animal mentation and the ethics of their treatment.

ANTHROPOMORPHISM AND ANIMAL ADVOCACY

The moral stance individuals take toward animals is related to their perceptions of the mental capacities of other species. Although "sentimental anthropomorphism" is a part of everyday lay perceptions of animal psychology, the claim that it is largely responsible for the rise of the modern animal protection movement (Jasper and Nelkin, 1992) seems overstated. While emotion and empathy motivate many animal activists (Herzog, 1993), the philosophical underpinnings of the movement are not based on sentimentality but cohesive and logical arguments. Animal "rights" philosophers use several different paths to reach the conclusion that our present moral relationship with other species is ethically untenable (Bekoff & Jamieson, 1991; Sapontzis, 1987). All these arguments rest on the assumption that at least some nonhuman animal species are capable of experiencing mental states that are also experienced by humans.

There are differences of opinion among these philosophers as to which mental states are necessary to entitle a species to moral consideration. Peter Singer (1975)—following the dictate of Jeremy Bentham (1789/1976), who wrote "the question is not, Can they *reason?* nor, Can they *talk?* but, Can they *suffer?*" (p. 130)—argues that the simple possession of sentience, the capacity to experience pain and pleasure, entitles a creature to moral consideration. In the 1975 edition of his well-known book *Animal Liberation,* Singer drew the proverbial line between species that are entitled to moral consideration and those that are not "somewhere between a shrimp and an oyster." (In the second edition of the book, Singer [1990] admits to being increasingly uncomfortable with drawing the line at this point, based in part on recent behavioral studies of cephalopods.)

Though often touted as the founder of the modern animal *rights* movement, Singer, being a utilitarian, does not believe that animals, or humans for that matter, have rights. Philosophers who do take the rights (deontological) route to animal liberation (e.g., Regan, 1983; Rollin, 1981) argue that the moral status of animals should be based on the premise that they possess certain rights. The question for them becomes, who and what is entitled to rights? Tom Regan's (1983) criteria for moral consideration is considerably more stringent than Singer's. He restricts right holders to creatures that are the "subjects of a life," that is animals having the following: beliefs, desires, preferences, emotions, self-consciousness, memory, a sense of the future, and an emotional life. Whereas Singer grants moral consideration even to advanced invertebrates, Regan feels that his criteria is clearly met only by mammals at least one year of age.

Once the issue of which mental attributes entitles a creature to moral status is resolved, a second question arises more germane to the topic of anthropomorphism: how can we recognize the relevant mental attributes in other species so that we can know which creatures rate our moral concern? Surprisingly, the answer often given by philosophers is simply *common sense.* For example, Rollin (1989) wrote, "In the world-view of ordinary common sense, animal consciousness is a fact, and the sort of fact which we experience directly and daily, just as we do human mentation" (p. 65). Regan (1983), Sapontzis (1987), Bekoff, Gruen, Townsend, and Rollin (1992) also invoke common sense as a major source of information about animal mental capacities. Thus personal intuition in the guise of common sense ultimately is a cornerstone underlying the philosophical arguments for animal protectionism.

Needless to say, the use of intuition and common sense to determine moral status can create problems. Rollin (1989), for example, argues that scientists possess common-sense attitudes toward animal consciousness that are quite different from those of laypeople. He traces conflicts between scientists and animal advocates over the use of animals in research to their different common-sense notions of the experiential worlds of nonhuman species. But, common sense tells different people different things. We are reminded of a cockfighter in North Carolina who, defending his illegal sport, told one of the authors, "A chicken is just too dumb to feel pain."

Thus, folk psychological beliefs about the animal mind are of more than theoretical interest. Differing views as to what animals experience as they go about their lives are related to disagreements about the ethics of their treatment. Identifying discrepancies in common-sense notions of animal mentality may be an important step in explaining the sometimes bitter disagreements over the use of animals within our culture.

With a few exceptions (Eddy, Gallup, & Povinelli, 1993; Rasmussen et al., 1993; Wuensch, Poteat, & Jernigan, 1991; Mitchell, this volume, chap. 13), there has been little study of how we attribute mental states to animals. In the research described here, we investigated common-sense beliefs about the mental lives of a number of species in a group of "intelligent laypeople" (college students). We sought to: (a) develop an instrument that would assess how people perceive animals, including beliefs about their mental capacities; (b) uncover, using principal components analysis, hypothetical dimensions underlying the perceptions of different species; and (c) examine the relationship between individual differences in attitudes toward animal welfare issues and perceptions of animal mentality.

THE STUDY

Assessing Beliefs about Animal Mentation

To assess beliefs about the mental capacities of nonhuman species, we developed the Attributions Questionnaire. The scale consists of 11 questions (Table 19-1). As we were interested in creating an instrument that would describe anthropomorphism among laypeople rather than professionals, we did not include some mental states and processes which interest cognitive ethologists but which might not be familiar to the public (e.g., levels of intentionality, deception, reflective and unreflective consciousness). Eight of the questions concerned the willingness of individuals to attribute mental capacities to other species (consciousness, emotions, suffering, ability to reason, self-awareness, intelligence, pain, and affection toward humans). We were also interested in factors other than perceived mental capacities that might predict individual

TABLE 19-1
Attributions questionnaire

1. Typically, how capable of experiencing pain are each of the following animals?
2. Typically, how intelligent do you think each of the following animals are?
3. Typically, how attractive do you find each of the following animals?
4. To what degree are the following animals typically capable of "consciousness"?
5. How much do you like each of the following animals?
6. To what extent do animals feel the types of emotion that humans experience such as joy, anger, and sadness?
7. Typically, how affectionate are each of the following animals toward humans?
8. Morally, how much consideration do each of the following species deserve in weighing their concerns against those of humans?
9. To what degree do the following animals have the capacity to reason?
10. To what degree do the following animals possess the capacity to suffer?
11. To what extent are each of the following animals self-aware?

differences in attitudes about animal welfare. Thus we included several items not related to anthropomorphism but which might be related to attitudes toward animals: how attractive the participants found the species, how much they liked the species, and how much moral consideration they felt the species deserved.

Each of the 11 items was followed by a series of five statements reflecting increased attributions. For example, item number 4 read: "To what degree are the following animals typically capable of 'consciousness'":

1. no consciousness
2. little consciousness
3. moderate consciousness
4. high degree of consciousness
5. humanlike consciousness

Eighteen types of animals were listed after each question, and the subjects were asked to indicate the statement that best reflected their thinking about the species. The list contained three invertebrates (worms, ants, spiders), five non-mammalian vertebrates (goldfish, frogs, snakes, turtles, pigeons), and eight species of mammal (mice, rats, bats, dogs, cats, pigs, chimpanzees, and dolphins). For an ontogenetic comparison, we also included puppies and kittens on the list.

Because we were interested in the relationship between the attribution of mental states to animals and attitudes toward their treatment, an instrument designed to measure attitudes towards animal welfare was included in the survey packet. This questionnaire, the Animal Attitudes Scale (AAS), was a slightly modified version of the instrument described in Herzog, Betchart, and Pittman (1991). It consists of 20 Likert-like scale items related to attitudes toward the use of animals in research, agriculture, education, and recreation. Sample items included, "It is morally wrong to hunt wild animals just for sport," and "Continued research with animals will be necessary if we are to ever conquer diseases such as cancer, heart disease, and AIDS."

Statistical Analysis

We were particularly interested in identifying underlying dimensions that would account for similarities in response patterns across the 11 attributes. These dimensions were identified using principal components analysis. This analysis was conducted by first collapsing the scores of each subject across all 18 animals on each of the 11 attributes. Thus Subject 1 was given a "pain" score indicating her willingness to attribute pain to animals as a group by summing her responses to the pain item over all of the species. The resulting scores were then subjected to principal components analysis. As described below, this procedure revealed that the 11 items on the questionnaire fell into three groups. We called these the "cognition factor," the "affect factor," and the "sentience factor." Factor scores were then generated for each of the subjects on each of the three dimensions. We then used multiple regression to examine the relationships between the cognition, affect, and sentience factors and individual differences in attitudes toward animal welfare issues as indicated by the AAS. Factor scores were also generated for each species and used to give a general picture of how the species were perceived in terms of the three underlying dimensions.

Subjects

The surveys were administered to 57 males and 112 females taking social psychology classes at the University of Tennessee in the spring of 1990. The average age of the subjects was 22. About half (82) of the students were psychology majors, with the remaining being predominantly business majors. It might be argued that the psychology majors would be more sophisticated in terms of their attitudes about animal mentality than the general student population. To assess this possibility, we compared the mean scores of the psychology majors and the non-majors on the Attributions Questionnaire dimensions. There were no significant differences in the willingness of students from the two groups to attribute any of the mental states to animals.

Results

Species Differences in Perceived Mental Life. Our respondents clearly believed that animals experienced all eight of the subjective states listed on the questionnaire. Not surprisingly, the degree that animals were perceived as experiencing the mental state depended both on the species and the state. This finding is illustrated in Figure 19-1 which presents profiles of three representative species (ants, mice, and dolphins) based on the percentage of subjects who felt that the species was capable of at least "moderate capacity" to experience each state (the percentage that rated the species as three or higher on the five-point scale).

There were dramatic differences in how the species were perceived. More than 90% of the subjects believed that dolphins experienced at moderate or higher levels all eight of the states. Though not shown on the figure, chimpanzees had an almost identical profile. The pet animals on the list (dogs, cats, puppies, and kittens) were also perceived by the majority of subjects as having at least a moderate capacity to experience all of the mental states.

While we expected that the respondents would think that animals such as dolphins and chimpanzees had mental experiences, we were surprised at the number who felt that these species were capable of

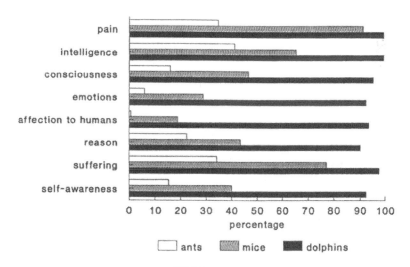

FIGURE 19-1
Percentage of subjects who felt that ants, mice, and dolphins
possessed moderate or greater levels of mental states

experiencing *humanlike* levels of the states (e.g., rated the species a five on the five-point scale). For example, 72% felt that chimpanzees possessed humanlike capacities to feel pain, 40% felt they had humanlike intelligence, and 42% thought they had humanlike ability to experience emotions.

In contrast, relatively few subjects thought that the invertebrates experienced the mental states. Figure 19-1 shows the profile of ants, which was typical of the invertebrates. Some participants did feel that invertebrates experienced forms of subjective states. Thirty percent, for example, thought that worms felt pain to at least a moderate degree, and 16% indicated that spiders possessed at least a moderate degree of consciousness. (Given their response toward worms, probably about a third of our subjects would have disagreed with Peter Singer's statement, "Most mollusks are such rudimentary beings that it is difficult to imagine them feeling pain or having other mental states" [Singer, 1975, p. 178].)

Some species showed a mixed pattern of perceived capacities. This pattern is illustrated by the profile for mice shown in Figure 19-1. About 90% of the subjects felt that mice had at least a moderate capacity to experience pain. On the other hand, only about 30% thought that mice experienced emotions.

Perception of Consciousness. Due to the current interest in animal awareness within ethological circles, we were particularly interested in how the participants would rate the species on the consciousness item. Figure 19-2 shows the percent of subjects who indicated that they felt the various groups of mammals were capable of at least a moderate degree of consciousness. We emphasize mammals because, with the exception of bats who were ranked lower than pigeons, all of the mammalian species ranked higher than all of the non-mammals. There was wide disparity among the eight mammalian species in the number of students who attributed consciousness to the species. Only 33% of the subjects thought that bats were at least moderately conscious, whereas 96% thought that chimps were at least moderately conscious.

That consciousness would be thought to be more characteristic of chimpanzees than bats is not surprising. Apparent inconsistencies, however, emerged in the profiles of other groups. The biggest jump in perceived consciousness occurred between pet and nonpet animals, a result also found by Eddy, Gallup, and Povinelli (1993). Pigs are thought by scientists who work with them to be fairly intelligent creatures, but only 59% of our subjects perceived pigs as being at least moderately conscious compared to 92% who felt that dogs were conscious. We know of no systematic research findings that would support

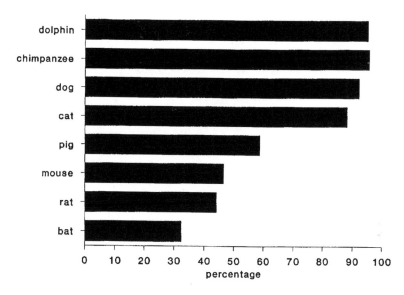

FIGURE 19-2
Percentage of subjects who felt the species
possessed moderate or greater levels of consciousness

the belief that dogs and cats are more conscious than pigs, but our subjects intuitively felt that their mental capacities were quite different.

Basic and Advanced Mental States. We examined the willingness of the subjects to ascribe each state to animals as a group by collapsing the attribution scores of the 8 states across all of the species. The mean scores for the states are shown in Figure 19-3. Pain and suffering were the most basic states in that they were perceived as being more characteristic of animals generally than more "advanced" states such as the ability to experience emotion, show affection toward humans, and reason.

Underlying Dimensions. The principal components analysis revealed that the 11 anthropomorphism items fell into three dimensions: one related to how the individuals felt about animals (referred to as the "affect dimension"), one related to pain/suffering/moral consideration ("sentience dimension"), and a third related to perceived capacity for higher mental processes ("cognition dimension"). The rotated loadings for each of the components are presented in Table 19-2. Only one item loaded significantly on more than one factor; the emotion item loaded

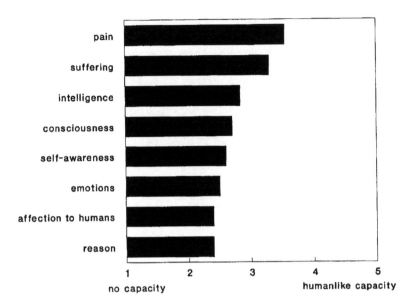

FIGURE 19-3
Mean capacity of 18 species to experience each mental state

TABLE 19-2
Item loadings from principal components analysis

Dimension	Factor 1 (Cognition)	Factor 2 (Affect)	Factor 3 (Sentience)
Cognition			
capacity to reason	.783	.155	-.012
self-awareness	.774	-.179	.325
experience emotions	.672	.515	.274
capable of consciousness	.667	.396	.250
intelligence	.523	.339	.310
Affect			
how much species is liked	.104	.871	.182
attractive	.050	.750	.301
affectionate to humans	.427	743	.042
Sentience			
capacity to suffer	.270	.087	.827
capable of experiencing pain	.195	.161	.796
deserve moral consideration	.060	.337	.646

on both the cognition and the affect factors. Of particular interest is the fact that pain and suffering formed a common factor along with moral consideration. This pattern is significant in light of the utilitarian argument that pain and suffering are the ultimate moral leveler. The fact that moral consideration fell into the same factor as pain and suffering seems to provide a psychological analog to Singer's (1975) contention that moral status ultimately rests on the capacity for sentience.

Subjective States and Animal Welfare Attitudes. Factor scores for each subject on the three dimensions were calculated and used to examine the relationships between the factors and individual differences in attitudes toward animal welfare issues as indicted by the AAS. There were significant correlations between AAS scores and the affect factor ($r = .309$) and the sentience factor ($r = .327$), indicating that individuals who rated animals more highly on these factors were more likely to be concerned about animal welfare issues. Interestingly, scores on the cognition factor were not significantly related to attitudes toward the use of animals ($r = .154$).

Multiple regression was used to examine the relationship between the sentience, cognition, and affect factors and the animal welfare attitudes of our subjects. Multiple regression is a statistical technique for analyzing the relationship between a set of independent variables and a single dependent variable. In this case, we were interested in the relative contributions of the three factors to the prediction of animal attitudes as measured by the AAS. Because gender is known to influence attitudes toward the treatment of animals (Herzog, Betchart, & Pittman, 1991; Driscoll, 1992; Gallup & Beckstead, 1988; Kellert & Berry, 1987), we also included it in the regression equation. The standardized regression weights (a measure of relative independent contribution of the factors to AAS scores) are shown in Table 19-3. The affect and sentience factors,

TABLE 19-3
Standardized regression weights describing how gender
and the anthropomorphism dimensions contribute
to the prediction of attitudes toward animal welfare

Variable	Standardized Regression Weight	Significance
affective factor	.305	.0001
gender	.277	.0001
sentience factor	.270	.0001
cognitive factor	.110	.0980

along with gender, contributed significantly to variation in AAS scores. Together these variables accounted for 30% of the variation in the AAS score (Multiple R = .55).

Gender Differences. Sex differences were found for two of the three factors. Women had significantly higher scores than men on the cognition factor (t = 2.16, df = 167, p = .032) and the sentience factor (t = 2.39, df = 167, p = .018). In both cases women were more willing to ascribe mental states to animals than were men. There was not a significant gender difference in affect factor scores (t = 0.25, df = 167, p = .80). As found previously (Herzog et al., 1991), there was a significant sex difference on the Animal Attitudes Scale with women having higher mean scores than men (t = 4.78, df = 167, p < .01), reflecting more sympathetic attitudes toward the treatment of animals.

Species Differences in Factor Scores. The means of the factor scores for each species are shown in Figures 19-4, 19-5, and 19-6. An examination of the figures indicates that phylogenetic status as popularly conceived is a major component determining how the species are perceived. Species commonly thought of as being "lower" tend to have low factor scores,

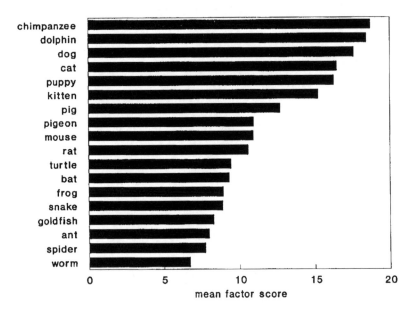

FIGURE 19-4
Mean factor scores on the cognition factor

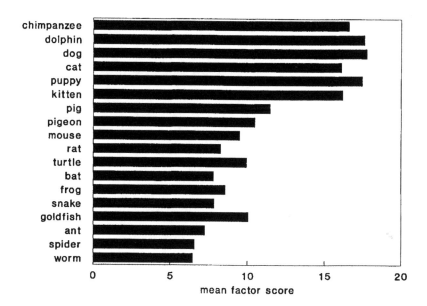

FIGURE 19-5
Mean factor scores on the affect factor

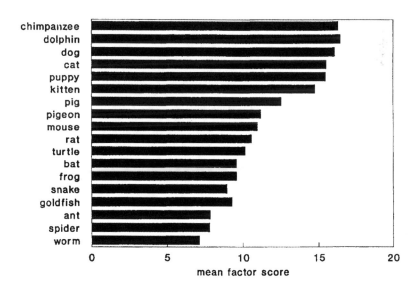

FIGURE 19-6
Mean factor scores on the sentience factor

whereas "higher" species were consistently given higher factor scores.

Species are listed in Figures 19-4, 19-5, and 19-6 in descending order as ranked on the cognition factor. The factor scores of the species on the three factors were highly interrelated, as shown by the high correlations between the factors for the various species (between the affect and cognition factors, $r = .97$; between the sentience and cognition factors, $r = .99$; between the sentience and affect factor, $r = .98$). Species that were perceived to have high cognitive abilities were also highly ranked on the other two factors. The high correlations were no doubt related to the effect of phylogenetic status on the perceived capacities of the species. We further examined the phylogenetic effect by comparing the mean factor scores for the invertebrates ($n = 3$), non-mammalian vertebrates ($n = 5$) and mammals ($n = 10$). MANOVA indicated a highly significant difference between these groups ($F = 7.039$, $df = 6$, $p < .001$). Univariate F tests ($df = 2$, 15) on all three variables, showing a highly significant effect of phyletic levels (cognition, $F = 11.035$, $p < .001$; affect, $F = 7.035$, $p < .01$; sentience, $F = 12.81$, $p < .001$).

Discussion

Many writers on animal rights assume that it is common sense that animals experience subjective mental states. If this is the case then we would expect to see agreement among human subjects on the degree to which animals of different species are thought to be conscious, feel pain, et cetera. Our results show that indeed there is a considerable agreement among the undergraduate students we queried on these issues. The idea common sense tells us that animals have mental experiences is to some extent confirmed. However, the subjects did not attribute the same degree of mental experience to all animals. They tended to accord mental experiences to species along phylogenetic lines that are familiar. Moreover, there were differences of opinion among subjects based on gender and other factors. Animals such as chimpanzees that are perceived to be closely related to humans and those with large brains such as dolphins were thought to have mental experiences at levels quite similar to humans. Pets were also thought by most subjects to experience emotions, consciousness, pain, etc. In contrast, many fewer subjects thought that invertebrates possessed these capacities.

In addition, some subjective states were perceived to be more widespread among animals than others. Pain and suffering were thought to be more common than states of mind such as emotions, reason, and affection toward humans. The principal components analysis indicated that there were three dimensions underlying attributions of mental states to animals: cognition, affect, and sentience. Finally, there were ethical consequences to the attribution of mental states to animals; individuals with high affect and sentience factor scores were more likely to show concern about how animals were treated.

Attitudes toward other species are influenced by a complex milieu of logical and emotional factors (e.g., Burghardt & Herzog, 1980; Galvin & Herzog, 1993). Superficially, our subjects' perceptions of mentality in different species were consistent with a phylogenetic-based logic; invertebrates were perceived as having the lowest levels of mentality, followed respectively by non-mammalian vertebrates and mammals, with the large-brained mammals (chimps and dolphins) on top of the hierarchy scale. Closer examination of rankings within taxa, however, revealed that less logical factors also influenced attitudes toward subjective states in animals. The best example of this phenomena is the attribution of mental states to pet and non-pet mammalian species. Dogs and cats were perceived to have sophisticated mental states by many more subjects than were non-pet mammals with roughly equivalent brain sizes. The reason for this finding is not clear. Perhaps people are simply more familiar with dogs, cats, puppies, and kittens than with bats, mice, rats, and pigs. It is also possible that the dogs and cats that we live with are in some senses our friends and that we attribute higher levels of mental processes to those creatures that we consider our friends (Serpell, 1992). These possibilities could be tested by comparing perceptions of the mental capacities of non-pet species of individuals who have extensive experience with the species with the perceptions of those whose experienced with the species is limited (i.e., Do pig farmers and laboratory technicians who work with rats attribute higher levels of mental prowess to these animals than do laypeople?).

Attitudes toward animals and their treatment are related to a host of variables including geography, religious beliefs, semantic categories, age, philosophical views, race, gender, socioeconomic status, and even belief in evolution (Burghardt, 1985a; Driscoll, 1992; Dunlop, 1989; Herzog et al., 1991; Gallup & Beckstead, 1988; Kellert, 1980; Rajecki, Rasmussen, & Craft, 1993). We would be foolish to imply that our results based on university undergraduates apply to all segments of a complex society. Our results do, however, indicate that an empirical approach to the investigation of intuitive beliefs about the minds of other species is feasible and even offers insight into the psychological underpinnings of differences of opinion about the use of animals in research and education.

ANIMALS AND SCIENTISTS

There is one group whose intuitive beliefs about the mental lives of other species affects the study of animal behavior, but about whom little is known: scientists. Rollin (1989) has argued that, with a few exceptions, scientists who study animals hold quite different beliefs about animal consciousness than laypeople. According to Rollin, most behavioral scientists are socialized into "the common sense of science," a counterintuitive Cartesian world in which conscious experiences in other species are conveniently denied. As he puts it, "It became clear to me that an integral part of becoming a scientist is learning to abandon ordinary common sense in a number of areas, including that of ascribing mentation and subjective experience to animals" (p. xii).

Rollin's portrayal of scientists may not be accurate. Certainly, the writings of the early ethologists (e.g., Lorenz, von Uexküll) reveal that some of the founders did not doubt that animals experienced subjective states. More recently, the dramatic upsurge of interest in cognitive ethology indicates that animal behaviorists are presently rethinking scientific approaches to animal mentation. However, there is no systematic information about what animal scientists *believe* about the minds of the creatures under study. Certainly the dry texts of professional journals offer little insight into the candid thoughts of the individuals who for one reason or another decided to spend their professional lives trying to figure out why other creatures do the things they do. Popular accounts and biographical sketches provide some insight into the minds of scientists (e.g., Crocker, 1984), but systematic information about the attitudes of animal scientists about their subjects is largely lacking.

What little evidence there is suggests that Rollin may be wrong in his assertion that, as a group, scientists deny the existence of consciousness in their animal subjects. We recently informally asked 14 biologists and psychologists if they felt that mice could experience pain and if they felt that mice were conscious. All the respondents thought that mice could experience pain and 12 believed them to be conscious. Phillips (1993) interviewed 27 scientists from 23 laboratories as part of an ethnographic study of laboratory culture. She reported, "There were no latter-day Cartesians among them who claimed that animals do not feel pain" (p. 76).

Animal activists often paint animal researchers as white-coated technocrats who are oblivious to the pain and suffering of other creatures and who perform trivial experiments in an ego-driven quest for fame and federal dollars. Even some sophisticated critics believe that the majority of animal researchers hold a simplistic Cartesian world view in which the existence of mental experiences in other species is denied. While this stereotype may be true of some scientists, it is probably not true of many ethologists, as was dramatically illustrated at a round table discussion on animal cognition at the 1991 International Ethological Congress. At the end of the session, one of the conveners asked the audience to indicate by a show of hands why they had chosen to devote their professional lives to the study of animal behavior. He then offered a list of fairly predictable reasons, such as the desire to investigate certain problems of theoretical interest in biology and simply liking animals. The last question was a surprise. "How many of you went into the field of animal behavior because you wanted to know what it was like to be a member of another species?" Even more surprising was the response: over half of the audience raised their hands.

20

Amending Tinbergen:
A Fifth Aim for Ethology

Gordon M. Burghardt

QUESTION: Do ants really know what they are doing and why?
ANSWER: This question has been argued for many years
among scientists. The truth seems to be that they could not
possibly build such well thought-out homes for themselves
and regulate their lives so intelligently unless they knew
what they were doing.

—Anonymous,
Uncle Milton's Ant Watcher's Handbook

In der Tierpsychologie gilt daher mehr denn sonst der Satz:
Das Maß aller Dinge aber ist der Mensch! [In animal psychology, therefore, the statement holds more than anywhere
else: The measure of all things is man himself.]

—Hempelmann,
Tierpsychologie vom Standpunkte des Biologen

This chapter had almost a 30 year gestation, since initially fertilized in the garden
of Eugene Gendlin, and a harried 3-month delivery. To mention all the people
who stimulated, corrected, clarified, and challenged my thinking over the years
is impossible, but William S. Verplanck was always there with his inimitable
comments, and the curmudgeonly ghost of Eckhard Hess was everpresent over
my shoulder. In the delivery of this specific neonate, however, I want to thank
Marc Bekoff, Frans de Waal, Don Dewsbury, Neil Greenberg, Hal Herzog, John
Malone, Larry Shapiro, and the three editors, Bob, Nick, and Lyn.

THE FOUR AIMS OF ETHOLOGY

The four aims of ethology, as delineated by Tinbergen (1963) 30 years ago in his seminal paper entitled "On Aims and Methods of Ethology," are to study causation (proximal mechanisms, control), ontogeny (development), evolution (phylogeny), and survival value (function, adaptive value). The parenthetical terms are alternatives used by some commentators. These four aims have been elaborated and discussed by many authors and their utility has been repeatedly stressed. Many ethologists, myself included, have claimed that together the four aims (sometimes framed as problems or questions to be asked) cover virtually every important issue in the study of behavior. Dewsbury (1992) has recently reviewed the history of this classification, reasserted its value, and provided a most useful clarification and extension (see also Lauder, 1986).

Although Tinbergen's 1963 paper drew explicit attention to, and formalized, the conception of the four aims, they were implicit in much of the early writings of ethology that Lorenz synthesized into what we now recognize as core or classical ethology (Burghardt, 1985b). Tinbergen's earlier *Study of Instinct* (1951) also mentioned them by name and, while emphasizing internal and external causation, included substantial chapters devoted to development, adaptiveness, and evolution (his terms). Survival value or *function*, originally used primarily for the role of a behavior pattern, morphological structure, or physiological process in the continued functioning of the individual, is often extended to include reproductive or inclusive fitness (cf. Lauder, 1986).

Now the conceptual separation of the four aims by most workers, reinforced by the increasingly specialized methods needed for their modern study, was not meant to suggest that they should be pursued independently of each other. Tinbergen earlier had acknowledged that "there is, of course, overlap among the fields covered by these questions," nonetheless asserting "that it is useful to both distinguish between them and to insist that a comprehensive, coherent science of Ethology has to give equal attention to each of them and to their integration" (Tinbergen, 1951, p. 411). Dewsbury (1992) emphasized that Tinbergen's emphasis on "equal" has been lost by some recent commentators, perhaps because of the current popularity of the proximate-ultimate dichotomy and the emphasis of 1970s- and 1980s-style sociobiology and behavioral ecology on demonstrating adaptation. Indeed, the "rediscovery" of the importance of mechanisms has been heralded as ushering in a new era in these fields (e.g., Stamps, 1991; Krebs & Davies, 1991). Regardless of fads and enthusiasms that tout one aim by demeaning or ignoring others, a balance among the four seems to be reasserting itself (e.g., Barlow, 1989; Dawkins, 1989). Table 20-1 lists the standard four aims, with control substituted for causation due to the frequent confusion between "proximate" and "ultimate" causation, with Tinbergen's causation being limited to the former.

IS IT NOW TIME TO FORMALLY RECOGNIZE A FIFTH AIM?

> We cannot understand animals fully without knowing what their subjective worlds are like. Until this is possible, and at present it is possible to only a very limited degree, we will remain unable to appreciate adequately either the nature of animals, or how we differ from them.
>
> —Griffin,
> *Animal Minds*

> Probably a majority of animal behaviour workers would now accept that, no matter how difficult they may be to study, it is no longer possible to deny the existence of true thought processes and even consciousness of some type in some mammals and birds.
>
> —Manning and Dawkins,
> *An Introduction to Animal Behaviour*

Old problems in the study of animal behavior and new directions in psychology and neuroscience suggest the time is right for the addition of a further aim. Darwin and the early comparative psychologists such as Romanes (1883) and Morgan (1894) were interested in the mental life of animals as well as behavioral study as an end in itself, an endeavor largely abandoned as behaviorism became dominant in the post World War I decades (Burghardt, 1978, 1985a; Rollin, 1990). But experimental psychologists are again willing to study cognitive events such as ideas, thoughts, memories, expectations, representations, images, and decisions, without subsuming them in behavioristic language, whether methodological or radical. Cognitive neuroscience is providing methods to explore activity occurring in the brain during specific perceptual or mental operations (e.g., Pellizzer & Georgopoulos, 1993). Emotion and motivation are again active research areas as the reciprocal interactions between physiological and psychological events are documented and explored (e.g., psychoimmunology, Martin, 1990). Computer-based neural net and connectionist models, often inspired by Hebb's (1949) cell assemblies, are ever more closely mimicking animal and human learning and behavior. In short, ethology, responsible for rehabilitating "instinct," is itself challenged by

TABLE 20-1
An amended five aims of ethology

Name	Other identifiers	Brief description
1. Control	Causation, Mechanism	Identify internal and external factors underlying behavioral performance.
2. Function	Adaptiveness, Survival Value	Identify contributions of behavior patterns to group, individual, reproductive, and inclusive fitness.
3. Ontogeny	Development	Identify patterns and processes in behavioral change during individual lifetimes.
4. Evolution	Phylogeny, Genetic and Cultural Inheritance	Identify historical patterns and processes in behavioral change across generations and taxa.
5. Private Experience	Personal World, Descriptive Mentalism, Subjective Experience, Heterophenomenology	Identify patterns and processes in life as experienced.

Note. Phenomena typically considered cognitive, emotional, and motivational may be incorporated into all 5 aims.

the rehabilitation of "mind" (Burghardt, 1978). There is, of course, debate as to whether these particular trends are advances or regressions (Skinner, 1990). For Skinner (e.g., 1974), addressing the phenomena labeled "private" is necessary, although he would definitely use more active gerunds (e.g., *thinking* rather than thought) and emphasize bodily experience rather than invoke a disembodied dualistic mind. But there is no question that the psyche, variously conceived, has openly reentered much scientific psychology. In spite of impassioned arguments, some in this book, that ethology and comparative psychology must remain aloof from any "New Mentalism," to do so will but set us at the periphery of what may prove to be the most scientifically and intellectually challenging and exciting issues of the next century. Indeed, ethology may be in a position to uniquely contribute to scientific and philosophical discourse and prevent the errors that befell the first systematic attempts in this area 100 years ago (Burghardt, 1985b). The present volume, a critical look at the role of anecdote and anthropomorphism in the study of animal lives and abilities, will be successful to the extent that it fosters a constructive forward-looking agenda that recognizes the new tools, possibilities, and demands to come, as well as the lessons to be learned from over a century of insights, promises, and failures.

This volume is also a testament to the impact of Donald Griffin, who became convinced of the importance of animal awareness and cognition as a field to be studied after thinking deeply about how bats echolocate and about the implications of Karl von Frisch's work on dance communication in honeybees (Griffin, 1976). Something more than stimulus-response mechanisms were at work and needed attention. Griffin's appeal was greatest outside traditional experimental psychology, where a methodological behaviorism reigned supreme. This behaviorism, although not denying subjective states or complex mental operations, considered them beyond the realm of scientific exploration. The cognitive revolution in psychology initially brought back the latter but not the former; Griffin used the increasingly complex behavior being documented in animals to suggest the need for complex mental operations, which themselves called for reopening discussion of consciousness, awareness, and other apparently subjective phenomena. Herein lies the problem this chapter attempts to clarify, if not resolve, by integrating the issues raised into an ethological framework.

It should be emphasized that all involved in these debates, especially Griffin (e.g., 1992), today accept that, however characterized, mental events and experiences are products of physiological, materialistic processes dependent on the nervous system. Thus any stated contrasts between physiological and psychological aspects of behavior refer to different levels of analysis, traditional dichotomies that probably still have their heuristic value in many contexts (such as cell biology and molecular biology) or the language, methods, and interests of the scientists themselves. That is, a classification useful in discourse does not need to reflect a fundamental metaphysical dichotomy.

WHAT THE FIFTH AIM SHOULD ENCOMPASS

. . . relationships have objective aspects that are apparent to an outside observer and subjective aspects that are specific to each participant, known in their entirety only to him or her, and shared only partially. Similarly, the objective aspects of the sociocultural structure may be partially codified in laws and customs, but the subjective aspects may be subtly different for each individual.

—Hinde, "Developmental Psychology
in the Context of Other Behavioral Sciences"

The affective world of the black bear does not mirror ours. Nor is it a parallel alien universe. If we have the privilege of being individually bonded to bears, their world touches us at their convenience, and we are given partial access to it through a darkened glass, but never face to face.

—Burghardt, "Human-Bear Bonding
in Research on Black Bear Behavior"

Should this fifth aim be cognition as in the cognitive ethology of Griffin (e.g., 1976) or the more restrictive (i.e., nonsubjective) comparative cognition of experimental psychology (Burghardt, 1985a)? There are category problems. For example, Sherman (1988) explicitly advocated the addition of a fifth "approach" for ethology dealing with cognitive processes inspired by some of the very developments noted above. But later in his paper he, perhaps inadvertently, showed that cognition was not a viable separate fifth aim at all. For now he included cognitive and physiological processes as subsets of the aim of *causation*. And he is right to do so, for cognition can be approached purely mechanistically without reference to the experience of the subject. So cognition can be, and currently largely is, studied as mental or brain processes inferred from empirical findings and amenable to computer simulation. Indeed, *comparative cognition* is a current popular term for the study of animal learning, aspects of which can be the target of any and all of the four Tinbergian aims. There is also a flourishing developmental comparative cognition movement (e.g., Mitchell, 1987; Parker & Gibson, 1990).

Another category problem with cognition as a separate aim is that the terms *cognitive* and *cognition* omit the other two major categories of psychological phenomena, emotion and motivation (Burghardt, 1991). However, as with cognition, these two phenomena can also be incorporated into all four aims. This tripartite division of psychological phenomena into cognitive, affective (emotion), and motivational aspects predates Darwin and was explicitly recognized by Romanes (1883) and McDougall (1923). Psychologists are again recognizing the value of achieving a balance and integration among all three aspects. The controversy on the role of cognition in emotion is but a representative example (e.g., Izard, 1993).

Much of the controversy engendered by Griffin's writings, beginning with his 1976 book, is related not so much to his emphasis on the complex abilities of animals (abilities that call for explanations involving cognitive processing more sophisticated than previously thought possible in animals), but his insistence on discussing them using attributions of awareness, feelings, consciousness, and other terms referring to the personal or subjective experience of organisms. The latter are frequently viewed as distinctly human phenomena or outside the realm of scientific study (compare the similar comments on this point in the reviews in *Science* of Griffin's first and latest books [Griffin, 1976; 1992] by Mason [1976] and Yoerg [1992] respectively; see also Bekoff & Allen, this volume). The conflation of the study of complex behavioral phenomena in animals with claims about their subjective experience (Burghardt, 1991) is behind much of the opposition to Griffin's views, even by those studying the very behavior Griffin emphasizes and whose research Griffin (1992) uses to support his approach (e.g., Wasserman, Kiedinger, & Bhatt, 1988).

Thus, although Griffin certainly means to include much of what I advocate including in the fifth aim, the term *cognitive ethology* is, in retrospect, not radical enough. Questions involving the experience by organisms of emotions, thoughts, decisions, feelings, intentions, consciousness, and awareness do merit a separate aim, but one once removed from their mechanisms, function, ontogeny, or evolution. Just as questions about ontogeny can be heuristically separated from those of evolution, so can questions about experience (variously labeled inner, subjective, phenomenal, self, or personal) be separated from those of causation or function and be asked about many diverse instances of behavior. Questions involving foraging need ultimately to incorporate the four aims; similarly, questions of these private experiences need to be integrated into the other four. But trying to combine analysis of the experience of cognitive events with the efforts of those studying cognition with a different agenda, and with no scientific interest in "life as experienced," is needlessly controversial and confusing.

It is thus time to return to the effort of von Uexküll (1909/1985) to attempt to understand the perceptual and inner worlds of other organisms, both human and nonhuman, and to try to gain some understanding of what it is like to *be* the animal, to make inferences about private experience, and to see what such understanding can contribute to studies in the traditional four aims. This will help to legitimate a rigorous focus on such phenomena and not treat them as epiphenomena or merely speculative "throwaway" inferences. In this new fifth aim they will be the focus of the scientific endeavor.

Although I have had a career-long fascination with Jacob von Uexküll's writings (e.g., 1909, 1934/1957) expressed in chapters beginning over 20 years ago (Burghardt, 1973), the quest to understand the self or phenomenal world of other species seemed doomed, a grabbing at the smiling grin of the Cheshire cat: so vivid but unattainable. The *Umwelt* concept of von Uexküll that had such an impact on the early ethologists (as in sign stimuli and releasers) is simply the environment as perceived and responded to by an organism. But von Uexküll viewed as equally essential for analysis and inference the *Gegenwelt* (inner or counterworld) companion concept. The *Gegenwelt* becomes more involved in animal behavior as the central nervous system puts more neural pathways between receptors and effectors. Its legacy in ethology was reduced to releasing mechanisms, schemata, and templates, while von Uexküll's broader and more profound ideas were forgotten and ignored in both ethology and psychology.

Is it not hard to gain accurate information on the experienced life of another human being, let alone another species? Surely it is. But just as stories, theater, poetry, art, cinema, even music are often attempts to portray the personal experiences of other people to an interested audience, so the primal drive for many persons fascinated with other species is to somehow enter into and understand their world (Burghardt, 1992b), even acknowledging that this knowledge may be as superficial as the reflection from a rusty, dented, mirror. This understanding is in many ways incomplete, and will remain so, being an inferential description of what is taken as important by the organism and how it is experienced. Consider the response of the audience survey at the animal cognition workshop at the International Ethological Congress in 1991 described by Herzog and Galvin (this volume), attended by both Herzog and me. The majority admitted that the primary motivation for their interest in the species they studied was to understand what it was like to be that species. Here I need to emphasize that we are attempting to understand the experience of others through a variety of necessarily analogical tools that we can apprehend (e.g., words, pictures, sounds, bodily experience).

Not surprisingly, this is far from a new idea. Besides von Uexküll, continental animal psychologists (e.g., Bierens de Haan 1940, 1947b), phenomenologists (e.g., Buytendijk and Merleau-Ponty, see Thinés, 1977; Shapiro, 1990b for a recent example), and experimental psychologists (e.g., Hebb, 1946; Skinner, 1961) have tried to evaluate the nature of experience in animals, although their ideas have never been thoroughly explored, compared, and tested. It is time that we carry out systematic studies of how far we can go and of what value such knowledge may be. Otherwise the endeavor to explore other species' lives is left to novelists, often imbued with a strong love of natural history, and plausible, even insightful and penetrating, narratives from the animal's perspective (e.g., Ernest Thompson Seton, Jack London), but with the same relation to ethology that Dostoyevsky's novels have to psychology. Indeed, recent writings based on real individual animals, carefully observed for long periods, are having a major effect on our views of animal mentality (e.g., Goodall, 1971; de Waal, 1982). Their powerful narrative format, often reinforced with photographs and videotape, has, however, been criticized as highly anthropomorphic (Peters, 1991).

By adding as a fifth aim the analysis of how other individuals and species experience events, the origin, evolution, control, or consequence of this experience is not prejudged. This aim is not limited to consciousness as typically conceived. Consciousness may be neither necessary for learning nor for action (Burghardt, 1985a; Yoerg, 1992). The recent fascinating work on "memory free" reasoning in children (Brainerd & Reyna, 1992) is further evidence that conscious processes may not be used or linked to intelligent behavior in the obvious ways supposed since Romanes (1883). The evidence has accumulated to the point that there is now discussion of implicit cognition, learning, and memory that can be collectively termed the *cognitive unconscious* and which even has been placed into an evolutionary context (Reber, 1992).

But that feelings, urges, and thought processes *exist* cannot be denied by us, and not to extend this to nonhumans is both anti-evolutionary and nonparsimonious, even if the extent and nature of the phenomenon and its occurrence in any given instance is uncertain. Although many have criticized Griffin for bringing back introspective analysis of a nonmaterial mindstuff, Griffin (1992) clearly views the experiences of animals as a product of physical events in the body no less than does his avowed opponent, B. F. Skinner (1989). Thus we need to set aside at the start discussion-stopping rhetoric about mind-body dualism and agree that there are phenomena of experienced life that we know exist to the extent that we are personally privy to such events ourselves. But the limits of our ability to interpret our own experience must be considered. Our privileged access to this information is far from being accurate or complete. Consider the conundrum of establishing intention or awareness in assessing guilt in wrongdoing and judicial trials in our own species (see Mitchell, this volume, chap. 13). The difficulties attendant on such endeavors would never justify our walking away from making such decisions; the functioning of any complex society is dependent upon the effort.

Recent events suggest, moreover, that our continuation as a scientific field will be deeply affected by our willingness to attend to issues of how animals experience their lives. Rollin (1989) argues, not entirely without justice, that scientists working with animals often deposit their common sense (folk wisdom?) at the laboratory door, especially in often ignoring issues of suffering. Assessment methods for a moral calculus are being seriously proposed in the most prestigious journals (e.g., Porter, 1992) as a response to laws and regulations involving animal use that specifically call for reduced suffering and enhanced well-being. Animal welfare and treatment concerns involving physical and psychological suffering are at base a concern for the quality of life and experience an animal is granted. But we don't quite know what to do as scientists in this debate. For example, Bateson (1991) effectively argued that the issues of suffering and pain cannot be ignored, but stumbled when it came to trying to legitimate their formal incorporation into ethology, as Lorenz (1963/1970) and I (1985a) did earlier. Tinbergen (1951) merely adopted the stance of methodological behaviorists; he had many other battles to fight and was opposed to what seemed a subjectivism beyond objective analysis. But whether, for example, animals housed singly or without companions do less well in some circumstances (e.g., using the Novak & Suomi [1988] typology) is a legitimate empirical question. Using terms such as unhappy or despondent can be warranted if we do not go too far from our data, as shown by Hebb (1946) many years ago. But when is departing too much from data equivalent to taking an intellectual vacation? Although this issue is far from trivial, de Waal (1991), among many others (e.g., Hebb, 1946; Bekoff, 1992; Bekoff, Gruen, Townsend, & Rollin, 1992), has pointed out the rigidity forced on us when concepts such as friendship, reconciliation, greeting, appeasement, even courtship, are objected to because they contain some interpretive content. That each concept reflects a heterogeneous category is a given, but such is also true of concepts such as eyes, wings, territory, play, and communication.

At stake here is the issue of whether to assume that animals are on an evolutionary continuum with us, or to begin by holding that there is a qualitative difference before even studying the relevant phenomena. Are such terms as friendship, for example, "inevitable when one studies acting agents instead of merely moving objects" (de Waal, 1991, p. 299)? Similarly, to consider our relationship and bonds with our animal subjects as an important aspect of our research (Davis & Balfour, 1992a) has proved controversial among academic scientists (Hank Davis, personal communication, January 1993). I suspect that the objectors are precisely those who have been inculcated with the necessity, *as scientists,* to at all times view animals as objects, just as data gathered from them must be public and objective. Does viewing animals as objects then mean that we should treat animals as mere objects? These scientists would most likely say, "Of course not, treat them well." But we will have no scientific basis to discuss why, unless we at least de facto accept the call for the fifth aim. In other words, insofar as we do not view other animals as mere objects, we need to inquire rigorously as to why not. I submit that the experiences animals may have provides the why not. Adding this fifth aim will return ethology to its prescient avant-garde roots by providing legitimacy for work in this area by those so interested and who think the enterprise valid. The results of such work, as on assessing the experience of suffering, may be really needed by all of us using animals in research. Philosophical reflection and analysis can play an important role in posing questions and clarifying concepts (Nagel, 1974; Russow, 1982; Bekoff & Jamieson, 1991), but science needs creative methods and concentrated systematic research. Before addressing such research, however, another matter intrudes.

WHAT IS THE MOST APPROPRIATE TERM FOR THE FIFTH AIM?

> I call this new individual world that has arisen in the central nervous system of the higher animals the counterworld of the animals. . . . The environment as reflected in the counterworld of the animal is always a part of the animal itself, constructed by its organization, and processed into an indissoluble whole with the animal itself.
>
> —von Uexküll, "Environment (Umwelt) and the Inner World of Animals"

> Radical behaviorism . . . can . . . consider events taking place in the private world within the skin. It does not call these events unobservable, and it does not dismiss them as subjective. . . . The position can be stated as follows: what is felt or introspectively observed is not some nonphysical world of consciousness, mind, or mental life but the observer's own body.
>
> —Skinner, *About Behaviorism*

I do not advocate cognition as the fifth ethological aim for two reasons. As noted above, these are that, first, cognition does not necessarily include emotional/affective or motivational/intentional aspects of psychological phenomena, and second, it may conflate the study of felt experience and the other four ethological aims. Cognitive ethology as a label has been immensely valuable in opening up study on questions most ethologists have previously ignored because of concerns about their scientific relevance, appropriateness, or respectability (Ristau, 1991b; Burghardt, 1992b). Other chapters in this volume detail some of the consequences, good and bad. In short, the explosion of research and interest has made the need for a fifth aim more timely.

But what to call this fifth aim now becomes a critical issue. The wrong term can lead to misunderstanding, polemical outbursts and treatises, and slow the scientific effort. The genius in the terms selected by Tinbergen was their relative clarity. Here I cannot discuss all the candidate terms I have uncovered or that were generously brought to my attention.

A main consideration was that the label be compatible with the four existing ethological aims and their common labels. Another was that it be a term without so much excess or controversial meaning that it would not be accepted readily. This precludes terms such as *consciousness* and *subjective*, which also are not inclusive of the proposed subject matter. The latter might be deemed the best apposition to "objective" (in the sense of publicly verifiable), but in fact can refer to subjective rather than objective methods, not just the subject of the subject's subjective life. Another constraint was that it not be overly identifiable with a specific school of thought. This eliminated such terms as *phenomenology* (even von Uexküll's phenomenal world) and the Skinnerian *private events*. The term also had to refer to the problem area and not to a methodology. Thus empathy, sympathy, intuition, and anthropomorphism were out of the running. I also thought the term should not be slanted toward a focus on either the individual or the species, gender, age, or population analysis of experiential events. Thus self-world was ruled out.

The most appropriate term may well be *private experience* and it is thus incorporated into the revised list of ethological aims (Table 20-1). *Experience* has a meaning in ethology and psychology where it refers to the conditions and stimuli presented to organisms, as in studies of "early experience." But this is an example of misplaced concreteness insofar as the experience is treated solely as an environmental manipulation independent of the animal to which the events are occurring. Appending the term *private* makes clear that the experience this aim investigates is that available to the organism. Complete knowledge of this lived experience is impossible to another organism. Private experience thus acknowledges those critics who have pointed out that the inner world of animals can never be directly known. Yet it isolates out this very area for study and analysis to determine what we can know or reasonably conclude. Partial understanding is obtainable (von Uexküll, 1934/1957) and is, in fact, essential to fulfilling our obligation to treat well our nonhuman research subjects, the other species living on our planet, and, of course, conspecifics.

Dogmatism here is not advisable, however. Privacy is, after all, relative, and can be invaded. And this goes beyond the taking of private (material) property. From Freudian slips to lies, secrets can be breached from analysis of words, facial expressions, brain scans, and electrical skin conductance. That no method is always correct is not a valid critique. It is also wise not to preclude what we will learn to do in the future. Already thinking in certain ways can be localized in the brain and people can be taught to produce reliable brain activity that can be externally detected and prompt computers. Some impressive virtual reality technology is already here that provides realistic experiences of such events as white-water rafting from one's living room. Creating such experiences for other species and monitoring behavior and physiology is not out of the question. Private experience, just as private phone lines, electronic mail, and correspondence, can be partially uncovered. Doing so may allow better inferences to be made about overt events and their underlying motivation and context. This has been known to biographers and historians for centuries. Still, while we may be able to understand and even imagine how Napoleon felt at Waterloo, we are limited to objective information, even if intuitively processed.

Some may argue that if complete knowledge in an area can never be obtained then the area is not amenable to scientific analysis at all. This view reflects a grammar school notion of what science is about. Scientific methods are certainly applicable to the study of events distant in time. The inability to "be there" or to recreate in the laboratory the events studied is a point the creationists are fond of making in claiming that evolutionary science is nothing but nefarious ideology, not an empirical science. Any historical reconstruction is similarly constrained, but no one would seriously suggest abolishing such fields as cosmology, paleontology, and archeology from the house of science. Any "proof" we are likely to have of the big bang, or the origin of amniotes, or the specific events and persons that triggered human agriculture has to be based on inferred evidence, typically fragmentary, and may deal with a one-time event that can never be recreated or repeated (no replication here). In this sense cosmologists and the searchers for Eve are in a much more precarious position than the interpreters of rare, but presently still occurring, events, or those indirectly experimentally manipulating private experience. In another realm, Hoffmann (1993) criticizes the view that only now that atoms can be "seen" are we really sure that they exist:

> How then do we know? Indirectly, but quite certainly. . . . we were dead sure of the existence of atoms and knew just how atoms connected up to form molecules, before those beautiful STM pictures. . . . Chemistry, like any human activity, proceeds simultaneously on many levels, with partial understanding, always incomplete, sometimes wrong, incredibly, mostly right. (p. 12)

We, of course, are far from being as exact a science as chemistry, but then they had their alchemy, phlogiston, and other detours also. Science is best conceived as a process for gaining improved understanding, a search for truth without any hope that we will ever fully attain it.

HOW CAN PRIVATE EXPERIENCE BE STUDIED?

> Since the difficulty in knowing the subjective experience is the same in our fellow man and the animals, the method of overcoming this difficulty must be the same in both cases. With the animals, therefore, our knowledge of their psychic life is also primarily based on sympathetic intuition. Let no one accuse us of unscientific phantasy or mysticism. Every one of us possesses this faculty and makes use of it in his everyday life
>
> —Bierens de Haan, *Animal Psychology*

> The question then is raised whether one might start with the intuitive categories of emotion and discover by analysis what behavioral relationships they are based on. In this way one might objectify the categorization and make it suitable for the purposes of a scientific comparative psychology.
>
> —Hebb, "Emotion in Man and Animal"

This is the most difficult issue and can only be partially addressed here. Clearly, we do not yet know the best ways to study private experience. But of course that is why I am proposing the fifth aim in the first place, to find out what we can find out. But adding to the promise of the new effort is the explosion of new techniques to assess activity in various parts of the brain in both intact humans and nonhuman animals. Some possible applications to accessing nonhuman private experience are developed elsewhere (Burghardt, 1994, 1995). New methods in cognitive science also allow finer distinctions and the isolation of processes formerly hidden from view (e.g., Edwards & Potter, 1993). Gains in our knowledge of sensory processes, perception, preferences, motor abilities and their nervous system linkages will thus provide exciting new possibilities barely imagined but a few years ago.

Although Skinner and the radical behaviorists claimed their approach, unlike methodological behaviorism, could accommodate private experience, it is clear that they have carried out little such research with animals. The following quote tells us why.

> Other species are also conscious in the sense of being under stimulus control. They feel pain in the sense of responding to painful stimuli, as they see a light or hear a sound in the sense of responding appropriately, but no verbal contingencies make them conscious of pain in the sense of feeling that they are feeling, or of a light or sound in the sense of seeing that they are seeing or hearing that they are hearing. (Skinner, 1974, p. 242–3)

By not being able to express feelings or ideas verbally, nonhuman animals cannot, for example, feel that they are feeling and are not 'conscious' in any way comparable to people. Skinner (1961) outlines four methods in which a "verbal community which has no access to a private stimulus may generate verbal behavior in response to it" (p. 276). Indeed, private experience is a product of learning in a language-based society. Thus, these behaviorists, as Watson before them (Burghardt, 1985a), are language chauvinists and apparently believers in as sharp a human-animal discontinuity as was Descartes.

But if the behaviorists are little help empirically, do we retreat to the approach of Bierens de Haan (1947) and the wisdom of D. O. Hebb (1946), who openly advocated a wide open exploration of animal mental life using almost folk psychology methods? These must be a part, but only a part, of our methodological toolbox. Inferences about the experienced life of other species can be approached through applying a critical anthropomorphism in which our statements about animal joy and suffering, hunger and stress, images and friendships, are based on careful knowledge of the species and the individual, careful observation, behavioral and neuroscience research, our own empathy and intuition, and constantly refined publicly verifiable predictions (Burghardt, 1985a; Morton, Burghardt, & Smith, 1990). "Descriptive mentalism" (Thompson, 1987) also captures much of what I think is involved. We formulate interpretations that can be tested in future observations or experiments with the same or other individuals. We ask animals questions to be answered by their behavior or physiology (not verbal or written reports), and these answers allow us to judge the extent and kind of private experience the animal is having. By posing questions anthropomorphically, but thinking about them critically, we can develop specific hypotheses and even the methods to test them, as shown in some of the most creative, influential early work in animal psychology (e.g., Köhler, 1917/1927; Tolman, 1932) and in the work of those who were influenced by such studies in their work with humans (e.g., Piaget, 1952; Werner, 1948). Although they do not discuss the issue at length, I think that Cheney and Seyfarth (1990) do an admirable job of using a critical anthropomorphism to pose novel questions and to develop simple but ingenious methods to answer them in their study of the world view of vervet monkeys. Unfortunately, these authors, as many inspired by cognitive ethology rather than cognitive or learning psychology, often blur the mechanism and private experience issues (as discussed above), even though they recognize the distinction (Burghardt, 1992b). The proposed fifth aim will help to keep this distinction at the fore, just as the aims of control and function led to the recognition of the confusion between proximate and distal (ultimate) processes in controversies of the past, such as those involving play (Burghardt, 1984).

Elsewhere I have discussed the early efforts of the first comparative psychologists, particularly Romanes (1883) and Morgan (1894), to gain knowledge of animal mental life (Burghardt, 1985a). Romanes used a method he called "ejectivism" and Morgan developed the "doubly inductive method." The essential feature of both methods is the use of objective knowledge of the animal's behavior along with subjective analogical inference (using our own introspected experience) and neural analogical inference (using the extent of the animal's neural development and physiology to infer possible mental abilities or states as in today's sleep/dream research). Although these methods failed to convince even sympathetic scientists (e.g., Jennings 1906/1976) early in this century, if carefully used and combined in a critical anthropomorphism they may yet be helpful. But although Morgan's doubly inductive method is a useful model, his certain faith in introspective psychology as the firm bedrock for ascertaining truth is, as pointed out above, not to be emulated today.

FROM ANECDOTE TO MATURE SCIENCE

... we want to try to give an account of which view, according to our present knowledge, best corresponds to the counterworld. This can happen only by means of suggestion and must necessarily remain very incomplete until more obser-

> vational material is collected. But in any case a visually clear
> conception is useful because, on the one hand, it aids in the
> clear statement of questions, and on the other hand, it gives
> us a feeling for the overall context. . . . The earthworm can
> serve as a starting point for our observations.
>
> —von Uexküll, "Environment (Umwelt)
> and the Inner World of Animals"

Anecdotes and anthropomorphism are often discussed together, as in this volume. Why? Historically this may be due to the fact that anthropomorphic human-based interpretations were tied to uncritical anecdotes and, as Washburn (1908) in her fine but generally ignored (Burghardt, 1973) analysis pointed out, the human tendency to want to tell a good story.

As an example of the linkage between the ill-considered anecdote and anthropomorphism consider the quote from Trivers (1991) on the head-cocking insect as evidence of consciousness.

> The image I have of a conscious animal is one in which a light is on
> inside the organism. In this sense, insects are certainly conscious: there
> is a light turned on inside them when you interact with them. For
> example, I may try to countersing with a male or make a series of little
> threatening moves or even friendly ones, pseudofriendly ones. You
> can certainly see the insect cock its head and try to get a fix on me
> from several angles to figure out what on earth I am doing. (Trivers,
> 1991, p. 179)

Now there are many problems with this passage, from the lack of giving the reader any indication of what species of insect was observed and other details to the kind of language the author used. Most likely the insect was a rather large one, at least of grasshopper size. It was also probably one with good vision, perhaps a mantid, but also one that sings.

Shortly after I read Trivers's comments I came across an article by Srinivasan (1992) on distance perception in insects. This brief paper reviews a series of studies on how insects estimate distance, an important question since their visual apparatus is so different from ours. Only mantids have stereopsis, for example. It seems that peering, or visual scanning, in locusts and other insects is a very simple but elegant means of determining whether an object is moving and in what direction, via motion parallax. Presenting targets that were stationary or moving with or against the peering movements of the insect "demonstrated clearly that the peering locust was estimating the range of the target in terms of the motion of its image on the retina. We now know that certain other insects, such as grasshoppers and mantids, also use peering to measure object range" (Srinivasan, 1992, p. 23). The author does not rule out mental events in insects, as in the discussion of a "mental image" of a landmark. What the experiments have done is give us some understanding of the visual world of some insects, rather than using "consciousness" merely as a heavily freighted synonym for responsiveness, as did Trivers. Trivers gave us a crude anecdote to make a point and provided the novice reader with an amusing picture of *Trivers* playing peekaboo with an insect. But the true nature and complexity of the behavior he saw was not conveyed. Indeed, this example is one of loose speculation in which the more fascinating current understanding of the behavior is not even regarded as worthy of scholarly exploration. This is the kind of crude anthropomorphism that a critical anthropomorphism should aid scientists in avoiding.

As scientists trained in and imbued with mechanistic or functional approaches ("inclusive behaviorists" according to Griffin, 1984) start, or are forced, to also contemplate the experienced world of other animals, it is understandable that there will be a "gee whiz" period in which uncritical anecdotal accounts and uncritical anthropomorphism abound. We now do seem to be at the stage where we can say that how animals perceive, experience, and comprehend their world is an important question. The lesson of ethology in reference to traditional psychology is that naturalistic approaches are more likely to be informative than using the animal as a convenient "model" to investigate human-behavior-derived questions that ignore the evolutionary, physiological, and ecological nature of the model species. Probably the best way of identifying those who fall into this error is to check whether closely related species are considered before making comparative extrapolations across taxa (Burghardt & Gittleman, 1990). The dropping of the modifier comparative from ethology seems to have encouraged this tendency even among ethologists and behavioral ecologists.

An analogous situation that also derives from a fascination with the behavior of animals is the widespread speculation and adaptationist just so stories about the evolution and function of all kinds of human and animal structures and behavior patterns. These became controversial in sociobiology, but were controversial before that as well. Such adaptationist stories have been widely and severely criticized. But evolutionary biologists do not then conclude that the adaptationist approach is impossible; rather they conclude that we need to use appropriate and currently developed methods to make inferences about both the current utility and phylogenetic history of such behavior patterns (Brooks & McLennan, 1990; Harvey & Pagel, 1991).

Whiten and Byrne (1988) have made the case that rare and adventitious observations can be an important part of the science of cognitive abilities in animals, such as deception by primates. They solicited, collected, and organized anecdotes to discover some common themes, species differences, contextual differences, and so on. A problem with this method is that the anecdotes were collected by different people studying different phenomena using different methods. The case could be made, of course, that any commonalities are thereby strengthened. Alternatively, all could have made similar interpretive errors. Hopefully, however, we can develop additional methods to evaluate interpretations based on collected anecdotes.

A systematic study of infants, children, and even human-animal interactions (Burghardt, 1985a; Davis & Balfour, 1992a) may provide rich material for identifying different kinds of private experiences and then testing the resulting hypotheses. The rich descriptions of dog-human play provided by Mitchell and Thompson (1991) are an example where experiential descriptions are eventually necessary for the most complete understanding.

Frans de Waal (1991) presents a hierarchical typology of complementary methods that is an excellent starting point for investigations of private experience as well as other topics (see also Silverman, this volume). These progress from (1) *qualitative observation* (careful natural history observations or behavioral anecdotes) which can raise issues but cannot conclusively deal with alternative explanations to (2) *quantitative observation* that systematically gathers data on the range of alternative explanations as well as compares alternatives. (3) *Controlled observation* compares behavior before and after a certain event occurs spontaneously and also compares results with matched controls in which the event did not occur. (4) *Experimentation* is deliberate manipulation and is most powerful in deciding among alternative explanations. However, due to the often artificial settings and human involvement, findings must also be evaluated in light of results gathered in the observational settings.

Another approach, developed for assessing psychological well-being of captive primates as mandated by recent U.S. legislation, is presented by Novak and Suomi (1988). Four aspects are considered, including physical health, ability to engage in species typical behavior, signs of distress or chronic stress, and ability to adapt to new situations. This approach can be useful in evaluating the life experience of the animals but is largely a descriptive statement of status for evaluating the appropriateness of captive management. Thus, without applying the methods outlined by de Waal, we gain little insight into novel psychological states. To be fully useful we need to know much more about the private experience of primates in order to develop a more accurate, reliable, and valid set of items for inspectors to check off on their rounds. Clearly, pursuing the fifth aim will have many applications in captive animal maintenance.

Anthropomorphism is an inherent human propensity that is many faceted (Fisher 1991). No one can avoid it; in fact, anthropomorphism probably had extraordinary adaptive value in dealing with animals as well as conspecifics throughout human evolution (Burghardt & Herzog 1989). Scholars have long recognized that anthropomorphic assessments often work in the sense of facilitating predictions about an animal's behavior (e.g., Morgan, 1894; Hebb, 1946). Those who think they have avoided all anthropomorphic terms (e.g., color) have had to do so with extreme deliberation, focusing on highly selected activities, and then they often risk being misunderstood by not using clear language. The challenge is to use our anthropomorphic propensity critically and with scientific and objective referents to both data and our propensity. Regardless of the fate of the term *critical anthropomorphism*, I am convinced that we will all be using some such concept more openly in the future and not be defensive or apologetic. As we discover how often research agendas and approaches are nonobjective, we will discover the value of working with, and not against, our human qualities in developing hypotheses. The fact that we all have different private experienced worlds will both enrich and delimit our progress. The process will be comparable to how consideration of cultural, gender, and age diversity puts the brakes on overly enthusiastic extrapolation of research findings on nutrition, medical care, and education to groups not studied.

BUT DOESN'T THIS REALLY MEAN . . . ?

This proposal for a fifth ethological aim centered on the analysis of private experience is a radical departure from current stipulated practice in ethology and I have found that a number of objections are frequently raised. The answers are, I believe, dealt within the body of this chapter, although each is worthy of longer treatment. Here I will only list and answer a few of them.

QUESTION: Isn't the study of private experience a return to body-mind dualism? Introspection?
ANSWER: No and no.

QUESTION: Is private experience just another term for consciousness?

ANSWER: There are too many typologies of consciousness out there for a simple answer, but in any event the call here is for us to go beyond general states that explain little to a focus on the experience of specific life events.

QUESTION: Does this fifth aim necessitate a class of special knowledge that cannot be known directly by anyone other than the experiencing organism?

ANSWER: It assumes that organisms may experience events in ways we have difficulty sharing and describing. But whether they have access to knowledge that we cannot appreciate, and whether or not their experiences provides accurate or useful information, are questions that should yield to scientific inquiry. The distinction between first person and third person knowledge is readily accepted in other fields (e.g., history).

QUESTION: Does removing private experiences from the other four ethological aims, such as control or ontogeny, imply they are mere epiphenomena?

ANSWER: Absolutely not. What it does do is remove the phenomena themselves from debates such as whether such experiences are causes (traditional mentalism), to be studied only through ontogeny (radical behaviorism), and so on, and focus on how we can identify and describe them so that we can study them effectively through the other four aims.

QUESTION: Tinbergen's four aims all refer to behavior. Doesn't the private experience aim refer to different phenomena that, if they are to be considered at all, just need to be looked at with the existing four aims?

ANSWER: A very good question. But the answer is no for at least three reasons. First, what ethologist and psychologists meant by behavior was relatively clear to all scientists, even if they chose to emphasize different aspects. We still need to characterize what will go under the rubric of the fifth aim and how to measure and study it. We need to understand how we can construct an 'ethogram' of private experience comparable to a behavioral one. Second, private experiences of some sort are postulated as potentially important aspects of any and all behavior in animals. They may have a variety of consequences that we can only dimly imagine. They may also be occurring even if the animal is not overtly behaving. Third, without special highlighting, the difficulty of even conceptualizing the issues of private experience will lead to continued formal suppression of the topic. This may also be contributing to the uncritical speculative mindtalk that even otherwise rigorous scientists are too often tempted to espouse.

QUESTION: Is not an incorporation of our own often unreliable and incomplete personal experience too weak a reed to secure a science?

ANSWER: There are two ways to answer. First, using our personal experience to originate hypotheses we may otherwise never consider is one thing, basing a science on it is another. Anthropomorphism and introspection may do both, but critical anthropomorphism only does the former. Second, we can ascertain which aspects of our own experience are common or universal and capable of being shared with conspecifics. This sharing may be cultural or gender specific. What can males really know about the experience of pregnancy and childbirth? How can we understand what it is like to be a member of another racial, social, or age class? If trying to answer such questions is without value in understanding other people, then I will agree that going across species boundaries to study life as experienced is futile.

QUESTION: While the goal here is admirable, I still say we lack useful methods to understand private experience and none will ever become available.

ANSWER: Statements of faith are not science. I prefer a faith that looks toward new horizons and understudied phenomena, and that recognizes incongruities indicating that not all is well with our traditional world view. Slight cracks in the walls that may signal serious faults in the entire edifice have indeed appeared. Do we keep applying dry wall spackling to cover up increasingly numerous fault lines? Or do we examine the foundation and see if a major renovation is necessary?

SUMMARY AND CONCLUSION

In recent years discussion has been heated and intense concerning aspects of human and nonhuman animal life that seem to lie outside of the Tinbergian four aims, however integrated or synthesized. I propose that a fifth aim be recognized that focuses on *private experience* and encompasses phenomena variously called conscious, aware, subjective, emotional, and phenomenal. This fifth aim is nothing less than a deliberate attempt to understand the private experience, including the perceptual world and mental states, of other organisms. The term private experience is advanced as a preferred label that is most inclusive of the full range of phenomena that have been identified without prejudging any particular theoretical or methodological approach. It is beyond this chapter to discuss why ethologists grew to avoid an open concern with subjective processes and anthropomorphism, or to relate this perspective to current philosophical discourse. It is important to maintain a balanced perspective that does not dismiss the potential role of individual personal experience in studying behavior. In the past this dismissal has occurred by treating such phenomena as outside science, by explaining them away as something else and thus effectively ignoring them, or by embracing uncritically a methodology and conceptual framework that may prove to be insular, ineffective, untestable, and ultimately incapable of being integrated with the standard four aims of ethology.

If I have been successful, the arguments and analyses have established that incorporating such a fifth aim is worthwhile, timely, and fascinating. If we do not deal with these issues we will find ourselves derailed from the advancing frontiers in behavioral and neural science, lacking the ability to attract the next generation of creative students, and excluded from full participation in societal policy debates on the treatment of animals, conspecifics included. Certainly progress may not be quick or easy, and solutions discovered may meet resistance from the scientific establishment. But ethology has a history of successfully challenging the scientific status quo concerning topics that conservative authorities claimed were incapable of scientific study, such as innate behavior, behavioral phylogenies, neurophysiology of instinct and communication, identifying maturation and experience in development, isolating genetic factors in complex social behavior, and applying evolutionary methods to understand human behavior and psychology.

21

A Phenomenological Approach to the Study of Nonhuman Animals

Kenneth J. Shapiro

The phenomenological method presented here uses kinesthetic empathy, social construction, and individual history to understand a dog, Sabaka.[1] Because the animal is nonverbal, empathic access is through posture, gesture, and behavior of the animal. I describe two vignettes of Sabaka's behavior in context, and reflect on the phenomenology of Sabaka's experience.

To help locate what is meant by kinesthetic empathy, the primary investigatory posture in the present method, I begin with a working example. Following the discussion of empathy, I further describe the two other moves, social construction and individual history, which comprise the method. Since a phenomenologically based method, particularly one applied to the study of a nonhuman animal, is probably unfamiliar to most readers, I will occasionally comment on the method in the results section as well. In a concluding section I will discuss the method more critically and in the context of some of the issues raised in this volume.

METHOD

I have of necessity spent much time in hospitals this past year, not as a patient but, I have discovered in some ways less fortunately, as a "visitor." I am watching a simple procedure—the doctor is giving the patient an injection, in the patient's bottom. After the fact, I find that my body had arched, pelvis forward as had the patient's body, the other's body, let us say, your body. I had not directly been aware that you had moved, let alone that I had. Now as I reflect on that bodily attitude, I sense that it embodies a refusal of the illness on your part. In the posture which I unwittingly imitated, you are closing yourself off—not just from the pain of the treatment but from the treatment itself, for it implies that you have the disease.

A kinesthetic empathy is possible which consists of the meaningful actual or virtual imitation or enactment of bodily moves. It is possible because we both have a living, mobile, intending body. Further, the particularity, richness, and confidence with respect to the meaning given in my empathic sense of your experience in any moment is a function of my knowledge of both general social construction of, here, disease metaphors and hospital situations in a given culture and, as well, knowledge of your history. Informed by such implicit prior understandings, the investigator can utilize sensitivity to his or her own body to explicate the experience of the individual who is the object of study.

Empathy is the direct apprehension of the intent, project, attitude, and experience of the other. It is direct in the sense that, phenomenologically, it is given as unmediated—as more akin to the forms of presentation of the perceptual and intuitive modes than of the deductive or inductive modes of discursive thought. That which is apprehended is not limited to "sensory feeling,"[2] as I wince when the door catches your or my dog's foot. Rather, it includes a more general sense of your situation, as, in the example, your refusal to accept your illness. More generally, I can also directly apprehend your or a nonhuman animal's project, purpose, or anticipated end.

Empathy is a general access to the intended world of another person. However, that access is not accomplished by inference. It is not of the form—that action in that context implies this intention. Nor is it gained through analogy. Again, it is not of the form—if I made that kind of move in relation to that kind of object, I would be having this kind of intention; and so you probably are too. More subtly, by empathy I do not refer to self-identification, to an act by which I put myself in your shoes. Nor, finally, does empathy refer to a bodily complementarity, as when your body invites mine to adopt a certain fit with yours—to support you while you lean, or be encompassed by your position. Of course, all of these postures are possible and potentially informative. However, here I take empathy to refer to a moment in which I, if only focally, forget myself and directly sense what you are experiencing. We all have at least fleeting moments of this. A contemporary approach to psychotherapy, Rogerian or client-centered therapy is predicated on the possibility and therapeutic power of consistently empathizing with the experience of another person and sharing a description of that experience with him or her (Rogers, 1951/1965).

However, in reflection we discover that the structure of the empathic moment is more complex than a momentary leave-taking of the self. The forgetting of the self is relative for implicit in the empathic moment are several features which point back to that momentarily passed-over self: a variable sense of confidence as to whether the empathically given meaning is accurate; a sense of control which allows me to stop experiencing what you are experiencing, to have some distance from it. More generally, then, there is the sense in empathy that this is the experience *of the other*—as direct and poignant as the given sense of refusal is, it is not *my* refusal, it is not *my* experience.

As imperfect as empathy is, how is it possible at all? Here I can only suggest an outline to an answer through a set of concepts adopted from several phenomenological thinkers. All experiencing is sensuous in that it involves bodily sensibility (Sokolowski, 1985, p. 23). Thought as well as perception, ideas as well as sights, states of mind as well as tickles can be "registered." As distinguished from a reported fact, a registered thought or a registered percept is "grasped" or "appropriated" or allowed to "sink in" (pp. 34–36). I can think that the house is abandoned and merely report it to myself or to you; or I can actually let that thought sink in, actually have some intuitive presentation of an abandoned house, a beached whale, or even a complex concept. This potential or implicit sensibility, bodily presence, or possibility of "taking up" is a helpful first step in thinking about how empathy is possible.

All experience is at least potentially embodiable. This is the case because an individual's lived body is his or her access to objects (Merleau-Ponty, 1962). To perceive is to live the object, to inhabit it through the potential mobility of the lived body (pp. 67–68). What is meant by this is perhaps clearer in examples of objects to be manipulated or dealt with physically—a hammer or reclining chair or jungle gym. When I look at the jungle gym, I can immediately see it, that is without yet reflecting on my perception, as a set of possible moves which it would allow my body to make were I actually over at it. For Merleau-Ponty all objects invite a motor intention with respect to them and this is the ground of our perception of them. To see an object is to virtually go over to it, walk around it, try it on. While immediately given in the moment, the meaningfulness of our perception is informed by past experience with and thought about related objects and situations.

Following Polanyi (1967), intentionality has a from/to structure. We attend *from* that lived body *to* the object of our experience. To perceive an object, then, is to focus primarily on it while my own body is present only as a background or "subsidiary" awareness (p. 10). In these terms, empathy is a second-order application of this notion of intentionality. Empathic experience involves appropriating a second body which then becomes my subsidiary focus. Through the potential mobility of my own body, I can virtually accompany yours as it perceives an object.

To concretize how this is possible, consider the following progression which shows how we extend our body to incorporate and inhabit other objects and, eventually, other lived bodies. We can readily extend our effective arm so that it becomes coterminous with our tennis racket. The tennis racket is integrated into our forgotten body for the effective return of service. However, I can extend the sensational as well as the manipulative power of my body. For example, I can directly know the obstacles in my path by extending my body into my blindperson's stick, which is then part of my auxiliary focus or "forgotten body." Finally, we can transform our body, so that we incorporate not only the width and length but also the complex mobile capacities of our automobile, which we then realize in the directly and immediately given sense that that truck heading toward us cannot be avoided.

Still, how can we understand the possibility of inhabiting another person's body? Following the critique found in much phenomenological literature, it is helpful to rethink the common Cartesian ontology which locates us holed up in our bodies as some mental stuff; and, instead, to recognize that we are out there in the world *through* our bodies. Our bodies do not encase us; rather we are our bodies. Given that identity, we are radically in touch with, immediately over at things, and, as well, over at other living bodies in the world. It is also helpful to recognize that we human animals have the same basic bodily possibilities—to move toward and away from, to effect, manipulate, gesture, posture, and the like. This is consistent with the familiar observation that an infant can immediately take up the postures and moods of others (Meltzoff, 1990). In a sense much of our later development is learning not to do so—not to be imitative and empathic so that we can attain the distance and externality necessary to make inferences and think logically.

With these briefly noted considerations, we can begin to understand the possibility of empathizing with or inhabiting the body of a being of another species. For, like us, nonhuman animals are intentional beings who move in purposeful ways, who run into barriers, reach for things and find other things unreachable; who also posture, gesture, effect, and manipulate. (For a discussion of the opposite thesis, that nonhuman animals are so "essentially alien" as to preclude any possible empathic understanding of their world, see Nagel, 1974, and Akins, 1990). The assumption that differentness and inaccessibility across species lines constitutes a more radical cleavage than that across culture, history, or even developmental periods is often an unexamined and speciesist prejudice.

Unlike us, however, nonhuman animals live more exclusively in a region where meaning is embodied and directly enacted but remains nonlinguistic. While signification for us more typically involves a complex dialectic between signifiers and the signified, we also live in and have access to a "non-" or at least prelinguistic bodily region, and, in fact, arguably must do so to experience meaning (Gendlin, 1962). If empathy can be conceptualized largely at the level of the lived body, as I am arguing here, then nonhuman animals are, likely, a more propitious object of study for a methodology featuring empathy than are humans.

A final argument for the possibility of empathizing with animals is from common experience. Most of us relate to nonhuman animals, particularly companion animals, as if we, at least at times, immediately apprehend their concerns, projects, and experience—at least until we are trained not to do so.

I have argued that empathy refers to a directly given sense of another being's experience. However, this claim does not imply that that immediately given apprehension of another's world is not influenced by the investigator's own history, biases, intended project, and the like. Our access to each other is not that of a transparency that allows a direct tracing. To the contrary, empathy must be understood in the context of an interpretative act. In the same way that, as a postmodernist, I reject the possibility of an unbiased objectivistic stance, particularly its usefulness when applied in the social sciences, I do not conceive of empathy as a transparent access, a pure mirroring of what we seek to understand. Rather, the empathic move is necessarily part of an interpretative approach. The understanding I bring to the object of study necessarily affects and is affected by that object such that I am involved in a progressive circle of further understanding—what is called the hermeneutic circle. While the empathic act aims to deliver just what the object of study is experiencing, that act is necessarily informed by my "preunderstanding" of him or her. My lived body is continually informed by the world and subsequently takes up that world, including the other's world, differently. Because of these imperfections in understanding gained through empathy, it is critical to the present method that the investigator critically evaluate the product of his or her kinesthetic empathy. I have distinguished two regions of such reflection: social construction and history.

I use "social construction" in the broad sense of that inclusive set of attitudes and beliefs that are co-constitutive of a particular social group's perception of reality (Berger & Luckmann, 1967). As investigator, then, I have the task of clarifying the social construction of that animal in the setting in which I and the animal as object of study are embedded. Clearly, I know Sabaka, the dog that is the object of the present study, through complex and sedimented layering of, among other categories, "animal," "pet," and "dog."

For example, one important although subtle contemporary attitude toward animals is that a nonhuman animal is, as it were, a species not an individual. Influenced by this construction, often when we refer to a particular dog we mean the reified, deindividuated, generic collectivity, "the dog" rather than that individual. As a pet, an example of a contemporary attitude is the pet as a "minimal animal"—as sanitized, neutered, neotenized, fixated at adolescence (Shepard, 1978). An interesting and only emerging social construction of "dog" is as an extension of human eyes or hands, as therapeutic companion, as stress reducer.

In addition to the social constructions of the popular culture, as investigator I am also embedded in those constructions which are the meanings of animal, pet, and dog in the scientific literature. This literature is itself pluralistic and constructs the object of study differently not only in substance, through findings, but, as well, through radically different methods which imply different forms of access and relationship to animals. Ethology, comparative psychology, the behaviorist program, physiological psychology, and an emerging cognitive psychology of animals, through their respective methods, define different ways of knowing, treating, and relating to animals.

The investigator must take these social constructions, both the popular and the scientific, into account. They cannot simply be subtracted or bracketed for many are formative of the dog's experience as well as of the investigator's approach to the study of that experience. We breed and socialize dogs to realize certain social constructions. When I have kinesthetically empathized with Sabaka these social constructions come into play and, as investigator, I must attempt to clarify their presence. I must judge whether any particular feature is part of the dog's present experience or is merely projected as part of my social construction of him.

Generally, social constructions bias our perception toward an emphasis on "typifications," toward features purportedly common to the *class* of individual objects under study (Wagner, 1970). Influenced by their operation, we are less apt to be open to the individuality of the object of study. This necessitates as a corrective the second region of reflection in the present method—individual history.

While empathy is a direct apprehension, unmediated in the moment, that which I apprehend when I empathize is necessarily informed by history. For example, my immediate sense of your grief is informed by my understanding of your other losses and how you grieved them. This is the case even when my emphasis is on kinesthetic empathy and even when the object of my empathy is a nonhuman animal. In addition to reflection on social constructions, then, the investigator must reflect on his or her understanding of the history of the individual under study.

History also informs the experience of the animal. This may seem self-evident but often we fail to give to specific major events the standing of "true historical particulars" (Gould, 1987), particularly when dealing with nonhuman animals. Instead of a concrete individual history, we limit the historical to the developmental, for example, to an account of the general developmental milestones of that species. Goodall's work (1986) is an exception and shows clearly that a given individual's personality and behavior is often largely intelligible in terms of historical events involving his or her immediate family group, the larger social group, and even that group's interaction with other social groups. In like manner, certain major events had significant impact on Sabaka—the fact of his abandonment at the dump, our adoption of him and of Elkie as his companion, the death of Elkie, and our trip abroad without him.

RESULTS: DESCRIPTION AND REFLECTION

In this first brief section, I select a few features from a more extended analysis (Shapiro, 1990b) of relevant social constructions of Sabaka and of his individual history to illustrate how they are taken into account in the primary investigative posture of kinesthetic empathy. The remainder of the presentation will be in the form of two vignettes describing Sabaka's behavior and my reflections on that behavior. In phenomenological terms, the description will move from concrete accounts of a particular dog's world to some of the structures of that *Umwelt*.

Sabaka (the name comes from the Russian for dog) is a five-year-old male dog, of mixed breed. He has been in our household since he was about four weeks old, at which time we got him at a local animal shelter. He had been abandoned at a town dump. Three months or so after he joined us, we obtained a second dog, a female collie mix, probably four years old, also from a shelter. We had decided on two dogs so that they would be company for each other. Undoubtedly influenced by the local Maine practice of keeping a dog outside even in winter, as a watchdog, we planned to raise the dogs primarily in an outdoor yard with a run joining a shed in the barn. Elkie, the collie, adopted Sabaka and the two were inseparable for about a year, when Elkie died. That event changed our relation to Sabaka significantly. A second major event in Sabaka's life, in his third year, was our absence from the house for a six-month period, during which time Sabaka was left with a house-sitter.

Vignette 1

When he is outside Sabaka spends much of his time lying in a certain spot, at the head of the drive. This place allows him an optimal view both down the driveway to the street and through the windows of the kitchen. It also allows him to be in the sun. From this spot, he can comfortably half-sleep while vigilantly smelling, listening, and watching, ready to bark, bay, and half-charge at passers-by. He also can watch family comings and goings. Other places within the house offer some of these features—under the couch in the playroom, on the second-floor landing, at the threshold between the dining room and the kitchen.

Currently, Sabaka sleeps overnight on the landing, although I had originally intended for him to sleep outside. When Elkie's premature death derailed that plan, slowly Sabaka moved to sleeping arrangements closer to us. First Sabaka slept in a shed attached to the main house. However, during our six-month stay abroad, we instructed the housesitter to let him sleep inside. She was away during the day, during which time he was outside, it was very cold at night, and we felt guilty at our absence from him. Also, in retrospect, it is clear that there was a conflicting construction of "pet" at work here vying with the Maine woodsman construction of outdoor watchdog—namely, that of dog as an integral member of the family. Apparently it took family absence to give that construction formative power. In any case, Sabaka now sleeps on the landing twelve feet from my bedroom.

During the day and early evening when in the house, he stays under the couch, sometimes to be away from us as when he has done something he should not have, and sometimes to be near as when I am on the couch. When we eat in the dining room, he remains on the threshold of that room, although over the years almost imperceptibly that threshold has gotten closer and closer to my soup, as the sleeping arrangement has gotten closer to my bed. At most any time of day or night he may, if given the chance, sleep on a second favored couch in my study. It would seem that I cannot (or perhaps do not really wish to) train him otherwise. While he generally takes a somewhat distant position of surveillance with respect to us, he will quickly occupy a bedspread or cushion left on the floor and when curled up next to one of us will immediately commandeer the apparent choice center of another family member's resting spot, even as he or she is setting it up or rolling over for a second to change the channel.

Reflection 1. Apparently, Sabaka lives space in various ways. Some of these have been described in ethological literature on dogs, and on their evolutionary ancestors, wolves. With respect to their instinctive behavioral patterns, both Scott and Fuller (1965) and Fox (1978) assert that dogs retain much in common with wolves. Most of the activities I have just described fit between a dog's territorial space, that space which is defended, marked, and tracked and, on the other hand, the "personal space" (Katz, 1937, p. 95) at the border of which and within which a dog performs numerous complex greeting, courting, dominance determination, and care-soliciting behaviors. Between, then, territorial and personal space is what I will refer to as "the space of place."

Within the literature on Canidae, instinctive patterns are described under the rubric of lair behavior—shelter and care seeking, building and maintenance behaviors (Scott & Fuller, 1965, pp. 64–65). Informed by these descriptions and their, typically, functional or evolutionary explanatory accounts, this investigator then returns to the animal under study and attempts to empathize kinesthetically with his or her lived sense of these activities. What is Sabaka's bodily experience of the space of place?

I want to appreciate directly Sabaka's bodily experience, his posture, attitude, incipient and actual moves and be carried along toward them as features of his own intended world. As I watch him in this way, I sense that he spends much time seeking and checking on previously established places. As he approaches a prospective place his bodily posture already begins to assume the contour and, as well, appreciate the lookout that the prospective place would offer. He begins to circle it and to curl and lower his body. There is more to understanding this than accepting it as the vestigial instinctive grass-flattening or snake-checking behavior of his wolf ancestors. In his bodily attitude I am aware of his sense of how this space could contain him. He is, as it were, trying it on for size. He is seeking a kind of space which he already knows bodily. It is an optimal resting place that provides a sense of the protection and lookout advantage given by a partial enclosure. It also allows comfort, the warmth of the sun, or the softness of the carpet. As a vantage point it is both a lookout or rather smelling station or listening post for detecting outside threat and, at the same time, it is a place that allows him to keep track of our presence and that gives him a sense of being with or close to us. In it he is in the family lair.

Once in that space of place he lives it in a certain bodily way. He curls his body in the recess for physical warmth and for closeness to the pack or family of which he is a member; he sighs and purrs at this contentment and security much like he does when petted; he lies oriented to keep watch for both strangers and for the possibility of even more access to the family hearth. But, again, he already assumes this posture as a kind of set, as the project of finding such places.

More generally, I sense that Sabaka's bodily experience intends objects in the world as possible sites of his inhabitation. He is looking for potentially secure places. One way Sabaka lives space is as to be appropriated, as to be incorporated so that it will serve as and become his point of view. He can assume a bodily attitude which intends complex configured objects as to-be-lived-in and lived-from, as optimal vantage points, as advantages.

He inhabits the spaces of place in at least two senses. He tries them out by virtually dwelling in them, that is in anticipation of the actual; and in that, once established, they are his habitations—etymologically from *habeo*, to have and hold (Jager, 1983, p. 156). The historical account of the journey of his primary sleeping place moving closer and closer to the family lair shows that this appropriation of place can be an ongoing project. Sabaka slowly, over a long period of time, whittled away at the boundaries of permissible sites to establish his preferred space of place.

While the emphasis on spatiality here admittedly may reflect the peculiar construction of pets in the Western world and/or the investigatory posture being utilized, speculatively, I would suggest that spatiality may ground the being of Sabaka in the way that it is often claimed that temporality grounds human being (Heidegger, 1962). This is not to say that there are not temporal structures operative in his experience. However and more particularly, I would suggest that place primarily grounds being for Sabaka. He belongs in the place and relates to others from and through that place. He can just lie there for hours because he is not primarily waiting, he is not primarily anticipating; he is already arrived, he is at home.

Correlatively, his is a spatial identity. In contrast to a reflective self that is constituted and developed as a unity through and over time, his is a self constituted through association with spaces. Sabaka's habitat, that space he inhabits, is his self in the sense that, while he is in it, it provides his point of view on the world. The space he has and holds is his appropriated self. In that he cannot disembed or reflect on his position (as we will demonstrate and discuss in the following vignette/reflection), that self is radically place-dependent.

Vignette 2

It is late in the day and I am trying to finish some work that has taken me considerably longer than I had anticipated. I become aware of Sabaka. From my study I can see him coming down the stairs in the hallway. He is moving with that slow, angled inward cautious placement of one foot at a time which is his gait when he has just woken up and is not yet fully mobilized. I recall now having walked past him three hours earlier while he was lying in his favorite spot on the landing. I had said to him, "You wait, Sabaka, I'll take the puppy"—a reference to our custom of going for a late afternoon walk together. My comment got little response as he continued to lie there eyes opened, attentive but expressionless. I patted him in passing without breaking stride and he took a slightly longer breath, settling in, becoming slightly less vigilant. I realize now with discomfort that I had failed to keep the promise.

He comes now almost to the threshold of my study and sits looking at me while he bends his head toward his right rear paw, which is raised motionless—possibly testing the activity of the fleas at the base of his ear. He does not scratch but goes down from the sitting position, slowly, to a passive crouch. As I now begin to close up shop, shutting down the computer, rustling papers, I am aware that he is watching me. At first he stays motionless with only his brown eyes following me as I move from desk to files. After a minute or two, he begins to change from this passive recline. He draws his stretched out forepaws in toward his face; he raises his haunches ever so slightly to a more action-readied position. Now he is posturally orienting not just his eyes but his whole body toward me as I move. His tail has some slight lift—incipient wag. As this develops, he is getting more restless, sighing, changing postures suddenly, almost jerkily. I feel the burden of his impatience and find myself geared more to readying the paraphernalia of our walk than to complying with my compulsive shop-closing protocol—so, walking boots, longsleeve shirt, dog biscuits, whistle.

Now as I wander more broadly around the first floor, he is following me, tail lifted and wagging, bright eyed, up on his toes, a spring in his gait and, as our walk now is clearly imminent, occasionally punctuating his movement by a play-invitation posture (down on his front legs, up on his rear, sporadic short barks). When my moves take me toward the back door, he continues toward it even when I veer away to go to the coat rack to get my hat; he orients himself toward the door, prancing and looking back at me. When we are finally back at the door, he leaps up toward the latch again and again.

Once in the adjoining shed, sometimes I reach for the leash before letting him out and he carries it in his mouth. By now he is wagging so vigorously that most of his body is as tail—it half circles me, half directs me down the drive. At other times, when the red squirrel has come in from the woods to check the spillings at the birdfeeder, I open the exterior door with a "Where's that squirrel?" and Sabaka explodes like a thoroughbred out of the starting cage and beelines, not for the presumptive presence of the squirrel at the feeder, but for the rear of the barn where he correctly anticipates the squirrel will make his leap from the roof to the trees for his escape.

After we cross into the field and are well away from the road, I take him off the leash and he sprints freely for a minute—showing flashes of his former inexhaustible two-year-old self. In this way he quickly fans out across the hills to inspect the two or three previously discovered woodchuck holes.

Reflection 2. This vignette lets me think through intention in a nonhuman animal, for in it it appears that Sabaka has the intention to go for his afternoon excursion. However, interestingly, this intention shows initially and most prominently in his apparent attention to my intentions. Does Sabaka have something akin to what has been called second-order intention—an intention about another intention (Dennett, 1978, p. 273; Byrne, this volume)? Here, does he have an intention about the intention of another being? For example, does he intends to see if I intend to take him for a walk? More generally, then, this vignette directs me to attempt a double reading, a reading of Sabaka's reading of me.

Sabaka comes and seeks me out so I will take him for a walk. In noticing that act, his seeking me out, I immediately understand it as I have just described. I ascribe to him an intentional act. The form of my description is an account of "what *further* [he is] doing *in* doing something" (Anscombe, 1957, p. 85, emphasis in original). He seeks me out in order to go for a walk. I ascribe to him and I explain his behavior in terms of intention.

How do I know his is an intentional act? There are various ways I know, each with its own contributing degree of confidence. I know because when I adopt an explanatory style which treats him as an "intentional system," to use Dennett's term (1983), I can generate an account which is cohesive and compelling. Further, it is one which allows me to make reasonably good predictions about his behavior. Still, a behaviorist could provide a nonintentional account with reasonably good predictive power. Well, then, I know because not only do I explain him as if he had an intention, I find myself actually living toward him as if he had such. But then at a very young age we impute soulful or animated or intentional being to virtually all objects. One further try: I know because in that moment his seeking me out pulls me out of my own project and makes me become aware of his project. In fact, in the first instant the effect on me of his act of seeking me out is that of Sartre's (1966) "look"—I become a being for him. With rare exceptions, celery stalks do not have that effect on me. Still, I am an "animal lover," and perhaps dogs have a stronger impact on me than on many other people.

Finally, one way I can know whether his is an intentional act is by empathizing with him. I can empathize with him in his act by focusing on his bodily comportment, posture, and action. I find I can gain a sense of his experience, here, of what further he is doing in seeking me out. In reflection, I understand that his act has the structure of an intentional act.

Now I wish to further intuit: What is his sense of me? Sabaka comes, seeks me out, and watches me. What does he see, what is he watching for?

Recall his behavior vis-à-vis the squirrel. I open the door and he makes a beeline to the place where the squirrel leaps to liberty. His action here is anticipatory only in a weak sense of that term. He anticipates the trajectory of the squirrel's escape much as he veers as an animal he is chasing veers. He does not need to have a sense of the squirrel's action as intentional to do this. Arguably, he would follow the squirrel's anticipated track in the same way if the squirrel accidentally fell rather than intentionally flew. And does he not act the same way as a pup in chasing a leaf? Here, then, while Sabaka's action to catch the squirrel is intentional, it is a first-order intention only. Our empathic sense from his kinetics is that he is chasing in order to catch the squirrel. Our sense is not that he is chasing in order to catch the squirrel who, in Sabaka's experience, is running in order to get away.

But now again consider the example of the slower, more extended, and more interactive affair which is his seeking me out and watching me at our presumptive walk-time. He is doing more than following the trajectory of my successive moves toward the boots, shirt, whistle, and leash. Particularly in the first moments he is watching to see if I intend to take him out. More than the simple extrapolation from the direction of my immediate movement, the range of his scrutiny is both broader and more articulate. Sabaka is sensitive to my bodily bearing, my attitude, my incipient movement, my gesture as embodiments of my intention—I am readying to take him for a walk. And he is sensitive to these in the context of his first-order intention to go for a walk. This, then, is a second-order intention—he wants me to want to take him for a walk. How do I know he is watching me to read that or any other intention?

First, two ways I know it indirectly, that is by inference. I have experimented with Sabaka by approximating, imitatively, those postures and incipient moves of mine that I suspect he scrutinizes in such moments. For example, I try to see how minimally I can direct myself to the preparation of taking Sabaka for a walk before he will show the excitement of his anticipated walk. In another context he is not fooled when I approach him as I would to be with him in a companionable way but say, "Bad, Sabaka," mimicking my gruffest tone. It takes very little: a certain stirring at my desk that to a less sensitive scrutinizer might as easily be read as a shift from one sore sedentary haunch to the other, he can read, correctly, as the beginnings of my intention to stop work. Of course, sometimes he is wrong—I mean sometimes, when I am not playing but am making some move within my work setting, he will misread a gesture on my part as an indication of my intention in regard to his walk.

At least I think he is wrong. At least I am aware of no such intention on my part in such a moment. But, and this is the second indirect way I am led to infer that his intention might be a second-order one, sometimes his obvious conviction that I am going to take him out promotes just that project in me. I do not make the further inference that his display of this conviction is an intended ploy on his part to manipulate my intention.[3] However, I do infer from Sabaka's obvious responsivity to approximations to the bodily comportment of my intention to take him out that a sensitivity to that intention is a feature of his scrutiny of me. Still, I admit it is not a necessary inference—in a behavioristic account he could be highly discriminatory to behaviors that regularly accompany that intention.

More directly, that is relying on kinesthetic empathy rather than inference, my sense of Sabaka is that he is sensitive to my intended action in a way that, while more sophisticated than the straightline extrapolation of the trajectory of my movement, stops short of the capacity to empathize with my intended action. He does not have an intuition of my intention "as lived"—as it is for me in my experience.

For him, I am not the subject of a world to which he can have access, as by contrast, it is the burden of this chapter to establish, I have access to his world. His sense of my intended action is, though, the relatively robust form of anticipation of the consequences of my intention. His is the practical knowledge or know-how that if I, his human companion, move this way, I will also make these other moves, the result of which will be a romp in the field for Sabaka.

In different terms, he cannot *take* me as an object in the full meaning of that predicate (Donceel, 1967, pp. 109–114). He can not objectify me or my concerns as mine. I am an object of his consideration in the more limited sense of a means to his practical ends. I am a focal feature of a field organized entirely with respect to his needs and intention. He cannot refer to me, but he can use me.

From most analytic philosophical points of view this is a sufficient form of objectification for the possibility of a higher order intention. Using Anscombe's language again, Sabaka acts intentionally on the basis of what further I am doing in doing something—that is on the basis of my intentional action.

However, from a phenomenological point of view, we can make a further distinction, following From (1971). In our perception of another individual, we distinguish a moment in which we perceive him or her as doing something for some further end (he is striking a match to light her cigarette) *and* one in which we empathize with his or her (perhaps here his romantic) sense of doing just that. In the first instance, the intention is implicit in the action sequence in and as the implicit narrative it is unfolding. In the second, we are focally aware of the intention as it is present to the acting person himself or herself (p. 69). In the terms of our discussion, the distinction is between a more practical, consequence-oriented and a more empathic understanding wherein we momentarily adopt the other individual's point of view. Here, we limit Sabaka to the former but take the position that even that practical orientation constitutes a perception of another's act as intentional. We conclude, then, that Sabaka is capable of second-order intention. Mitchell and Thompson's (1986b, p. 200) account of how "dogs can assess the project of the other" within the context of their own project of play behavior provides examples comparable to the one offered here.

Dennett (1978, pp. 269–270) lists six conditions of personhood in ascending order of importance. Higher order intention is the fourth; the fifth is verbal communication and the sixth is self-consciousness. Dennett (1983, p. 346) also argues that the capacity for second-order intention is the "crucial step." By his account then, Sabaka is a person by most accounting.

My reading, then, of this erstwhile person's, of Sabaka's, reading of people like me is that he, Sabaka, is not a behaviorist for he reads intentions. However, if he were a behaviorist, he would be Tolman not Skinner. On the other hand, Sabaka, in his personal style of reading, is not a phenomenologist for while he is sensitive to intention he is not attuned to the intention of the other as lived. But if he were a phenomenologist he is closer to an investigator utilizing kinesthetic empathy rather than hermeneutic empathy. He is sensitive to me as a living purposeful body; he cannot read me as if my action were a text.

CONCLUSION

This study of an individual dog utilizes an empirical phenomenological method designed for the investigation of nonhuman animals. It keys on an investigatory posture of bodily sensibility adopted to promote empathic access to the meaning implicit in an animal's postures, gestures, and behavior. The investigator's sense of the animal's experience is critically informed by reflection on (1) social constructions, both popular and scientific, that might affect researcher apprehension and/or the animal's actual experience and (2) understanding of the history of the individual(s) under study.

Two vignette/reflections provide the form of the presentation of results. In earlier work (Shapiro, 1990b), I describe Sabaka's concern with maintaining and monitoring a certain form of relationship with me. Here, I describe a predominant way in which Sabaka, as part of his own ongoing project, seeks out and lives spaces that provide both havens and points of view for him. In the second vignette/reflection, I show the presence of second-order intention in Sabaka's attention to and reading of my intention with regard to him. It is suggested that this capacity makes the maintenance and monitoring of relationships possible.

The present method is set in a philosophy of science which conceives of understanding as interpretative in nature. What is to be understood is thought of as a recalcitrant text that always remains subject to further readings or interpretations. This position replaces the positivistic notion that the object of understanding is an independently observable phenomenon consisting of eventually indivisible and unchanging universal components waiting there for our discovery. An interpretative science has its own criteria of validity and reliability (Shapiro, 1986; Wertz, 1986).

To anticipate several objections to the present method, what is the text here? Following Ricoeur (1981), any action can be taken as a meaningful text which can be more or less adequately interpreted. Surely, the leap from the original application of a traditional positivistic scientific method in the study of falling bodies to, say, achievement motivation or depression is as great as that from the original application of an interpretative method in the study of sacred texts to that of a sequence of action.

But even granting that action can be taken as a text, the objection might continue, does not the minimal condition even for an interpretative science require that that text be available to any and all interpreters? Indeed, the present method might be more compelling if a video of Sabaka's behavior were supplied. However, that would not provide an objective presentation of the object of study, Sabaka's behavior. For any video (or set of field notes) is already an interpretative act in that, both literally and figuratively, they are taken from a certain perspective. Further, as we have argued and demonstrated, any action or recording of that action requires interpretation grounded in an understanding of pertinent social constructions and individual history.

With respect to recent concern in comparative psychology regarding anthropomorphism as a possible methodological contaminant, how does the present method and the philosophy of science in which it is grounded understand that "error?" The position of this investigator is that with the application of the critical empathic investigatory style demonstrated here we can know with reasonable conviction the features of nonhuman animals' experience. Since, properly defined, the charge of anthropomorphism is limited to the "attribut[ion] of exclusively human characteristics to animals" (Noske, 1989, p. 88) and since we rarely find features that meet that criterion, we can confidently plead "not guilty."

More formally, from a phenomenological point of view this charge is typically overapplied in two ways in contemporary comparative psychology and cognitive ethology. Many behavioristically oriented investigators still take any attribution of "mental" life to nonhuman animals as necessarily anthropomorphic, thereby denying any subjectivity to them. More subtle than this neo-Cartesian ontological doubt, many investigators retain an epistemological skepticism regarding nonhuman experience. Arguing either according to the lights of traditional positivism or with the more sophisticated Wittgensteinian twist of Nagel (1974), their position is that we cannot know the mind of nonhuman animals. It then follows that any attribution of subjectivity to animals necessarily opens the investigator to the error of anthropomorphism.

Phenomenology is vulnerable to its own form of overapplication of the charge of anthropomorphism. For the phenomenologically oriented investigator, all understanding is necessarily perspectival for it consists of an object seen or considered from the point of view of a certain subject. Put another way, all experience consists of a subject/object relation. Since both poles of that relation together constitute it, any experience necessarily implicates the person of the investigator. In this sense, all understanding is anthropomorphic (from *anthropo*, meaning "man" and *morphe*, "form" or "shape") for it is partly shaped by the human investigator as subject. However, since this is a perspective or "bias" inherent in all experience, it is not an occasional attributional error to which we are particularly prone when we cross species' lines. It is a condition of science which prevents it from reaching certainty and, therefore, from supporting a positivistic philosophy.

Offsetting this particular overapplication of a concept of anthropomorphism, the present method builds on a feature of human experience derivative of Merleau-Ponty's (1962) conception of experience as embodied consciousness and my application of that to phenomenological psychology (Shapiro, 1985). Consciousness itself is not a particular shape but has the capacity to take or assume shapes. Further, we have the capacity to approximate the shape of another being's experience, including that of other species. When we critically apply this empathic ability, the product is not anthropomorphic error. We do not project our own shape, although we can. Rather, as we have described under the rubric of kinesthetic empathy, we can assume the shape of another being's experience and thereby gain in understanding of that experience.

On a final note, both the present method and its substantive results have bearing on the current debate over the use and treatment of animals in laboratory research. As an example of the latter, the peculiar ontological dependence on space described in the first reflection is a vulnerability which has ethical implications for the practice of housing animals in cages whether in laboratory research or shelter or kennel.

Regarding the method, as I have argued elsewhere (Shapiro, 1990a) the absence of a method utilizing the possibility of empathizing with animals is itself partly a product of a traditional Enlightenment or humanistic philosophy of science which, in its press to establish the individual autonomy and integrity of human being, denied these to nonhuman animals. The positivistic approach to science to which this philosophy gave rise and its applications to wildlife management, agricultural production, and laboratory animal research have systematically deindividualized animals by exclusively defining them and treating them in terms of species-specific behavior. A more recent development further reduces them beyond species identity to a generic identity. An individual animal is now an organism or preparation (Shapiro, 1989). By contrast, the present method is consistent with a definition of an animal as an individual in that he or she is understood and treated as a being with a meaningful world (an individual locus of subjectivity) and with a particular formative individual history (an individual locus of particular historical events).

NOTES

1. This is the second in a series of studies developing and applying this method to the study of an individual animal (see Shapiro, 1990b).

2. In his stratification of feelings, Scheler (1967, p. 26) distinguishes sensory feeling from vital feeling, the discrete localizable feeling of pain, for example, from the global but still bodily sense of comfort.

3. Dennett (1978, p. 275) discusses an example of a dog who, apparently deceitfully, scratched the door in order to get her human companion to get up to let her out in order to climb into the preferred chair he had been occupying.

22

Anthropomorphism, Anecdotes, and Mirrors

Karyl B. Swartz and Siân Evans

Despite efforts to maintain objectivity, scientists are often biased by their preconceptions and expectations. Anthropomorphism is especially likely to assist these biases, and anecdotes to confirm them, when the subjects of scientists' attention are nonhuman primates, particularly great apes. Because these species are expected to be so humanlike in their behavior and psychology as a result of their phylogenetic closeness to humans, anthropomorphic interpretation of anecdotes about their behavior often remain acceptable as evidence of their psychology (e.g., Mivart, 1898; Roberts, 1929; Griffin, 1981). Yet concerns about potential problems resulting from anthropomorphism and anecdotal evidence remain, regardless of phylogenetic closeness, especially since anthropomorphism and anecdotal evidence are problematic in interpreting *human* behavior (Caporael & Heyes, this volume; Mitchell, this volume, chap. 13).

In this chapter, we present a history of a research paradigm—mirror self-recognition (MSR)—which has been powerfully influenced by unacknowledged anthropomorphic interpretation of chimpanzee

We would like to thank Robert W. Cooper, Gilbert J. Harris, Robert W. Mitchell, Roger K. Thomas, and Roger K. R. Thompson for their comments on earlier versions of this manuscript.

Preparation of the manuscript was supported by Grant #GM08225 awarded to Lehman College.

behavior, projection from one's own (human) experience to great apes, overemphasis on anecdotes to accommodate beliefs, and overgeneralization from a few subjects to species-wide characteristics. An analysis of the evidence from studies of MSR raises questions about the impact (or lack of impact), on scientists' thinking, of single, few, or many instances which contradict generally accepted findings, theoretical accounts, or even a scientist's own account. The history presented here suggests that anthropomorphism and anecdotal evidence have not only propelled an intriguing phenomenon to scientific attention, but have also hindered our understanding of its nature and explanation. Because concerns about anthropomorphism and anecdotes are usually part of larger concerns about theory construction and evaluation, this history examines the difficulty scientists find in maintaining a relatively "objective" perspective toward evidence in relation to their theories.

THE PHENOMENON

Mirror self-recognition was originally described by Gallup (1970) based on the behavior of four chimpanzees (*Pan troglodytes*). Gallup reported that when presented with a mirror, his subjects initially responded with social behavior, consistent with the interpretation that the mirror image was perceived as an unfamiliar conspecific. Those social behaviors waned after two or three days of mirror exposure, followed by an increase in what Gallup called "self-directed behaviors." Self-directed behaviors involved use of the mirror to investigate areas of the body observable only with the mirror, for example,

> grooming parts of the body which would otherwise be visually inaccessible without the mirror, picking bits of food from between the teeth while watching the mirror image, visually guided manipulation of anal-genital areas by means of the mirror, picking extraneous material from the nose while inspecting the reflected image, making faces at the mirror, blowing bubbles, and manipulating food wads with the lips by watching the reflection. (Gallup, 1970, p. 86)*

Gallup suggested that these self-directed behaviors indicated that the chimpanzees recognized themselves in the mirror.

As a further test of self-recognition, Gallup (1970) anesthetized the chimpanzees and placed a red mark on one brow ridge and the

---

*Excerpted with permission from G. G. Gallup, Jr., "Chimpanzees: Self-recognition," *Science, 167*, p. 86. Copyright 1970 American Association for the Advancement of Science.

opposite ear, locations that could only be seen with the use of the mirror. When the chimpanzees recovered from the anesthesia, they were observed for a control period to determine the number of times they would touch their marks in the absence of the mirror. Then they were presented with the mirror. All subjects responded by touching the facial marks when they saw their marked mirror images. Following some touches to the mark, subjects visually inspected the fingers, and one animal sniffed the fingers following contact with the mark. A total of 27 mark-directed responses was reported for all four chimpanzees during the mirror presentation period, a sharp contrast to the single mark-directed behavior during the immediately preceding control period. Further, chimpanzees who had not had previous mirror exposure did not touch the facial mark under the same conditions. Gallup interpreted this behavior (what is now referred to as "passing the mark test") as indicating that chimpanzees recognize themselves in a mirror, and the phenomenon was established.

More than the frequencies of mark-directed, self-directed, and social actions toward the mirror or the minutes of looking time which Gallup (1970) reported, it was the description of chimpanzee behavior which supported claims of self-recognition. In the original report, Gallup wrote:

> On occasion, mark-directed behaviors also took the form of direct visual inspection of the fingers which were used to touch marked areas even though the dye had long since dried and was not transferable to the fingers. In *one particularly noteworthy instance* there was olfactory as well as visual inspection of the fingers which had been used to touch marked areas. (1970, p. 87, emphasis added)

In his next paper, the report of this behavior was similar:

> Occasionally a mark-directed response was also followed by direct visual inspection of the fingers which had been used to touch marked areas. In another case there were olfactory as well as visual attempts to inspect a finger which had made contact with a dyed spot. (1975, p. 327)

But as time went on, the report of the original observation changed:

> In addition to mark-directed responses, there were also *a number of noteworthy attempts* to visually examine and smell the fingers that had been used to touch marked areas of the skin, even though the dye had long since dried and was indelible. (Gallup, 1977b, p. 333, emphasis added; see also Gallup, 1979b, p. 109)

> In addition there were *several noteworthy attempts* to examine visually and smell the fingers which had been used to touch marked portions of the face. (Gallup, 1979a, p. 419, emphasis added)

> Upon reintroduction of the mirrors the chimpanzees repeatedly touched the marks, *often followed by* visual and olfactory inspection of the fingers which had touched the mark even though the dye was indelible. (Suarez & Gallup, 1981, p. 176, emphasis added)

> . . . *several noteworthy attempts* to visually examine and smell the fingers which had been used to touch these facial marks. (Gallup, 1982, p. 238, emphasis added)

> After touching marked portions of the skin, the animals would *frequently* look at and/or smell their fingers in an apparent effort to identify what was on their faces. (Gallup, 1994, p. 36, emphasis added)

Although only one of Gallup's (1970) subjects sniffed the fingers that touched the mark during the mark test, this observation came to be repeated in the literature numerous times with the implication that this is the standard way that chimpanzees respond to a mark on their face.

The trajectory of these varying depictions of the original data toward the characterization of chimpanzee MSR as a universal species-wide phenomenon should not be surprising: researchers studying memory have shown that we normally transform information in memory to make it more consistent with our beliefs (e.g., Bartlett, 1932; Neisser, 1982). Furthermore, in his analysis of the conditioning study with Little Albert by Watson and Rayner (1920), Harris (1979) showed that reports of a phenomenon can shift with repeated citation, even by the author of the original study.

The phenomenon of MSR, as it came to be described, was so compelling that it became difficult not to anthropomorphize. If chimpanzees *as a rule* not only touched the mark, but also looked at and smelled their mark-touching fingers, then they apparently had some ideas about how they should look and sought to understand the source of their new visual body-image. Indeed, when describing the chimpanzees' response to the mark, Gallup (1979a) remarked, "I suspect that most people would react in the same way if, upon awakening one morning, they saw themselves in the mirror with red marks on their face" (p. 419).

Note that the point here is not to deny that some chimpanzees smell their mark-touching fingers. We observed visual and olfactory inspection of the fingers that touched the mark by one chimpanzee in our study (Swartz & Evans, 1991), and Povinelli, Rulf, Landau, and Bierschwale (1993) reported similar observations. The point is that, rather than supplementing the data presented in the study, anecdotal aspects of these reports became important in the interpretation of the phenomenon because of its surface similarity to human behavior toward mirrors, and served to shape the image of MSR in chimpanzees as being a robust phenomenon typical of all normal chimpanzees, a perception that still exists (see Bradshaw & Rogers, 1993; Peterson & Goodall, 1993).

A more explicit anthropomorphism influenced the interpretation of MSR when Gallup (1977b, p. 335; 1983; Gallup, McClure, Hill, & Bundy, 1971) used Cooley's (1912) ideas that the development of the self-concept (in humans) is a result of other's appraisals of ourselves (and hence "is dependent upon social interaction with others" [Gallup et al., 1971, p. 73]) to interpret findings that only those chimpanzees raised with social interaction showed MSR. According to Gallup et al. (1971, p. 73), isolation-reared chimpanzees are without a self-concept, whereas mirror self-recognizing chimpanzees have a rudimentary self-concept. Yet the idea that chimpanzees are influenced by others' evaluative appraisals of them (in the sense intended by Cooley) was never supported by evidence; the existence of social interaction in chimpanzees was taken to support this anthropomorphic interpretation and thwarted the search for nonanthropomorphic interpretations (see Mitchell, 1993a, 1993c, 1994b). Unfortunately, over time the mere fact of MSR in chimpanzees came to be used to support the claim that chimpanzees had knowledge of other minds, instead of the reverse as Cooley's argument predicts (see Mitchell, 1993a). Rather than the self-concept developing out of social interaction with others and their consequent evaluative appraisal (which provides the "opportunity to examine one's self from another's point of view" and therefore a means of self-objectification—Gallup, 1977b, p. 335), now the self-awareness presumed to be indicated by MSR allowed for the modeling of others' minds (Gallup, 1982, 1985; Gallup & Suarez, 1986; Povinelli, 1993).

A FAILURE TO REPLICATE

We became interested in the phenomenon of MSR in 1984. During the fourteen years since Gallup's original (1970) article, the phenomenon had been widely discussed (Gallup 1975, 1977b, 1979a, 1979b, 1982, 1983), although few additional empirical investigations were made with chimpanzee subjects (Hill, Bundy, Gallup, & McClure, 1970; Gallup et al., 1971; Lethmate & Dücker, 1973; Suarez & Gallup, 1981). Other species of nonhuman primates were studied to determine which species were capable of demonstrating this capacity, and which ones were not (e. g., Gallup, 1977a; Gallup, Wallnau, & Suarez, 1980; Suarez & Gallup, 1981; Ledbetter & Basen, 1982; Eglash & Snowdon, 1983; see Anderson, 1984, for a complete review). The consistent story was that, of the great apes, chimpanzees and orangutans (*Pongo pygmaeus*) robustly demonstrated the phenomenon, bonobos (*Pan paniscus*) had never been tested, and gorillas (*Gorilla gorilla*) were incapable of demonstrating MSR with the use of the mark test. Other nonhuman primate species showed no evidence of MSR despite some very clever and persistent attempts by human researchers (see Anderson, 1984).

One of the most puzzling empirical findings at that time was the failure of gorillas to show evidence of MSR. This element of the story must be amended now because of evidence that some gorillas, when in front of a mirror, show self-directed behavior and others show both self-directed and mark-directed behavior (see Parker, 1994; Patterson & Cohn, 1994; Swartz & Evans, 1994). In the early to mid-1980s, however, the prevailing story was that no gorilla presented with a mirror had shown self-directed behavior or passed the mark test (Ledbetter & Basen, 1982; Suarez & Gallup, 1981), and the presumption was that all chimpanzees and orangutans with normal rearing histories would show evidence of MSR.

During the 1984–85 academic year, we (Swartz and Evans) were both at the Centre International de Recherches Médicales de Franceville (CIRMF) in Franceville, Gabon, Africa. We were puzzled by the result with gorillas and planned to study the phenomenon, hoping to develop a way to demonstrate that gorillas were able to show MSR. Our effort was unsuccessful (see Swartz & Evans, 1994). While studying the mirror behavior of Zoé, a five-year-old gorilla orphan living in the CIRMF colony, we felt we should observe the phenomenon in chimpanzees to compare with our observations of gorillas. We were confident about the prospect of obtaining MSR with chimpanzees, as we believed in the universality of this phenomenon with chimpanzees.

Our first chimpanzee subject was a four-year-old orphan, Makata, chosen because he was close in age to Zoé. Although Makata showed the "classic" social and subsequently some of the self-directed behaviors reported by Gallup (1970), he did not pass the mark test (Swartz & Evans, 1991). Our surprise at our failure to replicate Gallup's (1970) findings led us to study several chimpanzees. Because our primary objective was to investigate MSR in chimpanzees in order better to understand the phenomenon for our study with gorillas, we decided to investigate the relationship between behavior shown during mirror exposure and responses to the mark during the mark test. That is, although the increase in self-directed behavior was accompanied by a decrease in social behavior, and although chimpanzees passed the mark test following mirror exposure (Gallup, 1970), it was unclear (a) how much mirror exposure was necessary, and (b) what relationship, if any, existed between the frequency or quality of self-directed behaviors and the ability to pass the mark test. In order to investigate these questions, we conducted mark tests at predetermined intervals during mirror exposure.

Gallup's (1970) subjects had received a total of 80 hours of mirror exposure before the mark test. We conducted mark tests with two subjects before mirror exposure, two subjects following 2 hours, nine following 12 hours, six following 40 hours, eight following 80 hours, and one following 134 hours (see Swartz & Evans, 1991, for details). Individual animals received one to five mark tests during the course of the study. In only one case did we obtain mirror-guided mark-directed responses during the mark test. That behavior was shown by one 16-year-old female, Berthe, who showed great interest in the mirror, demonstrated high levels of self-directed behavior from the outset, passed the first mark test following just 13.5 hours of mirror exposure, and subsequently passed two additional mark tests. In total, we provided mirror exposure to 11 chimpanzees, with only one subject showing a positive result on the mark test. Two subjects were eliminated from the study following their first mark test at 12 hours mirror exposure, one (Gemini) because she appeared to be more interested in the human observers than in her own mirror image, and one (Edgar) for reasons unrelated to the study. A third subject showed no interest in the mirror throughout 134 hours of mirror exposure. The failures of the remaining seven subjects, including Makata, were quite puzzling because they showed interest in the mirror, and three (Makata, Henri, and Nestor) showed self-directed behaviors during mirror exposure.

We had considerable difficulty getting the report of these data published, partly because we were blinded by our uncritical acceptance of the robustness of this phenomenon and we offered no suggestions for why our subjects did not touch the mark. Not only were we hindered by our own blindness when writing the manuscript, we were also hindered in publishing these data because our reviewers appeared also to be blinded by their beliefs in the robustness of the phenomenon.

Our paper was rejected by two journals before it was accepted for publication by *Primates*. Quotations from our reviews suggest that the reviewers were responding on the basis of pre-existing assumptions about the robust nature of MSR in chimpanzees:[1]

> This paper reports that only 1 of 11 chimpanzees tested using Gallup's mirror recognition paradigm evidenced mirror recognition. As the author[s] note . . . , these results are contradictory to previous studies of mirror recognition in this species which have proved to be very robust. (Reviewer 1, second submission)

> A major problem is that the authors are unable to explain the discrepancy between their data and previously reported results. . . . The authors go to great lengths to demonstrate that procedural differences are not responsible for the differences in results between their and Gallup's studies. . . . The authors argue that different procedures from Gallup's cannot explain the different results, because one subject, Berthe, passed the mark test. Perhaps Berthe is a very quick ape, and the other [ten] individuals' data better reflect . . . performance using these different procedures. (Reviewer 2, first submission)

> Had self-recognition in chimpanzees only been reported once, or only by one investigator, then a failure to find this phenomenon might be informative. However, self-recognition has been reported in chimpanzees numerous times, by different investigators, working under different conditions in different parts of the world. (Reviewer 3, first submission)

In fact, at the time that this review was written (early 1987), the number of papers published reporting MSR in chimpanzees was only five (Gallup, 1970; Hill et al., 1970; Gallup et al., 1971; Lethmate & Dücker, 1973; Suarez & Gallup, 1981), four of which included Gallup as an author. One additional paper (Robert, 1986) reported a failure to obtain MSR in a very young chimpanzee who had had only 46.5 hours of mirror exposure.

The reviewers' responses quite easily bring to mind Maier's Law: if facts do not conform to the theory, they must be disposed of (Maier, 1960, p. 208; see also Dewsbury, 1993). Note, however, that one reviewer suggested that we consider alternative implications of our data, including questioning Gallup's framework, and suggested the view that MSR (in the form of passing the mark test) may be the exception rather than the rule in chimpanzees:

> it is Berthe who is the exception, not Makata and the nine others. It makes more sense to have one exception out of 11 subjects, rather than ten exceptions, as the authors propose. (Reviewer 2, first submission)

At the time that we submitted our manuscript in 1987, only 14 chimpanzees had been reported to demonstrate MSR with the mark test (4 in Gallup, 1970; 2 in Hill et al., 1970; 3 in Gallup et al., 1971; 3 in Suarez & Gallup, 1981; 2 in Lethmate & Dücker, 1973), yet the pervading perspective was that it was a universal phenomenon (see, e.g., Goodall, 1986; Griffin, 1984; Tuttle, 1986). The existing data for orangutans also showed some failure to find MSR. In 1987, there were two reports with orangutan subjects. Lethmate and Dücker (1973) reported positive mark tests with two orangutans, and Suarez and Gallup (1981) obtained self-directed behavior and a successful mark test with only one of two orangutans. Indeed, in 1994 Gallup still reported the orangutan data as a strong demonstration of the phenomenon, failing to note that one of four did not demonstrate the phenomenon.

How much emphasis one should put on a single instance of failure or success frequently depends upon the theoretical concerns of the evaluator. For example, whereas Gallup (e.g., 1982; Gallup & Suarez, 1986) for many years discounted evidence that at least one gorilla, Koko, recognized herself in the mirror because other gorillas failed the mark test, he ignored his own evidence that not all orangutans show MSR (Suarez & Gallup, 1981). Similarly, although Gallup has for years supported claims of intentional deception, pretense, and theory of mind in chimpanzees and orangutans using anecdotal evidence (see Gallup, Boren, Gagliardi, & Wallnau, 1977; Gallup, 1982, p. 244, Table 1; 1983, pp. 493–497; 1985, pp. 635–638; 1988; Gallup & Suarez, 1986), he eschews its use by others to support interpretations alternative to (or even the same as) his own (see Gallup & Povinelli, 1993; Mitchell, 1993c). From our perspective, we consider the single orangutan who failed (Suarez & Gallup, 1981) as significant in part because it parallels our claim that MSR is not a universal phenomenon in chimpanzees (Swartz & Evans, 1991). In addition, however, this single instance represents 20% of currently tested adult orangutans (including Chantek, Miles's [1994] sign-using orangutan). It would be less significant to us if 99 had passed and one had failed.

In the context of the failure to acknowledge the similarity between our results with chimpanzees and his with orangutans, Gallup (1994) continued to discount our results, first with misinterpretation of our discussion, and second, with criticism of our methods (see Gallup, 1994, and Swartz & Evans, 1994, for specifics). Gallup (1994) stated, "Swartz and Evans (1991) report finding self-recognition in only one out of eleven chimpanzees, and on that basis call into question the methodology I developed" (p. 40). However, we did not question Gallup's methodology; rather, we wrote:

> Although it is proposed here that procedural variables do not account for the discrepancy in results between Gallup's studies and the present results, if such methodological details as size of mirror or temporal spacing of mirror exposure affect the demonstration of this phenomenon, that poses a serious problem for Gallup's theory of self-recognition and mind in animals. . . . The results of this study, at the very least, call for a reassessment of mirror behavior in nonhuman primates. (Swartz & Evans, 1991, pp. 491, 495)

A more serious and unjustified criticism by Gallup (1994) was his claim that

> . . . not only were the animals marked repeatedly, but following successive mark tests they were returned to cages containing other chimpanzees that were also part of the same study. Thus, in some instances prior to even being tested for the first time, some of the animals had already received extensive exposure to these marks on cagemates, and to the extent that animals were mark tested and/or exposed to others with facial marks before learning to recognize themselves in mirrors, the failure to find "evidence" of self-recognition is hardly surprising. (p. 44)

These statements about extensive previous exposure to marked cagemates are exaggerations, and were not indicated in our presentation of method. In fact, four animals (Amélie, Koula, N'tébé, and Masuku) had been exposed to one cagemate (Henri) who had been marked once prior to their receiving mirror exposure and twice prior to their receiving their own mark tests; Henri saw marked cagemates following his third mark test; and Berthe was exposed to a marked cagemate following her first mark test but proceeded to pass two more such tests. None of the other animals were exposed to a marked cagemate, and all of these (Nestor, Henri, and Makata) failed the mark test. Gallup's claim that exposure to a marked cagemate inhibits a chimpanzee's tendency to touch a mark on his/her own head is apparently a red herring (but see Mitchell, 1993b, and Swartz & Evans, 1994, for further discussion of the effects of marks on others). Given his claims, it is surprising that as editor of the *Journal of Comparative Psychology*, Gallup accepted two manuscripts for publication (Lin, Bard, & Anderson, 1992; Povinelli et al., 1993) which contained procedural similarities to our study (e.g., exposure to marked cagemates) as well as additional problematic features (e.g., group mirror exposure; see Swartz & Evans, 1994).

The claim that repeated mark tests inhibit MSR is another red herring (Swartz & Evans, 1991, 1994). Like our chimpanzee Berthe, one of Lethmate and Dücker's (1973) orangutans was mark tested twice, and passed both times. Further, new data with human infants suggest that repeated mark tests neither hindered nor encouraged the developmental onset of MSR (Hart & Fegley, 1994). If chimpanzees are so different from humans as to be easily influenced to avoid exhibiting MSR by repeated mark tests, Gallup's claims about psychological similarities between humans and chimpanzees are not very believable.

Even with all of Gallup's criticisms, it should be noted that our findings have been replicated: (a) not all chimpanzees show self-directed behavior when given mirror exposure, (b) not all chimpanzees provided with mirror exposure will pass the mark tests, and (c) the occurrence of self-directed behaviors does not predict passing the mark test (Povinelli et al., 1993; Povinelli, 1994; Thompson & Boatright-Horowitz, 1994). It is unclear why Gallup still takes issue with our results.

WHEN THEORY MEETS DATA . . . AGAIN

The field of nonhuman primate MSR seems to be proceeding into another area of controversy in which single instances are at odds with claims, this time concerning the development of MSR in chimpanzees. This issue, although subject to empirical investigation, threatens to take on the same features that MSR itself manifested prior to the publication of our study (Swartz & Evans, 1991). Povinelli (1993; Povinelli et al., 1993) has suggested that MSR does not emerge until 6 to 8 years of age in chimpanzees, and that the ability declines after 15 years of age (Povinelli et al., 1993). Once again we find a situation in which a bold claim is at odds with widespread beliefs, but here empirical findings contradict the claim. Lin, Bard, and Anderson (1992) found evidence for MSR in chimpanzees as young as 2.5 years as well as in those who were 3, 4, and 5 years old. Early reports found evidence for MSR in three socially housed chimpanzees who were 3 to 6 years (Gallup et al., 1971) and two who were less than 2 years old at the time of testing (Hill et al., 1970). In another study, not only did two male chimpanzees aged 3 and 4 show evidence of MSR, they also showed retention of the phenomenon one year later (Calhoun & Thompson, 1988). If, indeed, chimpanzees do not begin to show MSR until 6 to 8 years of age, the widely cited study indicating that isolation-reared chimpanzees fail to show MSR needs to be replicated, as the isolation-reared chimpanzees studied were between 3 and 6 years of age (Gallup et al., 1971), and thus too young to be expected to show MSR. Finally, our own observations of the exuberant self-directed and mark-directed behavior of the 16-year-old Berthe is a single instance which contradicts the claim that MSR declines in chimpanzees after 15 years of age.

Although Povinelli (1993; Povinelli et al., 1993) dispenses with the above age-related data which contradict his claims by ignoring them or questioning them on the basis of various procedural grounds, it is difficult to discount these data entirely. Further, Povinelli et al. (1993) provide evidence against their own ideas when they report MSR in a three-year-old chimpanzee. In nominal support of their ideas, they designate this chimpanzee as "precocial," but this designation is based on no other evidence than the chimpanzee's exhibition of MSR.

Surprisingly, Gallup's original (1970) data have been used to support both those who believe MSR begins earlier than six years in chimpanzees and those who believe it occurs later. In our own study, Makata, a four-year-old chimpanzee, showed clear evidence of self-directed behavior but did not pass the mark test. We argued that his behavior could not be an age-related effect, as our strong impression was that Makata was the same age as Gallup's original subjects (Swartz & Evans, 1991, 1994), that is, about four years old. By contrast, Povinelli estimated that Gallup's (1970) subjects were six to eight years old (Povinelli, 1993) or five to eight years old (Povinelli et al., 1993). Given this contradiction, we decided to investigate the original subjects, recognizing the supreme irony that our investigation might show that the four original subjects whose behavior began all the interest in chimpanzee MSR were, by one account, too young to show it!

Gallup had previously provided the first author with summary sheets of his original data for the purpose of comparing his data with those we collected (see Swartz & Evans, 1994).[2] With the subject identification available on those sheets, it was possible to obtain some background information about the subjects. All were wild-born; one (Roxie) was brought into the Delta Regional Primate Research Center in 1964. At that time she was estimated to be about 1 year old. The other three were brought into Delta during 1967, and one (Mack) was estimated to be about 2.5 at the time. Information about estimated age at arrival for the other two has been lost (M. Rodriguez, personal communication, July 1993). Based on the assumption that Gallup conducted the study late in 1969, Roxie and Mack would have been approximately 6 and 5, respectively, at the time of the study. This assumption is based on the submission date of November 10, 1969; if the study was conducted earlier, the animals would have been younger.

It is possible, however, to make age estimates for all the subjects based on their weights at the time of the study. In contrast to the estimate of Povinelli et al. (1993) that Gallup's (1970) subjects were 5 to 8 years old, we estimate that they ranged in age from 3.6 to 5.7 years (using medians), with a possible range of 3.3 to 6.4 years if one uses the widest range possible. These estimates were derived in the following manner.

The population from which we obtained these estimates were wild-born chimpanzees living at CIRMF during the period from 1979 until 1986. These chimpanzees entered captivity because they were orphaned in the wild (Gabon), and they were behaviorally indistinguishable from comparable chimpanzees housed in U.S. colonies

(Robert W. Cooper, personal communication, November 30, 1993). We have monthly weights for nine males from August, 1982, until August, 1985, with some additional weights for some individuals dating before and after those dates. We also have weights obtained at irregular intervals for nine females across a period from 1979 until 1985. The age estimates were derived from tooth eruption (Nissen & Riesen, 1964) and were constantly revised as more data became available. Using the most recent (and hence, most accurate) age estimate available for each animal, age-weight charts were constructed for all individuals. To replicate the attributes of Gallup's (1970) subjects as closely as possible, age-weight data available only from chimpanzees in the CIRMF population who had been in the laboratory for at least two years and who had come into the facility before three years of age were used. That provided a pool of seven males and seven females.

This population seems, with one possible exception, to be an ideal comparison group for Gallup's (1970) original subjects. That one possible exception is subspecies. The animals at CIRMF were all *Pan troglodytes troglodytes*, which are smaller than *Pan troglodytes verus*. It is almost certain that Gallup's subjects were *Pan troglodytes verus*, as their origin was West Africa (Sierra Leone and Liberia; International Animal Exchange, personal communication, September 21, 1993), an area populated only by *P. t. verus* (Napier & Napier, 1967). As they almost certainly were *P. t. verus*, our age estimates should be adjusted downward; that is, Gallup's animals would be younger than the estimates we provide here.

To obtain specific estimates for each of Gallup's (1970) subjects, we determined the age at which individual same-sex CIRMF chimps attained the weight reported for each of his subjects. That provided a range of ages from which we calculated the mean and median. The median age estimates are 5.7 years for Marge (#1825, 22.5 kg), 5.6 years for Roxie (#355, 23.75 kg), 5.1 years for Mack (#1831, 20.0 kg), and 3.6 years for Raymond (#1854, 14.75 kg). The ages of six CIRMF animals were used to estimate the ages of Roxie and Mack. Seven were used for Marge's estimate. Unfortunately, the estimate for Raymond, the youngest subject, was based on only two CIRMF animals; however, there is additional evidence that he was younger than 5.

It is difficult to make age-weight comparisons between wild-born and laboratory-born chimpanzees; however, a gross estimate of the expected weight difference between these two populations would be that the wild-born animals might be 15–20% lighter than laboratory-born animals (Robert W. Cooper, personal communication, September 2, 1993). Age-weight charts are available for laboratory-born chimpanzees with known ages. Such a chart, drawn from a laboratory-born population similar to Gallup's wild-born subjects is provided by Gavan (1971). If we adjust the weights of Gallup's subjects by as much as 20%, and refer to the data provided in Gavan (1971), the age range estimates are 5.5–7.5 for Marge, 6.5–7.5 for Roxie, 5.5–6.5 for Mack, and 3.5–4.5 for Raymond. An adjustment of 15% provides about the same estimates. These estimates are consistent with the estimated birth dates provided for Roxie (1/1/64) and Mack (1/1/65) by the International Species Information Exchange (David R. Lukens, Jr., personal communication, July 12, 1993).

To further support their ideas that MSR begins from six to eight years of age in chimpanzees, Povinelli et al. (1993) reported that the two females in Gallup's (1970) study were "undergoing sexual swelling" (p. 32), which in their view indicated ages of at least six to eight years. They based this on the following information provided in Gallup (1975):

> Marge used the mirror for the first time to manipulate her genital area, which had begun to show signs of pubertal swelling. (After the original incidence of visually guided genital manipulation this behavior was seen repeatedly in both females in subsequent sessions.) (p. 324)

It is not unusual for a laboratory-living chimpanzee to show the beginning of sexual swellings as early as six years of age, but the range observed recently at the Yerkes Regional Primate Research Center was from four years, nine months, to seven years, one month, with an average of six years, four months (Sarah Phythyon, personal communication, August 17, 1993). This average age is consistent with our age estimates for Marge and Roxie, but is at the low end of Povinelli et al.'s (1993) suggested age range for Gallup's (1970) subjects. Note also that Gallup's subjects had not had previous mirror experience, which indicates that the age at which they showed MSR in Gallup's (1970) study need not indicate the earliest age at which they might have shown it.

CONCLUSIONS

During its brief history as an object of scientific scrutiny, the phenomenon of MSR in nonhuman primates has been influenced, in no consistent way, by anthropomorphism, anecdotal evidence, and a failure to analyze objectively data both supportive and contradictory to a particular theory. The over- and under-interpretation of anecdotes as a result of beliefs derived from theory, itself derived in some cases from anthropomorphic extrapolation, led to widely accepted but not completely accurate interpretations of the phenomenon and hindered the publication of data which ran counter to theory. Although Maier's Law—that if facts do not conform to theory they must be disposed of (Maier, 1960, p. 208)—has at times had a stranglehold on our understanding of MSR, that stranglehold is loosening. It is not by chance that a conference to evaluate new ideas about MSR was organized by individuals with theoretical orientations at odds with the standard view of MSR (see Parker, Mitchell, & Boccia, 1994). Still, however influenced by anthropomorphism and anecdotal evidence, Gallup's (1970) initial investigation of the phenomenon opened a fruitful and provocative area of inquiry to the field of comparative cognition.

NOTES

1. We do not intend, by the inclusion of these quotations from reviewers, to demean or question their hard work and judgment. Indeed, each revision of the manuscript was significantly strengthened by our taking into account the comments made by the reviewers. One reviewer of the second submission, who is not quoted here, chose to reveal his identity; the identity of all other reviewers remains unrevealed to us. We should also note that not all the reviews were negative, and two of the three reviewers of the first submission (including Reviewer #2 who is quoted here) recommended publication.

2. Gallup is aware of the use of his data to determine these ages. The first author wrote to him on July 23, 1993, providing him with her estimates of the ages. Those age estimates were preliminary and not based on as much data as are provided here, but they were consistent with those presented here. Gallup sent the letter to Povinelli (D. Povinelli, personal communication, August 3, 1993), who responded to Gallup in a letter dated August 13, 1993, providing a rationale for his age estimates and stating that he stood by his estimates. Gallup sent a copy of that letter to the first author, with an accompanying letter dated August 23, 1993, stating that he was satisfied with Povinelli's estimates. In a reply to Gallup dated October 27, 1993, the first author stated her intention to publish her age estimates along with the rationale for those estimates.

PART VII

Cognition

23

Cognitive Ethology:
Slayers, Skeptics, and Proponents

Marc Bekoff and Colin Allen

The interdisciplinary science of cognitive ethology is concerned with claims about the evolution of cognitive processes. Since behavioral abilities have evolved in response to natural selection pressures, ethologists favor observations and experiments on animals in conditions that are as close as possible to the natural environment where the selection occurred. No longer constrained by psychological behaviorism, cognitive ethologists are interested in comparing thought processes, consciousness, beliefs, and rationality in nonhuman animals (hereafter animals). In addition to situating the study of animal cognition in a comparative and evolutionary framework, cognitive ethologists also maintain that field studies of animals that include careful observation and experimentation can inform studies of animal cognition, and that cognitive ethology will not have to be brought into the laboratory to make it respectable. Furthermore, because cognitive ethology is a comparative science, cognitive ethological studies emphasize broad taxonomic comparisons and do not focus on a few select representatives of limited taxa. Cognitive psychologists, in contrast to cognitive ethologists, typically work on related topics in laboratory settings, and do

We thank Andrew Whiten, the TAMU Animal Behavior Study Group, Meredith West, and two anonymous reviewers for providing comments on a previous draft of this chapter, and also Gordon M. Burghardt, Susan E. Townsend, and Dale Jamieson for discussing many of the issues with which we are concerned.

not emphasize comparative or evolutionary aspects of animal cognition. When cognitive psychologists do make cross-species comparisons, they are typically interested in explaining different behavior patterns in terms of common underlying mechanisms; ethologists, in common with other biologists, are often more concerned with the diversity of solutions that living organisms have found for common problems.

Many different types of research fall under the term "cognitive ethology," and it currently is pointless to try to delimit the boundaries of cognitive ethology; because of the enormous amount of interdisciplinary interest in the area, any stipulative definition of cognitive ethology is likely to become rapidly obsolete. Currently, cognitive ethology faces challenges to its scientific status. Criticism is based on both the subject matter and the methods of cognitive ethology.

In this chapter we identify three major groups of people (among *some* of whose members there are blurred distinctions) with different views on cognitive ethology, namely, *slayers, skeptics,* and *proponents.* Our analyses are based on our reading of some published reviews of Donald Griffin's works in cognitive ethology (1976, 1978, 1981, 1984, 1992) and other clearly stated opinions concerning animal cognition, in the sense of attribution of mental states and properties such as beliefs, awareness, and consciousness. Two points need to be made clear at the outset. First, while Griffin's handling of issues such as anthropomorphism and the use of anecdote are often rather superficial (Griffin, 1977; Mitchell, 1986), and some even question whether or not his early works (at least) had much to do with ethology (Hailman, 1978; see Jamieson & Bekoff, 1993, for discussion), we are of the opinion that Griffin's (1976) rekindling of interest in the field that has come to be called cognitive ethology is responsible for the recent surge of interest in the comparative and evolutionary study of animal cognition. Thus our concentration on his work. Second, we ignore, but do not wish to downplay, the important experimental research on cognitive aspects of animal learning (see Galef, 1990; Kamil & Clements, 1990; Pepperberg, 1990a, 1990b; Timberlake, 1990; and Roitblat & von Fersen, 1992 and references therein), although those who do these types of studies may not call themselves cognitive ethologists; space does not allow us to discuss *all* views on cognitive ethology.

Categorizing views on cognitive ethology in the way we do here helps us to identify common themes, which in turn helps us to see to what extent genuine dialogue between critics and defenders is possible; analysis of both criticisms and confusions arising from this dialogue will help improve the science. We were surprised by the number and the strength of some of the attacks on cognitive ethology by ethologists, behavioral scientists, and evolutionary biologists (see also Bekoff & Jamieson, 1991). Our purpose is to bring attention to those attacks and to try to understand the sort of argument that underlies them. These arguments include the views that anthropomorphism is unscientific, anecdotes do not constitute legitimate data, attributing beliefs to nonhumans is impossible, and cognitive ethology is a soft science. We are concentrating on the critics because we are sympathetic to cognitive ethology, and categorizing their various views will allow those who are sympathetic to cognitive ethology to recognize common themes and provide appropriate responses. We do not agree with Epstein's (1987) claim "that our understanding of the behavior of organisms might advance ever so much faster if we spent more time studying it and less time philosophizing about it" (p. 21). We do, however, believe that the most effective philosophizing about cognitive ethology will occur only if careful attention is paid to the actual empirical work conducted by cognitive ethologists.

SOME VIEWS OF COGNITIVE ETHOLOGY: SLAYERS, SKEPTICS, AND PROPONENTS

For cognitive ethology, the major problems are those that center on methods of data collection and analysis, and on the description, interpretation, and explanation of behavior (Bekoff & Jamieson, 1990a, 1990b). Here we consider how different people's views of cognitive ethology appear to be informed by available evidence, social factors, or some of both. We attempt to locate different views into three major categories, slayers, skeptics, and proponents.

Slayers: Slayers deny any possibility of success in cognitive ethology. In our analyses of their published statements, we have found that they sometimes conflate the *difficulty* of doing rigorous cognitive ethological investigations with the *impossibility* of doing so. Slayers also often ignore specific details of work by cognitive ethologists and frequently mount philosophically motivated objections to the possibility of learning anything about animal cognition. Slayers do not believe that cognitive ethological approaches can lead, and have led, to new and testable hypotheses. They often pick out the most difficult and least accessible phenomena to study (e.g., consciousness) and then conclude that because we can gain little detailed knowledge about this subject, we cannot do better in other areas. Slayers also appeal to parsimony in explanations of animal behavior, but they dismiss the possibility that cognitive explanations can be more parsimonious than noncognitive alternatives, and they deny the utility of cognitive hypotheses for directing empirical research.

Skeptics: Skeptics are often difficult to categorize. They are a bit more open-minded than slayers, and there seems to be greater variation among skeptical views of cognitive ethology than among slayers' opinions. However, some skeptics recognize some past and present successes in cognitive ethology, and remain cautiously optimistic about future successes; in these instances they resemble moderate proponents. Many skeptics appeal to the future of neuroscience, and claim that when we know all there is to know about nervous systems, cognitive ethology will be superfluous (Bekoff, 1995; it should be noted that Griffin, 1992, also makes strong appeals to neuroscience, but he does not believe that increased knowledge in neurobiology will cause cognitive ethology to disappear). Like slayers, skeptics frequently conflate the difficulty of doing rigorous cognitive ethological investigations with the impossibility of doing so. Skeptics also find folk psychological, anthropomorphic, and cognitive explanations to be off-putting.

Proponents: Proponents recognize the utility of cognitive ethological investigations. They claim that there are already many successes and they see that cognitive ethological approaches have provided new and interesting data that also can inform and motivate further study. Proponents also accept the cautious use of folk psychological and cognitive explanations to build a systematic explanatory framework in conjunction with empirical studies, and do not find anecdotes or anthropomorphism to be thoroughly off-putting. Some proponents are as extreme in their advocacy of cognitive ethology as some slayers are in their opposition. But most proponents are willing to be critical of cognitive ethological research without dooming the field prematurely; if cognitive ethology is to die, it will be of natural causes and not as a result of hasty slayings.

Slayers

Let us first consider some of what the slayers have to say about cognitive ethology. We realize that we are presenting only snippets from longer statements, but we feel that our abbreviated quotations capture the flavor of the general stance of the people involved.

Heyes (1987a, 1987b) is concerned with what she perceives to be muddled thinking that pervades the area of cognitive ethology, and focuses her attacks on Griffin. She claims that

> Donald Griffin's . . . attempts to enrich the study of animal behaviour by making it cognisant of consciousness . . . deserves attention as an example of the muddled thinking that is increasingly influential in this area . . . (Heyes, 1987a, p. 107)

Many of her (and others') concerns center on methodological and explanatory problems in studies of animal cognition, issues to which Griffin actually gave little detailed attention. If Griffin contributed to the muddle, it was by his failing to consider these areas critically; this has been pointed out in many of the reviews of Griffin's major books (Mason, 1976; Huntingford, 1985; Whiten, 1992; Yoerg, 1992; Bekoff, 1993b).

Anthropomorphism can also be off-putting (e.g., Humphrey, 1977; Colgan, 1989), and many people have used their distaste of anthropomorphism to condemn cognitive ethology (Bekoff, 1995). In John S. Kennedy's recent book, *The New Anthropomorphism* (1992), a volume devoted to a grand-scale and rather superficial condemnation of anthropomorphism (see Ridley, 1992) and its supposed negative effect on the field of ethology in general, it is claimed that:

> Those who would have us go all the way back to traditional explicit anthropomorphism are still a minority but they show us the way things could go if we are not careful. . . . Anthropomorphism must take its slice of the blame for a sort of malaise that has lately afflicted the subject of ethology as a whole. . . . In conclusion, I think we can be confident that anthropomorphism will be brought under control, even if it cannot be cured completely. Although it is probably programmed into us genetically as well as being inoculated culturally that does not mean the disease is untreatable. (pp. 5, 55, 167)

Humphrey (1977) similarly writes: "Away with critical standards, tight measurements and definitions. If an anthropomorphic explanation *feels* right, try it and see; if it doesn't feel right, try it anyway" (p. 521). Kennedy and Humphrey are typical of critics who write as if the only alternatives are an unconstrained, fuzzy-minded use of anthropomorphism on the one hand, and the total elimination of anthropomorphism on the other (see also Estep & Hetts, 1992). But there is a middle position which they ignore (Fisher, 1990, 1991; Lehman, 1992). Anthropomorphism can be useful if it serves heuristically to focus attention on questions about animal behavior that might otherwise be ignored. Anthropomorphism might be used in a rigorous way to assist theory construction (Asquith, 1984) and to motivate empirical research projects (Burghardt, 1991; this volume).

Now, let's consider some of Kennedy's other claims. That there is a sort of malaise in the subject of ethology as a whole simply does not jibe with current interest in the field by many diverse people. A glance at any of the newer textbooks in the field as well as Griffin's (1992) *Animal Minds* shows that there is a lot of interest in ethology and the stimulating and difficult problems that ethological studies consider. Unlike Kennedy, some of Griffin's other strongest critics note that there has been, and still is, a great deal of interest in ethology. (However, they go on to condemn Griffin for providing merely natural history accounts of the behavior of animals; they fail to recognize that there has been a large amount of experimental work done in cognitive ethology. Griffin does, in fact, write about this research and he should be credited for inspiring much of it directly [Huntingford, 1985]). Kennedy also claims, in the total absence of any sort of data, that there is some sort of genetic predisposition to engage in anthropomorphism. This claim is out of character with the rest of his arguments that demand appeal to hard data. Perhaps Kennedy's closing sentences capture the essence of his views:

> If scientists, at least, finally cease to make the conscious or unconscious assumption that animals have minds, then the consequences can be expected to go beyond the boundaries of the study of animal behaviour. If the age-old mind-body problem comes to be considered as an exclusively human one, instead of indefinitely extended through the animal kingdom, then that problem too will have been brought nearer to a solution. (1992, pp. 167–168)

Kennedy's idea that the mind-body problem might be closer to solution if restricted to humans actually flies in the face of much recent work in the philosophy of mind, where a significant trend is toward naturalizing and even "biologizing" mental properties (e.g., Millikan, 1984). No doubt Kennedy would regard such moves by philosophers as misguided, but they cannot simply be dismissed by a wave of the hand; at best, his argument depends on a contentious philosophical position. This sort of argument is typical of many slayers, as we shall further illustrate below.

Some slayers use the "hard" (but not necessarily more difficult) sciences as a model for the criticism of "soft" cognitive ethology (Humphrey, 1977; Heyes, 1987a, 1987b; Kennedy, 1992; Williams, 1992), while others tend to be dismissive and conveniently ignore available empirical evidence of animal cognition in their sweeping claims (Ingold, 1988; Colgan, 1989; Zuckerman, 1991). They write:

> There can be no historical doubt that behaviourism has advanced ethology as a science, whereas the methods advocated by cognitivists have yet to prove their worth. Until mental concepts are clarified and their need justified by convincing data, cognitive ethology is no

> advance over the anecdotalism and anthropomorphism which characterized interest in animal behaviour a century ago, and thus should be eschewed. (Colgan, 1989, p. 67)

> A scientist's hunch is acceptable as a start, provided that it leads to a theory that can be rejected in the face of evidence. This has not been achieved in the field of animal cognition. (McFarland, 1989a, pp. 146–147)

A few comments are in order. First, it is notable that the ethologist Patrick Colgan's failure to give an adequate account of available literature has been challenged (McCleery, 1989; see also Houston, 1990), even by slayers (McFarland, 1989b). McCleery (1989) notes that "Brevity also characterizes . . . [Colgan's] discussion of cognition"; he continues on to say that Colgan's dismissal of cognitive ethology "scarcely seems warranted on the grounds given here" (p. 1092). McFarland (1989b) notes that Colgan's "section on cognition is poor. It fails to home in on the essential features of cognition, devotes too much space to fairly irrelevant historical matters, and offers no good examples of cognition in animals" (p. 170). Nonetheless, McFarland (1989a) seems to think that he has reviewed the field of animal cognition and that his conclusion (above quotation) is well-founded. In fact, his chapter entitled "Cognition in Animals" contains very few references to recent work in the field. Others who dismiss the field of cognitive ethology also make little effort to consider available evidence. Zuckerman (1991), in his review of Cheney and Seyfarth's book *How Monkeys See the World: Inside the Mind of Another Species* (1990), exemplifies this dismissive attitude: "Some of the issues they do raise sound profound as set out but, when pursued, turn out to have little intellectual or scientific significance" (p. 46). Heyes (1987a), who is a laboratory psychologist, advises cognitive ethologists to turn to laboratory research if they want to understand animal cognition. She writes: "It is perhaps at this moment that the cognitive ethologist decides to hang up his field glasses, become a cognitive psychologist, and have nothing further to do with talk about consciousness or intention" (p. 124). Thus, Heyes denies that evidence gained by observing animals in natural settings is particularly relevant to understanding animal minds. Other slayers, who claim that they need more convincing evidence from the field, rarely tell what evidence would be convincing. Heyes and these other critics generally simply assume that no evidence that could be collected from the field would provide convincing support for attributions of mental states.

Unlike Heyes, who thinks that animal cognition can at least be studied in the laboratory, some slayers argue against the study of animal cognition on the basis of a philosophical view about the privacy of the mental (for a well-developed counterargument see Whiten, 1993), or by the related "other minds" problem. These critics typically do not give specific critiques of actual empirical investigations carried out by cognitive ethologists; rather they dismiss such investigations on philosophical grounds alone. The renowned evolutionary biologist, George C. Williams, writes:

> Vitalism today can be recognized in the psycho-physical dualism of some neural and behavioral biologists (e.g., Griffin 1981), who claim that explanations must make use of explicitly mental factors in addition to the merely physical. Griffin's concession (1981, p. 301) that the mental depends on the physical is difficult for me to interpret. If he means total dependence there is no longer any reason to make use of mentalism in biological explanation . . . (Williams, 1992, p. 4)

In this quotation, Williams reveals his naiveté about contemporary philosophy of mind. In questioning Griffin's notion of dependence, Williams suggests that if the mind depends on physical facts, then it can be eliminated from scientific explanation. In fact, such a view is tenable if you assert that mental properties are strictly identifiable with physical properties (what philosophers call "type-identity" theory), which seems to be what Williams means by the phrase "total dependence." But there are forms of dependence other than type identity that do not support the view that mental states are irrelevant to the explanation of behavior. For example, functionalists like Fodor (1974) hold a token-identity theory that allows a particular mental state to be identified with a particular brain state (just like a particular dollar can be identified with a particular piece of paper) but does not allow the mental and brain state types to be identified because of the possibility that organisms with different nervous systems can be in the same psychological state (just as the type "dollar" cannot be identified with the physical type "piece of paper bearing such and such markings" since there are also silver dollars). Physical predicates and mental predicates appear to carve up the world differently. Whether there are scientific generalizations that require the divisions provided by mental vocabulary is an empirical question that is separate from the dependence of the mental on the physical. Token-identity theory is not the only way to defend the claim that the mental depends on but is not reducible to the physical (e.g., Kim, 1984). Williams's failure to consider alternate conceptions of dependence seriously undermines his argument against Griffin.

Williams (1992) also writes: "I am inclined merely to delete it [the mental realm] from biological explanation, because it is an entirely private phenomenon, and biology must deal with the publicly demonstrable" (p. 4). From this quotation, we can construct an argument like this:

1. Mental events are private phenomena.
2. Private phenomena cannot be studied biologically.
3. Therefore, mental events cannot be studied biologically.
4. Cognitive ethology is possible only if mental events can be studied biologically.
5. Therefore, cognitive ethology is not possible.

When analyzed in this way, Williams's argument is seen to depend on another contentious philosophical premise (no. 1). Slayers often base arguments on claims about the privacy of the mental or skepticism about other minds. But it is ironic that these premises, which can only be defended in nonempirical, philosophical fashion, are produced by critics who would typically regard themselves as hard-nosed empiricists. Cognitive ethologists do empirical work, yet slayers who argue on such philosophical grounds rarely analyze that empirical work to see what it is designed to show and whether it in fact shows what it is designed to show. Instead, they base their arguments on claims that are at least as fraught with interpretive difficulty as the cognitive conclusions they wish to deny. This unwillingness to engage in debate about the actual empirical work of cognitive ethologists gives the impression that many slayers simply barge in, declare victory, and get out without genuinely engaging cognitive ethologists in a dialogue about their work. Williams does not stand alone. McFarland (1989a, p. 146) wonders if we are designed by natural selection to assume that deceitful acts are intentional, and Kennedy claims that the sin of anthropomorphism is programmed into us genetically. If either or both is the case, how are we to know? These claims make for empirical questions that require detailed study.

Rosenberg is an example of a slayer who has paid careful attention to actual work in behavioral biology, specifically the study of play. He says:

> . . . we have lots of reasons to fear that there can be no evolutionary theory of intentionality. For there can be no scientific theory of intentionality. . . . Does a cat have the concept of mouse-hunting? . . . The reason we are reluctant to credit the cat with such concepts is that there seems to be no behavior it could engage in that is discriminating enough . . . This leads the tender-hearted among students of animal behavior to insist that though the animal may not have our concept, it has some concept or other . . . The kill-joy argues that short of behavior

discriminating enough . . . , attributing concepts at all is gratuitous
anthropomorphism. . . . When it comes to such abstract concepts as
those required to attribute playfulness, in the literal sense, to mammals
much below that of the monkey, the kill-joy's position seems hard to
deny. (Rosenberg, 1990, pp. 183–184)

In this passage, Rosenberg belittles the idea that it might be possible to
attribute human concepts to nonhumans. His argument against a science of intentionality is based on considerations that seem to conflate
the difficulty of specifying the content of intentional states with the
impossibility of doing so (Allen, 1992a; Allen & Bekoff, 1993). Specifically, Rosenberg appeals to his inability to imagine what organisms
could do to allow the attribution of certain concepts; the difficulty of
thinking of suitable experiments is a challenge, not necessarily a barrier
to concept attribution (Allen & Hauser, 1991).

Other tactics used by some slayers involve focusing their criticisms on very narrow issues. Cronin thinks that Griffin, a "sentimental
softy," and other cognitive ethologists are only concerned with demonstrating cleverness, and hence consciousness. In her recent review of
Griffin's *Animal Minds* (1992), she writes:

A Griffin bat is a miniature physics lab. So imagine the consternation
among behavioristic ethologists when Mr. Griffin came out a decade
ago, with "The Question of Animal Awareness," as a sentimental
softy. . . . For Mr. Griffin, all this.[cleverness] *suggests* consciousness.
He's wrong. If such cleverness were enough to *demonstrate* consciousness, scientists could do the job over coffee and philosophers could
have packed up their scholarly apparatus years ago. (Cronin, 1992, p.
14, our emphasis)

Even McFarland (1989a), whom we categorize as a slayer, recognizes
that there are indicators of cognition other than the ability to produce
clever solutions to environmental problems. Not only is Cronin wrong
about slaying the field of cognitive ethology because of the difficulty
of dealing with the notion of consciousness (think about all of the
other fields of inquiry that would suffer if it were appropriate to base
rejection of those fields on singling out their most difficult issues),
but she is also wrong to think that demonstrating cleverness is a simple matter. Certainly, the difficult work needed to demonstrate cleverness could not be done over coffee! Furthermore, Cronin conveniently slides from claiming that for Griffin, cleverness *suggests*
consciousness, to claiming that his view is that cleverness is "enough
to *demonstrate* consciousness" (our emphasis). Even Heyes (1987b)
notes that it is not Griffin's program to *prove* that animals are con-

scious. Cronin later goes on to claim that at least chimpanzees are conscious and tells us why. Cronin concludes her scathing review with the following statement:

> Well, I know that I am conscious, I know a mere 500,000 generations separate me from my chimpanzee cousin, and I know that evolutionary innovations don't just spring into existence full-blown—certainly not innovations as truly momentous as our hauntingly elusive private world. (p. 14)

Cronin places herself on a slippery slope here. Why did she stop with chimpanzees? After all, if evolutionary innovations do not spring into existence full-blown, where did chimpanzee consciousness come from? Her phylogenetic argument cannot be assessed directly; behavioral evidence is needed to help it along.

Slayers' arguments can be based on the issues discussed above, and some even throw the debate into the political arena. Thus, Garland Allen (1987) claims:

> . . . to argue that we are animals does not mean we are merely animals. . . . It is in this vein that I think that both sociobiology and the study of animal awareness can be seen as part of an evolutionary development of present-day capitalism toward fascism. In the abstract there is nothing inherently fascistic about asking whether animals "think" or have an awareness of their own existence. It is the asking of this question under the present social and economic conditions, and with no procedures for arriving at a rigorous or scientific answer that makes the whole enterprise part and parcel of a fascistic (albeit at this stage a protofascistic) social development. . . . Given the methodological problems inherent in the field of animal consciousness (as well as with sociobiology), pursuit of the problem in the ways exemplified by investigators such as Griffin, can only lead to the sort of confusion and false understanding of the biological nature of human beings, on which the future of fascism can be built. (pp. 158–160)

Slayers may also have something to say about animal welfare. Despite a very large data base demonstrating highly developed cognitive skills in many animals, there are those who ignore research on animal cognition, misinterpret data from studies on humans, and base their conclusions on the moral status of animals using an intuitionistic comparison of animal and human behavior (e.g., Carruthers, 1989; Leahy, 1991; for discussion of these views see Bekoff & Jamieson, 1991; Clark, 1991; Johnson, 1991; Griffin, 1992; Jamieson & Bekoff, 1992a; Singer, 1992).[1] Thus, Peter Carruthers (1989), who compares the behavior of animals with the behavior of humans who are driving while distracted and humans who suffer from blindsight, writes: "I shall assume that no one would seriously maintain that dogs, cats, sheep, cattle, pigs, or chickens consciously think things to themselves . . . the experiences of all these creatures [are of] the nonconscious variety" (p. 265). He then claims:

> Similarly then in the case of brutes: since their experiences, including their pains, are nonconscious ones, their pains are of no immediate moral concern. Indeed since all the mental states of brutes are nonconscious, their injuries are lacking even in indirect moral concern. Since the disappointments caused to the dog through possession of a broken leg are themselves nonconscious in their turn, they, too, are not appropriate objects of our sympathy. Hence, neither the pain of the broken leg itself, nor its further effects upon the life of the dog, have any rational claim upon our sympathy. (p. 268)

And finally, "it also follows that there is no moral criticism to be leveled at the majority of people who are indifferent to the pains of factory-farmed animals, which they know to exist but do not themselves observe" (p. 269). All in all, slayers conclude that the study of cognitive ethology is too fraught with difficulty to be worth pursuing, or they simply fail to address the results of cognitive ethological research.

Skeptics

Skeptics' views are more open-minded than those of slayers. Skeptics have some of the same worries about anthropomorphism, anecdote, folk psychological explanations, the inaccessibility of the mental, and the ability of field studies to shed light on animal cognition, but they are not as dismissively forceful. Two representative views are:

> . . . folk psychological theory pervades human thinking, remembering, and perceiving and creates a very subtle anthropomorphism that *can* corrupt the formation of a science of cognitive ethology. (Michel, 1991, p. 253, our emphasis)

> We argue that attempts to study these processes [intentionality, awareness, and conscious thinking], while revealing impressive behavioral complexity, have proven unsuccessful in establishing the importance of mental experiences in determining animal behavior primarily because of the intractability of the problem. (Yoerg & Kamil, 1991, p. 273)

Alcock (1992) is a notable example in that he does not find the inaccessibility of consciousness to be grounds for dismissing the study of animal cognition (see also Whiten, 1993). In his review of Griffin's *Animal Minds* (1992), he writes:

We need ways in which to test hypotheses in a convincing manner. In this regard *Animal Minds* disappoints, because it offers no practical guidance on how to test whether consciousness is an all-purpose, problem-solving device widely distributed throughout the animal kingdom. . . . And there are alternative approaches to consciousness not based on the behavioristic principle that thinking cannot be studied because it does not exist. (Alcock, 1992, p. 63)

In his attempt to gain more control in studies of animal cognition, Premack, like Heyes (1987a), wants to see more highly controlled laboratory work. He believes that:

. . . cognition is not exclusively a field phenomenon; it can take place quite nicely in the laboratory. Indeed, in the case of chimpanzees, advanced cognition would appear to be largely a laboratory phenomenon. For only the chimpanzee who has been specially trained—exposed to the culture of a species more evolved than itself—shows analogical reasoning. . . . advanced cognition, such as analogical reasoning, is confined to the laboratory. (Premack, 1988a, pp. 171–172)

One problem here is that gaining too much control over what *may* be important variables may result in an impoverished environment that does not allow the animals to use the *combination of stimuli* that enables them to make assessments of others' behavior or others' minds (Bekoff, Townsend, & Jamieson, 1994). Only careful observations under field and captive conditions will permit us to assess just what are the important variables that influence how individuals interact with their social and nonsocial worlds, how these different variables are used, and how they may be combined with one another. Along these lines, de Waal (1991) notes that "Certain social phenomena cannot be transferred to the laboratory" (p. 311), because it would be impossible to recreate the social and nonsocial environments that are responsible for producing and maintaining ongoing encounters, such as those observed in dominance interactions. McGrew (1992) also stresses that "No experiment has ever simulated any of the delicate probing tasks such as termite-fishing shown by wild chimpanzees" (p. 83).

The types of explanations that are offered in studies of animal cognition also enter into how cognitive ethological studies are viewed. For example, many slayers and some skeptics favor noncognitive explanations because they believe them more parsimonious and more accurate than cognitive alternatives, and less off-putting to others who do not hold the field of cognitive ethology in high esteem. Snowdon (1991) claims that:

It is possible to explore the cognitive capacities of nonhuman animals without recourse to mentalistic concepts such as consciousness, intentionality, and deception. Studies that avoid mentalistic terminology are likely to be more effective in convincing other scientists of the significance of the abilities of nonhuman animals. (p. 814)

Some just make simple claims about the supposed parsimoniousness of noncognitive explanations and move on. Thus, Zabel, Glickman, Frank, Woodmansee, and Keppel (1992) in their attempts to explain redirected aggression in spotted hyenas (*Crocuta crocuta*) *before* quantitatively analyzing it, are of the opinion that "One must be cautious about inferring complex cognitive processes when simpler explanations will suffice" (p. 129). However, they admit that the other noncognitive explanations they offer are questionable. The statement by Zabel et al. is a paraphrase of Morgan's (1894) Canon: "In no case may we interpret an action as the outcome of the exercise of a high psychical faculty, if it can be interpreted as the outcome of the exercise of one which stands lower in the psychological scale" (p. 53). It is possible that Morgan's Canon, which is concerned with the complexity of processes, should be distinguished from parsimony, which is concerned with the number of processes needed to explain a given behavior (as an anonymous reviewer noted). However, unless we are given an explicit standard for judging the complexity of a cognitive process or its place in "the psychological scale" (Sober, in press), appeals to the canon rest on nothing other than intuitions about relative complexity and need not be counted as scientific. Appeals to parsimony on a case by case basis do not take into account the possibility that cognitive explanations might help scientists come to terms with larger sets of available data that are difficult to understand and also help in the design of future empirical work. (For further discussion of the weakness of the idea that the simplest explanation is always the most parsimonious, see Bennett [1991]. For a comparison of different perspectives on cognitive "versus" more parsimonious explanations, see proponent de Waal's [1991] consideration of the view of skeptics Kummer, Dasser, & Hoyningen-Huene [1990] on parsimony.)

Beer (1992) also thinks that if cognitive ethology limited its claims for animal awareness to sensation and perception, a practice which could change the vocabulary used in cognitive ethological studies, then "even tough-minded critics would be more receptive" (p. 79). Griffin (1992, p. 11) actually agrees with this point, but even a consideration of simple forms of consciousness is contentious to many slayers and skeptics (see Bekoff, 1993b).

Some other skeptical views include:

Perhaps cognitive ethology will prove to be an advance in the study of behavior in consequence of resuscitating questions of animal awareness. How far back it will be necessary to go to find the more fruitful formulation of these questions remains open to doubt, however: to Darwin and Romanes who believed in animal awareness? or to Descartes, who did not? (Beer, 1992, p. 105)

I am afraid that I am somewhat more skeptical than your reviewers on the value of such an approach [cognitive ethology] . . . Well, these are just my thoughts. Clearly, the paper sparked my interest about the issue, as it did for your two reviewers . . . I want to see controversial subjects discussed . . . and I do not think that [you] would ever get complete agreement from reviewers on an article like this . . . (Jane Brockmann, personal communication, January 24, 1992)

Beer remains cautiously optimistic, but wants cognitive ethologists to pay close attention to problems that are raised by philosophers of mind; his essay is essential reading for those interested in animal cognition. The second quotation is notable, for it reflects the refreshing open-mindedness of an editor of a major journal in ethology.

Proponents

Let's now consider the proponents. A quotation from William Mason's (1976) review of Griffin's *The Question of Animal Awareness* (1976) is a good place to start with respect to proponents' views. Mason writes: "That animals are aware can scarcely be questioned. The hows and the whys and wherefores will occupy scientists for many years to come" (p. 931). Mason's claim is a strong one. Note that in his endorsement of the field, Mason does not qualify his statement by writing "That *some* animals are aware. . . ." However, he does recognize that animals may differ with respect to levels of development of their cognitive abilities, and at a later date notes that "On the basis of findings such as those reviewed in this paper, I am persuaded that apes and man have entered into a cognitive domain that sets them apart from all other primates" (1979, pp. 292–293). Mason's statement about animal awareness is typical of those who narrowly focus their attention only on primate cognition (Beck, 1982; Bekoff, 1995; Bekoff et al., 1994).

Proponents note that there are already many successes in cognitive ethology and are enthusiastic about the bright future for cognitive ethology (see Cheney & Seyfarth, 1990, 1992; Allen & Hauser, 1991; Burghardt, 1991; de Waal, 1991; Ristau, 1991a; Allen, 1992b; Bekoff & Allen, 1992; Bekoff 1993b, 1995; Bekoff et al., 1994; Jamieson & Bekoff,

1993; Whiten, 1993). Although proponents admit that cognitive ethology has had some failures, they do not single these out and use them to condemn the whole field (Dennett, 1987, p. 271). While skeptics are willing to state what would convince them that cognitive ethology is a viable science, they may be pessimistic about the prospects of cognitive ethology coming up with the goods. Often, however, it is difficult to differentiate between skeptics and moderate proponents who argue that if there is to be a science of cognitive ethology, we must develop empirical methods for applying cognitive terms and making talk about animal minds respectable (e.g., Dennett, 1983; 1987, p. 271; 1991, p. 446; Kummer et al., 1990). Jamieson and Bekoff's (1993) differentiation between weak cognitive ethology, where a cognitive vocabulary can be used to explain, but not to describe, behavior, and strong cognitive ethology, where cognitive and affective vocabularies can be used to describe and to explain behavior, highlights some of the difficulties of clearly differentiating between some skeptics and some proponents. Nonetheless, proponents are more willing than skeptics to rely on methods of classical ethology (Jamieson & Bekoff, 1993) to help them along, whereas skeptics and slayers generally want to see cognitive ethology become consumed by cognitive psychology or by neuroscience.

Proponents are more optimistic in their views about the contributions that the field of cognitive ethology, and its reliance on field work and on comparative ecological and evolutionary studies, can make to the study of animal cognition in terms of opening up new areas of research and reconsidering old data (de Waal, 1991; Bekoff, 1995). As Carolyn Ristau (1991b) notes in her attempts to study injury-feigning under field conditions, the cognitive ethological perspective ". . . led me to design experiments that I had not otherwise thought to do, that no one else had done, and that revealed complexities in the behavior of the piping plover's distraction display not heretofore appreciated" (p. 102). The challenge of using ethological ideas in the study of animal cognition is reflected in the following quotation:

> At this point . . . cognitive ethologists can console themselves with
> the knowledge that their discipline is an aspect of the broader field of
> cognitive studies and conceptually may not be in any worse shape
> than highly regarded, related fields such as cognitive psychology. We
> are a long way from understanding the natural history of the mind,
> but in our view this amounts to a scientific challenge rather that
> grounds for depression or dismissal. (Jamieson & Bekoff, 1992b, p. 81)

Premack's (1988, see above quotation) views about the absence of high-level cognition in the field and its being largely a laboratory phenomenon ignores the complexity of behavior that is found in wild animals (Griffin 1976, 1981, 1984, 1992; Byrne & Whiten, 1988b; de Waal, 1991; Ristau, 1991a). While proponents recognize that Griffin has not really made detailed suggestions for experimental studies, this doesn't discourage them from seeking ways to make ideas like Griffin's empirically rigorous (Allen & Hauser, 1991, 1993; de Waal, 1991; Ristau, 1991a; Allen, 1992a, 1992b; Whiten, 1992; Bekoff, 1993b, 1995).

Proponents also share some of the concerns of the slayers and skeptics with respect to problems associated with the use of anecdote, anthropomorphism, and folk psychological explanations. However, proponents claim that the careful use of anthropomorphic and folk psychological explanations can be helpful in the study of animal cognition, and they also maintain that anecdotes can be used to guide data collection and to suggest new experimental designs (Dennett, 1987; Fentress, 1992). Thus, other proponents write:

> Cognitive ethology, rescued from both behaviorism and subjectivism, has much to say about what the life of the animal is really like. It is silent on what it is like to have that life. (Gustafson, 1986, p. 182)

> Others have picked up where Griffin leaves us by using his collection of anecdotes, his discussion of empirical research, and his ideas, to motivate new and extremely innovative studies, the bases for which might not have been obvious before his work. (Bekoff, 1993b, p. 168)

> . . . I have advocated use of a critical anthropomorphism in which various sources of information are used including: natural history, our perceptions, intuitions, feelings, careful behavioral descriptions, identifying with the animal, optimization models, previous studies and so forth in order to generate ideas that may prove useful in gaining understanding and the ability to predict outcomes of planned (experimental) and unplanned interventions . . . (Burghardt, 1991, p. 73)

> . . . when a number of anecdotal examples, each with a possible alternative explanation, collectively point to the likelihood of intentional deception, and this is supported by more rigorous tests in the laboratory . . . I would argue that it adds up to a strong case. (Archer, 1992, p. 224)

Perhaps some would assume Griffin to be the strongest proponent of cognitive ethology. After all, his rekindling of interest in the area (Griffin, 1976, 1978, 1981, 1984) is usually given sole credit for restimulating modern interest in cognitive ethology. Furthermore, even in his most recent book, *Animal Minds*, he has unabashedly launched a number of polemical attacks on his opponents for being closed-minded about the nature of available evidence on animal cognition and on the field of cognitive ethology in general. These sometimes snide attacks are considered unnecessary and off-putting to his critics and supporters alike (Huntingford, 1985; Yoerg, 1992; Bekoff, 1993b). Toward the end of *Animal Minds* Griffin (1992) writes:

> Contrary to the widespread pessimistic opinion that the content of animal thinking is hopelessly inaccessible to scientific inquiry, the communicative signals used by many animals provide empirical data on the basis of which much *can reasonably be inferred* about their subjective experiences. (p. 260, our emphasis)

Note that Griffin counters some slayers' and skeptics' concerns about the inaccessibility of animal minds, but he does *not* make a very strong claim that he or others can ever know the content of animal minds (see italicized phrase above; recall Cronin's (1992) misrepresentation of Griffin's agenda above). Rather, Griffin, like other proponents, remains open to the possibility that we can learn a lot about animal minds by carefully studying communication and other behavior patterns. He and other proponents want to make the field of cognitive ethology more rigorous on theoretical and empirical grounds.

DISCUSSION

It is very useful for cognitive ethologists to engage in some introspection concerning how their fields of interest are viewed. As a result of our inquiry, different views on animal cognition have become clarified. As proponents we argue that there are many reasons for studying cognitive ethology from comparative and evolutionary perspectives.[2] Many models in ethology and behavioral ecology presuppose cognition, and it is useful to have an informed idea about the types of knowledge nonhumans have about their social and nonsocial environments and how they use this information (Yoerg, 1991; Real, 1992). The assumption of animal minds also leads to testable hypotheses about, and more rigorous empirical analyses of, behavioral flexibility and behavioral adaptation. From the applied (and perhaps political) side of things, views on animal minds are tightly linked to issues that center on animal welfare (Bekoff & Jamieson, 1991; Bekoff et al., 1992) and human dignity (Rachels, 1990; see also references in note 1). Contrary to G. Allen (1987, quotations above) who thinks that closing the gap between human and nonhuman animals will lower the value placed on humans, we think that closing the gap might raise the value placed on nonhumans.

Studying animal cognition is not easy. As Yoerg (1992, p. 831) notes: "It is isn't a project I'd recommend to anyone without tenure"

(p. 831). Clearly, proponents do not accept that cognition is a phenomenon associated only with captivity. While proponents are aware of the need to be critical, they also recognize that an extensive data base of cognitive ethological investigations will not be built rapidly, because of the demanding types of research that are required in the study of animal cognition, especially under field conditions. Patience is needed, as Bennett (1978) noted in his discussion of Griffin's (1978) earlier views:

> Just because I find G[riffin]'s campaign so sympathetic, and so many of his details interesting and persuasive, I would like to urge upon him the importance of circumspection—of a patient, continuous attention to conceptual foundations. (p. 560)

Future data from comparative analyses of animal cognition, along with existing information, should help us along in developing what some people think the field of cognitive ethology needs, namely, an integrative model or theory (Whiten, 1992) not concentrated solely on primates (Beck, 1982; Bekoff, 1995; Bekoff et al., 1994). Perhaps it is the lack of an integrative theory of cognitive ethology and the presence of one in evolutionary biology that is responsible for many people dismissing tenuous cognitive ethological explanations but accepting often equally tenuous evolutionary tales (Myers, 1990, pp. 211ff; Hurlbert, 1992; Jamieson & Bekoff, 1993).

Our analysis of criticism of cognitive ethology as a scientific discipline has also revealed the large extent to which critics depend on philosophical views about the nature of mind. Considering the open disdain that several of these critics have for philosophers, this is ironic indeed; see, for example, Kennedy's (1992) *ad hominem* against "the ethological philosopher Dennett, who seems perhaps to be embracing genuine anthropomorphism like Dunbar, albeit more obscurely and at prodigious length" (p. 92). By exposing the extent to which slayers and skeptics rely on contentious philosophical views uncritically adopted, we hope to warn against facile arguments to the effect that cognitive ethology is unscientific. We believe that scientific soundness can only be based on careful analysis of the specific empirical practices of ethologists. We also think it worth pointing out that arguments against cognitive ethology appear to operate at the level of the paradigm (Kuhn, 1962) rather than at the level of the ordinary scientific practices of cognitive ethologists. We suspect that the stands taken by many critics on the issues of mentalistic concepts, anthropomorphism, and parsimony are likely to be as much the product of socialization (e.g., graduate student training) as rational deliberation. Hence there may be support for Kuhn's views about the importance of sociological factors in the development of a young science.

There are no substitutes for careful and rigorous observational and experimental studies of animal cognition and detailed analyses of subtle behavior patterns that often go unnoticed. Cognitive ethologists are now able to exploit techniques like experimental playbacks of vocalizations to conduct controlled studies under field conditions (e.g., Seyfarth, Cheney, & Marler, 1980; Cheney & Seyfarth, 1990; for other examples see Allen & Hauser, 1991; Ristau, 1991a; Real, 1992); the range of experiments made possible by such techniques means that there can be no easy dismissal of modern cognitive ethology on the grounds that it is anecdotal or lacks empirical rigor. Thus, we do not think that modern cognitive ethology will suffer the same fate as pre-behaviorist cognitivism. People should not come to cognitive ethology with axes to grind. Interdisciplinary input is necessary for us to gain a broad view of animal cognition. Philosophers need to be clear when they tell us about what they think about animal minds and those who carefully study animal behavior need to tell philosophers what we know, what we are able to do, and how we go about doing our research. Cognitive ethologists should put their noses to the grindstone and welcome the fact that they are dealing with difficult, but phenomenally interesting, questions. We hope that all views of cognitive ethology will remain open to change.

NOTES

1. Of course, not only slayers have something to say about the relationship of cognitive ethology to animal welfare (e.g., Bekoff & Jamieson, 1991; Bekoff et al., 1992; Jamieson & Bekoff, 1992a; Lehman, 1992; G. W. Levvis, 1992; M. A. Levvis, 1992; Lynch, 1992; Bekoff & Gruen, 1993). Griffin (1992), in an uncharacteristically strong comment, notes: "No one seriously advocates harming animals just for the sake of doing so, although thoughtless cruelty is unfortunately prevalent in some circles" (p. 251). Unfortunately, he does not tell us where.

Another issue that bears on studies of both animal cognition and animal welfare concerns the naming of animals, for this practice is often taken to be nonscientific (Bekoff, 1993c). Historically, it is interesting to note the Jane Goodall's first scientific paper dealing with her research on the behavior of chimpanzees was returned by the *Annals of the New York Academy of Sciences* because she named, rather than numbered, the chimpanzees she watched. This journal also wanted her to refer to the chimpanzees using "it" or "which" rather than "he" or "she" (Montgomery, 1991, pp. 104–105; see also Myers, 1990, pp. 199ff; Peters, 1991; Davis & Balfour, 1992a, 1992b; Jamieson & Bekoff, 1993; Phillips, 1994). Goodall refused to make the requested changes but her paper was published anyway. As pointed out elsewhere (Bekoff, 1993a), the words "it" and "which" are typically used for inanimate objects. Given that the goal of many studies of animal cognition is to come to terms with animals' subjective experiences—the animals' points of view—making animals subjects rather than objects seems a move in the right direction.

Studies where individuals are named typically involve small numbers of animals, thus raising worries about sample size—that what is being presented is anecdotal evidence rather than data. But, some general points can be made concerning sample size—specifically single-subject research—in studies of animal cognition. If we want to come to a better understanding of the animals' points of view, then working on a limited number of individuals would facilitate these efforts because research on animal cognition is extremely time consuming and often tedious. Furthermore, providing appropriate care for certain species may be more difficult than for others, and concern for animal welfare might also enter into decisions concerning sample size. Regardless of the reasons, many studies of animal cognition involve very detailed analyses, based on observation and experiment, of the behavior of only one, or of a few, animals. However, understanding points of view also entails understanding differences among individuals, and not only the behavior of individuals who are assumed to be representative of their species. Inferences made from averaging the behavior of many organisms can be misleading, especially in species in which individual differences in behavior are the rule rather than the exception.

Questions that need to be considered include: (a) Why does it seem to be permissible—in the sense of being scientifically acceptable—to study a single ape, a lone parrot, a few monkeys, or a few dolphins, whereas studies involving a few dogs, cats, or rodents are generally frowned upon? (b) How does sample size relate to the goals of a given study? For example, if an evolutionary or ecological account of cognition is desired, would we be better off studying more animals in less detail to gain normative information in which would be contained data on species-typical ranges of behavior? If we want to learn more about the potential cognitive skills of a given individual or class of individuals, who might or might not represent her species, would we be better off studying fewer animals in great detail?

Often, one of us (M. B.) is asked why he concentrates on his companion dog, Jethro, when making general points about social play in canids. The reason is that while M. B. has considerably more data on other canids, using Jethro's behavior as an instance of some of the general characteristics of social play makes discussion of the phenomenon of social play more accessible to those who are not familiar with other canids or individuals belonging to other species in which play has been described. On one occasion M. B. asked what people would think if he had data *only* for Jethro; most people thought that this would make for weak arguments concerning the cognitive aspects of social play behavior. Then, when M. B. asked about the use of single subjects in different types of studies of animal cognition, people reconsidered their hasty response to the question of the use of a single (or a few) domestic dogs in studies of animal cognition. Questions concerning sample size are not easy to answer. The goals of a given study, the accessibility of the animals being used, the type of care that captive individuals require, and the nature of the questions being asked are among the variables that need to be considered in answers to the question of what constitutes an adequate sample.

2. One goal of these sorts of studies would be to collect data that can be analyzed applying the rigorous methods that have been used in comparative and evolutionary analyses of other phenotypes (e.g., Gittleman & Luh, 1993).

24

Animal Cognition
Versus Animal Thinking:
The Anthropomorphic Error

Hank Davis

Over 100 years ago Romanes (1882) suggested that overt behavior was the "ambassador of the mind," thus allowing one to draw inferences about the mental life of animals from observing their daily activities. Like most metaphors, Romanes's proposal contained a germ of truth. Unfortunately, this germ proved to be highly contagious and has contributed to the present epidemic of anthropomorphism.

I will begin by commenting on the prevalence of anthropomorphism, after which I will indicate the basis for my criticism of the practice. This will lead to a distinction between *animal cognition* and *animal thinking*, the latter of which I believe to be problematic in ways that go to the core of the anthropomorphic error. Throughout the following discussion, I will define anthropomorphism as the attribution of human thoughts, feelings, and motives to nonhumans, although my primary concern is with the first of these three human activities.

The preparation of this chapter was supported in part by Grant No. A1673 from the Natural Sciences and Engineering Research Council of Canada. A preliminary version of these arguments was presented at the June, 1989, meeting of the Animal Behavior Society. I am grateful to the following persons for their thoughtful criticisms of the present chapter: Timothy Eddy, Laurie Hiestand, Loraleigh Keashly, Hugh Lehman, Carl Porter, Susan Simmons, Allison Taylor, and Andrew Winston.

THE PREVALENCE OF ANTHROPOMORPHISM

Anthropomorphism is commonplace for a number of reasons. Most basically, it represents a form of intellectual laziness. In its most extreme case, anthropomorphism results from the failure to make species differentiations. In this regard, very young children or pet owners whose animals have become "family members" are likely candidates for such attributions. Carried to its extreme, anthropomorphism may be directed at plants or inanimate objects like faulty toasters and automobiles.

Indeed, there is nothing except "common sense" to limit the application of anthropomorphism. Unfortunately, common sense is a variable commodity. Historically, before the dawn of behaviorism, anthropomorphism was carried far beyond other primates or household pets (see Boakes, 1984). For example, Binet (1889) attributed higher mental processes to protozoa. Nearly a century later, Griffin (1984) has done the same for wasps, flies, ants, and spiders.

THE PROBLEM WITH ANTHROPOMORPHISM

While it has its endearing or Disneyesque side, anthropomorphism does far more harm than good. Indeed, Broadhurst (1964) has called it the "cardinal crime" of animal observation. Those who defend anthropomorphism often note that it is at worst a misguided analogy. I will argue that the consequences are far more insidious. The anthropomorphizer does more than draw inferences about the *occurrence* of human thought, feelings, and motives in nonhumans. Rather, anthropomorphism also draws inferences about the *function* of these entities. I will argue that such inferences are baseless.

The nature of anthropomorphism is rooted in a classic logical error called "affirming the consequent." Most of us are familiar with this form of reasoning: it is the staple of religious tracts. For example: (a) If there were a God, (b) then the world would be a beautiful place. One then affirms the consequent (b) in order to prove the antecedent (a). In short, because the world *is* a beautiful place, there must be a God. Without entering into a theological debate, it goes without saying that there are many reasons the world might be a beautiful place; a deity is only one of them.

In the case of anthropomorphism, the process involves slightly different premises. For example, one begins with implicit assumptions such as those shown in Figure 24-1. If (or when) I think, I scratch my head. There is nothing inherently wrong with this premise pair. Indeed, most premise pairs on which affirming the consequent is based are not

If I think
Then I scratch my head

*****

I am scratching my head
Therefore I must be thinking

FIGURE 24-1
The error of affirming the consequent

illogical. Rather, it is the illogical use to which they are put that contaminates our conclusions.[1] Thus, for example, one cannot affirm the
consequent (I *am* scratching my head) in order to establish the occurrence of thought. Such specious logic would fail to rule out other
antecedents such as dandruff.

When the faulty logic of Figure 24-1 is applied, the stage is set for anthropomorphism. Consider a visit to the zoo and the sight of a chimpanzee engaged in head scratching. At this point the logical error that was previously confined to humans has been extrapolated to chimps. This is essentially the way anthropomorphism proceeds: from logical error with humans to paradigm for field research with animals. Needless to say, a casual, if ill-conceived, observation at the zoo is not cause for concern. However, the problem is far more widespread.

Donald Griffin's book *Animal Thinking*, published in 1984, shows that anthropomorphism is neither a quaint nineteenth-century problem, nor confined to those without scientific training. Griffin is an eminent scholar whose landmark research on echolocation in bats serves as the basis of our understanding of this complex process. However, Griffin's speculations about the function of animal thinking follow another standard of scholarship. As van Rooijen (1983) argues, Griffin uses a wholly inappropriate logical system to prove the existence of "awareness" in animals. Van Rooijen notes that Griffin's reasoning belongs in the *Geisteswissenschaften* (spiritual sciences), rather than the *Naturwissenschaften* (natural sciences). At the risk of unfairly indicting a single source, I will focus on *Animal Thinking* because I can think of no more explicit body of anthropomorphism in the modern literature. Griffin's position is clear and persistent and he is unrepentant about it (Griffin, 1987). He is not a straw man.

The preface to *Animal Thinking* sets the tone for what follows. Griffin (1984) argues that "Animals make so many sensible decisions concerning their activities . . . that it has become reasonable to infer some degree of conscious thinking" (pp. 3–4). Plainly, Griffin is affirming a slightly different consequent here, but the illogical process is the same. His premises are the following: (a) If I engage in conscious thought, (b) then I will behave in a sensible manner. Again, there is nothing unreasonable about these premises. The problem lies in assuming that conscious thought is the sole basis for sensible behavior. As I shall argue, this is an unwarranted assumption to make about human behavior and the situation does not improve when we apply it to animals. Indeed, the fundamental point of my chapter is this: we do not know enough about the role of conscious thought in determining human behavior to extrapolate to any other species.

When I say "we," I am speaking of scientific psychology. The field devoted to understanding human cognition is in its infancy. The role of *thought* in determining behavior is the subject of considerable debate (e.g., Skinner, 1953; Velmans, 1991; Weiskrantz, 1988). This contentious and impoverished data base can hardly be the basis for speculation about the thoughts and behavior of another species. On the other hand, *folk* psychology, that branch of the discipline practiced by our unaccredited grandparents, bartenders, talk show hosts and cab drivers, has a rich and fully developed data base concerning human thought and behavior. It is folk psychology that offers credibility to the kind of anthropomorphizing practiced by Griffin (cf. Stich, 1983; Zuriff, 1985).

There is a second reason that anthropomorphism may be inappropriate. Even if the role of conscious thought in determining behavior were known in humans, the extrapolation to other species rests on the assumption of a cognitive continuum between humans and other animals. While this Darwinian position is widely held, there is also reason to question it. For example, Terrace (1984) has argued that the existence of language, unique to our species, may result in a cognitive discontinuity between humans and other animals. Premack (1988) has put the case bluntly: "If individuals who speak different languages actually differ in how they think (Whorf's hypothesis), how great then must be the difference between a group that has language and one that does not" (p. 46).

Terrace (1984) also notes that biological constraints (e.g., Seligman, 1970; Seligman & Hager, 1972) may so limit the cognitive abilities of other species as to render comparison with human cognition impractical. For example, performance on a conventional radial maze task is facilitated in rats by a species-typical "win-shift" tendency, whereas the performance of pigeons is impeded by a species-typical "win-stay" pattern. By implication, human performance, which is generally unconstrained by either tendency, may alone reveal something about "pure" problem solving.

THE ROLE OF THOUGHT

Within the well-documented behaviorist system of B. F. Skinner, thought functions as a *dependent* variable, an *effect* in the causal chain governing behavior. This position is counterintuitive from the point of view of folk psychology and is also a widely resisted notion among undergraduate students. Nevertheless, it is worth exploring, for if thoughts are not prerequisites for human behavior, then it is pointless to examine animal behavior for evidence of thinking (cf. Griffin, 1984).

Radical behaviorism does not dispute the existence of thought, per se. Rather, it holds that (a) thought is irrelevant to the analysis of human (or animal) behavior, and that (b) thought, like behavior itself, is the result of antecedent events in the organism's life. Put simply, a causal sequence may lead to two effects: One, an overt behavior; the other, covert. The latter is called *thought*.[2] Thought may occur simultaneously with behavior or may follow it. In the former situation, one may mistakenly assign a causal role to thought simply because of its proximity to the behavior. In this regard, folk psychology will uncritically confirm its own bias: that conscious thought was responsible for behavior.

In the latter case, where thought follows behavior, there is a clear parallel to the James-Lange theory of emotion, in which action precedes conscious thought or labeling (James, 1884). This view is counterintuitive and is not widely held. Not surprisingly, there is a persistent belief that because one can consciously reflect about behavior after the fact, this reflection must have been instrumental in causing the behavior to occur in the first place. Such speculative first-person reports are problematic for a number of reasons (Velmans, 1991). Perhaps most surprisingly, there is evidence that the human brain is organized so as to distort or even reverse our perception of the sequence in which events occur (Libet, 1985). If this is correct, then subjective reports about thought preceding and therefore causing behavior would be an unreliable source of evidence.

Consider the case of a person trying to decide which of two films to see. While it is undoubtedly true that considerable information processing will occur—as it does in all foraging decisions—it is unnecessary to assume that this processing sequence will either be (a) verbal in nature or (b) consciously accessible. It is no doubt true that one may consciously reflect on the pros and cons of the two films, waxing poetic about the merits of the director or actors. However, to assume that this conscious deliberation underlies the actual decision-making process is fallacious. Such a view is consistent with the Victorian notion that all human behavior (and decision making) is rational and consciously motivated (Houghton, 1957). More realistically, such metacognitions may be seen as a parallel process, another form of behavior resulting from the same causal events that led to the film-goer's decision. The notion that humans behave rationally, whether consciously or not, has itself become progressively less tenable. There is now considerable evidence that behavior in situations calling for logical decision making rarely if ever follows the formal rules of propositional calculus (Cosmides, 1989; Johnson-Laird, 1986).

This view presumes that some sophisticated and complex processing may occur without language and without awareness. This is precisely what I am proposing. This is not to say that the *results* of such "automatic" processing cannot also be achieved, probably in a far less efficient manner, by conscious verbal heuristics. This slower verbal process, which *would* be accessible to conscious monitoring, may lead to the erroneous conclusion that *all* information processing involves and depends directly on conscious thought.

Velmans (1991) has examined the role of consciousness in a host of human activities, ranging from simple motor tasks to creativity, and finds no compelling evidence that consciousness is required for any of them. Velmans is a cognitive psychologist, whose critique of the role of consciousness does not stem from the radical behaviorist tradition or its well-known biases. The assumption that conscious thought underlies complex behavior is rarely investigated directly in humans. The *Zeitgeist* gives us little reason to do so. The assumption is relatively inoffensive and it is consistent with widely held folk wisdom about human behavior. It is not until one affirms the consequent regarding complex behavior in order to prove the existence of antecedent thought in nonhumans that the logic is called into question. In this regard, Griffin's (1984) book has done us a service. In making lavish claims about the role of mental life in a variety of invertebrates, Griffin has made it clear that he is no *speciesist*. His book allows us to raise questions about the nature of human behavior in general and anthropomorphism in particular.

Diving Gannets

I will examine a case not included by Griffin in his survey of "animal thinking," but one which follows the essential paradigm of his book. Let us first approach the episode in Griffin's terms, then consider it in alternate terms which preclude anthropomorphism by suggesting an alternative nonverbal, nonthinking mechanism.

Gannets (*Sula bassana*) are large fish-eating birds whose foraging behavior involves a dramatic plummet from as high as 100 feet in the air. This descent, during which the gannet moves at speeds approaching 50 miles per hour, is a spectacular example of navigational prowess and physical coordination. To capture its prey, which is moving through water while the gannet dives, is a feat of stunning grace, the basis of which has remained unknown until fairly recently.

In Griffin's hands, this episode might prompt the following narrative: It is inconceivable that such highly precise, mathematically regulated behavior could occur without a guiding intelligence, a conscious thought process which monitors incoming information and makes minute adjustments in behavior in a manner that reflects, at the least, knowledge of calculus and/or trigonometry. Such mathematical sophistication, which a human might struggle to simulate in a computer program, surely cannot be held to exist without some form of conscious thought. Thus, in this highly sophisticated example of foraging behavior, we have once more uncovered evidence of thinking in the animal kingdom.

If this narrative sounds all too familiar, let us note that there is another possible account of the gannet's behavior, which has been confirmed by detailed analysis (Lee & Reddish, 1981). After painstaking study, these researchers concluded that the gannet's navigation, like a baseball player's hitting a 90-mile-per-hour fastball or a bird's remaining perched on a twig in a windstorm, allows precise visual-motor coordination by use of an *optic flow field*. Lee and Reddish analyzed filmed dives and, applying a simple parameter from the flow field, were able to account for all the navigational adjustments made by the diving gannets. These authors suggest that the same relatively simple visual processing mechanism may be involved in "complex" human motor skills such as high diving. "It is beginning to appear that the information afforded by this ubiquitous optical parameter is much exploited by visual systems. The information is needed for many activities and is available to any seeing organism" (1981, p. 294). In sum, just because behavior is describable in highly sophisticated verbal/quantitative terms, does not mean that such mechanisms were necessary to produce the behavior.

Although the evidence does not yet exist, I would like to believe that all of the examples that appear in *Animal Thinking* are amenable to precisely this kind of analysis. What, then, have Griffin's cognitive descriptions added to our understanding, and why are they so compelling to so many? Although the answer may be tied to our sentimental attitudes about animals, I believe it is largely rooted in misguided folk wisdom surrounding *human* behavior. If we did not hold a needlessly complex view of the importance of conscious (verbal) thought in the genesis of our own behavior, we would not even be tempted by the rampant analogizing of Griffin's book. It is the extent to which Griffin takes the analogy (e.g., thinking in insects) that calls the process into question for many readers. Ironically, these same critics rarely take issue with affirming the identical consequent when analyzing their own behavior.

ANIMAL COGNITION VERSUS ANIMAL THINKING

It is entirely possible to study the complex forms of behavior described in Griffin's book without sinking into the quagmire of anthropomorphism. Indeed, two related disciplines have brought complementary strategies to this work. Both have demonstrated that such research need not herald a resurgence of crude introspection and anthropomorphism. Despite Griffin's book, a new era of naive mentalism is not upon us.

For example, one may follow an ethological approach to issues such as kinship recognition and optimal foraging. Both of these topics hold the potential for rampant anthropomorphizing, but in the hands of competent scientists (e.g., Kamil & Yoerg, 1985; Walls, 1991), neither has succumbed to this treatment. Such research is in evidence in virtually any recent issue of the journal *Animal Behaviour*.

A second approach to complex animal behavior stems from the more experimental tradition of psychology, including the methodology of radical behaviorism (Shimp, 1989). Numerous laboratories are engaged in research on topics such as animals' use of space, time, and numbers (e.g., Davis & Albert, 1986; Davis & Bradford, 1986; Gibbon & Church, 1984; Olton, 1978; Roberts, 1979). In addition, much contemporary research is focused on issues such as (1) concept formation in animals (e.g., Blough, 1984; Herrnstein & Loveland, 1964; Pepperberg, 1990a; Thomas & Noble, 1988); (2) nonhumans' use of formal logic, such as transitive inference (e.g., Davis, 1992; von Fersen, Wynne, Delius, & Staddon, 1991); (3) the role of redundancy in information processing and how such information may be "blocked" by more salient cues (e.g., Kamin, 1968; Rescorla & Wagner, 1972); (4) the extent to which minor differences in the probability of stimulus occurrence affects the strength of conditioning (e.g., Rescorla, 1967); (5) the detection of complex visual patterns by animals (e.g., Blough, 1984). All of these topics, along with the study of memory (e.g., Honig, 1984), form the basis for what is globally known as *animal cognition*.

Animal cognition is now approached empirically, using the precepts of scientific method. Its practitioners address research questions directly by specifying properties of the physical environment and elucidating the processes by which information is processed and behavior is directed. Animal cognition does not imply, nor does it necessitate, speculations about thought, mentalism, or consciousness.

COMPARATIVE COGNITION AND ANTHROPOMORPHISM

Although animal cognition is typically studied within the framework of rigorous science, it shares an undeniable feature with anthropomorphism: its subject matter is guided by assumptions about how humans operate. Like comparative psychology itself, the study of comparative cognition assumes a continuum of abilities across animal species. Its essential mandate is to find rudiments of our own "higher" behavior in other species. Comparative psychology subscribes to the Darwinian notion that differences in the behavior of humans and other animals are essentially quantitative. Thus, we find discussions of the "roots of cognition" which explore analogs of various human behaviors in "simpler" species. The point here is that we are once again dependent upon notions about our own behavior to analyze performance in other species.

The case is nowhere plainer than in the study of numerical competence. The fundamental lexicon of numerical processes (e.g., counting, subitizing, estimation) derives directly from classification of human behavior. Our reluctance to apply the term "counting," for example, to an animal's performance (see Figure 24-2) stems from our difficulty in detecting all the fundamental attributes of counting (e.g., use of a cardinal chain; partitioning and tagging) that have been observed when human adults and children count (Davis & Pérusse, 1988; Gelman & Gallistel, 1978). The possibility that animals may use an alternative process to enumerate, that is altogether unknown to humans, may be

FIGURE 24-2
Rat correctly "noses" her way into fourth tunnel. But is she "counting"?
(After Davis & Bradford, 1986.)

acknowledged (e.g., Davis & Pérusse, 1988), but is rarely pursued systematically. We simply do not know what to look for because we do not have our own behavior to guide us.

Does this dependence on human behavior doom comparative cognition to the errors of anthropomorphism? I believe that it does not as long as our relevant knowledge of human behavior is based on scientific analysis. It is here that comparative cognition has a distinct advantage over anthropomorphism. Although both derive from assumptions about human behavior, these assumptions are generally better grounded in the study of cognition. For example, there is a large and systematic literature on human memory that can be applied to the analysis of animal memory. One may be criticized for assuming that animals use memory in a way similar to humans, but at least we have some knowledge of how humans function. Where our understanding of the human condition is more tenuous, such as our species' use of formal logic, our attempts to analyze animal behavior are correspondingly more problematic (Davis, 1992). Where knowledge of the human situation remains unresolved, such as the role of thought in guiding complex behavior, extrapolations to animals, such as those proposed by Griffin (1984), are futile.

THE UTILITY OF ANTHROPOMORPHISM

Some cogent arguments have been made for the utility of anthropomorphism, not just in elevating our sensitivity to the welfare of other species, but in aiding scientific endeavors as well (Gallup, 1983; Lockwood, 1985). Like anecdotal reports, which also curry disfavor with many scientists (e.g., Thomas, 1988), anthropomorphism may yield insights which can be transformed into testable hypotheses (Garcia, 1981). There is nothing objectionable about this strategy; indeed, one can argue that the *source* of a hypothesis is irrelevant. It is the manner in which it is stated and how (or whether) it is tested that determine its relevance to science.

There is no reason that between-species extrapolation must invariably stem from human behavior. Within psychology there is considerable tradition for using animal models to elucidate fundamental human processes. Although it is less frequently done, there may be a special virtue in studying "higher" human functions such as cognition in relatively simple or "unaware" animals. Such demonstrations would help divest human cognition of its excess dependence on verbal thought. The selection of "simple" subjects would be a relatively easy matter as there appears to be a continuum of species about which anthropomorphic attributions are likely to be made (Eddy, Gallup, & Povinelli, 1993; Herzog & Galvin, this volume). Rather than study "counting" in chimpanzees, for example, one might investigate its occurrence in garden slugs. If such a test were to succeed, we might reconsider the possibility that counting might be accomplished by a simpler process than previously believed.

Although a reappraisal of human counting would make sense in light of such evidence from garden slugs, such a reappraisal is unlikely to occur. It is more likely that our "simple" demonstration would be discounted, and the conclusion reached that the garden slug has taught us nothing about the human. Generalizations from "lower" animals that demystify human performance are often sabotaged by an intuitive notion that turns Morgan's Canon (1894) on its ear. Put simply, this belief holds that if a simple task can be approached in a complex manner, *that* must be how the human does it. Since humans have evolved all this extra brain tissue, they will use it whenever possible.

Not only does this logic confound our understanding of human behavior, but it ultimately contaminates our understanding of animal behavior as well. It maintains the very myths about the role of human thought that led to the kind of anthropomorphism practiced by Griffin and his antecedents (e.g., Binet, 1889; Witmer, 1909). It is only by letting go of our self-aggrandizing myths about human verbal thought that we will defuse the most naive kind of anthropomorphism.

Commenting on some of the virtues of anthropomorphism, Gallup (1983) has proposed that the practice may be an "inevitable byproduct of the ability to be aware of one's own mental states" (p. 503). This is an intriguing notion. However, if it is true that anthropomorphism ultimately stems from one of our highest mental faculties, then surely we would have the capacity to modulate its use. Although it may require mental discipline *not* to be anthropomorphic, no one has suggested that anthropomorphism is an unmodifiable reflexive behavior. Why, then, not subject it to the rigorous scrutiny of scientific method that, itself, is further evidence of our higher mental faculties?

SUMMARY

Eventually, some form of anthropomorphism may be an acceptable practice—not simply as a source of testable hypotheses, but also as a primary source of insight into how other species function. But before this occurs, we will need to satisfy two criteria: (1) Be convinced that there is indeed a continuum of mental life that includes both humans and the animals to which we wish to extrapolate, and (2) Achieve a sufficient understanding of the mental processes *in humans* that we wish to extrapolate to animals. What is the role of these thoughts, feelings, or other internal states in determining human behavior? To date, neither of these criteria has been satisfied.

NOTES

1. Affirming the consequent may actually be a useful adjunct to scientific discovery. The difference between its legitimate use, and the cases about which I am critical, lies in the extent to which we have ruled out other possible antecedents. In effect, the logical syllogism itself is blind to whether adequate steps have been taken so that the antecedent we are considering stands alone among causal possibilities. Our general ignorance about animal behavior (or, for that matter, about the origins of the universe) set the stage for the illegitimate use of this logical process as part of the discovery process.

2. I am not postulating the most extreme version of this position, under which thought is viewed solely as an after-the-fact *epiphenomenon*; that is, an event that has causes but no effects (Velmans, 1991). I concede that thought may, in special cases, serve as an independent variable, probably one of many, which influences subsequent behavior or decision making. It is worth stressing, however, that this is fundamentally different from saying that thought is (1) required to account for behavior in all cases, or (2) the *primary* cause whenever several factors are present.

25

Anthropomorphism Is the Null Hypothesis and Recapitulationism Is the Bogeyman in Comparative Developmental Evolutionary Studies

Sue Taylor Parker

RETURN TO ANTHROPOMORPHISM

Beginning in Darwin's time, students of animal behavior began to look at animal behavior through the lens of evolutionary theory. Struck by parallel abilities and developmental patterns in other species, such psychologists as Hall (1908) and Baldwin (1894/1903) related stages of child development to purported stages of animal evolution. Burgeoning interest in primate mentality expressed in the anecdotal approaches of Romanes (1882, 1883), however, was gradually overwhelmed by the growing momentum of the behaviorist movement.

The attendant emphasis on discovering universal laws of learning through controlled investigation of a few laboratory-bred species discouraged most animal psychologists from any anecdotal and anthropomorphic tendencies. A few romantic souls compared the development of mental abilities in home-reared chimpanzees with those of human children (Kohts, 1935; Kellogg & Kellogg, 1933; Hayes, 1951), but the mainstream psychologists avoided this approach. Although Köhler (1917/1927) investigated the mentality of chimpanzees from a Gestaltist perspective, most American investigators, like Yerkes and Harlow, emphasized mainstream learning paradigms. These paradigms discouraged interest in species differences or in behavior in natural environments even in the face of continuing challenges from the more evolutionarily oriented European ethologists. Beginning in the sixties, however, the cognitive revolution in American psychology began a revival of interest in mentality of humans and other animals. One outcome of this revival has been the emergence of a field known as cognitive ethology which, as the name suggests, focuses on the evolution of animal mentality from an ethological and cognitive perspective (e.g., Hinde, 1974; Griffin, 1978; Roitblat, Bever, & Terrace, 1984; Cheney & Seyfarth, 1990; Ristau, 1991b).

The emergence of comparative evolutionary developmental psychology (e.g., Jolly, 1972; Parker, 1990) has been another outcome of the cognitive revolution in psychology. This field, which began with ape language studies, contrasts with cognitive ethology in its use of models of human development to study animal cognition across a variety of domains: language development (e.g., Gardner & Gardner, 1975; Patterson, 1980; Miles, 1990; Greenfield & Savage-Rumbaugh, 1990); Piagetian and neo-Piagetian sensorimotor series stages (e.g., Chevalier-Skolnikoff & Poirier, 1977; Redshaw, 1978; Mathieu & Bergeron, 1983; Mignault, 1985; Doré & Dumas, 1987; Antinucci, 1989; Parker & Gibson, 1990); deception (e.g., Mitchell & Thompson, 1986b; Byrne & Whiten, 1988b); imitation (e.g., Mitchell, 1987; Whiten & Ham, 1992; Russon & Galdikas, 1993); "theory-of-mind" (Premack & Woodruff, 1978; Whiten, 1992; Cheney & Seyfarth, 1990); numerical competence (e.g., Boysen & Capaldi, 1993); and self-awareness (Lin, Bard, & Anderson, 1992; Povinelli, 1993; Parker, Mitchell, & Boccia, 1994), as well as play (e.g., Dolhinow & Bishop, 1970; Goodall, 1976a, 1976b, 1976c; Mitchell, 1990; Russon, 1990) and social-emotional development (e.g., Harlow, 1969; Russon, 1990; Johnson & Morton, 1991).

Comparative developmental and cognitive ethological approaches share four important perspectives which distinguish them from earlier learning theory approaches:

1. emphasis on species-typical abilities;
2. an interest in evolutionary reconstruction;
3. flexible methodologies combining results from the field, the colony, and the laboratory;
4. an interest in models from human psychology as well as evolutionary biology.

The two approaches differ in the following features:

1. the models they use (developmental vs. learning and cognitive theories);
2. the focal stages of life cycle (immature vs. mature); and
3. the range of taxa they embrace (primates vs. vertebrates as a whole).

As cognitive ethologists increasingly focus on development (e.g., Bateson, 1991) and especially on theory-of-mind research (e.g., Cheney & Seyfarth, 1990), however, the two approaches are coming closer together.

Although both approaches use models from human psychology, and hence could be viewed as anthropocentric, comparative developmental approaches seem to feel especially anthropomorphic. This feeling seems to arise from their use of stage models which raise the specter of recapitulation by implying that related species follow a common course of epigenesis. This specter has caused considerable resistance because much of modern evolutionary biology as well as psychology has devoted itself to expunging the nineteenth-century notion of recapitulation (Richards, 1992). Models that fly in the face of such sensibilities come under special scrutiny: How can such approaches be justified? Won't an epigenetic model developed on one species bias observations of other species? These are legitimate concerns, but concerns about biases cut across all approaches. Given that human instruments of perception and conceptualization are inherently biased both by nature and by culture (including various cultures of science) and given that all studies are theory-laden, the question might better be rephrased as follows: How can epigenetic models be tamed and used as tools? What are the costs versus the benefits of comparative developmental approaches (i.e., controlled epigenetic anthropomorphism) as compared to the alternative approaches of cognitive ethology?

THE NATURE OF DEVELOPMENTAL MODELS

Models of social and nonsocial cognition derived from developmental psychology offer a wide variety of frameworks for comparative studies of primate mentality. In addition to studies of physical and logical cognition (e.g., Piaget, 1952, 1954; Langer, 1980), and such related areas as comprehension of number (Gelman & Gallistel, 1978), these include frameworks in such domains as self-awareness (e.g., Lewis & Brooks-Gunn, 1979), perspective-taking, role-playing, and "theory-of-mind" (e.g., Flavell, Botkin, Fry, Wright, & Jarvis, 1968; Astington, Harris, &

Olsen, 1988), pretense (e.g., Leslie, 1988; Bretherton, 1984), empathy (e.g., Hoffman, 1983), and friendship and moral judgment (e.g., Likona, 1976; Turiel, 1983; Selman, 1980). Recent research on deception and "theory-of-mind" in nonhuman primates (Premack & Woodruff, 1978; Mitchell & Thompson, 1986b; Whiten & Byrne, 1988) is a good example of the comparative use of social models.

Their variety and comprehensiveness render models of human development attractive for comparative studies of primate mentality:

1. Because they follow development from birth, tracing the emergence of complex behaviors from their simple beginnings, they offer an appropriate beginning point for comparative studies of simpler behaviors in other primate species.
2. Because they address many areas of development, they offer models for development of a wide range of behavioral phenomena.
3. Because they focus on the development of the most complex behaviors from the simplest beginnings, they direct attention to nascent phenomena that might otherwise be missed in comparative studies.
4. Because they focus on the level of behavioral organization as opposed to the level of motor output, they focus attention on deeper structures.
5. Because they are epigenetic, they can be correlated with growth and development and with one another. In other words, in common with other good models, they allow investigators to discover relationships they would miss if they were using alternative models.

This comprehensiveness, however, repels many investigators precisely because it reflects the epigenetic nature of developmental models and hence raises the specter of recapitulation.

Most developmental theory today involves continuing testing and reformulation of the ideas of James Mark Baldwin (1894/1903), George Herbert Mead (1934/1974), Lev Vygotsky (e.g., 1930/1978), and Jean Piaget (1952, 1954, 1962), who at least implicitly relied upon recapitulation theory. This tradition has been enriched by theories of language acquisition pioneered by Roger Brown (1973), Jerome Bruner (1975), Patricia Greenfield (Greenfield & Smith, 1976), Elizabeth Bates (1976), Dan Slobin (1979) and others who have used Piagetian and Vygotskian concepts, and theories of socialization pioneered by Vygotsky and Mead. Even today these approaches are united by their epigenetic structures.

Although Piagetian theory has been widely criticized and revised, it provides the conceptual core of much of modern cognitive developmental psychology (e.g., Užgiris & Hunt, 1975; Selman, 1980; Case,

1985; Siegler, 1986; Langer, 1980), and even social psychology (e.g., Flavell et al., 1968; Astington et al., 1988). Many investigators have demonstrated that infants and children can succeed at simplified versions of Piagetian tasks at substantially earlier ages than Piaget described. This research has made important contributions in controlling variables that Piaget conflated and in parsing out elements that Piaget lumped together, but for this very reason it cannot "disprove" descriptions of stages based on more complex phenomena. Piagetian theory persists despite such "disconformations" because it has yet to be replaced by a more comprehensive and integrated model. Indeed it persists in the face of a more serious structural critique of Piaget's idea that representation arises at the end of the sensorimotor period (e.g., Mounoud & Vinter, 1981; Meltzoff & Moore, 1977; Mandler, 1988). Although representation is a broader phenomenon (both developmentally and taxonomically) than Piaget appreciated, he may still be correct about the emergence of the particular kind of representations he described: representation of novel trial-and-error actions and imitation as opposed to representation of simpler models (Parker, 1987).

As indicated above, Piagetian and neo-Piagetian stages and series are not simply descriptive, but rather constitute a model in the sense that they embody a set of theoretical postulates concerning the constructive nature of knowledge and the epigenetic nature of its development (what Piaget [1970] called "genetic epistemology"). The postulates can be summarized as follows:

1. Behavior is organized into mental schemes of action which are generalized instruments for assimilating stimuli and for accommodating to those stimuli as means for adapting to reality. The characteristics of the schemes therefore determine the potential nature of experiences. The grasping schema, for example, assimilates graspable objects within the child's reach and accommodates to the topological and gravitational and tactile/thermal properties of such objects.
2. The nature of these schemes changes systematically through the interplay of physical maturation and feedback generated through assimilation of objects and other stimuli and through accommodation to these objects and stimuli. During development sensorimotor schemes become more elaborate as they become internally differentiated and coordinated with other schemes. New more powerful schemes are progressively constructed through these processes, eventuating in new stages of development.
3. Piagetian sensorimotor and symbolic stages are epigenetic, schemes characteristic of each new stage within a series being constructed

from schemes in the preceding stage. These stages are universal among humans who have undergone normal development.

4. The epigenetic nature of stages implies a necessary evolutionary sequence.

5. The various developing schemes fall into the following Kantian categories of knowledge: objects, space, time, and causality, as well as such other categories as play and imitation.

6. The stages of development in these various aspects of knowledge tend to progress in parallel with one another.

7. Each new stage represents a new level of achievement of the child, but lower stages of adaptation also continue to manifest themselves situationally (e.g., Piaget, 1952, 1954, 1962).[1]

Piaget's (1962) imitation series provides a good example of these aspects of his developmental model. From birth through the second year of life infants traverse six stages of imitation, beginning with "contagious" crying in response to hearing other babies crying and ending with the capacity for deferred imitation of novel schemes. This sixth or final stage is necessarily preceded by the fifth stage, in which infants are first able to imitate novel actions through trial-and-error matching of their actions to those of the model. The ability to engage in trial-and-error matching emerges simultaneously with the fifth stage in the sensorimotor intelligence series and the so-called "tertiary circular reactions" of trial-and-error groping to discover new properties of objects. The sixth-stage achievement of mental representation of imitation leads to another related developmental series in pretend play during the preoperations period from two to five years of age. Although neo-Piagetians have challenged and revised aspects of the imitation series (e.g., Užgiris & Hunt, 1975; Bretherton, 1984; Meltzoff & Moore, 1977; Meltzoff, 1988), the basic framework survives.[2]

Indeed, application of Piaget's imitation series in studies of monkeys and apes has revealed certain distinctions of level, modality, and patterning which had been missed by investigators using other frameworks (e.g., Antinucci, Spinozzi, Visalberghi, & Volterra, 1982; Chevalier-Skolnikoff, 1976, 1977; Mathieu & Bergeron, 1983; Mignault, 1985; Parker, 1977, 1990, 1993). Its power to reveal greater detail arises from the fact that the stages distinguish multiple levels of imitation from infancy through childhood rather than simply distinguishing dichotomously between imitation and lack of imitation. It has the additional virtue of leading into pretend play and hence into theory-of-mind, providing developmental landmarks for these systems.

RESEARCH TACTICS IN
COMPARATIVE DEVELOPMENTAL STUDIES

Research involves at least two phases, the intuitive phase and the research design phase. Anthropomorphism plays a direct role in the intuitive phase of research into animal cognition. The identification of great apes with human infants in regard to capacities for gestural imitation, for example, stimulated the idea that these creatures could learn sign language (Gardner & Gardner, 1975). Likewise, anecdotal identification of the behaviors of dogs and primates with those of humans in regard to deception stimulated the idea that dogs, monkeys, and apes were capable of deception (de Waal, 1982; Mitchell & Thompson, 1986a; Byrne & Whiten, 1988b). Without the identifications, fruitful research programs would never have been initiated.

On the other hand, the design of comparative studies based on such intuitions goes beyond anecdotal anthropomorphic identification to a systematic process of hypothesis formulation and research design and testing. The critical point is that anthropomorphic intuitions stimulate hypotheses about the ways in which other species are similar to and yet different from human children. The null hypothesis in such studies is that the subject species display the same abilities, in all of the same series, developing in parallel, in the same developmental sequence, at the same time as they do in human children. At a deeper level the null hypothesis is that they are epigenetically constructed. The first task in research design comes in the selection of models appropriate to testing the null hypothesis.

Testing the null hypothesis requires systematic comparison of mental abilities and their development in many individuals of each closely related species of primates. In order to test the hypothesis, models used for this purpose must provide categories that encompass the appropriate levels of complexity of behavior to make them relevant to both human and nonhuman primate species. At the same time models must be sufficiently rich to capture the complexity of mental operations, sufficiently comprehensive to cover the full spectrum of mental abilities, sufficiently basic to cover the simplest behaviors, yet sufficiently discriminating to diagnose differences among species. Their frameworks must be sufficiently flexible to allow investigators to chart species-specific behavioral organization in various species. The greater the range of stage discriminations in a developmental model, the greater its potential value for disconfirming the null hypothesis.

Once specific developmental models have been selected, many tactical decisions regarding the design of the studies remain: These include the choice of species, the choice of settings, the choice of subjects, the length of the study, and the breadth of the study in terms of the number of cognitive domains included. Finally, after these tactical decisions have been made, other decisions about observational sampling techniques, translation of concepts into data categories and data analysis remain to be made.

Studies of animal mentality using human developmental frameworks can focus either on the terminal levels of achievement seen in adults of the subject species, or on the developmental sequences and their rates in the subject species (using longitudinal or cross-sectional methods or both). In Piagetian studies, for example, they can focus on the highest level of cognition achieved within each series, or on the rates and sequences of development within each series. While both kinds of studies offer important insights into primate mentality, studies of the developmental sequences and rates offer more information about species differences.

Likewise, studies of animal mentality can focus either on a single series or domain, or on many series and/or domains. In Piagetian studies, for example, they could focus solely on one series (e.g., object concept series), or on all the sensorimotor series (space, time, causality, sensorimotor intelligence, and imitation in addition to the object concept series). Because of the mosaic nature of evolution, that is, the fact that some abilities have evolved independently at different times and rates, and in different modalities, studying multiple series can give more information than studying a single series can. Likewise, distinguishing among various modalities of achievement within the same series increases the power of comparative developmental studies (e.g., Chevalier-Skolnikoff, 1977; Parker, 1977).

Finally, studies of animal mentality can use developmental concepts as a basis for identifying the achievement or lack of achievement of criterion abilities (category approach), or as a basis for inferring the organization of behavior in the subject species (pattern approach). While both categorical and pattern approaches provide important comparative data, the pattern approach provides a more comprehensive model for the mental organization of the subject species insofar as it tries to describe the mentality of subject species rather than sampling points which may or may not match those of human infants. In this sense data from a pattern approach offer a better test of the null hypothesis than data from a category approach.

The most powerful methodology for comparative studies would employ developmental studies of multiple behavioral series and domains combining categorical data with pattern data. Ideally, such studies should be done on many individuals in each of several of closely related species. It should be done in a variety of settings including laboratory studies of cross-fostered animals, colony studies of naturally fostered animals, field studies of wild animals, and cross-cultural studies of humans. Comparisons between humans and nonhumans should be based on the same methodologies. Finally, the most powerful methodology would place comparative developmental data on social and nonsocial cognition in a broader life-history perspective of stages of anatomical and physiological growth and life cycle stages.

Even the relatively simple approach of using categorical data on terminal levels of achievement in multiple series, however, has been useful for characterizing differences among great apes and monkeys that have been missed by more traditional learning studies (e.g., Rumbaugh & Pate, 1984). When Piagetian abilities of fifth- or sixth-stage imitation, tool use, and object permanence are mapped onto a primate phylogeny along with mirror self-recognition (MSR) data, for example, the resulting pattern reveals a perfect correlation between self-awareness and imitation but not between object permanence or tool use and imitation (and hence not between object permanence or tool use and self-awareness) (Parker, 1991). See Table 25-1. This analysis, along with comparative data on theory-of-mind, contributes to the emerging picture of cognitive correlates of MSR which were highly problematic before these emerging lines of research (e.g., Gallup, 1977b, vs. Gallup, 1982).[3]

Comparative studies of epigenetic levels of development of specific abilities across various domains and series provide a finer scale for discriminating species differences and similarities than ethology or traditional learning paradigms. Moreover, when such data are combined they reveal heretofore unrecognized distinctions among species which provide new data for reconstructing the evolutionary history of each particular ability.

RECONSTRUCTION OF THE EVOLUTION OF PRIMATE COGNITION

Comparative developmental studies become evolutionary studies when the data they generate are used for evolutionary reconstruction. By combining the methodologies of developmental psychology, ethology, and cladistics to compare the various forms of a behavior in closely related species, it is possible to reconstruct the probable behavior of an extinct common ancestor, and in some cases to infer the sequence of the evolution of the behaviors. The first step in such analysis is to map data on closely related species onto their family tree (see Figure 25-1). In

TABLE 25-1
Taxonomic distribution of mental abilities among primates based on a
terminal-level, multiple-series, categorical methodology

Fifth-stage Sensorimotor Mental Abilities:	*Humans*	*Great Apes*	*Old World Monkeys*	*Cebus*
Object permanence	+	+	+	+
Intelligent tool use	+	+	–	+
Imitation of novel behaviors	+	+	–	–
Mirror self-recognition	+	+	–	–

Note. In this analysis, humans and great apes are the ingroup, Old World monkeys are the outgroup, and cebus monkeys are a more distant outgroup with convergent characters. Object permanence is a shared character state, and imitation of novel behaviors and mirror self-recognition are shared-derived characters among the ingroup. This pattern indicates that the capacity for imitation of novel behaviors and the capacity for self-recognition must have evolved first in the common ancestor of the great apes and hominids. Note that, although initial expression of mirror self-recognition seems to occur during the fifth stage of the sensorimotor period, stable mirror self-recognition may depend upon sixth-stage sensorimotor abilities.

the case of comparative developmental evolutionary studies, these data are developmental.

Primate phylogeny, which has been constructed using genetic, molecular, and/or morphological data (e.g., Weiss, 1987), provides a framework for mapping each behavior or developmental series in the members of an ingroup (i.e., all the species in a genus or family that is the focus of study, such as humans and great apes) relative to behavior or development of members of the outgroup (i.e., the next most closely related group of species such as lesser apes and Old World monkeys). Specific characteristics must be treated independently because various characteristics of a species may evolve independently, each at its own rate (a phenomenon known as mosaic evolution). Thus, one characteristic of a species, for example, object permanence, may be ancient (shared with a large number of distantly related species) while another characteristic, for example, imitation, may be relatively recent (shared with only a few closely related species).

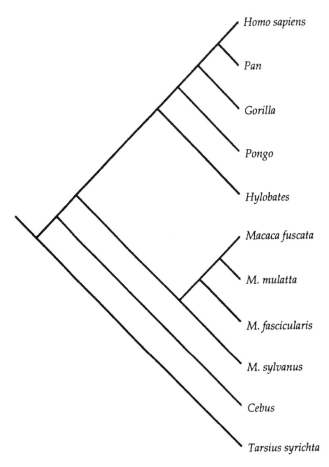

FIGURE 25-1
Family tree of primates based on molecular data
(created using "MacClade" by Maddison & Maddison, 1994)

The cladistic method of constructing phylogenies proceeds by mapping comparative data onto all logically possible trees to discover the most parsimonious alternative. The most parsimonious tree is the one that involves the fewest character-state transitions, that is, the one that shows the simplest pattern of inheritance through common ancestry. The method depends upon distinguishing among three classes of homologous characteristics (characters acquired through common ancestry):

1. shared characters which are present in all species in both the ingroup and the outgroup;
2. shared-derived characters which are present in most or all of the sister species of the ingroup;
3. derived characters which are present in only one of the sister species.

Only shared-derived characteristics denote common ancestry within the groups (see, e.g., Hennig, 1966; Ridley, 1984; Wiley, 1981).

By mapping species abilities onto existing phylogenies it is possible to directly reconstruct the evolutionary history of a specific ability without having to construct the family tree first: Through an analysis of the distribution of abilities it is possible to discover whether a given ability is a derived, shared-derived, or shared characteristics within the group in question, and it is possible to reconstruct its ancestry. Thus, for example, fifth-stage object permanence is probably a shared-derived characteristics among anthropoid primates, while fifth-stage imitation in the gestural modality is a shared-derived characteristic among the great apes, and elaborated pretend play is a derived characteristic in humans. This distribution implies that the capacity for fifth-stage object permanence arose in the common ancestor of monkeys and apes, while the capacity for sixth-stage imitation (in some modalities) and sixth-stage understanding of causality and space arose in the common ancestor of great apes and humans, and the capacity for pretend play arose in one of the ancestral species in the hominid lineage (see Figure 25-2).

Developmental sequences which, like Piaget's, are epigenetic in nature have special significance within a cladistic framework: They provide a basis for arranging abilities into an evolutionary series that implies a direction of evolutionary modifications with descent (i.e., a "polarized transformations series" in cladistic terminology) (Brooks & McLennan, 1990). Because of their unidirectionality, comparative data on epigenetic developmental sequences can be used to reconstruct the timing of the evolution of cognitive ontogeny (see Gould, 1977; e.g., McKinney & McNamara, 1991; Parker, 1996).[4]

CONCLUSIONS

Comparative developmental evolutionary studies of primate mentality may seem more anthropomorphic than cognitive ethological studies because their epigenetic models were stimulated by the recapitulationist theories of Haeckel (1874/1906) and Darwin (1871/1981). Although these associations place a greater burden of scrutiny on such studies, in

FIGURE 25-2

Mapping fifth and/or sixth stages of sensorimotor intelligence series
onto primate phylogeny to reconstruct common ancestry of particular abilities

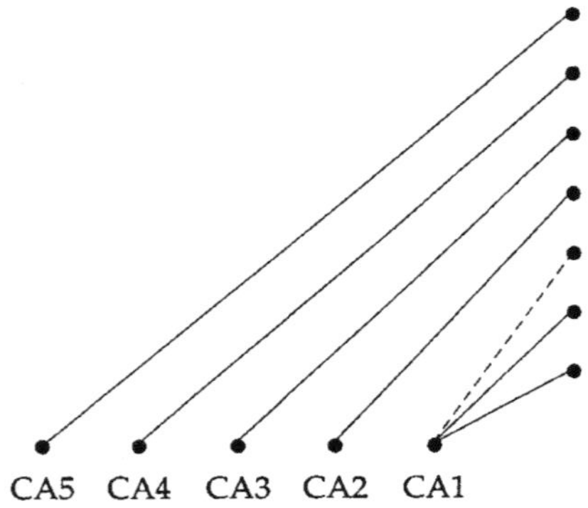

	Object Permanence	Tool Use	Novel Imitation	Self-Awareness
Cebus	+	+	–	–
Macaques	+	–	–	–
Gibbons	+	–	–	–
Orangs	+	+	+	+
Gorillas	+	+	+	+
Chimps	+	+	+	+
Humans	+	+	+	+

CA5   CA4   CA3   CA2   CA1

CA1 = common ancestor of humans and chimps and perhaps gorillas
CA2 = common ancestor of all great apes (Ingroup node)
CA3 = common ancestor of all apes (Outgroup node)
CA4 = common ancestor of Old World monkeys and apes (Outgroup node)
CA5 = common ancestor of New World monkeys and Old World monkeys (Outgroup node)

Note. According to this reconstruction, object permanence arose before the CA5 outgroup node, tool use arose twice (once immediately before the CA2 node, and again independently in cebus), and imitation of novel behaviors and mirror self-recognition arose immediately before the CA2 ingroup node.

fact, both approaches are anthropomorphic. In principle, moreover, anthropomorphism plays the same dual roles in all models drawn from human behavior, that is, as an intuitive generator of ideas about similarities between other species and humans, and as the testable null hypothesis that any given species is like the human species in any specific ability. The unresolved question is which approach provides the greatest heuristic power.

When anthropomorphism is the null hypothesis, developmental models provide a powerful set of heuristics for testing the nonanthropomorphic hypothesis. Specifically, this null hypothesis predicts that animals will traverse the same stages, in the same order, simultaneously in all the parallel series, at the same rate. In contrast to cognitive theory, it thereby provides not only a richer array of models for comparative studies of primate cognition, but a broader array of predictions. These models are useful because the epigenetic model allows investigators to notice phenomena they would otherwise miss and to discover relationships among these phenomena that they would otherwise overlook. The foregoing analysis suggests that the heuristic strength of comparative research is maximized by focusing on developmental studies (as opposed to terminal level studies) applied across multiple developmental series or domains (as opposed to single domain studies) using pattern analysis as well as categorical analysis.

Comparative developmental studies of mental abilities become evolutionary studies when data on closely related species of Old World monkeys, great apes, and humans are mapped onto existing phylogenies. Cladistic methodologies provide a means for reconstructing the patterns of ancestral species directly from comparative developmental data.

Ironically, in terms of its recapitulationist legacy, when anthropomorphism is the null hypothesis, comparative developmental approaches seem to reveal more that is unique to various primate species than the alternative approach has been able to. Of course, comparative studies are more revealing about our species than studies that have focused exclusively on humans: Both comparative developmental and cognitive ethological studies have begun to map out exactly what is unique to humans, what is shared among humans and other species, and when in phylogenetic history various shared and unique abilities arose. Perhaps humankind is not the only proper study of humankind after all.

NOTES

1. The epigenetic nature of stages of sensorimotor intelligence and stages of language development strongly suggest innate species-specific organiza-

tions. In the case of sensorimotor intelligence, another line of evidence for innateness is parallel occurrence in hunter-gatherer children (e.g., Konner, 1976). In the case of early stages of language development, multiple lines of evidence converge on this conclusion: familial language disorders, motor milestones, cognitive correlates, cross-cultural universals, and parallel courses of development in hearing and deaf children. Epigenesis, literally "upon the genes," is a phenomenon first identified by embryologists. It contrasts with other kinds of sequential acquisitions, such as calculus, based on previously acquired knowledge in that it occurs in all normal species members who develop in a normal environment.

2. Various investigators view Meltzoff's data on imitation from different perspectives: from Meltzoff's perspective it "disproves" Piaget's model, from other perspectives it elaborates the contents of his stages (Parker, 1993). Like most of the neo-Piagetian experiments published over the years, Meltzoff's studies describe much simpler schemes than those Piaget described in his stages, for example, the imitation of simple novel schemes Meltzoff elicited in 9- to 14-month-old infants did not involve the trial-and-error matching with the model's movements that Piaget described in 12- to 18-month old infants in the fifth stage.

3. Whether MSR depends upon fifth- or sixth-stage sensorimotor intelligence in one or more series is currently unresolved.

4. McKinney and McNamara define heterochrony as follows: "Literally, 'different time', more precisely, change in timing or rate of developmental events, relative to the same events in the ancestor. . . . Different kinds of heterochrony are recognized: dissociated, global, growth, differentiative, sequential, organizational" (McKinney & McNamara, 1991, p. 387). Neoteny and recapitulation are two widely known outcomes of heterochrony.

PART VIII

Language

26

Anthropocentrism and the Study of Animal Language

Judith Kiriazis and Con N. Slobodchikoff

In the 17th century, Descartes asserted that language was proof that humans alone have souls (Descartes, 1649/1927; Radner & Radner, 1989). This anthropocentric view of language has its philosophical roots in Descartes's notion of the relationship between the human body and the human mind. Descartes thought that the human body was analogous to an automaton, a machine that mechanically performed particular tasks. The human mind was separate from the body, the "Ghost in the Machine" (Ryle, 1949, pp. 15–16), and provided the body with a spark of consciousness. Human thought came from the mind, and was translated by the body machine into action. Nonhuman animals did not have a mind, but were instead merely unthinking automatons. What clinched this difference between human and animal, according to Descartes, was language: animals could never use words or other signs to convey any thoughts to others (Radner & Radner, 1989).

The assumption that language is a necessary condition of thought, and that both are exclusive properties of the human species, became a prevalent view among philosophers such as Hobbes, Herder, Wilhelm von Humboldt, and Hegel (Waldron, 1985). Linguists, biologists, and philosophers assume that language is what distinguishes the human species from other animals (Keyan, 1978; Crook, 1980; Chomsky, 1982; Waldron, 1985), and linguistics texts warn against assuming that animal communication and human language are even remotely analogous systems (e.g., Akmajian, Demers, & Harnish, 1985). Of all human capabilities, language remains *the* difference separating human beings from other creatures.

In an attempt to make clear what is specific to human language and what is present across species, Hockett (1960) developed a list of the design elements of human language that could be used to compare species' abilities (Thorpe, 1972, 1974a). Some of these elements are purely physical, dealing with the auditory channel and the mechanism of the transmission itself (Hockett & Altmann, 1968), such that they can be applied to a wide variety of vocal communication sources. Other elements relate specifically to aspects of human language: arbitrariness, semanticity, displacement of the message from an immediate place or time, syntax, and openness (where new messages are easily coined). Still other elements involve the ability to reflect upon, teach, and learn the communication system, and to use it to deliberately deceive the recipients of the signals.

Although implicit in Hockett's analysis is the expectation that particular attributes are present solely in the human domain, it is inappropriate to assume without evidence that the communication systems of other species cannot share elements of human language (Lieberman, 1975). According to Lieberman, failure to find syntax and other seemingly uniquely human elements in animal communication occurs because we do not ask the right questions. For example, many researchers attempt to define one single meaning of a particular animal signal by studying all the behavioral situations in which it occurs, and assigning a meaning based on something common to all these situations. The same method applied to human communication would lead to wholly inadequate results:

> If we looked for the common "meaning" that could be associated with most human vocal communications we would have to conclude that speech perhaps was a means whereby humans located each other—in other words, that one human made noise and the other human made a return noise. (Lieberman, 1975, p. 25)

Thus, it may be the methods of studying animal communication systems which are limited, rather than the communication systems themselves.

Recent findings have chipped away at the view that only humans have the cognitive capabilities for language (Parker & Gibson, 1990; Griffin, 1992). Other species have shown capacities for communicating using methods that contain some of the elements of human language. For example, some chimpanzees taught gestures of American Sign Language (ASL) communicate with these signs, as well as teach them to young (Fouts, Fouts, & Van Cantfort, 1989). In addition, chimpanzees and other primates taught ASL can assign at least two different semantic referents to a particular word, and differentiate those meanings through an application of syntax (Fouts, 1973; Gardner, Gardner, & Van Cantfort, 1989).

Even with the evidence of gestural abilities in apes, human uniqueness theorists such as Bickerton (1990) continue to argue that human language cannot have evolved directly from animal communication because the two systems are too different. Although animals as diverse as vervet monkeys (Cheney & Seyfarth, 1990), chickens (Gyger, Marler, & Pickert, 1987), and prairie dogs (Slobodchikoff, Kiriazis, Fisher, & Creef, 1991) can distinguish different types of predators and encode such information in their alarm calls, according to Bickerton these calls are limited to items that are directly relevant to animals' everyday experience, whereas human language can express ideas that are far beyond everyday experience. Even though alarm signals such as those produced by vervet monkeys can be used to identify the presence of an eagle or python (Cheney & Seyfarth, 1990), it is argued that these alarm calls cannot be broken down into parts the way that human sentences can (Bickerton, 1990).

Yet the information content of naturally occurring animal signals *can* be elaborately coded and structured in ways comparable to human sentences, but these comparisons need to be hypothesized explicitly rather than denied if they are to be discovered. For example, Gunnison's prairie dogs have demonstrated the ability to distinguish between individuals within a predator category, as evidenced by differences in their alarm calls (Slobodchikoff et al., 1991). These differences within a predator category appear to represent descriptors of the general size and shape of a predator, and with human predators, the color of clothes that each individual human is wearing. Prairie dogs seem to make cognitive assessments of the imminence of danger, and their alarms seem far more than just unintentional expressions of fear (cp. August & Anderson, 1987; cf. Morton, 1977).

While a calling animal might indeed produce something comparable to a sentence, its syntax and grammar might be unrecognizable to us because it evolved along a totally different pathway. If, as studies of alarm calls in prairie dogs have demonstrated, the semantic content of the call is generally descriptive and is given with the intent to communicate information about a predator, the possibility exists that such calls could include the syntactical construct of an actor or agent (perhaps a particular predator), a recipient (either the group or a specific animal that has been targeted by the predator), and an action indicating the type of attack (Slobodchikoff et al., 1991). If the animals can also encode information about the imminence of danger ("NOW!" or "Pretty soon"), perhaps this can be assigned, within the context of that species, a tense, according to some rule analogous to our rules of grammar. The precise forms of semantics, syntax, and grammar may be constrained by the ecological and evolutionary limitations of a particular species.

Testing for these ideas can be very difficult because, once again, we work from our anthropocentric need for a lexicon, or meanings encoded into discrete entities called words. Isolating such a lexicon in other animals might be impossible if we are not aware that their perceptions of their world, how they receive sensory information and interpret it, may be beyond the grasp of our senses. Animal language may elude our understanding because it encodes information about things and relationships beyond the realm of our immediate perceptual abilities.

Indeed, the evolutionary origins of human language itself are obscure. Some reconstructions of the vocal tracts of early hominids suggest that only some species had the bent supralaryngeal vocal tract that would have allowed them to physically produce the majority of sounds used to encode human speech (Lieberman, 1975); others suggest that there are no fundamental differences in the vocal tracts of mammals in general, and that the capacity to produce spoken language is due to differences in cerebral development (Wind, 1989). The evolution of tool use, the use of fire, bartering, social and kinship structures, and life in riparian habitats have all been suggested as possible factors leading to the evolution of language (Parker & Gibson, 1979; Grolier, 1989; Morgan, 1989).

Such explanations make a key assumption that human language adapted through ecological constraints. For language to evolve, it must have facilitated an understanding or manipulation of the world in ways that humans could exploit to increase their fitness (Dawkins & Krebs, 1978). Socioculturally, language might have empowered members of groups to deal more efficiently with changes in their environment, and to share this information with their kin. It also might have allowed some individuals to have power over others. Additionally, the complex patterns of behavior associated with early human culture, such as tool-making, hunting, and gathering of plant products, might have placed a selective advantage on linguistic systems that enhanced transmission of knowledge about these skills (Parker & Gibson, 1979; Crook, 1980). Without the needs of adjusting to changing ecological conditions within the context of a social system, there might have been little use for an elaborate vocalized communication system.

These ideas about the evolution of human language have implications about animal languages. If we try to keep an open mind about differing design features of animals that live in different ecological circumstances, we may find that many more species than we currently recognize have a language that has evolved to meet their specific needs, just as our language has evolved to meet our own ecological needs.

27

Pinnipeds, Porpoises, and Parsimony: Animal Language Research Viewed from a Bottom-up Perspective

Ronald J. Schusterman and Robert C. Gisiner

Animal Language Research (ALR) includes a variety of experimental studies of complex learning and cognition by nonhumans in which human language serves as a model for experimental design and data interpretation. For example, dolphins, who learned to react either to human sign language or to computer-generated sounds, were said to "give evidence of both semantic and syntactic processing. . . . The responses of the dolphins were consistent with the meaning of the words in a sequence and with the constraints imposed by word-order, and also appeared to take account of the state of the world" (Herman & Morrel-Samuels, 1990, p. 297). Without further analysis the above account may be more complicated and anthropomorphic than is necessary. The principle of parsimony states that simple explanations are usually better than complicated ones. Therefore some aspects of language comprehension by dolphins (and humans as well) might be

This research was supported by Contract N00014-85-K-0244 from the Office of Naval Research and a grant from the Center for Field Research to Ronald J. Schusterman. We thank Murray Sidman for giving us a framework for the study of marine mammal cognition and we thank Emil Menzel for reinforcing the idea that many aspects of "language learning" in animals depend on principles of discrimination learning.

described and explained more effectively by straightforward variants of discrimination learning principles (Catania, 1992).

The human linguistic model in ALR has indeed generated new ideas about nonhuman cognitive abilities as well as produced experimental demonstrations which were quite unexpected. However, a strong use of the human linguistic model in ALR has sometimes led to a reification of constructs, concepts, and terminology without further questioning and analysis of such linguistic concepts as "word," "meaning," "syntax," and "sentence." Thus, the top-down approach of adapting terminology from the study of a very complex set of learned skills (linguistics) to considerably less complex performances (ALR) has not been successful in (a) defining the learning abilities required for normal human language performance and (b) defining the quantitative or qualitative differences (if any) between learning abilities of different animals, including humans.

We have instead adopted a bottom-up approach: starting with operantly conditioned motor acts and moving up through an increasingly complex network of conditional relations emerging from and consistent with earlier discrimination learning. This gives us an operational model, or "how-to" guide, for investigating the role of reinforcement-based learning skills in complex cognitive performances, including human language. We describe the learning abilities required for appropriate responses to a simple artificial language taught to bottle-nosed dolphins (Herman, Richards, & Wolz, 1984) and a pinniped species—California sea lions (Schusterman & Krieger, 1984). These include the ability to learn (frequently by "exclusion") one-way sign-referent (conditional) relations, the ability to classify or categorize signs into functional categories, and the ability to learn generalized "rules" for the integration of multiple signs into an appropriate response (Gisiner & Schusterman, 1992; Schusterman & Gisiner, 1988, 1989; Schusterman & Krieger, 1984, 1986; Schusterman, Gisiner, Grimm, & Hanggi, 1993). To the degree that these processes and equivalence class formation are related to nonverbal behavior, they may give us a clue to the origins and evolution of human language (Catania, 1992; Sidman, 1990)

TRAINING STAGES AND LEARNING SKILLS IN SEA LIONS' ALR

Training Stages

Since 1981 three California sea lions (*Zalophus californianus*)—Rocky, Bucky, and Gertie—have been trained in two artificial language comprehension formats (Schusterman & Krieger, 1984; Schusterman &

Gisiner, 1988). The ALR with sea lions was designed to parallel a similar program on dolphins (Herman et al., 1984). This chapter is mostly
concerned with Rocky, since this female sea lion has undergone the
longest and most intensive analysis of all our sea lions in the ALR program.

The basic signaling repertoire and initial training procedures for
nonrelational (single object) instructions are described in Schusterman
and Krieger (1984). Training procedures for the relational (two object)
instructional form are described in Schusterman and Gisiner (1988).
The training stages between 1982 and 1988 are listed in Table 27-1. In
Stages I and II paired relations between gestural cues and actions (stimulus-response relation) and gestural cues and objects (stimulus-stimulus
relation) were established using standard operant conditioning techniques with food reinforcement. In Stage III a cue designating an object
and a cue designating an action were combined, the required response
being the performance of the specified action on the specified object
only (out of two or more available objects). In Stage IV signs designating
object brightness (white, black) and size (large, small) were added to the
object-action combinations, requiring the subject to integrate the information from multiple cues when selecting a response object. Finally,
in Stage V the sea lion was trained to conditionally modify its "fetch"
response (bringing the designated object back to the signaler) to a "rela-

TABLE 27-1
Training stages in the artificial language taught to Rocky

Stage	Training procedure
I. Actions (A)	Shape response and place under control of a gestural sign.
II. Objects (O)	Pointing orientation to an object shape under control of a gestural sign.
III. Integration (O+A)	Combine separate behavioral repertoires from I and II.
IV. Modifiers (M)	A. Add conditional cues for brightness. B. Add conditional cues for relative size. C. Combine brightness and size modifiers in either order.
V. Relational (O-p)	A. Shape response (take Object A to Object B). B. Put response under control of gestural signs by adding an object sign designating Object B to the combination Object A + FETCH.

tional fetch" response (bringing the object to another object) by placing a sign designating a destination object before the OBJECT + FETCH combination. The instruction was called a relational instruction because the sequential relationship between the two object signs determined which was fetched (the second) and which was the destination (the first).

These training procedures resulted in the formation of over 7,000 different instructions composed of 2 to 7 members out of a repertoire of 23 members (13 object shape signs, 4 modifier signs, and 6 action signs). The types of standard combinations and an example of each are shown in Figure 27-1. The standard sign combinations given to Rocky all followed conventions for the sequential arrangement of signs as shown in the figure. Table 27-2 expresses these sequential ordering conventions as pairwise sequential relations between two events. These conditional rules for sequential pairing of signs produced all standard combinatorial forms taught to Rocky.

Learning Skills: Sign-Referent Relations

In a procedure similar to arbitrary match-to-sample, conditional discrimination training was first used to direct a pointing response to one of two or more choice stimuli in the presence of an arbitrary gestural sign (stimulus-stimulus pairings): This procedure generated the "objects" and "modifiers" of the subsequent sign combinations (see Table 27-1). As training proceeded, sea lions would respond to an undefined or "unnamed" object immediately and persistently if it had been paired with a novel sign and if the other available comparison objects each had already been established or experimentally defined by a previous relation (Schusterman, Gisiner, Grimm, & Hanggi, 1993). Our training of the sea lions in this errorless manner was quite similar to the way dolphins had been trained to relate novel signs to unnamed objects

TABLE 27-2
Sequential pairings of signs in Rocky's standard combinations
(refer to Figure 27-1)

	Sign A	→	Sign B	(Sequence segment)
I.	Start	→	MODIFIER, OBJECT	(M-, O-)
II.	MODIFIER	→	MODIFIER, OBJECT	(M-M, M-O)
III.	OBJECT	→	Pause, ACTION	(O-p, O-A)
IV.	Pause	→	MODIFIER, OBJECT	(p-M, p-O)
V.	ACTION	→	Release	(-A)

FIGURE 27-1

Path diagram for development of repertoire of standard combinatorial forms

Training Stage (refer to Table 27-1)				Complete Combination	Number	Example
IV	V	IV	III			
			O-A	O-A	76	PIPE OVER
		M →	O-A	M-O-A	528	SMALL BALL FLIPPER
		M-M →	O-A	M-M-O-A	240	BLACK LARGE BALL TAIL-TOUCH
	O-p →		O-A	O-p-O-A	130	BAT, DISC FETCH
M →	O-p →		O-A	M-O-p-O-A	313	WHITE RING, CUBE FETCH
M-M →	O-p →		O-A	M-M-O-p-O-A	384	WHITE SMALL CUBE, CAR FETCH
	O-p →	M →	O-A	O-p-M-O-A	376	PERSON, BLACK RING FETCH
M →	O-p →	M →	O-A	M-O-p-M-O-A	956	LARGE CONE, BLACK CLOROX FETCH
M-M →	O-p →	M →	O-A	M-M-O-p-M-O-A	1,192	BLACK SMALL CONE, SMALL CLOROX FETCH
	O-p →	M-M →	O-A	O-p-M-M-O-A	464	WATER, SMALL WHITE CONE FETCH
M →	O-p →	M-M →	O-A	M-O-p-M-M-O-A	1,192	WHITE FOOTBALL, LARGE WHITE CUBE FETCH
M-M →	O-p →	M-M →	O-A	M-M-O-p-M-M-O-A	1,216	WHITE LARGE CUBE, SMALL BLACK BALL FETCH
					7,067	

Note. The path diagram on the left shows how the complete repertoire of standard combinatorial forms was produced by the training stages listed in Table 27.1. M = modifier signs, O = object (shape) signs, A = action signs, p = pause. The total number of instructions of each form and an example are listed to the right of each completed combination.

(Herman et al., 1984). However, without further analysis, Herman et al. took this errorless training technique to be the process by which the signs acquired meaning for these marine mammals and concluded that "It was sometimes sufficient to pair a new signal with an unnamed object for the dolphin to associate the two immediately. Successful association was indicated by the dolphin continuing to respond appropriately to the previously unnamed object in the presence of the new signal and to other objects . . ." (p. 157). Herman et al. go on to summarize their results on reference as follows: "The concept that signs stand for referents seems to come easily to the dolphins" (p. 207). Indeed, without questioning, analyzing, or experimenting further we would have arrived at the same conclusion about our sea lions. For example, following 716 trials, Rocky had learned to relate the signs A_1 and A_2 to the objects "pipe" (B_1) and "ball" (B_2) respectively. Afterward, Rocky, like the dolphins, would select a new object like a "ring" (B_3) in the presence of a novel signal A_3 rather than choosing one of the already defined comparison objects—pipe or ball. Had Rocky, like the dolphins, also come easily to the concept that signs stand for referents, or is there a simpler explanation for Rocky's behavior? Could the sea lion Rocky or, as a matter of fact, could the dolphins, prior to learning the relation "if sign A_3 then object B_3," simply have selected the novel object (B_3) by using a "rule of thumb"? That is, could they have excluded the defined or so-called named comparisons (B_1 and B_2) in the presence of the new sign (A_3)? The answer appears to be yes.

From the viewpoint of experimental psychology, signs and their referents can be seen as stimulus-stimulus relations that are established by conditional discrimination training or "if . . . then" rules, particularly in match-to-sample formats. In order to illustrate how a subject might *seemingly* associate a new signal with an unnamed object immediately, consider the following experiment—the subject in this case is taught to match geometric shapes to the number of fingers being held up by the trainer, and, after a period of arduous training, has indeed learned the following conditional discriminations: if the sample is one finger, then the correct comparison shape is square; if the sample is two fingers, then triangle is correct. Note that there are two relations being taught simultaneously. In contrast to two relations being taught simultaneously, a transfer test might be set up in the following manner: the novel sample is three fingers and the novel shape is circle, and the incorrect comparison is either square or triangle—the already defined comparisons. The subject is likely to respond correctly the first time to the circle-shaped comparison; however, it is not clear that the sample three fingers is the controlling factor in this instance. The subject is probably responding to "not square" or "not triangle" when presented with a sample (three fingers) which is likewise unrelated to a triangle or a square. The only way a cogent transfer test can be arranged is to train two novel relations simultaneously. Under these conditions, only if a subject learns the two sample/comparison relations immediately—or following a single information trial—can the investigator validly conclude that the concept that signs stand for referents come easily to the subject.

Laboratory demonstrations of exclusion are quite common with human subjects, some with severe and prolonged mental retardation, who have not as yet, for example, learned to relate Greek symbols with their printed names or numerals with their printed names (see e.g., McIlvane & Stoddard, 1985). Recently we have conducted experiments with sea lions showing that they also use exclusion during the initial phase of learning to relate a signal to an object or in arbitrary match-to-sample procedures, a sample to a comparison stimulus (Schusterman et al., 1993). Thus, selection of a particular comparison in the presence of a particular sample or signal does not necessarily indicate a controlling relation between the two (Dube, McIlvane, & Green, 1992; Kastak & Schusterman, 1992). In the experiments on sea lions, test trials were used in which novel signals like A_3 or A_4 were presented along with novel alternatives B_3 and B_4. Such test trials afford no basis for exclusion, and learning of the $A_3 \rightarrow B_3$ and $A_4 \rightarrow B_4$ relations by California sea lions took about 100 to 200 reinforced pairings using an errorless "exclusion" training technique (see Schusterman et al., 1993). These results cast considerable doubt on the ability of dolphins to readily acquire a concept that signs stand for referents (Herman et al., 1984). More likely dolphins do what some humans and sea lions do in similar situations— they respond "not B_1" or "not B_2" in the presence of the new A_3 signal and thus select object B_3. It is only later in the learning process, in contexts where the basis for exclusion is no longer available, that a novel conditional performance occurs. This is indeed the most parsimonious explanation of how dolphins learned sign-referent relations.

Functional Classes and Ordinal Relations

Novel sign sequences were introduced in a variety of ways and included new combinatorial structures (see the path diagram on the left of Figure 27-1). After a limited number of available signs were used in training and a sea lion achieved a high percentage of correct responses to the training set, the signs that had been held out of training were introduced to produce novel instructions. All three sea lions, including Rocky, were able to respond correctly to these new instructions on the very first exposure to them. This demonstrates that a sea lion is capable of learning something about signal class membership, allowing the animal to extend what it has learned with the training set to all other signs within the appropriate class. Three examples listed below show how Rocky was able to respond perfectly to novel sign sequences based on her previous reinforcement history.

1. After Rocky was trained by shaping procedures to jump over any object floating in a pool and the behavior was placed under the control of a gestural cue, she was able to perform perfectly the first time she was given the sign sequences BALL OVER and CUBE OVER.
2. Prior to training Rocky to relate a gestural sign to a cone-shaped object, she had been trained by our manipulation of reinforcement contingencies to respond appropriately to brightness and size cues as they were added to or conjoined with cues she had already learned to relate to different shaped objects (see Table 27-2). After Rocky learned to relate a gestural cue (CONE) to a floating cone-shaped object, she performed perfectly the first time she was given the sign sequences WHITE CONE MOUTH and BLACK CONE MOUTH.
3. Relational instructions ("take Object A to Object B") were formed by the addition of a second object sign (with optional modifiers) to a FETCH object-action pairing sequence. Again, this was done by manipulating reinforcement contingencies. Thus, we conditionally changed a single-object (O-A), nonrelational instruction like RING FETCH ("fetch the ring" to the signaler) to a relational instruction (O-O-A) like PIPE, RING FETCH ("take the ring to the pipe") or RING, PIPE, FETCH ("take the pipe to the ring"). The instruction was identified as a relational one because the ordinal relations between the two object signs determined which was the transported item (the second) and which was the goal item (the first). Another way of stating this is that in a relational instruction, depending on their ordinal positions, first or second, the same object signs (with optional modifiers) may serve as conditional cues for entirely different sea lion or dolphin performances. Thus, even such novel reversed, five-sign instructions like WHITE WATER-WING (a float), SMALL WHITE BALL FETCH and SMALL WHITE BALL, WHITE WATER-WING FETCH were performed perfectly by Rocky the very first time she was exposed to them.

In the artificial language taught to sea lions and dolphins, the fact that all signs in a given class (e.g., object signs) could be freely substituted in the appropriate place constitutes evidence of the formation of *functional* classes. Goldiamond (1966) has defined a functional stimulus class as a set of discriminative stimuli controlling the same behavior. Vaughn (1988) showed that pigeons were capable of forming what he called *functional equivalence sets* to photographic slides of trees divided into arbitrary sets of 20 slides each. After several reversals in the reinforcement contingencies for responses to stimuli in one or the other of the two sets, the pigeons began shifting their pecks from one set to the other after being exposed to only a few stimuli of a given set. Basically, the ability to extend relationships learned from one member of the set to other members of the set permits the subject to extend relationships learned from one member of the set to all other members of the set without additional differentially reinforced experience. The principle of functional equivalence may be the most parsimonious explanation of immediately correct responses by pinnipeds and porpoises to novel sign combinations.

Equivalence Relations

The concept of *equivalence relations* was derived from a study on reading and auditory-visual equivalences (Sidman, 1971). In this research, Murray Sidman applied a match-to-sample format to train conditional discriminations in a mentally retarded boy. The subject matched pictures of objects, like a cat, to spoken words as well as matching printed words to the corresponding spoken words. Subsequently, the subject showed that he could spontaneously relate the printed word cat to the picture of a cat and vice versa, even though the printed word and picture had not previously been explicitly paired but had only been related to the spoken word.

Experimental demonstration of equivalence involves testing for the properties of reflexivity, symmetry, and transitivity in the associations formed between three or more stimuli, in the context of a match-to-sample procedure. If these three properties can be demonstrated, the stimuli comprising the conditional relations are referred to as equivalent members or elements of a class.

Reflexivity, or identity matching, is demonstrated when an animal that has been trained to relate various identical stimuli can do so immediately and accurately when presented with completely novel stimuli (Kastak & Schusterman, 1992). Symmetry is demonstrated when an association between nonidentical stimuli is shown to be reversible (i.e., when trained to match A_1 [sample] with B_1 [comparison], the subject is able to match B_1 [sample] with A_1 [comparison]). Finally, transitivity can be exhibited by the ability of the subject to relate stimuli which share an intermediate stimulus, yet have never been presented together in the match-to-sample context (i.e., when trained to match A_1 to B_1 and B_1 to C_1, the subject immediately perceives a relationship between A_1 and C_1). If the subject of this experiment can immediately match C stimuli with the appropriate A stimuli (a combination of symmetry and transitivity) then a combined test for equivalence has been passed.

Recently, Sidman (1990) has distinguished between functional classes or sets and equivalence relations, stating that they are "closely related" but not the same. Sidman and his colleagues (Sidman & Tailby, 1982; Sidman, Wynne, Maguire, & Barnes, 1989) have shown that it is possible for differentially reinforced training of one-way sign-referent relationships ($A \rightarrow B$, $B \rightarrow C$) to produce in human subjects additional reflexive ($A \rightarrow A$, $B \rightarrow B$, $C \rightarrow C$), symmetric ($B \rightarrow A$, $C \rightarrow B$), and transitive ($A \rightarrow C$) relationships which together form a set of stimulus equivalence relations ($C \rightarrow A$) between the various signs and referents. They have pointed out the obvious significance of these learning abilities for language: "Stimulus classes formed by a network of equivalence relations establish a basis for referential meaning" (Sidman & Tailby, 1982, p. 20), and note for the equivalence relations (as we do for the conditional relations between functional classes—see below) that "It is not correct to assume that the new . . . performances emerged without a reinforcement history" (p. 20).

Stimulus equivalence learning abilities have recently been demonstrated with a single California sea lion (Schusterman & Kastak, 1993). Such a demonstration has considerable importance for the topics of interest to ALR: (a) defining the learning abilities required for language and (b) defining the differences in learning abilities of different animal taxa, including humans. Sidman et al. (1989) have made a similar point: "A continued search with nonhuman subjects may yet provide the key to the problem of what is primary, equivalence or language" (p. 273). Our recent finding suggests that equivalence relations are not mediated by language but may be a prerequisite for linguistic competence.

Sequential Conditional Relations

Gisiner and Schusterman (1992) presented Rocky with *anomalous* (unfamiliar combinations of) signs created by reordering (e.g., A-O instead of O-A), deleting (e.g., M-A instead of M-O-A) or adding signs (e.g., O-A-A instead of O-A). The results of this study provide perhaps the most compelling data in a nonhuman demonstrating the formation of functional equivalence relations within a variety of complex contexts. The sea lion Rocky had learned relations which were consistent with reinforcement contingencies established during her experience with the arbitrarily structured artificial language shown in Figure 27-1. For example, of 70 anomalous sign sequence probes embedded into a standard balanced series given over a period of 18 months, Rocky gave 62 orienting responses (prior to her release from station) which did *not* correspond to the class of sign actually given, but instead corresponded to a class of sign that would normally appear in a familiar, standard combination (see Table 27-2). For example, the anomalous combination of an action sign followed by an object sign (A-O) contains two sequence differences from the standard object-action (O-A) combination: first, the sequence begins with an action sign, whereas standard, familiar combinations always began with either a modifier or object sign; second, the action sign is followed by an object sign when normally it would be followed by a release from station (refer to Table 27-2). Rocky was given 9 different A-O anomalous combinations and either made an object orientation to the action sign (three times) or made an action orientation when given the object sign (six times). Such results show that Rocky had learned to use one class of signs to predict the class or classes of signs that would appear next in sequence. Another way of stating this is that each class of signs acted as a conditional cue for a subsequent class. We called this set of conditional relations between classes *sequential conditional relations* (Schusterman & Gisiner, 1988, 1989; Gisiner & Schusterman, 1992).

Although Rocky demonstrated sensitivity to fixed sequential relationships and learned to rely on them strongly during her processing of sign combinations, she was also able to learn to integrate elements that did not stand in fixed sequential relationship to each other. Double modifiers were trained and maintained in free sequential order; both brightness-size and size-brightness modifier pairs were used with equal frequency (e.g., both LARGE BLACK and BLACK LARGE were used equally often to indicate the larger and darker of four objects of the same shape). As a result modifier sign order did not affect Rocky's ability to select the appropriate response object (see Gisiner & Schusterman, 1992).

Similarly, another sea lion (Gertie) trained in a different artificial language format was exposed to a relational instructional form ("take Object A to Object B") that provided both sequence cues (first object sign, second object sign) and position cues (left object sign, right object sign) to indicate the destination (B) and transported (A) objects, respectively. Tests with anomalous trials showed that she had learned to use the position relationships rather than the sequential relationships to indicate the relative roles of the two object signs (Gisiner & Schusterman, unpublished data). Thus it would appear that the sequential relationships which Rocky learned were not absolutely necessary for processing multiple signs into a response, nor were sequential relationships necessarily the only type of relationship between combined elements that a sea lion might learn.

Hierarchical Conditional Relations

Gisiner and Schusterman (1992) also showed that Rocky had learned another set of relations for integrating information from individual signs into a final response, which they termed *hierarchical conditional relations*.

The relationships were hierarchical in the sense that a previously trained part of the combination had to be present before subsequently trained additional signs were secondarily incorporated into the response (see Table 27-1 and Figure 27-1 for the sequence of training stages). The first trained and hierarchically primary combination was the object-action (O-A) sign pair: Rocky's responses showed that she would not form a response unless an object sign and action sign were paired in that order (O-A) somewhere in the combination. Next trained, modifier signs secondarily modified selection of an object, but only if followed by an object sign. Modifier information was not incorporated into an object search until an object sign was given, and if no object sign was given Rocky did not use the modifier information alone to select a response object. By comparison, she *did* respond to O-A combinations in contexts that normally required modifier signs (more than one object of the same shape), thus illustrating the secondary and optional role of modifier in response formation. The last trained step in response formation was the addition of a second object sign to the beginning of an O-A sequence containing a FETCH action sign (Training Stage V, Table 27-1). The presence of the second object sign conditionally altered a fetch response into a relational response (fetching one object to another). When the action sign was not FETCH then the hierarchically secondary object sign was not integrated into the response and only the hierarchically primary step (O-A) was carried out (Schusterman & Gisiner, 1988; Gisiner & Schusterman, 1992).

CONCLUSION

Some investigators started into ALR in a relatively cautious manner and talked, for example, about a gesture *as if* it were analogous to a word or about a sequence of gestures *as if* it were analogous to a sentence. However, a strong use of the linguistic model in ALR led these investigators to lose their caution, take a top-down approach and begin to assume rather than to question, analyze, and demonstrate symbolism, meaning, syntactics, et cetera, in the animals under study. In contrast, our bottom-up approach attempts to define symbolism and syntax operationally in terms of equivalence relations (see Sidman, 1990) and demonstrates that straightforward discrimination procedures with California sea lions and dolphins can produce complex cognitive performances with relevance to both syntactic and semantic learning in human language (Schusterman & Krieger, 1984, 1986; Schusterman & Gisiner, 1988, 1989; Gisiner & Schusterman, 1992; Schusterman & Kastak, 1993, 1995; Schusterman et al., 1993).

In a search for the origins and evolution of language, the discovery of learned relationships emergent from and consistent with reinforcement contingencies, such as functional equivalence and stimulus equivalence, provides ALR with powerful new techniques for the investigation of the relationship between language and cognition.

28

Anthropomorphism, Apes, and Language

H. Lyn Miles

Joni, Gua, Viki, Washoe,
Sarah, Lana, Koko, Nim,
More than apes yet not quite human,
Emulating seraphim.

—Thomas Sebeok, "Apespeak"

INTRODUCTION

Among the many criticisms leveled at ape language research is the charge of anthropomorphism. In the poem "Apespeak," which appeared in *The Sciences* as a letter to the editor in response to Herbert Terrace's essay, "Why Koko Can't Talk," the famous linguist and harsh critic of ape language research, Thomas Sebeok (1981), wryly suggested

The support of the Yerkes Regional Primate Research Center, National Institutes of Health, National Science Foundation, and UC Foundation are gratefully acknowledged. I would also like to thank Bob Mitchell and Stephen Harper for their excellent comments and editorial assistance with this chapter. The help of Don Elrod, Roger Ling, and the members of Project Chantek at the Uni-

that these subjects are "more than apes, but not quite human." In the last stanza, it becomes clear that this suggestion is an ironic understatement, when he casts the apes as merely "an epigonic, revenant of Clever Hans," the horse who appears to count but was really being subtly cued by his trainer. Pundits accuse ape language researchers of over-interpreting the language abilities of apes, and engaging in both self-deception and deliberate deception of others, as they create personalities out of animal subjects. Thus, in this view an anthropomorphic fallacy is the primary reason that people are duped into believing that language is not an ability unique to humans.

Rather than focusing on the anthropomorphic bias of the animal researchers, other critics have concentrated on the ways in which the linguistic output of apes does not match human language, especially in terms of syntax (Terrace, Petitto, Sanders, & Bever, 1979; Wallman, 1992). Although this approach does not preclude the belief that language is exclusively human, it also does not exclude the position that some of the cognitive traits of humans, including language, are homologous with apes. In keeping with this position, some scientists, such as Cheney and Seyfarth (1990), have been open to the idea that monkeys and apes, along with children and artificial intelligences, have "almost minds."

Also brought into question are the stories or narrative anecdotes that have been created about all of the animals who have been cross-fostered with humans and taught human symbols. Critics charge that some of the anecdotes, while consistent perhaps with the factual events, improperly imply cognitive abilities and motives to the animals through elaboration of the contextual conditions in which the utterances take place. The utterances of these animals, then, are simply overinterpreted. Yet stories and anecdotes are a very old and primary means by which humans convey information, persuade, or construct identities. We rely on them to create concrete mental explanatory pictures to aid understanding (see Mitchell, this volume, chap. 13), and they are scientifically useful in understanding animal cognition when they are based on solid empirical evidence interpreted in context and consistent with carefully recorded longitudinal data.

There are, in fact, many ways that anthropomorphism, as well as anthropocentrism, has affected language studies with apes and other

versity of Tennessee at Chattanooga is also very sincerely appreciated. This chapter is dedicated to the memory of Beatrix Gardner who so consistently and eloquently stressed that we were not teaching signs to an ape but that we were cross-fostering another species, and who was unfailingly kind, generous, and supportive of students of ape language.

animals. Since the turn of the century, this has been true with regard to the choice of species, the research questions asked, how the questions are framed, and other aspects of animal research. For example, the search for syntax in ape communications was essentially an anthropocentric focus on the unique aspects of human language. However, since language abilities have their foundations in human biological and cultural adaptation, anthropomorphism and anthropocentrism in this research should not be surprising.

Indeed, the methodology of ape language research is, in part, to anthropomorphize your subject and then observe how far the envelope of human psychological boundaries can be pushed. Sorting out ape and human similarities and dissimilarities in a nonanthropomorphic way is a complex process, but it is problematic and paradoxical because inevitably home-reared apes are raised as human beings. How can you raise an ape in a human environment, provide human experiences, and encourage human behavior, such as language acquisition or comprehension, to make the ape more humanlike, and then succeed in being nonanthropomorphic in your interpretations? Ape and human aspects are blurred, and the "natural" ape is recast as an interspecies ape-human cultural hybrid for which our models come from fiction (Frankenstein or Caliban) and science fiction (Data or Spock from *Star Trek*).

It is certainly possible and desirable to eliminate an unconsidered anthropocentric perspective, but I will argue that it may be impossible or self-defeating to try to conduct this research without what I call "pragmatic anthropomorphism." Pragmatic anthropomorphism is treating a nonhuman organism as human so that the organism will gain human qualities and abilities. This method addresses the challenge of how to scientifically characterize our similarities with our closest biological relatives, the great apes, without attributing nonexistent humanlike qualities. I argue that it is only through a contextually grounded, observational and developmental life history method—used primarily in anthropology and biology—that we can approach the question of ape language abilities, and then make comparisons with developmental studies of human *children*.

By focusing on children, instead of adult humans, it is immediately apparent that in order to teach language to children we must behave *as if* they will learn the appropriate symbol codes for speech, facial expressions, gestures, dress, et cetera. That is to say, we behave anthropomorphically towards our own neonates (see Russell, this volume). In fact, this anthropomorphism is likely an essential element for the emergence of language, whether one takes the position that language is innate in our species and merely acquired, or whether it is gradually learned through forms of apprenticeship and scaffolding. Any comparative study with apes, then, must take the same position, or fail because of faulty theory and inadequate methodology. Because of the unique socialization process of humans, serious ape language research must be based on a pragmatic anthropomorphism.

Further, any comparative effort to teach language to apes must take note of the facts that language is based on human cultural models and does not exist in isolation, and that language acquisition occurs within dynamic social contexts, involving many different social roles. We do not just teach language; we teach a world view, social roles, and a vision of what constitutes reality. We describe this process as enculturation, by which a child becomes a part of the shared culture. From this perspective, it appears that much of the interpretation and critique of ape language research has been narrowly focused on a single dyadic relationship: the researcher/subject role. Alternatively, a richer understanding of both humans and apes may be developed by observing the totality of perspectives that are present in the language-learning experience, including human/animal, caregiver/child, and teacher/pupil. Thus, I will argue in this chapter that ape language research requires pragmatic anthropomorphism to provide a framework for language learning and enculturation. However, this explicit anthropomorphism is methodological and must be accompanied with careful contextually based developmental studies in order to answer the question, What language abilities can apes comprehend and express?

Because our closest biological relatives, the primates, share biological and behavioral similarities with us, the issue of anthropomorphism is especially complex. In cultures where humans live with or near other primates, the similarities between monkeys, apes, and humans are readily recognized and reflected in world views which stress biological and psychological continuity (Linden, 1974). For example, the Dyak of Malaysia respect orangutans as intelligent forest-dwelling persons (Galdikas-Brindamour, 1975; Freeman, 1979; Maple, 1980, p. 213), and ancient Greek and Egyptian cultures made pets, spiritual beings, and household helpers of primates (Morris & Morris, 1966). By contrast, most of the Western intellectual world lacked close everyday experience with primates and thus made strong distinctions between humans and other animals and elaborated these with philosophical justifications, such as Aquinas's characterization of animals and nature as deistically designed and uncontaminated by human qualities (Lovejoy, 1936/1978), and Descartes's (1649/1927) claim that animals are "machines" incapable of thoughts or concepts. Both perspectives separated animals from the human realm of mind and reason, and are still implicit in much scientific, philosophical, and popular discourse about animals.

When Western scholars finally came into contact with the great apes in Descartes's time, they marveled at their humanlike form and abilities, and many early descriptions were highly anecdotal and anthropomorphic, including speculations by Samuel Pepys that they might be taught to speak (Pepys, 1661/1970). Growing understanding of the evolutionary continuities between humans and other animals in the 18th and 19th centuries was reflected in books on apes by primatologists, with titles such as *The Speech of Monkeys* (Garner, 1892), *The Ape Within Us* (MacKinnon, 1978), and *Almost Human* (Yerkes, 1925; Strum, 1987). But, despite their proximity to us, much of what we know about apes still remains sketchy, and the true intricacies of ape life are still recorded as anecdotes (Jolly, 1985) which are challenged by many scientists.

When scientists raise charges of anthropomorphism, they usually refer to attributing human qualities that are thought to be unique to our species or lineage. Because some similarities between apes and humans are homologous, that is, rooted in our common evolutionary past, it can be questioned whether descriptions of humanlike behavior in apes based on appropriate descriptions of hominoid-hominid continuities are truly anthropomorphic because these behaviors are shared by the two and thus are not unique to humans. Modern researchers are presented with a challenge: how to scientifically characterize our similarities with our closest biological relatives, the great apes, without attributing nonexistent humanlike qualities. This challenge is made even more difficult when apes are deliberately taught aspects of human culture and sometimes language.

ENCULTURATED AND LANGUAGE-TRAINED APES

Apes who have been taught aspects of human culture can be referred to as "enculturated" (some prefer the term "encultured" to designate a lack of full enculturation and integration into the society). In this century, many apes were kept as pets, or enculturated as family members or companions, and their fascinating behaviors in captivity were observed and recorded (Kearton, 1925; Jacobsen, Jacobsen, & Yoshioka, 1932; Raven, 1932; Kellogg & Kellogg, 1933; Kohts, 1935; Hoyt, 1941; Lintz, 1942; Hess, 1954; Harrison, 1962; Temerlin, 1975). Enculturated apes are raised in a human cultural setting and, unlike most pets, are encouraged to adopt many of the aspects of human experience, including daily activities, use of human toys and tools, world view, communication, and sometimes manners and clothing. However, to say an ape has been enculturated does not imply that the ape has the full cultural understanding or status of a normal human. A few apes even became bicultural, that is, they led natural lives before their captivity, or they were returned to natural settings following captivity. Anthropologists themselves are not yet agreed upon the meaning of the term "culture." Thus, to call these animals "bicultural" is not to suggest that their culture is based on the same symbolic meanings as human culture, but rather to recognize the learned social and technological traditions in ape society, in an attempt to understand their subjective social experiences in both worlds. That apes could adopt aspects of human culture is remarkable, but, despite the devotion that human caregivers often showed to their apes, the role of enculturated ape was often unfortunate, and many early accounts of human-reared apes have sad outcomes for the animals involved.

The earliest attempts to teach language to apes were based on spoken language and focused on the production, rather than comprehension, of speech sounds and words. Garner (1900) taught a chimpanzee to produce four sounds, including a breathless "mama," and Witmer (1909) also taught a chimpanzee to say "mama." Furness (1916) trained an orangutan to produce four word sounds, and in the 1950s Keith and Cathy Hayes home-reared a chimpanzee, Viki, who could produce "mama," "papa," "cup," and "up" (Hayes, 1951). Keith Laidler (1980) obtained similar results when he replicated the vocal studies with an orangutan from the London Zoo. In terms of comprehension, the Kelloggs (1933) reported that their home-reared chimpanzee Gua reacted correctly to many words and phrases, but this may have been based on nonlinguistic cues and no claim for language ability was made. This limited success led to the conclusion that apes were incapable of producing speech or learning language.

It is now known that the vocal tract of apes lacks the right angle bend that permits the full range of human speech sounds, and this is likely the cause for their failure (Lieberman, 1975). The earlier scientists, as well as others, had noted the gestural propensity of apes, and in the 1960s, researchers turned to using several nonvocal means to communicate with apes, including gestural signs used by the deaf (Gardner & Gardner, 1969), plastic tokens (Premack, 1971, 1976), and computer lexigrams (Rumbaugh, Gill, & von Glasersfeld, 1973). The chimpanzees (Washoe, Sarah, and Lana) in these different experiments learned vocabularies of over 150 symbols. While these apes could combine these symbols in various ways to express rudimentary meanings at the level of two- to three-year-old human children, the conclusion that these animals had learned language or even languagelike skills was severely challenged on the grounds that the animals had merely mimicked their trainers, as had the nineteenth-century horse Clever Hans (Sebeok & Umiker-Sebeok, 1980; Sebeok & Rosenthal, 1981).

Joining the Sebeoks' critical perspective were Herbert Terrace (1979) and his colleagues, who concluded that the chimpanzee Nim had failed to demonstrate spontaneous and properly structured sentences, and thus had not learned language at all. This was based on an anthropocentric focus on syntax, and Nim's failure was later understood in terms of his unsatisfactory training conditions (Miles, 1983). In the 1970s, despite much criticism, this research was extended to additional chimpanzees by the Gardners (Gardner, Gardner, & Van Cantfort, 1989) and Fouts and Couch (1976) with an emphasis on peer language use and cultural transmission. The three other ape species were now included in language studies by Savage-Rumbaugh (1984) with the pygmy chimpanzees (bonobos) Kanzi and Matata, Patterson (1978) with the gorillas Koko and Michael, and Shapiro (1978) and Miles (1980a) with the orangutans Princess and Chantek, respectively. Shortly thereafter in Japan, Matsuzawa (1986) replicated the computer lexigram studies with a chimpanzee Ai. Savage-Rumbaugh radically altered the training methods that she had used previously and turned to a more interactive and nurturing enculturation model. More than the other researchers, Patterson focused on the humanlike quality of her gorillas and their language ability, making her research available to nonscientific audiences through her Gorilla Foundation and newsletter. Miles (1990) emphasized enculturation, contextual analysis, and the evolutionary implications of ape language abilities, suggesting a more developmental and ethological approach, and a move away from proving that apes were learning adult human language. Shapiro (1978) examined orangutan language learning with a bicultural orangutan in a natural setting in Borneo.

However, attention continued to focus not on the acquisition of these abilities for their own sake, or the relation of these abilities to the animal's evolutionary history, but on whether or not chimpanzees had acquired adult human language, that is, the focus remained on human uniqueness. What ensued was a game of linguistic "Yes, but . . . ," in which the standard was raised as to what constitutes language ability as claims were made that apes showed more skills (Miles, 1983). Premack remarked on this transition:

> Ten or so years ago when I first gave colloquia on the question, "Do
> chimpanzees have language?", we used to be booed and hissed. The
> colloquia had the character of a revival meeting. Then, after about five
> years, people were asking me, "Do your chimps do so and so?" and I'd
> say, "No," and they'd say, "Other people's do. You must be some sort
> of a dummy." (Bazar, 1980, p. 4)

The extent of ape language abilities remains only slightly less controversial (Wallman, 1992), and we still know relatively little about how
apes process the skills they exhibit, whether or not this qualifies as language. However, the number of full and partial studies of ape language
abilities in this century represents the first time that it has been shown
that an animal can use a productive symbolic code, even at a rudimentary level, with defined, shared meanings. All of these factors serve to
create a unique anthropomorphic research context, which has consequences for both the methodology and conclusions of language studies
with apes.

ANTHROPOMORPHISM AND ANTHROPOCENTRISM

Due to our efforts to teach apes language, a human ability, there are a
number of anthropomorphic and anthropocentric methodological issues
inherent in ape language research. First, the very choice of apes for
these studies is appropriate, but not without anthropocentric aspects.
Not surprisingly, our characterization of the apes has been affected by
nonscientific factors. For example, chimpanzees were believed to be
the "smartest" of the apes, perhaps because of their higher activity levels, as well as their biological proximity to us. Discussions of "ape language" are often only focused on chimpanzees, even though three other
species are now involved (Miles, 1993b). The language ability of nonprimate species has been largely ignored; for example, at approximately
the same time as the early studies, the dog Fellow showed, under properly controlled conditions, that he comprehended over 120 spoken
English words when uttered as commands, even with varying syntax
(Warden & Warner, 1928). This work was generally ignored in favor of
attempting to find humanlike behavior in our closest animal relatives,
and evaluation of dogs' linguistic receptive capacities has only recently
been taken up again, this time by someone who has studied ape language (Sally Boysen, personal communication, 1994). Language
research has also been carried out with an African gray parrot (Pepperberg, 1981). The expansion of the species which have received language training should provide a better understanding of the evolution

of language and cognition, as well as place the focus away from whichever species is deemed to be biologically closest to us at the moment.

Second, the fact that apes raised in human cultures are engaging in the same activities as humans serves to make their behaviors appear more humanlike than they may actually be. The very similarities between apes and humans, especially when apes are in human settings, provides more opportunities for inappropriate anthropomorphic interpretations or assumptions that the behaviors are homologous. For example, an ape ordering a hamburger at a fast food restaurant or cleaning up a bedroom to receive an allowance may all enhance the impression of humanlike processing of these experiences. The situation may be heavily caregiver-supported and the ape may only sign one or two words which can be given greater meaning when placed in the context of the caregiver's interpretation or repeated questioning. The researcher may not intentionally distort the situation, but the situation itself makes it possible to use more anthropomorphic language when reporting it.

Third, ape language studies utilize a nonvocal human communication system with conventionalized signs, symbols, and referents. In fact, ignoring the apes' own natural communication systems and natural environment and instead attempting to teach aspects of human culture to them, including language, is again an example of unconscious anthropocentrism (see Kiriazis & Slobodchikoff, this volume). Rumbaugh, Gill, Von Glasersfeld, Warner, and Pisani (1975) argued that "the chimpanzee has cognitive competence far beyond that required in . . . the jungle" (pp. 629–630). However, more recently, with the discovery of more complex social behavior and communication in monkeys and apes in natural and captive settings (e.g., de Waal, 1982, 1986; Cheney & Seyfarth, 1990), investigators suggested re-evaluating whether or not apes' natural environments are unchallenging (Miles, 1993a; Savage-Rumbaugh, Murphy, Sevcik, Brakke, Williams, & Rumbaugh, 1993; Miles & Harper, 1994).

Apes could actually be taught a select set of speech sounds that would not replicate any known human language, and apes could also learn a spoken communication system based on consonantal and some vocalic sounds forming a subset of the range of sounds used in human speech (Miles, 1978b). However, a series of popping and guttural sounds would appear even less humanlike and less appealing than present nonvocal methods of communication, and interest in such methods will probably remain low. This bias should not be surprising because language studies with humans historically excluded deaf culture and those with disabilities. Possession of language was denied to deaf humans who used gestural signs of the American Sign Language (ASL) for the deaf. It was believed that a gesture language was somehow inherently primitive or less linguistic than a spoken one, until Klima and Bellugi (1979) and others showed that ASL had the same complex grammatical features as spoken language.

Cognitive studies of primates in natural settings depend upon interpretations of primate behavior and signals based on human goals, categories of action, and motivations (Cheney & Seyfarth, 1990). But in language studies with apes, the animal actually uses a form of human communication, such as sign language, or a meaning system created by humans such as computer lexigrams or other symbols. The structure and meanings of the conventionalized symbols are already established. The results of language experiments, although they remain controversial, are in the form of human communication, similar to the utterances and signing of children in the very earliest stages of language acquisition. These communications are usually no longer than two or three words, show structural simplicity, and indicate greater comprehension than production (Klima & Bellugi, 1979). Direct comparisons can be made with the development of human language and other forms of communication with human children, and we can examine an ape and a child in a similar context with similar learning opportunities and levels of enculturating experiences with supportive caregivers.

Fourth, the human communication system is directly taught to the apes, and it is expected that the ape will respond in a meaningful way. This direct instruction represents a deliberate attempt to encourage humanlike behavior, not simply a passive lending of human interpretation to a natural animal behavior. Apes learn signs by molding, then later by imitation; or, in the case of one ape, Kanzi, almost exclusively by imitation (Savage-Rumbaugh & Lewin, 1994). Ironically, in order to teach language we must assume our conclusion, that is, it is not easy to learn language from someone who does not have the expectation that your communications will be meaningful since language learning depends on redundancy, scaffolding, and apprenticeship. The very act of signing with apes, then, requires us to act "as if" they can acquire language because this is the way that we learn language ourselves (Miles, 1990).

Fifth, researchers can directly query a language-trained ape by asking, for example, "Which one would you like?", "Do you see the truck?", and "Where does it hurt?" Because the ape responds in a concrete code there is less ambiguity about his or her message; the researcher is not lending a human interpretation to the communication—it has an already established meaning. There is still the important issue of validly interpreting the animal's execution of the sign and whether it is intended as a communication, but this is the same problem we have in communicating with a child or each other. When we interpret an action as the use of a human symbol and gloss it as a human word with a culturally defined meaning we believe that the communicator, whether ape or human, has intentionality and is using the symbol system meaningfully. The ape in language studies is an active agent, not a passive subject. Apes do not just respond passively to the experimenter's task of problem solving or learning, but can spontaneously initiate their own communications. Apes can even disagree with the researcher and express this through symbols or other behavior.

Sixth, although a symbol-using ape can initiate, an ape in a language study is also a communicative partner in a two-way exchange. The exchange does not consist of two discrete elements, but forms a blend of both participants' interactions. Much research has focused on only one half of this partnership—the ape's—and has failed to notice that the researcher's communications are intertwined with and help to shape the ape's use of symbols. For example, an analysis of the chimpanzee Nim's communications showed that they were part of a larger interaction that was affected by the researcher's interruptions (Miles, 1983). Terrace assumed that Nim's communications were controlled only by him, but they were actually affected by the researcher's actions. Thus, when we measure the ape, we are also measuring the effect of the human.

Finally, because the ape is being taught to use a human form of communication, a key factor impinging on anthropomorphism is that the researcher subjectively experiences the ape as a person with agency, intentionality, feelings, et cetera. Because of our own simulative propensities (Mitchell, 1994b), we readily engage in animism and lend personality to puppets, cartoon characters, and inanimate objects such as boats and automobiles (see Guthrie, 1993, this volume). In this research, the investigator not only projects sadness in the expression of a primate's face, but experiences the primate actually saying "me sad," "me good," or "me bad." The linguistic nature of the interaction and the long-term shared experiences within a human cultural context all foster this subjective experience and make communication with the animal seem more ordinary and acceptable, much like the way a child's communications are received. In the world of ape language researchers, apes "talk," have personalities and preferences, and have similar basic emotions. Subjectively, the ape enters a new semantic domain of symbol-users. This domain excludes pets, but includes apes and very young children (and perhaps animated figures, cartoons, puppets, and other animate or inanimate objects that appear to talk).

SOCIAL ROLES IN APE LANGUAGE RESEARCH

A highly underexamined aspect of ape language research has been the researchers' training of the apes and the culture of communication that is established in the research setting (Miles, 1978b). Both the researcher and the ape participate in roles that characterize the research in particular ways. These roles affect the communication dyad and frame the characterization or dominant theme of the research. The culture of communication and roles played have varied in these studies but consist primarily of: human and ape, researcher and subject, teacher and student, parent and child, and enculturator and enculturatee. Although many roles simultaneously operate in every research effort, projects have tended to emphasize one or two roles over others. These roles interface with anthropomorphism in several ways. A dominant role frames the research question, governs the researcher's behavior, affects the terminology used, influences the linguistic skills of the ape, and helps to shape the eventual research results. Further, while some research efforts have made these roles explicit and have discussed research philosophy in their protocols, in others the roles have remained implicit, or perhaps even unnoticed. These roles have even changed over time, as we gain more experience in what conditions foster even rudimentary language abilities in nonhuman species. Much of the controversy of ape language research, as well as relative charges of anthropomorphism, can be attributed to the interface of these roles. This is because the world view of the researcher and questions asked, both explicit and implicit, drive the theory and method of these studies. For example, if an ape is treated and encouraged to respond like a human child, which is likely necessary for human language to be acquired, the interpretation will be rich and the issues of anthropomorphism will be more serious. If the animal is objectified as a research subject, some of the elements of natural language learning may not be developed. And if an ape is seen as a participant in a cultural scene in which it may share some meanings but not others, the question of whether it has acquired (adult) human language will not be as important as an exploration of the systems of meanings it is able to comprehend and express—regardless of whether that qualifies as language.

Ape as Animal

The first basic set of descriptive roles in ape language research is at the species level: human versus ape, with a sense of the ape as an animal "other." This distinction is clearly seen in the treatment of Meshie, one of the first enculturated chimpanzees, who was taken from her mother in Africa in the 1920s, and purchased by Harry Raven who raised her in the sophisticated intellectual and political subculture of New York City (Preston, 1981). With Meshie there is also a secondary role of parent and child, as she lived like a child in the Ravens' home, and was treated as a sibling by Raven's children. Raven became quite attached to her and she dined in restaurants with politicians and celebrities, and rode her tricycle like a child around the floors of the American Museum of Natural History. Meshie lived as a child in a human setting and learned to eat with a knife and fork, untie knots, feed a human baby with a spoon, and light her human father's cigarette. Eventually, in her role as ape, she became too hard to control at home. She was transferred to a zoo in another city where she apparently led a stressful and cruel life, and died in childbirth. She was returned to the museum where, ironically, as a dead chimpanzee, she lost her former childlike status. Although Raven seemed greatly conflicted about it, her body was stuffed and placed on display in the Hall of Primates where Meshie had once freely ridden her tricycle, an atrocity unthinkable for a true child. She did not cross the boundary from ape to human and reverted to her animal status at her death. Her body can still be seen today on the third floor of the museum, with her chin in her hand in a reflective pose, perhaps a visual metaphor for her intelligence and enculturated status (Haraway, 1989), or an icon of our anthropomorphic folly.

Ape as Research Subject

As scientists became more interested in studying the similarities between apes and humans, the roles of researcher and subject were added, and cognitive studies were conducted to determine how humanlike the apes are, at the same time that paleoanthropological studies were carried out to discover how apelike humans are. Deliberately raising a child as an ape was rejected on ethical grounds, but apes, as research subjects and not pets, were exposed to human culture in restricted ways.

For example, Winthrop and Louise Kellogg sought to understand the relationship of heredity and environment, and to "obtain an infant anthropoid ape, as young as possible, and rear it in every respect as a child is reared" (Kellogg & Kellogg, 1933, p. 11). The Kelloggs reared the chimpanzee Gua for 9 months with their own infant son Donald, and compared them on tests of motor skills and thought processes. Gua was exposed to speech, but the Kelloggs noted that although in advance of Donald on most motor skills, she lagged behind in the areas of language and imitation (Kellogg & Kellogg, 1933).

Although Gua was raised as a child, a film of their work (Kellogg & Kellogg, 1932) shows that the researcher and subject roles predominated and were also applied to their own son Donald. The film depicts extensive and sometimes frightening tests with Gua and Donald, whom they refer to as "the chimpanzee" and "the child" in an objectified normative stance that compares types, not individuals (Kellogg & Kellogg, 1932, 1933). The young "subjects" are tied up, swung around, or made anxious when sacks are placed on their heads in the interest of science. The Kelloggs seem genuinely interested in both Donald and Gua, giving empathic hugs between scenes, but instead of having anthropomorphized Gua, they appear to have objectified both Gua *and* Donald. This depersonalized research style seems characteristic of the times: see, for example, M. C. Bateson's (1984) descriptions of how her parents Margaret Mead and Gregory Bateson carried out scientific tests on their newborn infant. This also reflected a great interest in ways to perfect and control human behavior, whether from the perspective of heredity, in the eugenics movement, or the environment, through behaviorism.

The Premacks also emphasized researcher and subject in their study of intelligence with the chimpanzee Sarah (Premack, 1971, 1976; Premack & Premack, 1983). Sarah was a passive subject who was able to answer questions with multicolored and shaped plastic tokens, but who did not initiate communications herself. In this sense, Sarah was an animal model for human cognition, not language, as the Premacks sought to tease out the nature of animal minds.

Rumbaugh and his associates modified the Premacks' approach and were initially interested in learning and grammar. They taught a chimpanzee, Lana, to communicate through a computer using lexigrams made up of geometric shapes. Lana spent much time in a clear plastic enclosure making requests of her machine. The grammar even required her to be polite, for example, "please machine give. . . ." Although Lana's communication was two-way, via the computer, it was recognized that a social peer element was missing from this method which might have affected her linguistic performance and prevented it from being referential (Rumbaugh, 1977).

To make this experiment more social, two other chimpanzee subjects were studied, Sherman and Austin, who were raised together and taught to categorize tools and foods (Savage-Rumbaugh, 1979, 1986; Savage-Rumbaugh, Rumbaugh, & Smith, 1980). The Rumbaughs, guarding their role as researchers, became strong proponents for strict methodological rigor. They were critical of users of other methods and argued that sign research with apes was less controlled and resulted in uninterpretable random hand movements or natural gestures (Savage-Rumbaugh, Rumbaugh, & Boysen, 1978). In contrast, the computer lexigrams used in their studies were not a natural human language, and were permanently recorded by a *machine*, which eliminated the more organic gestural problems of sign language interpretation and removed the researcher even further. However, with their more objectified methodology based on lexigrams they were able to conclude "there is now strong evidence that the great apes are capable of acquiring symbolic, language-like skills" (Savage-Rumbaugh & Rumbaugh, 1978, p. 265).

Ape as Student

Another important role in ape language research is that of teacher, since apes may require educational support to learn their symbols. Herbert Terrace (1979) experimented with various teaching techniques: raising the chimpanzee Nim with a human family where he could be taught by his human siblings, exposing Nim to a tedious teaching method that had been developed for retarded children, and ultimately raising him with an extremely diverse group of at least 60 "teachers" who took turns spending time with Nim at a "classroom" at Columbia University. Terrace (1979) was originally optimistic about training Nim to use gestural signs and speculated that Nim might be an informant or "translator" for wild chimpanzees, or might be coaxed to reveal his dreams. Nim at first appeared to learn approximately the same language skills as the other apes in these studies, but, in the late 1970s, Terrace et al. (1979) re-analyzed the tapes of Nim's discourse and concluded that Nim had not acquired language skills but was merely imitating the communications of his "teachers," a charge that he then generalized to the other ape studies. Nim had done too well as a compliant and imitative student, and not well enough as an independent initiator of his own communications.

If a single-subject approach with Washoe or Lana could not "make" the ape language case, how could the failure of one method with one ape, Nim, "break" it? The answer lay in the fact that Nim seemed to prove the anthropocentric case, that humans were unique in their language ability. Analyses of Nim's communications by other investigators showed that Nim had in fact succeeded in imitating his teachers in an unexpected way—Nim was interrupted twice as much as apes in other studies, and thus had learned to interrupt himself, resulting in a discourse style of frustrated communications and mutual inter-

ruptions. Under different, more relaxed, conversational conditions
Nim's communications were indeed spontaneous and meaningful
(Miles 1983; O'Sullivan & Yeager, 1989). Terrace had engaged in a well-
documented study, conducted extensive and careful analysis, and given
unusual detail in the behind the scenes aspects of his research, but
because only the student was scrutinized, and not the teachers, their
teaching method was essentially ignored. The lack of attention by ape
language critics to the researchers themselves, and to the sociocultural
context of language production, suggested that additional research
might provide different results, as Terrace (1979) himself suggested.

Ape as Child

In contrast to those pursuing more experimental methods, Alan and
Beatrix Gardner focused on a language system already used by humans,
the American Sign Language for the deaf (Gardner & Gardner, 1978). To
replicate the human language learning experience with the chimpanzee
Washoe, they adopted a parent/child model based on language acquisition in children, and placed an emphasis on meaningful two-way
communication with chimpanzees which would then be compared with
the language development of, especially, deaf children (Gardner et al.,
1989). They described their method as "cross-fostering," and carefully
raised their chimpanzee Washoe in a house trailer on their property
with a "family" of signing companions, some fluent in ASL; in a second
phase of their research, which is now continued by Roger Fouts, they
expanded their study to several other chimpanzees. Although they followed the Kelloggs' example, they did not objectify Washoe or the other
chimpanzees and maintained throughout their research that less naturalistic methods would not permit the kinds of comparisons with children that would ultimately be necessary. In many ways their early ideas
were foreshadowing changes in ape language research that were to
come much later.

The parent/child model was most extensively employed by
Francine "Penny" Patterson with the gorillas Koko and Michael. Patterson not only cross-fostered Koko and Michael, but created symbolic
environments for the gorillas that most replicated a childlike home setting. The gorillas not only learned signs but their vocabularies were
claimed to be the largest, up to 1,000 signs in some accounts (Patterson
& Linden, 1981). Koko and Michael also recently provided important
evidence of gorilla mirror self-recognition (Patterson & Cohn, 1994).
Koko was strongly encouraged to produce signs, and total agency and
personhood were attributed to her, while Patterson's task was to inter-

pret Koko's "messages." Koko played with dolls, had tea parties and a
pet kitten, and sat up and waited for Santa (Patterson & Linden, 1981;
Anne Southcombe, personal communication, 1986). Patterson claimed
that Koko not only used her signs in meaningful ways, but that the
gorillas had a human low normal IQ, and could joke, swear, make
rhymes (based on speech and signs), and describe the circumstances
of their own capture (which occurred at a very young age).

There has been some scientific neglect of Patterson's research
(Davidson & Hopson, 1988) perhaps in response to the extreme par-
ent/child roles that she has adopted with the gorillas. When writing
down conversations with Koko, Patterson or another companion frequently adds elaborate interpretation to Koko's sign use which must be
carefully examined. The resulting edited dialogue creates a narrative
with such strong supports for interpreting the "story" from the human
companion that the conversation may appear more humanlike than
when looking at the ape signs alone. Patterson claims that the gorillas
engage in storytelling themselves and can illustrate these stories with
partial representational drawings (Weller, 1980). However, when caregiver interpretations are eliminated from her reports of conversations
with the gorillas, it is clear that Koko and Michael are signing in ways
similar to the other signing apes, and additional discourse analysis and
contextual information within a developmental context would help
clarify more complex use of signs from errors or overinterpretation
(Miles, 1986). It is inappropriate to hold Patterson's methodology to
standards that she does not seek, such as the objectification of her subject. Her method is one of immersing herself in interactions with her
animal based on the assumption that her animal's communications are
meaningful; her method yields answers to questions that are not being
asked by her critics. Patterson's method, whatever its difficulties, has
resulted in a high frequency of signing by Koko, and it may provide the
basis for the type of scaffolding and support that a definitive ape language experiment would require.

Ape as Enculturated Individual

Anthropologists, in particular, stress language learning within a sociocultural context and take the position that this both reflects the social
context and helps to construct it. The process, discussed above, is called
enculturation. This approach was first used to describe the signing chimpanzees Ali and Booee (Miles, 1976; 1978b, p. 85), and formed the basis of
my study with Chantek who was taught gestural signs in a form called
pidgin Sign English (Miles, 1980a). Although utilizing parent/child roles,

this approach emphasizes the relationship between language and culture, including such dimensions as generational transmission, peer interaction, tool use, et cetera. This model has been used either explicitly or implicitly by three language projects. Fouts (1973) adopted this method in his continuation of the work of the Gardners, and based it on a model of language learning and transmission through social peer interaction. Initially, Fouts used an experimental learning model and referred to the chimpanzees as "subjects" who were engaged in "training procedures." But Fouts's success with ape language research was based on an empathic approach that stressed each animal's individual personality. For example, because "just a bowl of raisins was a special delight to Booee" (Fouts & Couch, 1976, p. 148), Fouts utilized Booee's gentleness and passion for food to pique his curiosity and motivate him to sign. On the other hand, Cindy was dependent upon praise and emotional support, as shown in Fouts's remark "she demanded not only raisins but compliments" (Fouts & Couch, 1976, p. 149). Fouts also commented "it is not surprising that Bruno was given the sign for proud as his name sign" (Fouts & Couch, 1976, p. 150), and indeed Bruno was an energetic and dominant chimpanzee. It should be noted that all apes have variable personalities; however, Fouts's methodology explicitly emphasized and capitalized on individual differences, and he often admonished his assistants to "listen to their music" as they were introduced to the apes (Fouts, personal communication, August 5, 1973). He referred to apes as "nonhuman beings" (Fouts & Budd, 1979, p. 375), and attributed behavior such as joking and insults to the chimpanzees (Fouts, 1974; Fouts & Couch, 1976).

In contrast to earlier work, the pygmy chimpanzees Kanzi, Matata, and others studied by Savage-Rumbaugh (1984, 1986) were raised under more enriched conditions with emphasis on enculturation and direct comparisons with human children, much like the Gardner work that had been criticized earlier. Savage-Rumbaugh considered the pygmy chimpanzee to be our closest relative and stressed that they have "a greater capacity for language" and are "brighter than other apes," and "can acquire language skills without explicit training" (Savage-Rumbaugh, 1986, pp. 385, 386; Zihlman, Cronin, Cramer, & Sarich, 1978). Savage-Rumbaugh and her colleagues (Savage-Rumbaugh et al., 1993) strove to combine careful controlled methods and an enculturation model with emphasis on "untrained" referential language use, language comprehension—not production—and tool use. Savage-Rumbaugh provided extensive trails on which the pygmy chimpanzees could spend their days playing in the woods, making campfires, seeking out locations, experiencing "surprising" events—in short a combination of a normal child-rearing context and a replication of a food-gathering stage of human evolutionary cultural transmission.

My own work with the orangutan Chantek, which was begun in 1978, was affected by my earlier work with Fouts and the chimpanzees under his care, and was a continuation of his empathic style, combined with my own anthropological method of participant observation. As a result, my research was explicitly founded on the enculturation model, and was a direct response to Terrace's failure (which I personally observed in 1976). Extensive comparisons were made between Chantek's environment and linguistic output and Nim's, supporting the argument that context and caregivers' linguistic style affect an ape's communications (Miles, 1978a, 1983, 1990). I was particularly interested in cultural models and the evolution of human language and cognition, but stressed an encultured environment where Chantek could learn symbols that were meaningful *to him*. Early work had assumed that signs mapped adult usage, but I found that contextual information showed that, like the language of children, the meanings of Chantek's signs changed over time, that is, the semantic domain of signs shifted to match more closely the expectations of his environment. From the beginning I focused on enculturating Chantek and adopted a posture of critical or reflective anthropomorphism (Burghardt, 1985a, this volume). I viewed Chantek as an informant and myself as an anthropologist who would chart the reinvention of culturally based language (Lock, 1980) with socioculturally shaped cognitive scaffolding (Vygotsky, 1930/1978).

Ape as Subject of Narrative (Anecdotes)

Scientists sometimes utilize anecdotal narratives, that is, stories, expressed in human psychological terms, to interpret their data and make the information more broadly accessible, for example, in reconstructing human evolution (Landau, 1991; Collins & Pinch, 1993; Mitchell, this volume, chap. 13). Anecdotes and other forms of storytelling may be our primary means of understanding ourselves as individuals and as a species (McAdams, 1993), so it may be difficult to create scientific interpretations of animals that do not have strong narrative or mythic elements (Candland, 1993). Stories are appropriate since narratives are the primary way we commonly convey information, from biblical narratives, to news stories on television, to the stories we tell about ourselves and others.

Recognizing this, Patterson has very effectively cast the gorillas in her stories as children: the story of Koko's kitten, Koko wondering about where you go when you die, Michael as the budding artist, both of them as siblings waiting for Santa, and Michael as the foundling who remembers the events of his capture (Patterson & Linden, 1981). This role matured as Koko grew up and developed infatuations with various human males. A "dating service" was engaged for her to find a gorilla mate (actually the Species Survival Plan computer bank of gorillas in captivity). These stories appear in popular media and in the newsletter of the Gorilla Foundation, in which Patterson acknowledges that events and discourse reports are edited to create a narrative (Davidson & Hopson, 1988). These anecdotes have not been discussed in the scientific literature, but they have been effective ways to make the gorillas accessible to a wider audience, and to aid in fund raising. Sometimes the stories are created by the media, as when Chantek picked a lock, escaped from his yard, wandered into the provost's office on campus, and was found attempting to make a phone call by campus security officers (we frequently tested Chantek's comprehension of speech without opportunity for cueing on the telephone, so he was quite familiar with listening to a voice and then translating or replying in signs while he was observed by another caregiver, who then translated back into English). This became a national news item and some stories portrayed Chantek as ET, an intelligent alien "other" trying to "phone home."

Narratives are not only a part of the media aspect of ape language research—they can also affect the science. Kanzi, informally dubbed by some as "the wonder chimp," was presented in a Japanese television broadcast as frequently bipedal and was shown making a fire, and hiking up a hill, standing upright and carrying a backpack. The image of an early hominid (or hominine) was a clear inference. Kanzi also learned to manufacture a stone tool, under apprenticeship with Nicholas Toth, a noted archaeologist and flint knapper. Similarly, Chantek learned to steer a car, use over a dozen workman's tools, cut leather, punch holes, and weave leather strips to make pouches.

Hopefully, the anecdotes told by the researchers themselves illuminate the science, not fabricate it. These anecdotes can create cult personalities of Washoe, Chantek, Koko, and Kanzi, and distort their actual behaviors, or they can be used wisely to illuminate the science, interpret the results for the general public, and foster interest in great ape conservation (Patterson, 1986).

PRAGMATIC ANTHROPOMORPHISM AS A HYPOTHESIS

The research described in this chapter places apes in a human environment, immerses them in human culture, attempts to teach them a human form of communication, and then compares the results with language development in children. Each research project involved various roles and models based on experimentation, parenting, or cultural transmission, and this emphasis affected the extent to which humanlike skills were attributed to the apes. It is an irony of ape language research that, as stated above, one must assume and attribute language abilities to apes in order to produce those abilities. This is a posture of what I have called pragmatic anthropomorphism. Under ordinary circumstances, one cannot prove a conclusion with an assumption, but this paradox is due to the language acquisition process itself. This assumption is not a sufficient condition, but it is a necessary one. To communicate with other than the intention that the recipients will understand and respond in a meaningful way will prevent their acquisition of linguistic skills. Thus, language research with animals must begin from an enlightened posture of reflective anthropomorphism if it is to possibly succeed.

Given this somewhat unusual state of affairs, it is also important to recognize that exceptional claims require exceptional evidence. The standards for data collection must be exceptionally high, and care must be taken to make the most parsimonious interpretation. In addition, the underlying communicative and cognitive processes must be sought so that we can understand if these are in fact homologous abilities in apes and humans. We must also investigate production as well as comprehension. In theories that language is acquired, that is, innate, and not learned, linguistic comprehension is emphasized because it is believed that there is a "universal language" or grammar that our species uniquely possesses and that children merely acquire it with very little learning. However, this leaves out the very important fact of language usage and the interface between language, culture, and society. The way symbols are used is as important as the symbols themselves. It is partly by attention to only the symbols themselves that some research efforts stumbled. For example, while Terrace (1979) describes in great detail the social context in which the chimpanzee Nim's language development took place, including changes in his environment and teachers, and various stresses that clearly had an effect on Nim's linguistic output, these descriptions do not appear in scientific journals because the sociocultural factors are not deemed important. As I have argued here, they may be the most important factors of all and may have a dramatic effect on the linguistic output of the participants in language experiments.

Similarly, investigations should not be limited to anthropocentric issues such as studying only the species believed to be biologically closest to humans, or focusing only on syntax, or some other aspect of grammar believed to be uniquely human. To focus only on proposed unique human traits is to stay locked into the loop of whom to let into our cognitive "club." I suspect, as in the past, our answers to this question when framed this way will always be only ourselves. In order to place both ape and human abilities into evolutionary perspective it will be necessary to explore the abilities of many species, including nonprimates, and to look much more closely at the natural gestural and signaling systems, and other communicative behaviors of apes and other animals (Kiriazis & Slobodchikoff, this volume), as Savage-Rumbaugh has begun to do with pygmy chimpanzees (bonobos).

Taking a developmental approach allows one to compare the language progress of apes with that of human children rather than adults. There is no evidence that apes have acquired adult human language, and to continue that comparison is to continue to be distracted by human uniqueness. Bates (1993) and other child language specialists accept that apes are communicating with symbols in rudimentary but meaningful ways. Bates recognizes that the proper comparison for sign-using apes is children's language, and not adult human language, and the details of exactly what is or is not in their repertoire of skills remains open. Simultaneously, any additional ape language research should be done under enriched enculturated conditions with fluent caregivers. Finally, ape language researchers themselves will have to refrain from ahistorical presentations of research and acknowledge that we use anecdotes, narratives, hero myths, and personality stories in order to convey the subjective aspect of language research with animals. Such an approach might please both critics and enthusiasts, since its foundation would be in a conservative yet open interpretation of data that examines both linguistic and sociocultural factors.

PART IX

Comparing Perspectives

29

Anthropomorphism and Anecdotes:
A Guide for the Perplexed

Robert W. Mitchell

This chapter is intended as a guide for those who are perplexed by the array of arguments and attitudes presented in this volume about the meaning and significance of "anthropomorphism" and "anecdotes" in our interpretations of animals. In this chapter, similar meanings and arguments are compared, and different interpretations are evaluated. I first discuss different types of anthropomorphism; next examine ideas about anecdotes as evidence for psychological states; and finally offer my integration of the varied perspectives on the significance of anthropomorphism and anecdotes for understanding animals. Despite my attempts at neutrality, I sometimes disagree with or question authors' ideas. So, to be fair, I recommend that readers use this guide to make connections among ideas in other chapters and elsewhere, so as to develop their own arguments about the topics.

At the risk of making simple what many authors have been at pains to detail, I suggest that definitions of anthropomorphism in this volume can be placed under three overlapping rubrics: global, inaccurate and subjective. Some chapters examine or mention all three views, but usually their focus is only one. Global anthropomorphism (which includes the other two forms) is an expectation (perceptual or theoretical) that things in the world are like human beings or are caused by human beings or humanlike entities. Inaccurate anthropomorphism is an erroneous depiction of animals as having (uniquely) human characteristics (usually psychological). And subjective anthropomorphism is attribution of mental states or other psychological characteristics to animals (whether accurate or not). This last rubric—mental state attribution—might seem inappropriate to a discussion of anthropomorphism per se, in that the mental states attributed might not be *human* mental states. As Millikan suggests, "there are many intermediate possibilities" between having human mental states and having none at all, some complications of which are imagined (quite differently) by Povinelli. However, with the rise of behaviorism, *any* psychological characterization of animal behavior came to be suspect, and "anthropomorphic subjectivism" specified where the error resided, even if the psychological characterization was not humanlike (Hull, 1943, pp. 23–24). (Note that I distinguish types of anthropomorphism per se, rather than types of ideas about how anthropomophism originated or can be explained, which are depicted briefly in the first chapter and extensively in other chapters.)

GLOBAL ANTHROPOMORPHISM

Global anthropomorphism is most elaborately explored in the chapters by Guthrie and Knoll. The global anthropomorphism depicted by Guthrie is a basic element of human perception and cognition, whereas that depicted by Knoll is part of a methodological attitude toward animals.

Guthrie paints a grand canvas, attempting to come to grips with the human tendency to respond to many different phenomena with expectations of humanness. Not only animals, but animated characters, clouds, sounds in the night—in fact almost any object of experience has been responded to or depicted by someone as having characteristics redolent of humanity. Anthropomorphism results because we experience the world in terms of what is most important to us—humanness. For Guthrie (1993), global anthropomorphism is the basis for religion, in that all religions depict anthropomorphic entities. By contrast, Gallup, Marino, and Eddy view theism as a by-product of mental state attribution—a projection of one's own attributes to make death less fearful. Guthrie (1993, this volume) elaborately argues against this idea, noting that many aspects of deities are quite fear-inducing, providing little comfort either before or after death.

Knoll's canvas fits a smaller frame with her in-depth focus on Darwin's and Romanes's uses of anthropomorphism as part of a method, untainted by concerns about accuracy or inaccuracy, of seeing human traits and qualities in animals. Contrary to other biological writers at the time, Darwin did not perceive the principle organizing evolution as anthropomorphic, although he used the anthropomorphic term "selection" metaphorically, denuded of its mentalistic implications, to designate this principle. The anthropomorphic ambiguity inherent in the selection metaphor helped his theory to be accepted by theologically minded thinkers as well as materialists (Young, 1985). Although few modern descriptions of animal behavior are so credulous or "quaint" as those of Darwin and Romanes appear to us today, much of the Darwinian tradition remains strong: as several authors mention, the evolutionary ideas propounded by Darwin lend credence to the idea that animals, particularly those most closely related to us, should have human or humanlike qualities. Indeed, this anthropomorphism remains the expected (null) hypothesis in Darwinianly inspired comparative developmental evolutionary psychology (Parker), and thinking of nonhuman primates as human is "theoretically justified anthropomorphism" because it is necessary for collecting the sort of data needed to evaluate claims from evolutionary biology (Quiatt).

Although the other authors in Parts I and II (and elsewhere) also touch on global anthropomorphism, they do so to bring into focus one of the other two forms. Caporael and Heyes use "anthropomorphism" to mean mental state attribution (so-called "folk psychology") regardless of its accuracy, whereas Asquith and Cenami Spada use "anthropomorphism" to mean inaccurate mental state attribution or inaccurate expectations about animal psychology which are based on ideas about human beings. Even though authors are not strict in maintaining a separation between inaccurate and subjective anthropomorphism throughout the volume, the focus generally remains on one or the other: those largely content to focus on the use of mental state attribution to understand animal or human psychology (Gallup et al., Povinelli, Millikan, Moynihan, Byrne, Herzog & Galvin, Shapiro, Parker) are generally a different group from those concerned with evaluating the accuracy of attributing mental states to animals and other human beings (de Waal, Russell, Rollin, Mitchell, Silverman, Beer, Quiatt, Bekoff & Allen, Swartz & Evans, Davis, Kiriazis & Slobodchikoff, Schusterman & Gisiner). Still, a few (Guthrie, Lehman, Burghardt, Miles) attend extensively to both.

INACCURATE ANTHROPOMORPHISM

Inaccurate anthropomorphism includes at one extreme the types called "crude" (Burghardt) and "uncritical" (de Waal), sometimes found in books for popular audiences. Some of these books (e.g., Barber, 1993; Marshall Thomas, 1993) misrepresent scientific attitudes toward anthropomorphism and anecdotes, and at times describe animals' behaviors in quite odd anthropomorphic ways (as de Waal as well as Mitchell [1995] state about Marshall Thomas), yet they sometimes offer intriguing interpretations. But popular books need not be uncritically anthropomorphic, and some provide philosophical critiques of scientific paradigms and elaborate competing perspectives (e.g., Hearne, 1986).

A useful distinction between types of inaccurate anthropomorphism distinguishes categorical anthropomorphism from situational anthropomorphism (Fisher, 1991). Categorical anthropomorphism refers to inaccuracy in psychological characterization which occurs because the species so characterized does not and cannot have the psychological characterization (e.g., "chimpanzees understand Einstein's theory of relativity"). Situational anthropomorphism refers to inaccuracy in psychological characterization about a particular instance of behavior, even though the same psychological characterization is appropriate for members of that species in other instances (e.g., that grimacing chimpanzee is "smiling"). Although several authors mention special problems of situational anthropomorphism (see, e.g., Caporael & Heyes on dog training), most authors writing about inaccurate anthropomorphism are concerned with the categorical variety.

Inaccurate anthropomorphism is always recognized in hindsight. Asquith notes that detection of inaccurate anthropomorphism depends on the detector's beliefs about animals, humans, and the relation between them, where the detector is (usually) someone other than the anthropomorphizer. These beliefs are, in her view, typically about "our perceived place in nature." Cenami Spada similarly argues that anthropomorphism occurs because unquestioned assumptions which presuppose aspects of human psychology are used to come to grips with animal psychology, and detection of anthropomorphism results only after the dominant paradigm that is the basis for these assumptions is questioned. For Guthrie (and for one account of Caporael & Heyes), detecting inaccurate anthropomorphism is an empirical process—we simply find out that we are wrong. But finding out we are wrong in the case of animal psychology is not always possible at present because we don't have methods for gaining knowledge of some aspects of animal psychology (Fisher, 1991). All these authors agree that inaccurate anthropomorphism is, contra Kennedy (1992) and others, not something you can "guard against" in the sense of providing procedures for avoiding it, because you cannot know ahead of time what will be considered inaccurate of human beings or animals.

Concerns about inaccuracy resulting from the "subjectivity" present in applying mental state terms are overvalued, because *any* observation statement is dependent upon the observer's subjective experience. Cenami Spada and Shapiro posit that subjectivity is an essential part of observation, in that any observer observes from a particular point of view. This point of view is part of human knowledge, not something which can be separated from it, leaving behind "objective" knowledge. To separate what is subjective from what is objective would require another perspective detailing how things "really are"—an impossible perspective untainted by any point of view, which Cenami Spada calls "amorphism." Rather, any observation is dependent upon an "already conceptualized field of experience" (Cenami Spada, p. 42); "any and all of our descriptions . . . are given in terms of frames of reference that are posited . . . by the language we choose to use" (Russell, p. 120). Although Cenami Spada and Russell present similar arguments, they differ in the details. Whereas Russell argues that mentalistic terms "belong to a particular frame of reference, and belong more to the system of description than to the things described" (p. 119), Cenami Spada contends that to distinguish the description from the object described requires someone who knows what the object truly is, which again presumes an amorphic perspective. For both Russell and Cenami Spada, scientists are left with competing frames of reference or descriptive systems, and must sort out which are the most compelling for their purposes. Scientists can presumably "test" divergent frames of reference or descriptive systems by presenting animal behavior within these different perspectives and finding which is most compatible with evidence (see discussion below about anecdotes). One sometimes gets the sense that one interpretation is more accurate than another when the interpretative frame changes, even by something as simple as slowing down a videotape to examine an animal's behavior more closely than under normal viewing conditions (Mitchell). Decisions that some description or attribution is anthropomorphic, then, probably rely on comparisons *between* perspectives, rather than with an amorphic perspective.

The frames of reference, or beliefs and assumptions, which generate or detect inaccurate anthropomorphism are thought to derive from a variety of overlapping contextual or experiential sources: culture and historical period (Knoll, Asquith, Caporael & Heyes, Russell), scientific paradigm or model (Knoll, Cenami Spada, Quiatt, Miles, Rollin, Swartz & Evans), socialization or training (Shapiro, Caporael & Heyes, Bekoff & Allen), self-examination (Gallup et al.), and anthropocentrism (Kiriazis & Slobodchikoff). By contrast, others assert that our anthropomorphic propensities either (a) are there from the beginning as part of a human perceptual "strategy," derived from natural selection, "that usually is out of our awareness and always is out of our control" (Guthrie, p. 58); or (b) are otherwise inextricably linked with humanness, being innate (Kennedy, 1992), inherent (Burghardt), inescapable and almost irresistible (Knoll), ineliminable and (virtually) unavoidable (Guthrie, Beer), and inevitable (Moynihan). Still, as Davis (p. 346) notes, "no one has suggested that anthropomorphism is an unmodifiable reflexive behavior," and if anthropomorphism were innate and unavoidable in the sense that we could not do otherwise, Cenami Spada avers, we would never be able to recognize it as an error.

Whether inaccurate anthropomophism results from inherent perceptions and attitudes or from experientially based beliefs and assumptions, determining the accuracy of any anthropomorphic equation requires three components: knowledge of humans, knowledge of animals, and knowledge of the relation between the two. Several authors have pointed out that neither our knowledge of human beings (Caporael & Heyes, Guthrie, Davis) nor of animals (Cenami Spada, Lehman, Asquith) is always particularly adept. Oftentimes, criticism of anthropomorphic language occurs prior to any established knowledge about animal psychology; as Asquith (p. 34) states, often it is "opinion and conjecture" that stokes the critical fire, a premise which Bekoff and Allen have elaborated in their depictions of slayers and skeptics of cognitive ethology. While the question of how we are to gain general knowledge of human psychology seems stuck in philosophical dilemmas about the accuracy of folk psychology in comparison with other systems (e.g., forms of behaviorism, functionalism, materialism, connectionism), the question of the accuracy of folk psychology in understanding animals is difficult to answer (Beer). However, we can use anthropomorphism (or the reverse extrapolation from animals to humans) to derive hypotheses which can then be tested. Anthropomorphic hypotheses for understanding animal cognition can be developed from areas of study in which we are knowledgeable about human cognition, such as memory and cognitive development (Davis, Parker); and scientists studying pain and depression are content to extrapolate to humans from their knowledge about mechanisms and circumstances leading to these experiences in animals (Rollin), although there are many discrepancies in the extrapolation (Silverman). Of course, if our knowledge of animals and humans were without error, concerns about the accuracy or inaccuracy of anthropomorphism would be completely obviated because the need for potentially inaccurate anthropomorphic language would vanish.

Often the inaccuracy of anthropomophism is already decided by its definition as the attribution of "*uniquely* human" characteristics.

How is one to decide that an attribute is uniquely human? Presumably that requires knowledge of all species, a practical impossibility. (Similar concerns have been voiced about the term "species-specific," which also presumes knowledge of all other species; "species-typical" is perhaps a more accurate though less exciting description of our knowledge.) As Asquith notes, designation of some characteristic as "unique" to humans is usually not made on the basis of evidence. In addition, what is considered "human" is culturally normative, such that extrapolation from one set of humans to another can be inaccurate (Russell, Mitchell; see Howes, 1991). Even if other species share characteristics with humans, application of this characterization to species inappropriately would still be considered anthropomophism: labelling actions of salamanders as "pretending" would be anthropomorphic even though pretending is not unique to humans. Shapiro wryly argues that, since we rarely find attributes that meet the criterion of being uniquely human, anthropomorphism is rarely a problem. And if *human* anthropomorphic attributions themselves are uniquely human and hence unattributable to other animals, then these animals may themselves have attributional systems which are unique to their species (Caporael & Heyes, Gallup et al.), thereby potentially posing a difficulty in elaborating "psychological diversity" across species (Povinelli).

Human uniqueness was what Darwin and Romanes sought to explain and erode by detailing comparable psychological examples in nonhuman animals, and to achieve this task they and their intellectual descendents sought evidence of indentical or *simpler* forms of human psychology in animals (Knoll, Davis). Morgan's (1894) response to Romanes' inaccurate anthropomorphism was to support a more parsimonious attitude toward animal mental states, unwittingly making the *simplest* psychological attributions (independent of accuracy) seem the best, at least when behaviorism was the only game in town (Mitchell, 1986). Such "cognitive parsimony" needs to be tempered with "evolutionary parsimony" (de Waal): the simplest means of explaining similar behaviors when animals are known to be closely related is that these behaviors are homologous, rather than derived from convergent evolution. If such behaviors clearly indicate a complex psychology in one species (such as humans), they probably indicate the same complex psychology in its close relatives (such as great apes) (de Waal, Parker). Error can be present here, however, as when the homological presumption that chimpanzee mirror self-recognition is *identical* to human mirror-self-recognition led to an overestimation of chimpanzees' abilities (Swartz & Evans). Evolutionary homologies between great apes and humans make exceedingly difficult the task of deciding which anthropomorphic descriptions of these animals are accurate and which are inaccurate, particularly in areas such as language in which great apes do not share every element of a complex behavioral pattern (Miles).

Anthropomorphism proposes a relationship between human and animal psychology, but how to characterize this relationship is unclear. For some, anthropomorphism results from reasoning by analogy or, as discussed above, reasoning from homology. Analogical reasoning involves extrapolating from a well understood base domain (in this case, knowledge of human beings) to a less understood target domain (knowledge of animals), and in some cases reasoning from homology does as well. However, reasoning to homology can occur by claiming that two domains (great ape and human cognition, for example) are identical, in that they derive from a common origin.

For both analogical and homological forms of reasoning there seem to be two types of arguments which are labelled anthropomorphic: from humans to animals, and from one's own experience to a non-human other's experience (see Thompson, 1994). In addition, there seem to be two attitudes toward the accuracy of the conclusions of analogical reasoning: some, such as Rollin, seem ready to accept their accuracy (as in ordinary common sense), whereas others, such as Lehman, explicitly denounce the use of analogy for obtaining knowledge. But Lehman's devastating critique is directed to extremely simple analogical relations, rather than to more structured analogies in which base and target domains share multiple interconnected elements which suggest further common relations between elements (see Gentner, 1989). Miles points out that the base domain must be well chosen if the analogy (or homology) is to be useful, noting that many critics make analogy to ape language from adult human language, rather than from the more appropriate base domain of children's language.

Most authors find analogy and homology useful for discovering hypotheses which can then be tested. About the analogy from humans to animals, Parker and Davis seem sanguine about its usefulness in forming hypotheses when the base domain (human cognition) is well understood. Indeed, Davis (p. 345) argues that without knowledge of human psychology, "We simply do not know what to look for because we do not have our own behavior to guide us." But this argument seems inadequate, given that, for example, scientists learned about echolocation in bats and dolphins even without their own behavior to guide them (Nagel, 1974). Schusterman and Gisiner believe that, although using human language as a base domain for understanding human communication with dolphins has provided fruitful hypotheses, more parsimonious explanations are even more fruitful.

About the analogy from one's own experience to a nonhuman other's experience, authors are divided. Again, several authors acknowledge that intriguing hypotheses can be derived from such analogies, but that is all. A few authors seem sympathetic to the idea that subjective inference by analogy provides us with knowledge about animal (and human) experience (Cenami Spada, Rollin, Gallup et al.). Arguing against this idea, Davis imagines a scenario in which he depicts Griffin's (1984) anthropomorphic analogical subjectivism at full tilt, but this imagined scenario is not taken directly from Griffin's books. In fact, perhaps none of the subjects in Griffin's depictions of animal subjective experience are as superhumanly knowledgeable as Davis's (straw) gannets, which seem to know and experience more than any human could. Although some might argue that empathy involves analogical subjective inference, in Shapiro's view it derives instead from our ability to directly apprehend another's bodily experiences through our own body (see discussion below on subjective anthropomorphism).

An idea similar to anthropomorphism as analogy is anthropomorphism as metaphor. Initially suggested by Asquith (1984) and supported tentatively by Beer, this idea is now judged inadequate by Asquith. Metaphors apply a term (the vehicle) to an object (the referent) which does not fall under the literal meaning of the vehicle. In a metaphor, characteristics of the vehicle are attributed to a referent to capture some striking qualities shared by both vehicle and referent. But the vehicle is taken to be like the referent in a nonliteral sense only. For example, calling someone a "cow" suggests that he or she is sluggish, ungainly and perhaps stupid because these are striking aspects of cows (at least in American urban culture) that might be applied to a person. The metaphor is not suggesting that the person has four legs and eats grass; "people" and "cows" are distinctly different. Indeed, the fact that the person called a "cow" is known not to have some essential elements of cows indicates that the term is used metaphorically. So-called anthropomorphic terms share some qualities of metaphor, in that the salient aspects of a base (the vehicle) and a target (the referent) are compared: for example, when using a term such as "begging" to describe a chimpanzee, one may be noting that the actions of the animal have the same function (and/or appearance) that begging in humans has. To say that this use of the term "begging" is metaphorical requires that some essential elements of the literal meaning of the term "begging" are lacking, for example, that begging presumes intentionality in humans, but no such intentionality is present in chimpanzee begging. However, as Asquith argues, it is unknown whether or not such elements of the literal meaning are present, because we don't *know* whether chimpanzees exhibit intentionality in begging. Hence, anthropomorphism is not metaphor, because metaphor requires the known absence of the potentially present essential elements.

Of course we don't know if humans *always* exhibit intentionality in begging, and most observers think that chimpanzees act intentionally in begging and elsewhere. For some, the attribution of intentions to animals is only the beginning of the work necessary to understand the representations required to produce these intentions in action (Millikan), and the extensive concern about potential inaccuracy in attributing intentions to animals is overwhelmingly belied by the ease and usefulness of the attribution (Moynihan, de Waal, Quiatt). The use of anthropomorphic language in sociobiology *is* metaphorical, but even when explicitly denied the mentalistic implications of that language influence our interpretations and acceptance of sociobiological ideas (Beer, Quiatt).

SUBJECTIVE ANTHROPOMORPHISM, OR MENTAL STATE ATTRIBUTION

Other authors argue that use of anthropomorphic (psychological) terms is in fact *not* based on analogy; rather, these terms are defined in ways that could include animals as well as extraterrestrials or, perhaps, machines (Asquith, Cenami Spada, Lehman, Mitchell; see Wittgenstein, 1953). Thus, using psychological terms to describe animals need not be viewed as anthropomorphism, and could instead be viewed as an appropriate use of language (Thompson, 1987; Mitchell & Hamm, in press; see also below). For example, in humans "grief" refers to the experience of sadness and despair over the death of a loved one, and "friendship" refers, in part, to a relationship between two people in which they seek each other out to enjoy extended time together. As such, "grief" can easily be applied to animals, and "friendship" can be also, simply by replacing the term "people" with "animals" in the definition. But to make these terms scientifically respectable in discussing either humans or animals, the evidence used needs to be more specific. For example, "grief" could be defined as a decrease in activity and self-care by an individual following the death of a familiar organism with whom the animal had frequently been physically close (e.g., Smuts, 1985, pp. 230–231), and "friendship" could be defined as a relationship between two not closely related individuals in which they spend more time with each other—grooming and otherwise remaining nearby—than they do with other individuals (approximately Smuts's [1985] sense). Although these definitions are not identical to those used normally in relation to humans, and although the terms "grief" and "friendship" pull for subjective emotional meanings, the new scientifically respectable definitions are applicable to both humans and nonhumans even without our knowledge of these organisms' subjective states (Asquith, Lehman). (Indeed, one easy way to avoid the problem of affirming the consequent as depicted by Davis is simply to define one premise in terms of the other.) In some cases, to provide a definition requires characterizing animals with anthropomorphic (mentalistic) terminology and then detailing the specific behaviors which appear indicative of this mentalistic characterization (Hebb, 1946; de Waal, 1982, 1991). Once a definition is stated, in Lehman's view, the failure to apply the term to nonhumans when the definition is satisfied is an instance of anthropomorphism. In this way, Lehman turns the table on those who are too ready to use anthropomorphism as a pejorative to deny psychological characteristics to animals.

By contrast with Lehman's definitional view, Russell disbelieves that psychological terms can be ostensively defined (by pointing). Instead he argues that psychological terms and sentences "summarize," "abbreviate," and "entitle" a variety of manifestations which can be included under them. Because psychological terms summarize a whole range of behaviors and underlying assumptions and expectations, they cannot be (ostensively) pointed to. Rather, psychological terms might be used because of their usefulness in understanding behavior, or because they confer a value on the organism so described (Caporael & Heyes, Russell). Similarly, and contrary to another author "who thinks that closing the gap between human and nonhuman animals will lower the value placed on humans," Bekoff and Allen (p. 330) write that they "think that closing the gap might raise the value placed on nonhumans" (see also Cavalieri & Singer, 1993). Presumptions about the use of anthropomorphic language to increase the value of animals or devalue humans may be accurate for Westerners, but not for all cultures. For example, among the Mende in West Africa, who share their environment with chimpanzees and view them as extremely humanlike, extermination of chimpanzees is sometimes believed to be better for humans than conservation, in part *because* chimps are too much like humans (Richards, 1995).

Our normal uses of psychological terms commonly refer to or entitle the circumstances in which animals act (such as the spatiotemporal relations between the animals), the bodily actions of the animals, or both. Asquith notes that an animal may threaten another animal in a manner utterly unlike a human but the threat is still recognized by humans, and Caporael and Heyes suggest that if images of animals were replaced by triangles, circles, and squares moving in relation to each other in the same dynamic manner as did the original animal images, we would have no problem recognizing many psychological attributes depicted by the geometric shapes. In addition, facial and bodily movements may define interactions in some species, and may be helpful in finding appropriate psychological characterizations when circumstances are similar. Indeed, if psychological terms were not dependent upon patterns in behaviors, interactions and circumstances, how would we humans ever develop an understanding of psychological terms to apply to other people, let alone animals?

Consider the position of a child learning to use appropriately words and sentences about the emotions, intentions, and thinking of other people: the child must be able to recognize which behaviors in which circumstances the words label, as Lehman argues (see also Austin, 1946/1979; Wittgenstein, 1953; Quine, 1960). In describing why it is not surprising that a man, after claiming that he is angry, can be convinced by others that in fact he is not angry, Austin (1946/1979) notes about this man that

> he, like all of us, has primarily learnt to use the expression 'I am angry' of himself by (*a*) noting the occasion, symptoms, manifestation, &c., in cases where other persons say 'I am angry' of *themselves* (*b*) being told by others, who have noted all that can be observed about *him* on certain occasions, that 'You are angry', i.e. that he should say 'I am angry'. (p. 110)

In addition, this same man (like other people) learned to use the expressions "He is angry" or "She is angry" by noting when other people say these words of him and of others, and how these others respond to these statements. (Note that learning about the point of view required to use "I" and "you" appropriately may be more complex than Austin imagines—see Loveland, 1984.) Given that our anthropomorphic psychologizing is an expectation based on observing people's statements about other people (including ourselves), subjective anthropomorphizing is not a projection of one's own mental states. Rather, it is the result of other people teaching us, and our discerning, the requirements for conventional usage (but see below). Therefore, using such mentalistic terms as "angry" does not require one to "introspect" another's experience (Austin, 1946/1979, p. 116) or even one's own. If laypeople and scientific researchers anthropomorphize human infants to mold them into human beings (Russell), it should not be surprising that researchers teaching apes and parrots to use signs to communicate start off with the pragmatic assumption that they are interacting with humanlike creatures so as to transform them as much as possible into humanlike creatures (Miles, Silverman). My own experience as an undergraduate working with Irene Pepperberg on the parrot Alex's developing ability to produce meaningful utterances with English word sounds showed me that Irene's ability to hear English words, in what I took to be indistinguishable sounds emanating from Alex, resulted in Alex's eventually producing English words.

The view that our use and understanding of mental state terms derives from observations of appropriate uses of these terms has intriguing consequences for a child learning to talk about and understand mental states. This child learns about terms for the mental states of others by connecting these terms with intercorrelated bodily, facial, and circumstantial manifestations perceived visually and perhaps acoustically, olfactorily, et cetera, whereas he or she learns about terms for his/her own mental states by connecting these terms largely with intercorrelated kinesthetic, somatosensory, and other internal experiences of his/her own body and face. (I ignore here comparisons between visual mental images and vision, and between auditory mental images and audition, but see Mitchell, 1993b; 1994a.) To use terms appropriately about both self and other, the child must have a means of making sense of the connection between the two diverse sets of "evidence" for the same term. Although the terms themselves to some degree forge a connection, the child could maintain two different senses of the same term, much as adults do when they use the word "buck" to refer to a male deer and to a dollar bill. One thing that allows the child to recognize that the same word (e.g., "angry") refers to the same thing when used in relation to himself and to others is intersensory matching (Meltzoff, 1990), most importantly (but not exclusively) between kinesthesis and vision (Mitchell, 1993b; 1994a). The result is that the two forms of evidence are simply assumed to be mutually cohering; that is, the self and the other are expected to have most experiences in common when any salient manifestation—visual, kinesthetic, somatosensory, et cetera—indicative of a particular mental state term is presented by either. In this way, the other's behavior is imbued with experiences comparable to those of the self, even though the self has no direct access to those experiences; and the self understands how his or her behavior appears to others, and experiences that appearance as also imbued with subjective aspects for others. Even without language, intersensory matching may provide animals such as great apes with comparable expectations of similarity between their own experience and that of others (Mitchell, 1993b, 1993c, 1994a, 1994b).

A consequence of the view described in the preceding paragraphs is that human children (and some other organisms) learn about other minds initially without any analogy or extrapolation from their own case; rather, experiencing others as experiencing beings is simply a presupposition which derives from simplifying the extraordinary wealth of material that perceptual experiences of self and other provide. By contrast, Gallup et al. argue that organisms who are self-aware (as indicated by mirror-self-recognition) are so because they have access to and can reflect on their own mental states; and that, *because* they can reflect on them, they make inferences from their own mental states to others' mental states. Anthropomorphism in this view results from these inferential subjective attributions when made by humans; zoomorphism, when they are made by animals. Accepting these ideas, Povinelli argues that only some animals—those who recognize themselves in mirrors—will provide evidence of zoomorphism, and these attributions may be different among species. Against Guthrie's claim that global anthropomorphizing (which includes mental state attribution) is not a product of "conscious intellection," Gallup et al. posit that anthropomorphic attributions to animals are consciously made, and are different from attributions to machines and plants in that people believe that the attributions to animals are accurate.

The ideas that mental state attribution results from extrapolation from one's own case, and that mirror self-recognition indicates the ability to reflect on one's own mental states, are problematic (Swartz & Evans, Mitchell; see also Mitchell, 1993a, 1993b, 1993c). That young children, in the middle of their second year when they first recognize themselves in mirrors, *think about* their own mental states, as adults might do, to come to an understanding of others' mental states is unlikely, given that children fail to understand their own and others' mental states until much later (Butterworth, Harris, Leslie, & Wellman, 1991; Perner, 1991; Mitchell, 1993b, 1993c; Lewis & P. Mitchell, 1994). And if, as several authors (Thompson, 1994; Davis, Caporael & Heyes, Guthrie) note, our knowledge of our own psychology is typically not only inadequate but riddled with falsehoods, it would be difficult to extrapolate accurately from our own experience to that of nonhumans or even other people. Indeed, Swartz and Evans note paradoxically that proponents of the idea that organisms extrapolate from their own experience to understand others may themselves have too readily extrapolated from their own experience to that of chimpanzees. If mental state attribution is indeed dependent upon access to one's own mental states, then our knowledge of other organisms' psychological attributions may be exceedingly difficult to discern in that we do not have access to their mental states (Povinelli); but if mental state attribution is dependent upon perceptual access to others' behaviors and circumstances (as discussed above), then our knowledge of other organisms' psychological attributions is likely to be more readily apparent.

The study of private mental states is perhaps the most contentious of all concerns of cognitive ethology (Bekoff & Allen). Lehman's argument against the "private episode view" of animal mental states does not indicate that animals *cannot* have private mental states, but rather that, whether or not animals have private mental states, their behavior can still be used as evidence for psychological terminology. Some authors propose methods for the scientific study of the private mental states of nonhuman organisms, either through "critical anthropomorphism" applied via the ethological method (Burghardt) or through phenomenological empathy (Shapiro). Burghardt acknowledges that analogy from one's own experience to that of animals may be useful in fashioning hypotheses about animals' subjective experience, but he views it as only one of many variables, including "careful knowledge of the species and the individual, careful observation, behavioral and neuroscience research, our own empathy and intuition, and constantly refined publicly verifiable predictions" (p. 264), a methodological inventory comparable to Shapiro's phenomenological approach. These suggestions for a new methodology reappropriate "ordinary common sense" but with additional skepticism, and violate what Rollin calls "scientific common sense"—the assumptions many scientists have about what is acceptable scientific practice and belief. (The reverse process, of accepting as believable only scientific ideas about social and ethical problems, is viewed with concern by Caporael and Heyes, and by Silverman.) But ordinary common sense tells us that we know our own minds, when evidence frequently suggests that we do not (Nisbett & Wilson, 1977).

ANECDOTES

For Rollin, ordinary common sense accepts anecdotes as important evidence for understanding animals. In fact, anecdotes are one of our primary ways of interpreting human behavior, although with humans anecdotes are not used to discern whether or not humans have intentions or other psychological states, but rather what those intentions or psychological states might be (Mitchell). In the study of animal behavior and psychology, anecdotes are narrative depictions of animal behavior interpreted psychologically (Mitchell). Although the term "anecdotes" is often used in a dismissive sense, implying untrustworthy, casual observations, the anecdotes of current concern to scientists depict behavioral patterns detected repeatedly and independently by experienced and well-trained observers (Byrne). Note, however, that the existence of these repeated, independently observed patterns is usually not what is being contested (as might, e.g., be anecdotal sightings of the Loch Ness monster); rather, what is contested is their causal (psychological) interpretation, which is not itself repeatedly and independently observed (Russell).

Anecdotes of complex behavior, reported by careful observers knowledgeable about their subjects, are appealing as evidence for an animal's psychological abilities, in part because of the apparent spontaneity of the animal's behavior, which suggests that the behavior is generated from the animal's thought processes. But apparent spontaneity of behavior need not indicate spontaneous thinking, as Lloyd Morgan (1900, pp. 144–145) pointed out almost a century ago when he described his observations of the slow, trial-and-error development of a dog's apparently thoughtful act of lifting a latch to open a gate. In other cases, the sheer numerousness of reports of complex behavior by some species contrasts sharply with the lack of similar anecdotes for other species (Byrne & Whiten, 1990), suggesting a difference in psychological abilities: as de Waal notes, after years of observation no one has yet reported an instance of ambush by a rat, but reports of ambush in chimpanzees abound. But so also do reports of ambush and apparently calculated hiding in coyotes, wolves, and bears, supporting the idea that these animals share a complex psychology comparable to that of chimps and humans (E. Mills, 1919/1976; McMahan, 1978; Peters, 1978; McNamee, 1982). Evolutionary parsimony would support an interpretation of chimpanzee and human ambushes as based on similar or identical psychological processes. But what about ambushing by canids and ursids? Apparently, similar predatory strategies evolved independently in the common ancestor of ursids and canids, and in the common ancestor of humans and chimpanzees (Peters, 1978), much as tool use evolved independently in cebus monkeys and in the common ancestor of great apes and humans (Parker). Unfortunately, such instances of convergent evolution offer support for similarity of function in distantly related species, but not for similarity of cause or psychological experience (Hobhouse, 1901, 1915). Animals closely related to humans appear to have a privileged place with regard to anthropomorphic extrapolation, but even here there can be problems, as Swartz and Evans claim about erroneous assumptions of psychological identity for self-recognition in chimpanzees and humans.

How far scientists are willing to accept anecdotes as scientifically useful evidence for any proposition depends upon the scientists, their purposes, and their training. Silverman acknowledges that individuals hoping to control and predict animal behavior or to develop ethical theories may find anecdotes useful for these purposes, but that those building formal theories of animal psychology should find them less compelling. Whereas Rollin believes that scientists are opposed to anecdotes only from nonscientists, in fact many scientists are opposed to anecdotes from anybody, including other scientists (see commentary following Whiten & Byrne, 1988), and some accept anecdotes as scientifically necessary for describing rare events or complex interaction (Byrne & Whiten, 1988b; Moynihan) or for suggesting hypotheses. Others view anecdotes from scientists as more compelling than those from carefully observant pet owners (Byrne), the reverse of Rollin's view. Still others point out that scientists have for years ignored the anecdotes of native people, who often already know what these scientists take years to figure out (Kellert & Wilson, 1993), and that animal keepers and farmers who utilize anecdotes are far more successful at managing their interactions with their animals than are scientists who do not (Hebb, 1946; Rollin, Silverman, de Waal). And as Byrne, Swartz and Evans, and others (e.g., Mitchell, 1993c) note, scientists are often far more accepting of their own anecdotes than they are of others'. Indeed, in some cases entire species may be depicted as having the characteristics of one animal anecdotally described, even when the characterizer disavows the use of anecdotes (Swartz & Evans).

Some authors provide anecdotes to give readers some of the flavor of watching their animal subjects (Smuts, 1985) or to illustrate established behavior patterns of a species or an individual animal they know very well (Bekoff & Allen), but such flavorful or illustrative anecdotes can sometimes take on a life of their own, creating an image of the animal (or species) that is more complex and humanlike than is warranted by the animal's behavior patterns (Byrne, Miles). Again, in the study of animal behavior the issue is usually not whether the behaviors depicted in anecdotes happened, but rather how they are to be interpreted (Silverman). For their interpretation, anecdotes depend upon storytelling. Stories capture aspects of complex social interaction that cannot be otherwise adequately described at present (Moynihan). However, situating animal behavior in stories constrains readers to conceptualize an animal's behavior within the story structure (i.e., to use the story as the frame of reference for evaluating the behaviors) and to look to consistency among all the elements of the story, rather than to the behaviors of the animal, to decide if the story is accurate. One problem is that the same behavior may be equally consistent with two entirely different stories, only one of which is presented (Mitchell). Indeed, researchers can become so caught up in the story they wish to present that they fail to take in discordant information (Swartz & Evans, Mitchell).

Against these problems, Rollin and Byrne argue that plausibility with what we know of the animal depicted in an anecdote should stop us from accepting unwarranted ideas. But if to be believed the anecdotes must be plausible versions of what we already know, why bother collecting them? If the purpose of collecting anecdotes is to learn something new about animals' mental abilities, the criterion of plausibility makes fulfilling this purpose impossible because new anecdotally derived ideas will not cohere with previous ideas and will therefore be judged implausible. Note that this argument does not proscribe using psychological terms to describe instances of behavior; rather, the problem is deriving complex psychological explanations from anecdotal descriptions. Such explanations require other evidence, derived from developmental history and experiment (Dennett, 1983; Mitchell, 1986, 1987, 1990). But even if such evidence is unavailable, descriptions of complex behavior by animals should make scientists more attentive to the idea that cognitive explanations of such behavior need to be tested, and not simply dismissed as anthropomorphic projection (de Waal).

Fashioning animal behavior into a story induces a psychological interpretation, in that animals within a story are perceived as intentional, perceiving agents. When asked to rank animals on a scale of how humanlike they are on various psychological abilities, people describe nonhuman primates, pets, and dolphins as not quite human, but certainly more like humans than are other mammals (Gallup et al., Herzog & Galvin). But when mammals (including humans) are described as agents in a story depictive of a complex psychology, they are ranked as having essentially the same psychological characteristics as humans, whether these mammals are nonhuman primates, pets, or neither primates nor pets (Mitchell). Stories are clearly powerful tools for anthropomorphizing, and primatologists use that power to make sense of other primates (Quiatt, Byrne). Contrarily, stories can be viewed, not as inducing anthropomorphism, but as providing or organizing evidence for psychological terms which are applicable across species (Lehman, Byrne; Mitchell & Hamm, in press). If stories are accepted as methodologically useful for understanding animal mentality, then researchers must not only examine the same behaviors from within a variety of different stories (i.e., from a variety of perspectives) to find maximum consistency (Byrne)—a process which may result in two stories which are equally consistent—but must also imagine that animals may look at the world from a perspective very different from that of the "ordinary person," such that what is perceived as inconsistent by researchers may not be perceived as problematic by the animals (Mitchell). Even though the point of view of nonhuman animals can and should be "managed" by ethologists (Quiatt), how far one can go in understanding alien experiences is uncertain (Nagel, 1974; Burghardt).

Although Byrne argues convincingly that the extensive interactions most field researchers have with their subjects help to narrow the range of possible interpretations of anecdotes, other scientists worry that just this extensive interaction with an animal creates a bond which stops the scientist from being "objective." But as Rollin, Burghardt, Quiatt, and Shapiro note, this bond may in fact provide the scientist (or any long-term watcher of animals) with intuitions which allow him or her to understand the animal better than someone without such an interaction history (see Davis & Balfour, 1992a). In addition, long-term interaction with the same animal provides extensive evidence of the animal's developmental history, which can be examined for new developments in the animal's understanding (W. Mills, 1898; Mitchell, 1986) and compared across closely related species (Burghardt, Parker).

Single subject studies receive some of the same criticism as do anecdotes, in that it may not be appropriate to generalize the observations or results to other members of the species. Indeed, one animal's behavior can have varying significances for different theoretical orientations (Bekoff & Allen). Consider the case of Nim, the chimpanzee who "failed" to learn sign language. His inadequate signing (though actually it was inadequate only in particular circumstances) lent credence to the idea that all sign-using apes were not really signing. Yet Nim was only one of many apes taught sign language, most of whom signed in a manner similar to that of human children (Miles). On the other hand, even though it has been known for years that at least one gorilla fulfilled all criteria of mirror self-recognition, it has been denied for almost as many years that gorillas were capable of it (Swartz & Evans, Quiatt). How many animals must pass or fail a test for one to acknowledge that animals of this species, or animals of this age class, or animals with this developmental history, have or do not have the capacity evaluated by the test remains problematic. The number of animals required to provide evidence for or against a hypothesis depends upon such factors as the ease or cost of testing the species, how many animals have been tested before, and people's expectations about the ability of the species in question (Bekoff & Allen, Swartz & Evans, Mitchell). When Swartz and Evans first tried to publish their finding that 10 of 11 chimpanzees they tested failed to show mirror self-recognition (eventually published as Swartz & Evans, 1991), they encountered difficulty because it flew in the face of the belief that all chimps must recognize themselves in mirrors, even though only 14 chimpanzees had actually been tested at the time (and not all of those showed mirror self-recognition). Here they concern themselves with the birthdates of the four chimpanzees who provided the original evidence for chimpanzee mirror self-recognition (Gallup, 1970) to argue against a recent interpretation (Povinelli, Rulf, Landau, & Bierschwale, 1993) which, if true, would predict that these animals had been too young to recognize themselves in mirrors.

CONCLUSION

The discussion of these ideas leaves one with some degree of consensus. Global anthropomorphism is a pervasive aspect of human perception and cognition, the use of which as a methodological tool to understand animals must be examined critically and empirically. Inaccurate anthropomorphism is by definition an error, but how one is to determine when one has erred is problematic: the method suggested is to use diverse perspectives (translated into hypotheses) to evaluate data, where these perspectives vary in their degree of anthropomorphism (from none to a great deal). Subjective anthropomorphism—mental state attribution—is just one perspective on how humans and perhaps animals understand other beings, and evidence for it and the conditions under which it occurs need to be critically examined. Subjective experience of both humans and animals should be studied empirically (if possible), and methods adequate to the task may be developing. Anecdotes by reliable observers can be accepted as believable evidence that specific *behaviors* occurred (that is, that *this* animal did *this* action and *that* animal did *that* action) when these behaviors can be accurately observed under the viewing conditions employed, anthropomorphic terminology can reasonably be used to label or entitle these behaviors, but the interpretation of these behaviors as scientific evidence for mental state attribution demands the same method used to evaluate the accuracy of anthropomorphism: data should be examined using multiple hypotheses varying in the sorts of mental states they propose (if any). As with any science, the accurate interpretation of evidence depends upon the hypotheses and perspectives available; and the available evidence may support multiple hypotheses and perspectives. Note, however, that people concerned with ethics toward animals or control of animal behavior, or those just wanting to interact pleasantly with animals, might reasonably use anecdotes and ordinary common sense to fulfill purposes associated with these concerns and desires.

The analyses of anthropomorphism and anecdotes presented in this volume suggest that the methods scientists use to understand animal behavior and psychology are far less rigid and foolproof than many scientists and philosophers would have us believe. Animals present an "embarrassing problem," as Cenami Spada (p. 49) states, and scientists seem willing to try potentially "embarrassing" methods—methods which may not lead to complete solutions—in hopes of gaining insight about animals' behaviors, lives, and experiences. As usual, Darwin is a beacon for ethologists and comparative psychologists, because even though his methods of thinking about animal psychology are, today, delightfully embarrassing, they have led to remarkably novel and productive ideas about the nature of biological life, including the nature of human beings. This volume might be considered an evaluation of the legacy that Darwin and his friend Romanes left to scientists, a legacy which is still provoking us with new puzzles to engage our minds.

List of Contributors

COLIN ALLEN, associate professor of philosophy at Texas A&M University, has published a number of articles about the application of philosophical ideas about intentionality and function to the practice and concerns of cognitive ethologists, and is co-authoring a book (with Marc Bekoff) about theoretical issues involved in attributing mental states to animals.

PAMELA J. ASQUITH, associate professor of anthropology at the University of Alberta at Edmonton in Canada, is co-editor of *The Monkeys of Arashiyama: Thirty-five Years of Research in Japan and the West*, and *Images of Japanese Nature: Cultural Perspectives*. She has co-translated from Japanese the scientist Imanishi Kinji's *Seibutsu no Sekai* [World of Living Things], and has written articles on anthropomorphism in Japanese and Western primatology, cultural influences on Japanese science, and the history of primatology.

COLIN BEER, professor of psychology at Rutgers University, Newark Campus, is the co-author of *A Dictionary of Ethology*, co-editor of *Function and Evolution in Behavior: Essays in Honor of Professor Niko Tinbergen, F.R.S.*, and has written on the complexities of Laughing Gull communication, and conceptual issues in ethology.

MARC BEKOFF is professor of environmental, population, and organismic biology at the University of Colorado, Boulder. He was awarded a John Simon Guggenheim Memorial Foundation Fellowship for his work on the social ecology of coyotes, and has recently been elected a fellow of the Animal Behavior Society. His main research interests center on behavioral ecology, cognitive ethology, and animal welfare.

GORDON M. BURGHARDT, professor of psychology and of ecology and evolutionary biology at the University of Tennessee at Knoxville, has published extensively on the behavior of reptiles, play, and the evolution of behavior. He is the editor of *Foundations of Comparative Ethology*, and a Fellow of the Animal Behavior Society, the American Psychological Association, and the American Psychological Society.

RICHARD W. BYRNE, reader in psychology at the University of St. Andrews, is the author of *The Thinking Ape* and co-editor of *Machiavellian Intelligence: Social Intelligence and the Evolution of Intellect in Monkeys, Apes, and Humans*. His field research on chimpanzees, gorillas and baboons in Africa has included the study of manual skill acquisition, vocal communication, foraging, and social organization.

LINNDA R. CAPORAEL is an associate professor of psychology in the Department of Science and Technology Studies at Rensselaer Polytechnic Institute in Troy, New York. She has written on the evolution of human cooperation, social cognition, and human interaction with artifacts. Her work has appeared in *Behavioral and Brain Sciences, Journal of Social Issues*, and *Research in Philosophy and Technology*. Currently, she is completing a book, *The Mind's Natural Environment: Evolution, Sociality, and Identity*.

EMANUELA CENAMI SPADA, researcher in comparative psychology and philosophy at the Universitá Statale di Milano in Italy, is the author of articles on cognitive and social capabilities of marine mammals and primates.

HANK DAVIS, professor of psychology at the University of Guelph, is the author of articles on numerical and logical competence in animals, and co-editor of *The Inevitable Bond: Examining Scientist-Animal Interactions*.

TIMOTHY J. EDDY, research associate in the behavioral biology division at the University of Southwestern Louisiana's New Iberia Research Center, has published articles dealing with the behavioral, cognitive, and physiological concomitants of human interactions with other species.

SIÂN EVANS is a British primatologist with degrees in both zoology and anthropology. She has studied the social behavior of primates in large naturalistic enclosures in three continents, and is currently conducting research on cognition in New and Old World monkeys and apes in the Monkey Jungle zoological park in Florida.

GORDON G. GALLUP, JR., professor of psychology at State University of New York at Albany, developed the first experimental test of self-recognition in nonhuman primates, and has written on self-recognition, anthropomorphism, and tonic immobility.

SHELLEY GALVIN, instructor of psychology at Mars Hill College, is the author of articles on personality correlates, ethical ideology and animal rights activism.

ROBERT C. GISINER, program director for Marine Mammal Science at the Office of Naval Research in Arlington Virginia, studies acoustics and cognition of marine mammals.

STEWART ELLIOT T GUTHRIE, professor of anthropology at Fordham University, is author of *Faces in the Clouds: A New Theory of Religion.*

HAROLD A. HERZOG, professor of psychology at Western Carolina University in Cullowhee, North Carolina, has written extensively on the topic of human-animal interactions.

CECILIA M. HEYES, reader in psychology at University College, London, has authored papers on animal social cognition in journals such as *Animal Behaviour, The Quarterly Journal of Experimental Psychology, Animal Learning and Behavior, Mind and Language, Behaviorism, The Journal of Comparative Psychology,* and *Biology and Philosophy.* She is co-editor of *Social Learning in Animals: The Roots of Culture.*

JUDITH KIRIAZIS, a part-time instructor in biology at Northern Arizona University, studies animal communication.

ELIZABETH KNOLL, who earned her Ph.D. in conceptual foundations of science at the University of Chicago, is senior editor for trade science books at W. H. Freeman & Co.

HUGH LEHMAN, professor of philosophy at the University of Guelph in Ontano, Canada, is the author of *Rationality and Ethics in Agriculture* and *Introduction to the Philosophy of Mathematics,* as well as articles in philosophy. He is co-editor of *The Pesticide Question: Environment, Economics, and Ethics,* as well as an editor of the *Journal of Agricultural and Environmental Ethics.*

LORI MARINO is visiting assistant professor of biology at Emory University and research associate in behavioral biology at the Yerkes Regional Primate Research Center in Atlanta. She studies self-awareness and the relation between intelligence and encephalization in cetaceans and primates.

H. LYN MILES, U. C. Foundation professor of anthropology at the University of Tennessee at Chattanooga, is Director of Project Chantek, a developmental study of the linguistic and cognitive abilities of the orangutan Chantek. She has written about Chantek's psychological development and our conceptions of apes. She is a commentator and associate for *Behavioral and Brain Sciences*, and is co-editor of the forthcoming *Mentalities of Gorillas and Orangutans*.

RUTH GARRETT MILLIKAN, professor of philosophy at the University of Connecticut at Storrs and the University of Michigan, has written *Language, Thought, and Other Biological Categories*, and *White Queen Psychology and Other Essays for Alice*, as well as journal articles on language and mind. She has been president of the Society for Philosophy and Psychology, and has held a variety of offices in the American Philosophical Association.

ROBERT W. MITCHELL, associate professor of psychology at Eastern Kentucky University, is co-editor of *Deception: Perspectives on Human and Nonhuman Deceit*, *Self-Awareness in Animals and Humans*, and the forthcoming *Mentalities of Gorillas and Orangutans*, and is writing a history of scientific attitudes toward anthropomorphism and animals.

MARTIN H. MOYNIHAN is senior scientist at the Smithsonian Tropical Research Institute in Panama. He has written *Communication and Noncommunication in Cephalopods* and *The New World Primates: Adaptive Radiation and the Evolution of Social Behavior, Languages, and Intelligence*. He has also co-authored *The Behavior and Natural History of the Caribbean Reef Squid Sepioteuthis Sepioidea*, and is currently working on a book concerning the social regulation of competition and aggression.

SUE TAYLOR PARKER, professor of anthropology at Sonoma State University, has co-edited several books, including *Language and "Intelligence" in Monkeys and Apes*, *Self-Awareness in Animals and Humans*, *Reaching into Thought*, and the forthcoming *Mentalities of Gorillas and Orangutans*.

DANIEL J. POVINELLI, director of the Division of Behavioral Biology at the University of Southwestern Louisiana New Iberia Research Center and the USL Center for Child Studies, is an associate editor of the *British*

Journal of Developmental Psychology and a former editorial board member of the *Journal of Comparative Psychology*. He has published articles on the self-recognition abilities of children, chimpanzees and other species, as well as the evolution and development of mental state attribution in primates.

DUANE QUIATT, professor of anthropology at the University of Colorado, Denver, is co-author of *Primate Behaviour: Information, Social Knowledge, and the Evolution of Human Culture.* He edited an early collection of papers on primatological method and theory, *Primates on Primates,* co-edited a volume on primate and human biocultural evolution, *Hominid Culture in Primate Perspective,* and is the author of articles on primate behavior and paleoanthropology.

BERNARD E. ROLLIN, professor of philosophy, professor of physiology, and director of bioethical planning at Colorado State University, is the author of ten books, including *Animal Rights and Human Morality, The Unheeded Cry: Animal Consciousness, Animal Pain and Science, The Experimental Animal in Biomedical Research* (2 vols.), *The Frankenstein Syndrome,* and *Farm Animal Welfare.*

ROBERT L. RUSSELL, professor of psychology at Loyola University in Chicago, has edited two books, *Language in Psychotherapy* and *Reassessing Psychotherapy Research,* and co-authored *Change Processes in Child Psychotherapy: Revitalizing Treatment and Research.* He has published articles on psychotherapy theory, research methods, narrative, and language interaction.

RONALD J. SCHUSTERMAN is professor emeritus in psychology and biology at California State University at Hayward, research marine biologist at University of California at Santa Cruz, and senior scientist at the Aquarium for Wildlife and Conservation in New York City. He is co-editor of *Dolphin Cognition and Behavior: A Comparative Approach,* and has written on marine mammal cognition and behavior.

KENNETH J. SHAPIRO, executive director and cofounder of Psychologists for the Ethical Treatment of Animals (PSYETA), is founding editor of *Society and Animals* and founding co-editor of *Journal of Applied Animal Welfare Science,* two journals in the emerging field of animal studies. He is co-author of *The Experience of Introversion,* and author of *Bodily Reflective Modes* and the forthcoming *Animal Models of Human Psychology: Critique of Science, Ethics and Policy.*

PAUL S. SILVERMAN, professor of psychology at the University of Montana, is a developmental psychologist with interests in the "construction" of mind, children's understanding of emotions and other mental concepts, and psychopathology.

CON N. SLOBODCHIKOFF, professor of biology at Northern Arizona University, is editor of *The Species Concept* and *The Ecology of Social Behavior*, and co-editor of *A New Ecology: Novel Approaches to Interactive Systems*. He studies the communication of ground squirrels.

KARYL B. SWARTZ, associate professor of psychology at Lehman College and the Graduate Center, both of the City University of New York, has written publications on learning, memory, and self-recognition in primates.

NICHOLAS S. THOMPSON, professor of ethology and psychology at Clark University, is series editor of *Perspectives in Ethology* and co-editor of *Deception: Perspectives on Human and Nonhuman Deceit*. He has written articles on dog behavior, bird song, babies' cries, and philosophical ethology.

FRANS B. M. DE WAAL, professor in psychology at Emory University and research scientist at the Yerkes Regional Primate Research Center in Atlanta, was trained in European ethology and zoology before moving to the United States in 1981. He is the author of *Chimpanzee Politics*, *Peacemaking Among Primates*, and *Good Natured*, and studies the social behavior of nonhuman primates in captivity and under semi-natural conditions.

References

Agassi, J. (1973). Anthropomorphism in science. In Philip P. Wiener (Ed.), *Dictionary of the history of ideas* (Vol. 1, pp. 87–91). New York: Charles Scribner's Sons.

Akins, K. (1990). Science and our inner lives: Birds of prey, bats, and the common (featherless) bi-ped. In M. Bekoff & D. Jamieson (Eds.), *Interpretation and explanation in the study of animal behavior* (Vol. 1, pp. 414–427). Boulder, CO: Westview Press.

Akmajian, A., Demers, R., & Harnish, R. (1985). *Linguistics: An introduction to language and communication.* Cambridge, MA: MIT Press.

Alcock, J. (1992). Consciousness-raising III. *Natural History, 101*(9), 62–65.

Alexander, R. D. (1989). Evolution of the human psyche. In P. Mellars & C. Stringer (Eds.), *The human revolution* (pp. 455–513). Princeton: Princeton University Press.

Allen, C. (1992a). Mental content. *British Journal for the Philosophy of Science, 43,* 537–553.

Allen, C. (1992b). Mental content and evolutionary explanation. *Biology and Philosophy, 7,* 1–12.

Allen, C., & Bekoff, M. (1993). Intentionality, social play, and definition. *Biology and Philosophy, 9,* 63–74.

Allen, C., & Hauser, M. D. (1991). Concept attribution in nonhuman animals: Theoretical and methodological problems in ascribing complex mental processes. *Philosophy of Science, 58,* 221–240.

Allen, C., & Hauser, M. D. (1993). Communication and cognition: Is information the connection? *Philosophy of Science Association, 2,* 81–91.

Allen, G. (1987). Materialism and reductionism in the study of animal consciousness. In G. Greenberg & E. Tobach (Eds.), *Cognition, language, and
consciousness: Integrative levels* (pp. 137–160). Hillsdale, NJ: Lawrence Erlbaum.

Altmann, J. (1974). Observational study of behaviour: Sampling methods.
Behaviour, 49, 227–265.

Altmann, S. A. (1962). A field study of the sociobiology of rhesus monkeys,
Macaca mulatta. Annals of the New York Academy of Sciences, 102, 338–435.

American Psychiatric Association. (1987). *Diagnostic and statistical manual of
mental disorders* (3rd ed., rev.). Washington, DC: American Psychiatric
Association.

Andersen, S. M. (1984). Self-knowledge and social inference: II. The diagnostic-
ity of cognitive/affective and behavioral data. *Journal of Personality and
Social Psychology, 46,* 294–307.

Andersen, S. M., & Ross, L. (1984). Self-knowledge and social inference: I. The
impact of cognitive/affective and behavioral data. *Journal of Personality
and Social Psychology, 46,* 280–293.

Anderson, J. L. (1989). A methodological critique of the evidence for genetic
similarity detection. *Behavioral and Brain Sciences, 12,* 518–519.

Anderson, J. R. (1984). Monkeys with mirrors: Some questions for primate psychology. *International Journal of Primatology, 5,* 81–98.

Andrews, P. L. R., Messenger, J. B., & Tansey, E. M. (1983). The chromatic and
motor effects of neurotransmitters in intact and brainlesioned *Octopus.
Journal of the Marine Biology Association, U. K., 63,* 355–370.

Anonymous. (1980). *Uncle Milton's ant watcher's handbook.* Culver City, CA:
Uncle Milton Industries.

Anonymous. (1989, 11–17 March). Thank your lucky stars. *The Economist,
310*(7593), 90.

Anscombe, G. E. (1957). *Intention.* Ithaca: Cornell University Press.

Antinucci, F. (Ed.). (1989). *Cognitive structure and development in nonhuman primates.* Hillsdale, NJ: Lawrence Erlbaum.

Antinucci, F., Spinozzi, G., Visalberghi, E., & Volterra, V. (1982). Cognitive
development in a Japanese macaque (*Macaca fuscata*). *Annali dell' Istituto
Superiore di Santa', 18,* 177–184.

Archer, J. (1992). *Ethology and human development.* London: Barnes & Noble.

Arluke, A. B. (1988). Sacrificial symbolism in animal experimentation: Object or
pet? *Anthrozoös, 2,* 97–116.

Asquith, P. (1984). The inevitability and utility of anthropomorphism in description of primate behaviour. In R. Harré & V. Reynolds (Eds.), *The meaning of primate signals* (pp. 138–174). Cambridge, UK: Cambridge University Press.

Asquith, P. (1986). Anthropomorphism and the Japanese and Western traditions in primatology. In P. Else & R. Lee (Eds.), *Primate ontogeny, cognition and social behaviour* (pp. 61–71). Cambridge, UK: Cambridge University Press.

Asquith, P. (1996). Japanese science and Western hegemonies: Primatology and the limits set to questions. In L. Nader (Ed.), *Naked science: Anthropological inquiry into boundaries, power, and knowledge* (pp. 239–256). New York: Routledge.

Astington, J., Harris, P., & Olsen, D. (Eds.) (1988). *Developing theories of mind.* New York: Cambridge University Press.

August, P. V., & Anderson, J. G. T. (1987). Mammal sounds and motivation-structural rules: A test of the hypothesis. *Journal of Mammalogy, 68,* 1–9.

Austin, J. L. (1962). *Sense and sensibilia.* Oxford: Clarendon Press.

Austin, J. L. (1979). Other minds. In *Philosophical papers* (3rd ed.). (pp. 76–116). Oxford: Oxford University Press. (Original work published 1946.)

Avis, J., & Harris, P. (1991). Belief-desire reasoning among Baka children: Evidence for a universal conception of mind. *Child Development, 62,* 460–467.

Ayer, A. J. (1936). *Language, truth and logic.* London: Gollancz.

Ayer, A. J. (1959). *Logical positivism.* New York: Free Press.

Bacon, F. (1960). *The new organon and related writings.* New York: Liberal Arts Press. (Original work published 1620.)

Baldwin, J. M. (1903). *Mental development in the child and the race* (2nd ed.). London: Macmillan Co. (Original work published 1894.)

Barber, T. X. (1993). *The human nature of birds.* New York: St. Martin's Press.

Barlow, G. W. (1989). Has sociobiology killed ethology or revitalized it? In P. P. G. Bateson & P. H. Klopfer (Eds.), *Perspectives in Ethology* (Vol. 8, pp. 1–45). New York: Plenum.

Bartlett, F. C. (1932). *Remembering.* Cambridge, UK: Cambridge University Press.

Bates, E. (1976). *Language and context: The acquisition of pragmatics.* New York: Academic Press.

Bates, E. (1993). Commentary: Comprehension and production in early language development. In E. S. Savage-Rumbaugh, J. Murphy, R. A. Sev-

cik, K. E. Brakke, S. L. Williams, & D. M. Rumbaugh, *Language comprehension in ape and child* (pp. 222–242). *Monographs of the Society for Research
in Child Development, 58*(3–4, Serial No. 233).

Bateson, M. C. (1984). *With a daughter's eye: A memoir of Margaret Mead and Gregory Bateson.* New York: William Morrow.

Bateson, P. (1991). Assessment of pain in animals. *Animal Behaviour, 42,* 827–839.

Bateson, P. (Ed.). (1991). *The development and integration of behaviour.* Cambridge,
UK: Cambridge University Press.

Bazar, J. (1980). Catching up with the ape language debate. *APA Monitor, 11*(1),
4–5, 47.

Beaver, B. V. (1989). Environmental enrichment of laboratory animals. *ILAR
News (Institute of Laboratory Animal Resources), 31*(2), 5–11.

Beck, A. T. (1974). The development of depression: A cognitive model. In R. J.
Friedman & M. M. Katz (Eds.), *The psychology of depression* (pp. 3–28).
Washington, DC: Winston.

Beck, B. B. (1982). Chimpocentrism: Bias in cognitive ethology. *Journal of Human
Evolution, 11,* 3–17.

Beer, C. G. (1969). Laughing Gull chicks: Recognition of their parents' voices.
Science, 166, 1030–1032.

Beer, C. G. (1970). On the responses of Laughing Gull chicks to the calls of
adults II. Age changes and responses to different types of call. *Animal
Behaviour, 18,* 661–677.

Beer, C. G. (1975). Multiple functions and gull displays. In G. Baerends & C.
Beer (Eds.), *Function and evolution in behavior* (pp. 16–54). Oxford: Clarendon Press.

Beer, C. G. (1976). Some complexities in the communication behavior of gulls.
Annals of the New York Academy of Sciences, 280, 413–432.

Beer, C. G. (1979). Vocal communication between Laughing Gull parents and
chicks. *Behaviour, 70,* 118–146.

Beer, C. G. (1980). Perspectives on animal behavior comparisons. In M. C. Bornstein (Ed.), *Comparative methods in psychology* (pp. 17–64). Hillsdale, NJ:
Lawrence Erlbaum.

Beer, C. G. (1991). From folk psychology to cognitive ethology. In C. A. Ristau
(Ed.), *Cognitive ethology: The minds of other animals* (pp. 19–33). Hillsdale,
NJ: Lawrence Erlbaum.

Beer, C. G. (1992). Conceptual issues in cognitive ethology. *Advances in the Study
of Behavior, 21,* 69–109.

Beer, T., Bethe, A., & von Uexküll, J. (1899). Vorschläge zu einer objektivieren-den Nomenklatur in der Physiologie des Nervensystems [Proposals for an objectivist nomenclature for the physiology of the nervous system]. *Biologisches Zentralblatt, 19,* 517–521.

Bekoff, M. (1992). Scientific ideology, animal consciousness, and animal protection: A principled plea for unabashed common sense. *New Ideas in Psychology, 10,* 79–94.

Bekoff, M. (1993a). Experimentally induced infanticide: The removal of females and its ramifications. *Auk, 110,* 404–406.

Bekoff, M. (1993b). Review of Griffin's *Animal Minds. Ethology, 95,* 166–170.

Bekoff, M. (1993c). Should scientists bond with the animals who they use? Why not? *Psycoloquy, 3*(37), 1–8.

Bekoff, M. (1995). Cognitive ethology and the explanation of nonhuman animal behavior. In J.-A. Meyer & H. Roitblat (Eds.), *Comparative approaches to cognitive science* (pp. 119–150). Cambridge, MA: MIT Press.

Bekoff, M., & Allen, C. (1992). Intentional icons: Towards an evolutionary cognitive ethology. *Ethology, 91,* 1–16.

Bekoff, M., & Gruen, L. (1993). Animal welfare and individual characteristics: A conversation against speciesism. *Ethics and Behavior, 3,* 163–175.

Bekoff, M., Gruen, L., Townsend, S. E., & Rollin, B. E. (1992). Animals in science: Some areas revisited. *Animal Behaviour, 44,* 473–484.

Bekoff, M., & Jamieson, D. (Eds.). (1990a). *Interpretation and explanation in the study of animal behavior: Vol. 1. Interpretation, intentionality, and communication.* Boulder, CO: Westview Press.

Bekoff, M., & Jamieson, D. (Eds.). (1990b). *Interpretation and explanation in the study of animal behavior: Vol. 2. Explanation, evolution, and adaptation.* Boulder, CO: Westview Press.

Bekoff, M., & Jamieson, D. (1991). Reflective ethology, applied philosophy and the moral status of animals. In P. P. G. Bateson & P. H. Klopfer (Eds.), *Perspectives in ethology* (Vol. 9, pp. 1–47). New York: Plenum Press.

Bekoff, M., Townsend, S. E., & Jamieson, D. (1994). Beyond monkey minds: Towards a richer cognitive ethology. *Behavioral and Brain Sciences, 17,* 571–572.

Beloff, J. (1973). *Psychological sciences: A review of modern psychology.* London: Crosby Lockwood Staples.

Benedict, H. (1976, June). Language comprehension in nine fifteen-month-old infants. Lecture presented at University of Stirling Psychology of Language Conference.

Bennett, J. (1971). *Rationality: An essay towards an analysis.* London: Routledge & Kegan Paul. (Original work published 1964.)

Bennett, J. (1978). Commentary on "Cognition and consciousness in nonhuman species." *Behavioral and Brain Sciences, 1,* 559.

Bennett, J. (1991). How to read minds in behaviour: A suggestion from a philosopher. In A. Whiten (Ed.), *Natural theories of mind: Evolution, development and simulation of everyday mindreading* (pp. 97–108). Cambridge, MA: Basil Blackwell.

Bennett, W. L., & Feldman, M. S. (1981). *Reconstructing reality in the courtroom: Justice and judgment in American culture.* New Brunswick, NJ: Rutgers University Press.

Bentham, J. (1976). A utilitarian view. In T. Regan & P. Singer (Eds.), *Animal rights and human obligations* (pp. 129–130). Englewood Cliffs, NJ: Prentice-Hall. (Original work published 1789.)

Berger, P. L., & Luckmann, T. (1967). *The social construction of reality: A treatise in the sociology of knowledge.* Garden City, NY: Doubleday Anchor.

Berry, D. S., Misovich, S. J., Kean, K. J., & Baron, R. M. (1992). Effects of disruption of structure and motion on perceptions of social causality. *Personality and Social Psychology Bulletin, 18,* 237–244.

Bickerton, D. (1990). *Language and species.* Chicago: University of Chicago Press.

Biddle, B. J., & Marlin, M. M. (1987). Causality, confirmation, credulity, and structural equation modeling. *Child Development, 58,* 4–17.

Bierens de Haan, J. A. (1940). *Die Tierischen Instinke und ihr Umbau durch Erfahrung* [Animal instincts and their modification through experience]. Leiden: E. J. Brill.

Bierens de Haan, J. A. (1947a). Animal psychology and the science of animal behaviour. *Behaviour, 1,* 71–80.

Bierens de Haan, J. A. (1947b). *Animal psychology.* London: Hutchinson.

Binet, M. (1889). *The psychic life of micro-organisms.* Chicago: Open Court.

Bitterman, M. E. (1975). The comparative analysis of learning. *Science, 188,* 699–709.

Black, M. (1962). *Models and metaphors: Studies in language and philosophy.* Ithaca, NY: Cornell University Press.

Blough, D. S. (1984). Form recognition in pigeons. In H. L. Roitblat, T. G. Bever, & H. S. Terrace (Eds.), *Animal cognition* (pp. 227–289). Hillsdale, NJ: Lawrence Erlbaum.

Boakes, R. (1984). *From Darwinism to behaviourism*. Cambridge, UK: Cambridge University Press.

Booth, W. C. (1978). Metaphor as rhetoric: The problem of evaluation. *Critical Inquiry, 5*, 49–72.

Boring, E. G. (1950). *A history of experimental psychology* (2nd ed.). New York: Appleton-Century Crofts, Inc.

Bornstein, M. H. (1980). On comparison in psychology. In M. C. Bornstein (Ed.), *Comparative methods in psychology* (pp. 1–7). Hillsdale, NJ: Lawrence Erlbaum.

Boucher, D. H. (1985). The idea of mutualism, past and future. In D. H. Boucher (Ed.), *The biology of mutualism* (pp. 1–28). New York: Oxford University Press.

Bouissac, P. (1989). What is a human? Ecological semiotics and the new animism. *Semiotica, 77*, 497–516.

Boyd, R. (1979). Metaphor and theory change: What is 'metaphor' a metaphor for? In A. Ortony (Ed.), *Metaphor and thought* (pp. 356–408). Cambridge, UK: Cambridge University Press.

Boysen, S. T., & Capaldi, E. J. (Eds.). (1993). *The development of numerical competence*. Hillsdale, NJ: Lawrence Erlbaum.

Bradshaw, J. L., & Rogers, L. J. (1993). *The evolution of lateral asymmetries, language, tool use, and intellect*. New York: Academic Press.

Brainerd, C. J., & Reyna, V. F. (1992). Explaining "memory free" reasoning. *Psychological Science, 3*, 332–339.

Bretherton, I. (1984). Introduction: Piaget and event representation in symbolic play. In I. Bretherton (Ed.), *Symbolic play* (pp. 3–41). New York: Academic Press.

Breuggeman, J. A. (1978). The function of adult play in free-ranging *Macaca mulatta*. In E. O. Smith (Ed.), *Social play in primates* (pp. 169–191). New York: Academic Press.

Bridgeman, P. W. (1927). *The logic of modern physics*. New York: Macmillan.

Bridgeman, P. W. (1938). Operational analysis. *Philosophy of Science, 5*, 114–131.

Broadhurst, P. L. (1964). *The science of animal behavior*. Baltimore: Penguin Books.

Brooks, D. R., & McLennan, D. A. (1990). *Phylogeny, ecology, and behavior*. Chicago: University of Chicago Press.

Brown, R. (1973). *A first language: The early stages*. Cambridge, MA: Harvard University Press.

Bruner, J. (1975). The ontogenesis of speech acts. *Journal of Child Language, 2,* 1–19.

Burghardt, G. M. (1973). Instinct and innate behavior: Toward an ethological psychology. In J. A. Nevin & G. S. Reynolds (Eds.), *The study of behavior* (pp. 321–400). Glenview, IL: Scott, Foresman.

Burghardt, G. M. (1978). Closing the circle: The ethology of mind. *Behavioral and Brain Sciences, 1,* 562–563.

Burghardt, G. M. (1984). On the origins of play. In P. K. Smith (Ed.), *Play in animals and humans* (pp. 5–41). Oxford, UK: Basil Blackwell.

Burghardt, G. M. (1985a). Animal awareness: Current perceptions and historical perspective. *American Psychologist, 40,* 905–919.

Burghardt, G. M. (Ed.). (1985b). *Foundations of comparative ethology.* New York: Van Nostrand Reinhold.

Burghardt, G. M. (1991). Cognitive ethology and critical anthropomorphism: A snake with two heads and hognosed snakes that play dead. In C. A. Ristau (Ed.), *Cognitive ethology: The minds of other animals* (pp. 53–90). Hillsdale, NJ: Lawrence Erlbaum.

Burghardt, G. M. (1992a). Human-bear bonding in research on black bear behavior. In H. Davis & D. Balfour (Eds.), *The inevitable bond* (pp. 365–382). Cambridge, UK: Cambridge University Press.

Burghardt, G. M. (1992b). Looking inside monkey minds: Milestone or millstone? *Behavioral and Brain Sciences, 15,* 150–151.

Burghardt, G. M. (1994). Evolution and the analysis of private experience. *Psycoloquy, 5*(73).

Burghardt, G. M. (1995). Brain imaging, ethology, and the nonhuman mind. *Behavioral and Brain Sciences, 18,* 339–340.

Burghardt, G. M., & Gittleman, J. L. (1990). Comparative behavior and phylogenetic analysis: New wine, old bottles. In M. Bekoff & D. Jamieson (Eds.), *Interpretation and explanation in the study of animal behavior* (Vol. 2, pp. 192–225). Boulder, CO: Westview Press.

Burghardt, G. M. & Herzog, H. A., Jr. (1980). Beyond conspecifics: Is "Brer Rabbit" our brother? *BioScience, 30,* 763–767.

Burghardt, G. M., & Herzog, H. A., Jr. (1989). Animals and evolutionary ethics. In R. J. Hoage (Ed.), *Perceptions of animals* (pp. 129–151). Washington, DC: Smithsonian Institution Press.

Burke, K. (1966). *Language as symbolic action: Essays on life, literature, and method.* Berkeley: University of California Press.

Butterworth, G. E., Harris, P. L., Leslie, A. M., & Wellman, H. M. (Eds.). (1991). *Perspectives on the child's theory of mind*. Oxford: Oxford University Press.

Buytendijk, F. J. J. (1961). *Pain: Its modes and functions*. Chicago: University of Chicago Press. (Original work published 1943.)

Byne, W., & Parsons, B. (1993). Human sexual orientation: The biological theories reappraised. *Archives of General Psychiatry, 50*, 228–239.

Byrne, R. W. (1995). The ape legacy: The evolution of Machiavellian intelligence and anticipatory interactive planning. In E. Goody (Ed.), *Social intelligence and interaction* (pp. 37–53). Cambridge, UK: Cambridge University Press.

Byrne, R. W., & Whiten, A. (1985). Tactical deception of familiar individuals in baboons (*Papio ursinus*). *Animal Behaviour, 33*, 669–673.

Byrne, R. W., & Whiten, A. (Eds.). (1988a). *Machiavellian intelligence: Social expertise and the evolution of intellect in monkeys, apes and humans*. Oxford: Clarendon Press.

Byrne, R. W., & Whiten, A. (1988b). Toward the next generation in data quality: A new survey of primate tactical deception. *Behavioral and Brain Sciences, 11*, 267–271.

Byrne, R. W., & Whiten, A. (1990). Tactical deception in primates: The 1990 database. *Primate Report, 27*, 1–101.

Byrne, R. W., & Whiten, A. (1991). Computation and mindreading in primate tactical deception. In A. Whiten (Ed.), *Natural theories of mind* (pp. 127–142). Oxford: Basil Blackwell.

Byrne, R. W., & Whiten, A. (1992). Cognitive evolution in primates: Evidence from tactical deception. *Man, 27*, 609–627.

Byrne, R. W., Whiten, A., & Henzi, S. P. (1987). One-male groups and intergroup interactions of mountain baboons. *International Journal of Primatology, 8*, 615–633.

Caine, N. G., Earle, H., & Reite, M. (1983). Personality traits of adolescent pigtailed monkeys (*Macaca nemestrina*): An analysis of social rank and early separation experience. *American Journal of Primatology, 4*, 253–260.

Caldwell, R. L. (1986). The deceptive use of reputation by stomatopods. In R. W. Mitchell & N. S. Thompson (Eds.), *Deception: Perspectives on human and nonhuman deceit* (pp. 129–145). Albany: SUNY Press.

Calhoun, S., & Thompson, R. L. (1988). Long-term retention of self-recognition by chimpanzees. *American Journal of Primatology, 15*, 361–365.

Callicott, J. B. (1982). Traditional American Indian and Western European attitudes toward nature: An overview. *Environmental Ethics, 4*, 93–118.

Candland, D. K. (1993). *Feral children and clever animals: Reflections on human nature.* New York: Oxford University Press.

Caporael, L. R. (1986). Anthropomorphism and mechanomorphism: Two faces of the human machine. *Computers in Human Behavior, 2,* 215–234.

Caporael, L. R., Dawes, R. M., Orbell, J. M., & van de Kragt, A. J. C. (1989). Selfishness examined: Cooperation in the absence of egoistic incentives. *Behavioral and Brain Sciences, 12,* 683–739.

Carruthers, P. (1989). Brute experience. *Journal of Philosophy, 86,* 258–269.

Cartmill, M. (1993). The Bambi syndrome. *Natural History, 102*(6), 6–12.

Case, R. (1985). *Intellectual development from infancy to adulthood.* New York: Academic Press.

Catania, A. C. (1992). *Learning* (3rd ed.). Englewood Cliffs, NJ: Prentice-Hall.

Cavalieri, P., & Singer, P. (Eds.). (1993). *The great ape project: Equality beyond humanity.* London: Fourth Estate.

Cenami Spada, E. (1994). Animal mind—human mind: The continuity of mental experience with or without language. *International Journal of Comparative Psychology, 7,* 159–193.

Chandler, M., Fritz, A. S., & Hala, S. (1989). Small-scale deceit: Deception as a marker of two-, three-, and four-year-olds' early theories of mind. *Child Development, 60,* 1263–1277.

Cheney, D. L., & Seyfarth, R. M. (1980). Vocal recognition in free-ranging vervet monkeys. *Animal Behaviour, 28,* 362–367.

Cheney, D. L., & Seyfarth, R. M. (1990). *How monkeys see the world.* Chicago: University of Chicago Press.

Cheney, D. L., & Seyfarth, R. M. (1992). *Précis of How monkeys see the world: Inside the mind of another species. Behavioral and Brain Sciences, 15,* 135–182.

Cheney, D. L., Seyfarth, R. M., & Smuts, B. (1986). Social relationships and social cognition in nonhuman primates. *Science, 234,* 1361–1366.

Chevalier-Skolnikoff, S. (1976). The ontogeny of primate intelligence and its implications for communicative potential: A preliminary report. *Annals of the New York Academy of Sciences, 280,* 173–211.

Chevalier-Skolnikoff, S. (1977). A Piagetian model for describing and comparing socialization in monkey, ape and human infants. In S. Chevalier-Skolnikoff & F. Poirier (Eds.), *Primate biosocial development: Biological, social and ecological determinants* (pp. 159–187). New York: Garland.

Chevalier-Skolnikoff, S., & Poirier, F. (Eds.). (1977). *Primate biosocial development: Biological, social and ecological determinants.* New York: Garland.

Chihara, C. S., & Fodor, J. A. (1981). Operationism and ordinary language: A critique of Wittgenstein. In J. A. Fodor (Ed.), *Representations* (pp. 35–62). Cambridge, MA: MIT Press. (Original work published 1965.)

Chomsky, N. (1982). Discussions of Putnam's comments. In M. Piattelli-Parmarini (Ed.), *Language and learning: The debate between Jean Piaget and Noam Chomsky* (pp. 310–324). Cambridge, MA: Harvard University Press.

Churchland, P. M. (1981). Eliminative materialism and propositional attitudes. *Journal of Philosophy, 78,* 67–90.

Churchland, P. M. (1988). *Matter and consciousness.* Cambridge, MA: MIT Press.

Churchland, P. S. (1986). *Neurophilosophy: Toward a unified theory of mind.* Cambridge, MA: MIT Press.

Clark, S. R. L. (1991, December). Not so dumb friends. *The Times Literary Supplement, 13,* 5–6.

Colgan, P. (1989). *Animal motivation.* New York: Chapman & Hall.

Collins, H., & Pinch, T. (1993). *The Golem: What everyone should know about science.* New York: Cambridge University Press.

Cooley, C. H. (1912). *Human nature and the social order.* New York: Charles Scribner's Sons.

Cosmides, L. (1989). The logic of social exchange: Has natural selection shaped how humans reason? *Cognition, 31,* 187–276.

Coussi-Korbel, S. (1994). Learning to outwit a competitor in mangabeys (*Cercocebus t. torquatus*). *Journal of Comparative Psychology, 108,* 164–171.

Craik, K. J. W. (1943). *The nature of explanation.* Cambridge, UK: Cambridge University Press.

Crider, C., & Cirillo, L. (1992). Systems of interpretation and the function of metaphor. *Journal for the Theory of Social Behaviour, 21,* 171–195.

Crocker, D. R. (1984). Anthropomorphism: Bad practice, honest prejudice? In G. Ferry (Ed.), *The understanding of animals* (pp. 304–313). Oxford: Basil Blackwell & New Scientist.

Cronin, H. (1992, November 1). Review of Griffin's *Animal Minds. New York Times, 142*(49137), sect. 7, 14.

Crook, J. H. (1980). *The evolution of human consciousness.* Oxford: Clarendon Press.

Darwin, C. (1872). *The expression of the emotions in man and animals*. London: J. Murray.

Darwin, C. (1896). *The descent of man and selection in relation to sex* (2nd ed.). New York: D. Appleton & Co. (Original work published 1874.)

Darwin, C. (1968). *The origin of species by means of natural selection; or, the preservation of favoured races in the struggle for life*. Harmondsworth, Middlesex, UK: Penguin. (Original work published 1859.)

Darwin, C. (1981). *The descent of man; and selection in relation to sex*. Princeton: Princeton University Press. (Original work published 1871.)

Dasser, V., Ulbaek, I., & Premack, D. (1989). The perception of intention. *Science, 243*, 365–367.

Davidson, K., & Hopson, J. (1988, April 10). Gorilla business. *San Francisco Chronicle, Image section*, 14–18, 33–36.

Davis, H. (1992). Logical transitivity in animals. In W. K. Honig & G. Fetterman (Eds.), *Cognitive aspects of stimulus control* (pp. 405–429). Hillsdale, NJ: Lawrence Erlbaum.

Davis, H., & Albert, M. (1986). Numerical discrimination by rats using sequential auditory stimuli. *Animal Learning and Behavior, 14*, 57–59.

Davis, H. & Balfour, D. (Eds.). (1992a). *The inevitable bond*. Cambridge, UK: Cambridge University Press.

Davis, H., & Balfour, D. (1992b). The inevitable bond. In H. Davis & D. Balfour (Eds.), *The inevitable bond: Examining scientist-animal interactions* (pp. 1–5). New York: Cambridge University Press.

Davis, H., & Bradford, S. A. (1986). Counting behavior by rats in a simulated natural environment. *Ethology, 73*, 265–280.

Davis, H., & Pérusse, R. (1988). Numerical competence in animals: Definitional issues, current evidence and a new research agenda. *Behavioral and Brain Sciences, 11*, 561–616.

Davis, W. (1989). Finding symbols in history. In H. Morphy (Ed.), *Animal into art* (pp. 179–189). London: Unwin Hyman.

Dawes, R. M. (1988). *Rational choice in an uncertain world*. New York: Harcourt Brace Jovanovich.

Dawkins, M. S. (1989). The future of ethology: How many legs are we standing on? In P. P. G. Bateson & P. H. Klopfer (Eds.), *Perspectives in Ethology* (vol. 8, pp. 47–54). New York: Plenum Press.

Dawkins, R. (1976). *The selfish gene*. Oxford: Oxford University Press.

Dawkins, R. (1981). In defence of selfish genes. *Philosophy, 56,* 556–573.

Dawkins, R. & Krebs, J. R. (1978). Animal signals: Information or manipulation? In J. R. Krebs & N. B. Davies (Eds.), *Behavioural ecology: An evolutionary approach* (pp. 282–309). Oxford: Blackwell.

Dennett, D. C. (1978). *Brainstorms.* Montgomery, VT: Bradford Books.

Dennett, D. C. (1983). Intentional systems in cognitive ethology: The "Panglossian paradigm" defended. *Behavioral and Brain Sciences, 6,* 343–390.

Dennett, D. C. (1987a). *The intentional stance.* Cambridge, MA: MIT Press.

Dennett, D. C. (1987b). Reflections: Interpreting monkeys, theorists, and genes. In *The intentional stance* (pp. 269–286). Cambridge, MA: MIT Press.

Dennett, D. C. (1988). Why creative intelligence is hard to find. *Behavioral and Brain Sciences, 11,* 253.

Dennett, D. C. (1991). *Consciousness explained.* Boston: Little, Brown & Co.

Descartes, R. (1927). Letter to Henry More. In R. M. Eaton (Ed.), *Descartes: Selections* (pp. 358–360). New York: Charles Scriber's Sons. (Original work published 1649.)

Descartes, R. (1955). *The philosophical works of Descartes* (2 vols.). New York: Dover Books. (Original works published after 1633.)

Dettwyler, K. A. (1991). Can paleopathology provide evidence for "compassion"? *American Journal of Physical Anthropology, 84,* 375–384.

DeVore, I. (1985). Foreword to B. B. Smuts, *Sex and friendship in baboons.* New York: Aldine.

Dewsbury, D. A. (1992). On the problems studied in ethology, comparative psychology, and animal behavior. *Ethology, 92,* 89–107.

Dewsbury, D. A. (1993). On publishing controversy. *American Psychologist, 48,* 869–877.

Dickens, C. (1964). *Bleak House.* New York: New American Library. (Original work published 1853.)

Dickens, C. (1980). *David Copperfield.* New York: New American Library. (Original work published 1850.)

Doherty, M. E., Tweney, R. D., & Mynatt, C. R. (1981). Null hypothesis testing, confirmation bias and strong inference. In R. D. Tweney, M. E. Doherty, & C. R. Mynan (Eds.), *On scientific thinking* (pp. 262–267). New York: Columbia University Press.

Dolhinow, P., & Bishop, N. (1970). The development of motor skills and social relationships among primates through play. *Minnesota Symposium on Child Psychology, 4,* 141–198.

Donceel, S. J. (1967). *Philosophical anthropology.* Kansas City: Sheeds, Andrews & McMeel.

Doré, F. Y., & Dumas, C. (1987). Psychology of animal cognition: Piagetian studies. *Psychological Bulletin, 102,* 219–233.

Dretske, F. (1988). *Explaining behavior.* Cambridge, MA: MIT Press.

Driscoll, J. W. (1992). Attitudes toward animal research. *Anthrozoös, 5,* 32–39.

Dube, W. V., McIlvane, W. J., & Green, G. (1992). An analysis of generalized identity matching-to-sample test procedures. *Psychological Record, 42,* 17–28.

Dunbar, R. I. M. (1984). *Reproductive decisions.* Princeton: Princeton University Press.

Dunlop, J. (1989). Moral reasoning about animal treatment. *Anthrozoös, 5,* 245–258.

Eddy, T. J., Gallup, G. G., Jr., & Povinelli, D. J. (1993). Attribution of cognitive states to animals: Anthropomorphism in comparative perspective. *Journal of Social Issues, 49,* 87–101.

Edwards, D., & Potter, J. (1993). Language and causation: A discursive action model of description and attribution. *Psychological Review, 100,* 23–41.

Eglash, A. R., & Snowdon, C. T. (1983). Mirror-image responses in pygmy marmosets (*Cebuella pygmaea*). *American Journal of Primatology, 5,* 211–219.

Eibl-Eibesfeldt, I. (1983). The comparative approach in human ethology. In D. W. Rajecki (Ed.), *Comparing behavior: Studying man studying animals* (pp. 43–65). Hillsdale, NJ: Lawrence Erlbaum.

Ekman, P. (1973). Cross-cultural studies of facial expression. In P. Ekman (Ed.), *Darwin and facial expression: A century of research in review* (pp. 169–222). New York: Academic Press.

Elstein, A. S., Shulman, L. S., & Sprafka, S. A. (1978). *Medical problem solving.* Cambridge, MA: Harvard University Press.

Empson, W. (1947). *Seven types of ambiguity.* New York: New Directions.

Epstein, R. (1987). Reflections on thinking in animals. In G. Greenberg & E. Tobach (Eds.), *Cognition, language, and consciousness: Integrative levels* (pp. 19–29). Hillsdale, NJ: Lawrence Erlbaum.

Essock-Vitale, S., & Seyfarth, R. M. (1987). Intelligence and social cognition. In B. Smuts, D. L. Cheney, R. M. Seyfarth, R. W. Wrangham, & T. T. Struhsaker (Eds.), *Primate societies* (pp. 452–461). Chicago: University of Chicago Press.

Estep, D. Q., & Hetts, S. (1992). Interactions, relationships, and bonds: The conceptual basis for scientist-animal interactions. In H. Davis & D. Balfour (Eds.), *The inevitable bond: Examining scientist-animal interactions* (pp. 6–26). New York: Cambridge University Press.

Etienne, B. (1989). *La France et l'Islam* [France and Islam]. Paris: Machette.

Faust, D. (1984). *The limits of scientific reasoning*. Minneapolis: University of Minnesota Press.

Feigl, H. (1945). Operationism and scientific method. *Psychological Review, 52,* 250–259.

Feinberg, R. R., Miller, F. G., & Ross, G. A. (1981). Perceived and actual locus of control similarity in friends. *Personality and Social Psychology Bulletin, 7,* 85–89.

Fentress, J. C. (1992). The covalent animal: On bonds and their boundaries in behavioral research. In H. Davis & D. Balfour (Eds.), *The inevitable bond: Examining scientist-animal interactions* (pp. 44–71). New York: Cambridge University Press.

Feyerabend, P. K. (1962). Explanation, reduction and empiricism. *Minnesota Studies in the Philosophy of Science, 3,* 28–80.

Feyerabend, P. K. (1975). *Against method*. Atlantic Highlands, NJ: Humanities Press.

Fisher, J. A. (1990). The myth of anthropomorphism. In M. Bekoff & D. Jamieson (Eds.), *Interpretation and explanation in the study of animal behavior* (Vol. 1, pp. 96–116). Boulder, CO: Westview Press.

Fisher, J. A. (1991). Disambiguating anthropomorphism: An interdisciplinary review. In P. P. G. Bateson & P. H. Klopfer (Eds.), *Perspectives in Ethology* (Vol. 9, pp. 49–85). New York: Plenum.

Fisler, G. F. (1967). Nonbreeding activities of three adult males in a band of free-ranging rhesus monkeys. *Journal of Mammalogy, 48,* 70–78.

Flavell, J. H., Botkin, P., Fry, C., Wright, J., & Jarvis, P. (1968). *The development of role-taking and communication skills in children*. New York: Wiley.

Fodor, J. A. (1974). Special sciences. *Synthese, 28,* 77–115.

Fodor, J. A. (1975). *The language of thought*. Cambridge, MA: Harvard University Press.

Fodor, J. A., & Lapore, E. (1992). *Holism: A shopper's guide.* Oxford: Blackwell.

Forguson, L. (1989). *Common sense.* London: Routledge.

Fossey, D. (1983). *Gorillas in the mist.* Boston: Houghton Mifflin.

Fouts, R. S. (1973). Acquisition and testing of gestural signs in four young chimpanzees. *Science, 180,* 978–980.

Fouts, R. S. (1974, July). Artificial and human language acquisition in the chimpanzees. In D. A. Hamburg & J. Goodall (Chairs), *The Behavior of Great Apes,* Wenner-Gren Foundation for Anthropological Research, New York, NY.

Fouts, R. S., & Budd, R. L. (1979). Artificial and human language acquisition in the chimpanzee. In D. A. Hamburg & E. R. McCown (Eds.), *The great apes* (p. 375–392). Menlo Park, CA: Benjamin/Cummings.

Fouts, R. S., & Couch, J. B. (1976). Cultural evolution of learned language in chimpanzees. In M. Hahn & E. Simmel (Eds.), *Communicative behavior and evolution* (pp. 141–161). New York: Academic.

Fouts, R. S., Fouts, D. H., & Van Cantfort, T. E. (1989). The infant Loulis learns signs from cross-fostered chimpanzees. In R. A. Gardner, B. T. Gardner, & T. E. Van Cantfort (Eds.), *Teaching sign language to chimpanzees* (pp. 280–307). Albany: SUNY Press.

Fox, M. W. (1978). *The dog: Its domestication and behavior.* New York: Garland.

Fox, M. W. (1985). Empathy, humanness and animal welfare. In M. W. Fox & L. D. Mickley (Eds.), *Advances in animal welfare science, 1984* (pp. 61–73). Boston: Martinos Nijhoff.

Fraser, A. F. (1974). *Farm animal behaviour.* London: Balliere Tindall.

Fraser, A. F. (1980). Ethics and ethology. *Applied Animal Ethology, 2,* 211–212.

Frayer, D. W., & Montet-White, A. (1989). Commentary on Gargett's "Grave shortcomings." *Current Anthropology, 30,* 180–181.

Freeman, D. (1979). *The great apes.* New York: Putnam.

Freeman, K. (1948). *Ancilla to the pre-Socratic philosophers. A complete translation of the Fragments in Diels, Fragmente der Vorsokratiker.* Oxford: Basil Blackwell.

Freud, S. (1964). *The future of an illusion.* Garden City: Anchor Books. (Original work published 1927.)

From, F. (1971). *Perception of other people.* New York: Columbia University Press.

Furnam, A. (1988). *Lay theories: Everyday understandings of problems in the social sciences*. Oxford: Pergamon.

Furness, W. H. (1916). Observations on the mentality of chimpanzees and orangutans. *Proceedings from the American Philosophical Society, 55*, 281–290.

Galdikas-Brindamour, B. (1975). Orangutans, Indonesia's "people of the forest." *National Geographic, 148*, 444–473.

Galef, B. G., Jr. (1990). An adaptationist perspective on social learning, social feeding, and social foraging in Norway rats. In D. A. Dewsbury (Ed.), *Contemporary issues in comparative psychology* (pp. 55–79). Sunderland, MA: Sinauer.

Gallup, G. G., Jr. (1970). Chimpanzees: Self-recognition. *Science, 167*, 86–87.

Gallup, G. G., Jr. (1975). Towards an operational definition of self-awareness. In R. H. Tuttle (Ed.), *Socioecology and psychology of primates* (pp. 309–342). The Hague, Netherlands: Mouton.

Gallup, G. G., Jr. (1977a). Absence of self-recognition in a monkey (*Macaca fascicularis*) following prolonged exposure to a mirror. *Developmental Psychobiology, 10*, 281–284.

Gallup, G. G., Jr. (1977b). Self-recognition in primates: A comparative approach to the bidirectional properties of consciousness. *American Psychologist, 32*, 329–338.

Gallup, G. G., Jr. (1979a). Self-awareness in primates. *American Scientist, 67*, 417–421.

Gallup, G. G., Jr. (1979b). Self-recognition in chimpanzees and man: A developmental and comparative perspective. In M. Lewis & L. A. Rosenblum (Eds.), *The child and its family* (pp. 107–126). New York: Plenum Press.

Gallup, G. G., Jr. (1982). Self-awareness and the emergence of mind in primates. *American Journal of Primatology, 2*, 237–248.

Gallup, G. G., Jr. (1983). Toward a comparative psychology of mind. In R. L. Mellgren (Ed.), *Animal cognition and behavior* (pp. 473–510). Amsterdam: North-Holland.

Gallup, G. G., Jr. (1985). Do minds exist in species other than our own? *Neuroscience and Biobehavioral Reviews, 9*, 631–641.

Gallup, G. G., Jr. (1988). Toward a taxonomy of mind in primates. *Behavioral and Brain Sciences, 11*, 255–256.

Gallup, G. G., Jr. (1991). Toward a comparative psychology of self-awareness: Species limitations and cognitive consequences. In J. Strauss & A. Goethals (Eds.) *The self: Interdisciplinary approaches* (pp. 121–135). New York: Springer-Verlag.

Gallup, G. G., Jr. (1992). The reflective mind: An alternative approach to animal cognition. *International Journal of Comparative Psychology, 6,* 106–108.

Gallup, G. G., Jr. (1994). Self-recognition: Research strategies and experimental design. In S. T. Parker, R. W. Mitchell, & M. L. Boccia (Eds.), *Self-awareness in animals and humans* (pp. 35–50). New York: Cambridge University Press.

Gallup, G. G., Jr., & Beckstead, J. W. (1988). Attitudes toward animal research. *American Psychologist, 43,* 474–476.

Gallup, G. G., Jr., Boren, J. L., Gagliardi, G. J., & Wallnau, L. B. (1977). A mirror for the mind of man, or will the chimpanzee create an identity crisis for *Homo sapiens? Journal of Human Evolution, 6,* 303–313.

Gallup, G. G., Jr., McClure, M. K., Hill, S. D., & Bundy, R. A. (1971). Capacity for self-recognition in differentially reared chimpanzees. *Psychological Record, 21,* 69–74.

Gallup, G. G., Jr., & Povinelli, D. J. (1993). Mirror, mirror on the wall which is the most heuristic theory of them all? A response to Mitchell. *New Ideas in Psychology, 11,* 326–334.

Gallup, G. G., Jr. & Suarez, S. D. (1983). Overcoming our resistance to animal research: Man in comparative perspective. In D. W. Rajecki (Ed.), *Comparing behavior: Studying man studying animals* (pp. 5–26). Hillsdale, NJ: Lawrence Erlbaum.

Gallup, G. G., Jr., & Suarez, S. D. (1986). Self-awareness and the emergence of mind in humans and other primates. In J. Suls & A. G. Greenwald (Eds.), *Psychological perspectives on the self* (Vol. 3, pp. 3–26). Hillsdale, NJ: Lawrence Erlbaum.

Gallup, G. G., Jr., Wallnau, L. B., & Suarez, S. D. (1980). Failure to find self-recognition in mother-infant and infant-infant rhesus monkey pairs. *Folia Primatologica, 33,* 210–219.

Galvin, S. L., & Herzog, H. A., Jr. (1992). Ethical ideology, animal rights activism, and attitudes toward the treatment of animals. *Ethics and Behavior, 2,* 141–149.

Galvin, S. L., & Herzog, H. A., Jr. (1993). The ethical judgment of animal research. *Ethics and Behavior, 2,* 263–286.

Garcia, J. (1981). Tilting at the paper mills of academe. *American Psychologist, 36,* 149–158.

Gardner, R. A., & Gardner, B. T. (1969). Teaching sign language to a chimpanzee. *Science, 165,* 664–672.

Gardner, R. A., & Gardner, B. T. (1975). Early signs of language in child and chimpanzee. *Science, 187,* 752–753.

Gardner, R. A., & Gardner, B. T. (1978). Comparative psychology and language acquisition. *Annals of the New York Academy of Sciences, 309,* 37–76.

Gardner, R. A., Gardner, B. T., & Van Cantfort, T. E. (Eds.). (1989). *Teaching sign language to chimpanzees.* Albany: SUNY Press.

Gargett, R. H. (1989). Grave shortcomings: The evidence for Neanderthal burial. *Current Anthropology, 30,* 157–190.

Garner, R. L. (1892). *The speech of monkeys.* New York: Charles L. Webster & Co.

Garner, R. L. (1900). *Apes and monkeys: Their life and language.* Boston: Ginn & Co.

Gavan, J. A. (1971). Longitudinal postnatal growth in chimpanzee. In G. H. Bourne (Ed.), *Behavior, growth, and pathology of chimpanzees* (Vol. 4, pp. 46–102). Basel, Switzerland: Karger.

Gelman, R., & Gallistel, C. R. (1978). *The child's understanding of number.* Cambridge, MA: Harvard University Press.

Gendlin, E. T. (1962). *Experiencing and the creation of meaning.* Toronto: Free Press of Glencoe.

Gentner, D. (1989). The mechanisms of analogical reasoning. In S. Vosniadou & A. Ortony (Eds.), *Similarity and analogical reasoning* (pp. 199–241). Cambridge, UK: Cambridge University Press.

Gibbon, J., & Church, R. M. (1984). Sources of variance in an information processing theory of timing. In H. L. Roitblat, T. G. Bever, & Terrace, H. S. (Eds.), *Animal cognition* (pp. 465–488). Hillsdale, NJ: Lawrence Erlbaum.

Gibson, K. R. (1981). Comparative neuro-ontogeny: Its implications for the development of human intelligence. In G. Butterworth (Ed.), *Infancy and epistemology* (pp. 52–82). Brighton, UK: Harvester Press.

Gibson, K. R. (1990). New perspectives on instincts and intelligence: Brain size and the emergence of hierarchical mental construction skills. In S. T. Parker & K. R. Gibson (Eds.), *"Language" and intelligence in monkeys and apes* (pp. 97–128). New York: Cambridge University Press.

Gibson, K. R. (1991). Review of Wallman's *Aping language. Anthropological Linguistics, 33,* 332–334.

Gisiner, R., & Schusterman, R. J. (1992). Sequence, syntax and semantics: Responses of a language-trained sea lion (*Zalophus californianus*) to novel sign combinations. *Journal of Comparative Psychology, 106,* 78–91.

Gittleman, J. L., & Luh, H.-K. (1993). Phylogeny, evolutionary models, and comparative methods: A simulation study. In P. Eggleton & D. Vane-Wright (Eds.), *Pattern and process: Phylogenetic approaches to ecological problems* (pp. 241–298). London: Academic Press.

Gladwin, T., & Sarason, S. B. (1953). *Truk: Man in paradise.* Chicago: University of Chicago Press.

Glass, A. J., & Holyoak, K. J. (1986). *Cognition.* New York: Random House.

Goldiamond, I. (1966). Perception, language, and conceptualization rules. In B. Kleinmuntz (Ed.), *Problem solving* (pp. 183–224). New York: John Wiley.

Goldman-Rakic, P. S., & Preuss, T. M. (1987). Wither comparative psychology? *Behavioral and Brain Sciences, 10,* 666–667.

Goodall, J. (1967). *My friends, the wild chimpanzees.* Washington, DC: National Geographic Society.

Goodall, J. (1971). *In the shadow of man.* Boston: Houghton Mifflin.

Goodall, J. (1975). The chimpanzee. In V. Goodall (Ed.), *The quest for man* (pp. 131–170). New York: Praeger.

Goodall, J. van Lawick- (1976a). Early tool use in chimpanzees. In J. Bruner, A. Jolly, & K. Sylva (Eds.), *Play* (pp. 222–225). New York: Basic Books.

Goodall, J. van Lawick- (1976b). Mother chimpanzees play with their infants. In J. Bruner, A. Jolly, & K. Sylva (Eds.), *Play* (pp. 262–266). New York: Basic Books.

Goodall, J. van Lawick- (1976c). Sibling relations and play among wild chimpanzees. In J. Bruner, A. Jolly, & K. Sylva (Eds.), *Play* (pp. 300–310). New York: Basic Books.

Goodall, J. (1986). *The chimpanzees of Gombe: Patterns of behavior.* Cambridge, MA: Harvard University Press.

Goodall, J. (1988). *In the shadow of man* (rev. ed.). Boston: Houghton Mifflin.

Goodman, N. (1978). *Ways of worldmaking.* Cambridge, MA: Hackett.

Gopnik, A., & Graf, P. (1988). Knowing how you know: Young children's ability to identify and remember the sources of their beliefs. *Child Development, 59,* 1366–1371.

Gordon, W. J. J. (1974). Some source material in discovery-by-analogy. *Journal of Creative Behavior, 8,* 239–257.

Gould, S. J. (1977). *Ontogeny and phylogeny.* Cambridge, MA: Harvard University Press.

Gould, S. J. (1987). Animals and us. *New York Review of Books, 34*, 11, 20–25.

Graham, P., & Rutter, M. (1985). Adolescent disorders. In M. Rutter & L. Hersov (Eds.), *Child and adolescent psychiatry: Modern approaches* (pp. 351–368). New York: Blackwell Scientific.

Gray, P. H. (1963). The Morgan-Romanes controversy. *Proceedings of the Montana Academy of Science, 23*, 225–230.

Greenfield, P., & Savage-Rumbaugh, S. (1990). Grammatical combination in *Pan paniscus*: Processes of learning and invention in the evolution and development of language. In S. T. Parker & K. R. Gibson (Eds.), *"Language" and intelligence in monkeys and apes* (pp. 540–577). New York: Cambridge University Press.

Greenfield, P., & Smith, J. (1976). *The structure of communication in early development.* New York: Academic Press.

Griffin, D. R. (1976). *The question of animal awareness. Evolutionary continuity of mental experience.* New York: Rockefeller University Press.

Griffin, D. R. (1977). Anthropomorphism. *BioScience, 27*, 445–446.

Griffin, D. R. (1978). Prospects for a cognitive ethology. *Behavioral and Brain Sciences, 4*, 527–538.

Griffin, D. R. (1981). *The question of animal awareness* (rev. ed). New York: Rockefeller University Press.

Griffin, D. R. (1984). *Animal thinking.* Cambridge, MA: Harvard University Press.

Griffin, D. R. (1987). *How can animal thinking be studied scientifically?* Paper presented at Animal Behavior Society Meeting, Williamstown, MA.

Griffin, D. R. (1990). Foreword. In M. Bekoff & D. Jamieson (Eds.), *Interpretation and explanation in the study of animal behavior* (Vol. 1, pp. xii–xviii). Boulder, CO: Westview Press.

Griffin, D. R. (1992). *Animal minds.* Chicago: University of Chicago Press.

de Grolier, E. (1989). Glossogenesis in endolinguistic and exolinguistic perspective: Paleoanthropological data. In J. Wind, E. G. Pulleyblank, E. de Grolier & B. H. Bichakjian (Eds.), *Studies in language origins* (Vol. 1, pp. 73–138). Amsterdam: John Benjamins.

Gruber, H. E. (1974). *Darwin on man: A psychological study of scientific creativity.* (Together with Darwin's Early and Unpublished Notebooks.) London: Wildwood House.

Guralnick, M. J., & Paul-Brown, D. (1977). The nature of verbal interactions among handicapped and non-handicapped preschool children. *Child Development, 48,* 254–260.

Gustafson, D. (1986). Review of Griffin's *Animal thinking. Environmental Ethics, 8,* 179–182.

Guthrie, S. E. (1980). A cognitive theory of religion. *Current Anthropology, 21,* 181–203.

Guthrie, S. E. (1993). *Faces in the clouds: A new theory of religion.* New York: Oxford University Press.

Gyger, M., Karakashian, S. J., & Marler, P. (1986). Avian alarm calling: Is there an audience effect? *Animal Behaviour, 34,* 1570–1572.

Gyger, M., Marler, P., & Pickert, R. (1987). Semantics of an avian alarm call system: The male domestic fowl, *Gallus domesticus. Behaviour, 102,* 15–40.

Haeckel, E. (1906). *The evolution of man: A popular scientific study* (5th ed.). New York: The Truth Seeker Co. (Original work published 1874.)

Hafez, E. S. E. (Ed.). (1969). *The behaviour of domestic animals.* London: Balliere, Tindall & Cassell.

Hailman, J. P. (1978). Review of D. R. Griffin's *The question of animal awareness. Auk, 95,* 614–615.

Hall, G. S. (1908). A glance at the phyletic background of genetic psychology. *American Journal of Psychology, 19,* 149–212.

Hamilton, W. D. (1964). The genetical theory of social behavior, I, II. *Journal of Theoretical Biology, 7,* 1–52.

Hamilton, W. D. (1971). Geometry for the selfish herd. *Journal of Theoretical Biology, 31,* 295–311.

Handler, P. (Ed.). (1970). *Biology and the future of man.* New York: Oxford University Press.

Hanlon, R. T. (1982). The functional organization of chromatic and iridescent cells in the body patterning of *Loligo plei* (Cephalopoda: *Myopsida). Malacologia, 23,* 89–119.

Hansen, B. (1986). The complementarity of science and magic before the scientific revolution. *American Scientist, 74,* 128–136.

Hanson, N. R. (1958). *Patterns of discovery.* Cambridge, UK: Cambridge University Press.

Hanson, N. R. (1969). *Perception and discovery*. San Francisco: Freeman, Cooper.

Haraway, D. (1989). *Primate visions: Gender, race, and nature in the world of modern science*. New York: Routledge.

Harcourt, A. H., & de Waal, F. B. M. (Eds.). (1992). *Coalitions and alliances in humans and other animals*. Oxford: Oxford University Press.

Hardy, A. (1956). *The open sea, Its natural history: Part 1, The world of plankton*. London: Collins.

Harlow, H. (1969). Age-mate or peer affectional systems. In D. Lehrmann, R. A. Hinde, & E. Shaw (Eds.), *Advances in the study of behavior* (Vol. 2, pp. 287–334). New York: Academic Press.

Harris, B. (1979). Whatever happened to Little Albert? *American Psychologist, 34,* 151–160.

Harris, R. (1984). Comment, Rejoinder 1. In R. Harré & V. Reynolds (Eds.), *The meaning of primate signals* (pp. 174–175). Cambridge, UK: Cambridge University Press.

Harrison, B. (1962). *Orang-utan*. London: Collins.

Harrison, R. (1964). *Animal machines: The new factory farming industry*. London: Vincent Stuart.

Hart, D., & Fegley, S. (1994). Social imitation and the emergence of a mental model of self. In S. T. Parker, R. W. Mitchell, & M. L. Boccia (Eds.) *Self-awareness in animals and humans* (pp. 149–165). New York: Cambridge University Press.

Harvey, P. H., & Pagel, M. D. (1991). *The comparative method in evolutionary biology*. Oxford: Oxford University Press.

Hauser, M. D. (1988). How infant vervet monkeys learn to recognize starling alarm calls: The role of experience. *Behaviour, 106,* 187–201.

Hausfater, G., & Hrdy, S. B. (Eds.). (1984). *Infanticide: Comparative and evolutionary perspectives*. Hawthorne, NY: Aldine.

Hawking, S. W. (1988). *A brief history of time*. New York: Bantam Books.

Hayes, C. (1951). *The ape in our house*. New York: Harper.

Hearne, V. (1986). *Adam's task: Calling animals by name*. New York: Alfred A. Knopf.

Hebb, D. O. (1946). Emotion in man and animal: An analysis of the intuitive process of recognition. *Psychological Review, 53,* 88–106.

Hebb, D. O. (1949). *The organization of behavior*. New York: John Wiley & Sons.

Hediger, H. (1955). *Studies in the psychology and behaviour of animals in zoos and circuses.* London: Buttersworth.

Heidegger, M. (1962). *Being and time.* New York: Harper & Row.

Heider, F. (1958). *The psychology of interpersonal relations.* New York: Wiley.

Heider, F., & Simmel, M. (1944). An experimental study of apparent behavior. *American Journal of Psychology, 57,* 243–259.

Heinroth, O. (1911). Beiträge zur Biologie, namentlich Ethologie und Psychologie der Anatiden [Contributions to the biology, namely, the ethology and psychology of Anatids]. *Verhandlungen des V Internationalen Ornithologen-Kongresses* (1910) Berlin, 589–702.

Heinroth, O., & Heinroth, K. (1924–1933). *Die Vögel Mitteleuropas, in allen Lebens—und Entwicklungsstufen photographisch aufgenommen und in ihrem Seelenleben bei der Aufzucht vom Ei ab beobachtet* [The birds of central Europe photographed in all their life and development, and observations of their inner life in the raising of their hatchlings]. (4 vols.). Berlin: H. Bermuhler.

Hempelmann, F. (1926). *Tierpsychologie vom Standpunkte des Biologen* [Animal psychology from the standpoint of biology]. Leipzig: Akademische Verlagsgesellschatt M. B. H.

Hennig, W. (1966). *Phylogenetic systematics.* Urbana: University of Illinois Press.

Herman, L. M., & Morrel-Samuels, P. (1990). Knowledge acquisition and asymmetry between language comprehension and production: Dolphins and apes as general models for animals. In M. Bekoff & D. Jamieson (Eds.), *Interpretation and explanation in the study of animal behavior* (Vol. 1, pp. 283–312). Boulder, CO: Westview Press.

Herman, L. M., Richards, D. G., & Wolz, J. P. (1984). Comprehension of sentences by bottlenosed dolphins. *Cognition, 16,* 129–219.

Herrnstein, R. J., & Loveland, D. H. (1964). Complex visual concept in the pigeon. *Science, 146,* 549–551.

Herzog, H. A., Jr. (1988). The moral status of mice. *American Psychologist, 43,* 473–474.

Herzog, H. A., Jr. (1993). "The movement is my life": The psychology of animal rights activism. *Journal of Social Issues, 49,* 103–119.

Herzog, H. A., Jr., Betchart, N. S., & Pittman, R. B. (1991). Gender, sex role orientation and attitudes toward animals. *Anthrozoös, 4,* 184–191.

Hess, L. (1954). *Christine the baby chimp.* London: Bell.

Hesse, M. (1988). The cognitive claims of metaphor. *The Journal of Speculative Philosophy, 2,* 1–16.

Heyes, C. (1987a). Cognisance of consciousness in the study of animal knowledge. In W. Callebaut & R. Pinxten (Eds.), *Evolutionary epistemology* (pp. 105–136). The Netherlands: D. Reidel.

Heyes, C. (1987b). Contrasting approaches to the legitimation of intentional language within comparative psychology. *Behaviorism, 15,* 41–50.

Heyes, C. (1993). Anecdotes, training, trapping and triangulating: Do animals attribute mental states? *Animal Behaviour, 46,* 177–188.

Hill, S. D., Bundy, R. A., Gallup, G. G., Jr., & McClure, M. K. (1970). Responsiveness of young nursery reared chimpanzees to mirrors. *Proceedings of the Louisiana Academy of Sciences, 33,* 77–82.

Hinde, R. A. (1974). *Constraints on learning.* Cambridge, UK: Cambridge University Press.

Hinde, R. A. (1992). Developmental psychology in the context of other behavioral sciences. *Developmental Psychology, 28,* 1018–1029.

Hirsch, E. D., Jr. (1985). Derrida's axioms. In K. Miller (Ed.), *London reviews* (pp. 27–36). London: Chatto & Windus.

Hobhouse, L. T. (1901). *Mind in evolution.* London: Macmillan & Co.

Hobhouse, L. T. (1915). *Mind in evolution,* 2nd ed. London: Macmillan & Co.

Hockett, C. F. (1960). Logical considerations in the study of animal communication. In W. E. Lanyon & W. N. Tavolga (Eds.), *Animal sounds and communication* (pp. 392–430). Washington, DC: American Institute of Biological Sciences.

Hockett, C. F., & Altmann, S. A. (1968). A note on design features. In T. A. Sebeok (Ed.), *Animal communication: Techniques of study and results of research* (pp. 61–72). Bloomington: Indiana University Press.

Hodges, B. H., & Baron, R. M. (1992). Values as constraints on affordances: Perceiving and acting properly. *Journal for the Theory of Social Behaviour, 22,* 263–294.

Hodos, W., & Campbell, C. B. G. (1969). Scala naturae: Why there is no theory in comparative psychology. *Psychological Review, 76,* 337–350.

Hoffman, M. L. (1983). Empathy, guilt, and social cognition. In W. F. Overton (Ed.), *The relationship between social and cognitive development* (pp. 1–52). Hillsdale, NJ: Lawrence Erlbaum.

Hoffman, R. R. (1985). Some implications of metaphor for philosophy and psychology of science. In W. Paprotte & R. Dirven (Eds.), *The ubiquity of metaphor: Metaphor in language and thought* (pp. 327–380). Amsterdam: John Benjamins.

Hoffmann, R. (1993). For the first time, you can see atoms. *American Scientist, 81,* 11–12.

Honig, W. K. (1984). Contributions of animal memory and the interpretation of animal learning. In H. L. Roitblat, T. G. Bever, & H. S. Terrace (Eds.), *Animal cognition* (pp. 29–44). Hillsdale, NJ: Lawrence Erlbaum.

Houghton, W. E. (1957). *The Victorian frame of mind.* New Haven: Yale University Press.

Houston, A. I. (1990). Review of Colgan's *Animal motivation. Quarterly Review of Biology, 65,* 383.

Howes, D. (Ed.). (1991). *The varieties of sensory experience.* Toronto: University of Toronto Press.

Hoyt, A. M. (1941). *Toto and I: A gorilla in the family.* New York: Lippincott.

Hrdy, S. B. (1977). *The langurs of Abu: Female and male strategies of reproduction.* Cambridge, MA: Harvard University Press.

Hull, C. L. (1943). *Principles of behavior: An introduction to behavior theory.* New York: Appleton-Century Crofts.

Hume, D. (1957). *The natural history of religion* (H. E. Root, ed). Stanford: Stanford University Press. (Original work published 1757.)

Hume, D. (1960). *A treatise of human nature.* Oxford: Oxford University Press. (Original work published 1739.)

Humphrey, N. K. (1976). The social function of intellect. In P. P. G. Bateson & R. A. Hinde (Eds.), *Growing points in ethology* (pp. 303–321). Cambridge, UK: Cambridge University Press.

Humphrey, N. K. (1977). Review of Griffin's *The question of animal awareness. Animal Behaviour, 25,* 521–522.

Humphrey, N. K. (1983). *Consciousness regained.* Oxford: Oxford University Press.

Humphrey, N. K. (1986). *The inner eye.* London: Faber & Faber.

Humphrey, N. K. (1988). *Lies, damned lies, and anecdotal evidence. Behavioral and Brain Sciences, 11,* 257–258.

Huntingford, F. (1985). Review of Griffin's *Animal thinking. Animal Behaviour, 34,* 1905–1906.

Hurlbert, E. M. (1992). Equivalence and the adaptationist program. *Ecological Modeling, 64,* 305–329.

Huxley, J. S. (1923). Ils n'ont que de l'âme: An essay on bird-mind. In J. S. Huxley, *Essays of a biologist* (pp. 105–129). London: Chatto & Windus.

Inagaki, K. (1989). Developmental shift in biological inference processes: From similarity-based to category-based attribution. *Human Development, 32,* 79–87.

Inagaki, K., & Sugiyama, K. (1988). Attributing human characteristics: Developmental changes in over- and underattribution. *Cognitive Development, 3,* 55–70.

Ingold, T. (1988). The animal in the study of humanity. In T. Ingold (Ed.), *What is an animal?* (pp. 84–99). London: Unwin Hyman.

Izard, C. E. (1993). Four systems of emotion activation: Cognitive and noncognitive processes. *Psychological Review, 100,* 68–90.

Jacobsen, C., Jacobsen, M. M., & Yoshioka, J. G. (1932). Development of an infant chimpanzee during her first year. *Comparative Psychology Monographs, 9*(41), 1–94.

Jacobson, J. L., Boersman, D., Fields, R., & Olson, K. (1983). Paralinguistic features of adult speech to infants and small children. *Child Development, 54,* 436–442.

Jager, J. (1983). Theorizing and the elaboration of place: Inquiry into Galileo and Freud. In A. Giorgi, A. Barton, & C. Maes (Eds.), *Duquesne studies in phenomenology* (Vol. 4, pp. 153–180). Duquesne: Duquesne University Press.

James, W. (1884). What is emotion? *Mind, 9,* 188–205.

James, W. (1890). *The principles of psychology* (Vol. 1). New York: Henry Holt.

Jamieson, D., & Bekoff, M. (1992a). Carruthers on nonconscious experience. *Analysis, 52,* 23–28.

Jamieson, D., & Bekoff, M. (1992b). Some problems and prospects for cognitive ethology. *Between the Species, 8,* 80–82.

Jamieson, D., & Bekoff, M. (1993). On aims and methods of cognitive ethology. *Philosophy of Science Association, 2,* 110–124.

Jamison, W. V., & Lunch, W. M. (1992). Rights of animals, perceptions of science, and political activism: Profile of American animal rights activists. *Science, Technology, and Human Values, 17,* 438–458.

Jasper, J. M., & Nelkin, D. (1992). *The animal rights crusade.* New York: The Free Press.

Jennings, H. S. (1976). *Behavior of the lower organisms*. Bloomington: Indiana University Press. (Original work published 1906.)

Johnson, C. N., & Wellman, H. M. (1980). Children's developing understanding of mental verbs: Remember, know, and guess. *Child Development, 51,* 1095–1102.

Johnson, E. (1991). Carruthers on consciousness and moral status. *Between the Species, 7,* 190–193.

Johnson, M. H., & Morton, J. (1991). *Biology and cognitive development: The case for face recognition*. Cambridge, UK: Blackwell.

Johnson-Laird, P. N. (1986). Reasoning without logic. In T. Myers, K. Brown, & B. McGonigle (Eds.), *Reasoning and discourse processes* (pp. 13–49). London: Academic Press.

Jolly, A. (1966). Lemur social behavior and primate intelligence. *Science, 153,* 501–506.

Jolly, A. (1972). *The evolution of primate behavior*. New York: MacMillian.

Jolly, A. (1985). A new science that sees animals as conscious beings. *Smithsonian, 15*(12), 66–75.

Jolly, A. (1991). Conscious chimpanzees? A review of recent literature. In C. Ristau (Ed.), *Cognitive ethology: The minds of other animals* (pp. 231–252). Hillsdale, NJ: Lawrence Erlbaum.

Jones, E. E. (1990). *Interpersonal perception*. New York: W. H. Freeman.

Jones, E. E., & Davis, K. E. (1965). A theory of correspondent inferences: From acts to dispositions. In L. Berkowitz (Ed.), *Advances in experimental social psychology* (Vol. 2, pp. 219–266). New York: Academic Press.

Kahneman, D., Slovic, P., & Tversky, A. (Eds.). (1982). *Judgment under uncertainty: Heuristics and biases*. New York: Cambridge University Press.

Kahneman, D., & Tversky, A. (1982). Variants of uncertainty. In D. Kahneman, P. Slovic, & A. Tversky (Eds.), *Judgment under uncertainty: Heuristics and biases* (pp. 509–520). Cambridge, UK: Cambridge University Press.

Kamil, A. C. (1984). Adaptation and cognition: Knowing what comes naturally. In H. L. Roitblat, T. G. Bever, & H. S. Terrace (Eds.), *Animal cognition* (pp. 533–544). Hillsdale, NJ: Lawrence Erlbaum.

Kamil, A. C., & Clements, K. C. (1990). Learning, memory, and foraging behavior. In D. A. Dewsbury (Ed.), *Contemporary issues in comparative psychology* (pp. 7–30). Sunderland, MA: Sinauer Associates, Inc.

Kamil, A. C., & Yoerg, S. (1985). Effects of prey depletion on patch choice by foraging blue jays. *Animal Behaviour, 33,* 1089–1095.

Kamin, A. (1968). Attention-like processes in classical conditioning. In M. R. Jones (Ed.), *Miami symposium on the prediction of behavior: Abusive stimulation* (pp. 9–31). Miami, FL: University of Miami Press.

Kant, E. (1991). *Critique of judgement.* Oxford: Clarendon Press. (Original work published 1790.)

Kastak, D. A., & Schusterman, R. J. (1992). Comparative cognition in marine mammals: A clarification on match-to-sample tests. *Marine Mammal Science, 8,* 414–417.

Katz, D. (1937). *Animals and men: Studies in comparative psychology.* London: Longmans, Green & Co.

Kearton, C. (1925). *My friend Toto: The adventures of a chimpanzee and the story of his journey from the Congo to London.* London: Arrowsmith.

Kellert, S. R. (1991). Japanese perceptions of wildlife. *Conservation Biology, 5,* 297–308.

Kellert, S. R. (1993). Attitudes, knowledge, and behavior toward wildlife among the industrial superpowers: United States, Japan, and Germany. *Journal of Social Issues, 49,* 53–69.

Kellert, S. R. (1980). American attitudes toward and knowledge of animals: An update. *International Journal for the Study of Animal Problems, 1,* 87–119.

Kellert, S. R., & Berry, J. K. (1981). *Knowledge, affection and basic attitudes toward animals in American society.* Washington, DC: United States Government Printing Office.

Kellert, S. R., & Berry, J. K. (1987). Attitudes, knowledge, and behaviors toward wildlife as affected by gender. *Wildlife Society Bulletin, 15,* 363–371.

Kellert, S. R., & Wilson, E. O. (Eds.). (1993). *The biophilia hypothesis.* Washington, DC: Island Press.

Kelley, H. H. (1971). *Attribution in social interaction.* Morristown, NJ: General Learning Press.

Kellogg, W. N., & Kellogg, L. A. (1932). *Child and chimpanzee.* Philadelphia: University of Pennsylvania Press.

Kellogg, W. N., & Kellogg, L. A. (1933). *The ape and the child: A study of environmental influence on early behavior.* New York: McGraw-Hill.

Kennedy, J. S. (1992). *The new anthropomorphism.* New York: Cambridge University Press.

Keyan, R. (1978). *The evolution of language.* New York: Philosophical Library.

Kim, J. (1984). Epiphenomenal and supervenient causation. *Midwestern Studies in Philosophy, 9,* 257–270.

Kitchell, R. L., & Erickson, H. H. (Eds.). (1983). *Animal pain: Perception and alleviation.* Bethesda, MD: American Physiological Society.

Kitchener, R. F. (Ed.). (1989). *The world view of contemporary physics.* Albany: SUNY Press.

Klima, E., & Bellugi, U. (1979). *The signs of language.* Cambridge, MA: Harvard University Press.

Klopfer, P. H., & Hailman, J. P. (1967). *An introduction to animal behavior: Ethology's first century.* Englewood Cliffs, NJ: Prentice-Hall.

Köhler, W. (1927). *The mentality of apes.* New York: Vintage Press. (Original work published 1917.)

Kohts, N. (1935). *Infant ape and human child* [in Russian, with English summary]. *Scientific Memoirs of the Darwin Museum* (Moscow), *3,* 524–591.

Konner, M. (1976). Maternal care, infant behavior and development among the !Kung. In R. B. Lee & I. DeVore (Eds.), *Kalahari hunter-gatherers* (pp. 218–244). Cambridge, MA: Harvard University Press.

Krebs, J. R. & Davies, N. B. (1991). Preface. In J. R. Krebs & N. B. Davies (Eds.), *Behavioural ecology: An evolutionary approach* (3rd ed., pp. ix–x). Oxford: Blackwell.

Krebs, J. R., & Dawkins, R. (1984). Animal signals: Mind-reading and manipulation. In J. R. Krebs & N. B. Davies (Eds.), *Behavioural ecology: An evolutionary approach* (2nd ed., pp. 380–420). Oxford: Blackwell.

Kuhn, T. S. (1962). *The structure of scientific revolutions.* Chicago: University of Chicago Press.

Kuhn, T. S. (1977). *The essential tension.* Chicago: University of Chicago Press.

Kummer, H., Dasser, V., & Hoyningen-Huene, P. (1990). Exploring primate social cognition: Some critical remarks. *Behaviour, 112,* 84–98.

Laidler, K. (1980). *The talking ape.* New York: Stein & Day.

Lakoff, G., & Johnson, M. (1980). *Metaphors we live by.* Chicago & London: University of Chicago Press.

Landau, M. (1991). *Narratives of human evolution.* New Haven: Yale University Press.

Langer, J. (1980). *The origins of logic.* New York: Academic Press.

Langer, J. (1993). Comparative cognitive development. In K. R. Gibson & T. Ingold (Eds.), *Tools, language and cognition in human evolution* (pp. 300–313). New York: Cambridge University Press.

Latour, B. (1987). *Science in action.* Cambridge, MA: Harvard University Press.

Lauder, G. V. (1986). Homology, analogy, and the evolution of behavior. In M. H. Nitecki & J. A. Kitchell (Eds.), *Evolution of animal behavior* (pp. 9–40). New York: Oxford University Press.

Lawrence, E. A. (1990). The tamed wild: Symbolic bears in American culture. In R. B. Browne, M. W. Fishwick, & K. O. Browne (Eds.), *Dominant symbols in popular culture* (pp. 140–153). Bowling Green, OH: Bowling Green State University Popular Press.

Leahy, M. P. T. (1991). *Against liberation: Putting animals in perspective.* New York: Routledge.

Ledbetter, D. H., & Basen, J. A. (1982). Failure to demonstrate self-recognition in gorillas. *American Journal of Primatology, 2,* 307–310.

Lee, D. N., & Reddish, P. E. (1981). Plummeting gannets: A paradigm of ethological optics. *Nature, 293,* 293–294.

Lehman, H. (1992). Scientist-animal bonding: Some philosophical reflections. In H. Davis & D. Balfour (Eds.), *The inevitable bond: Examining scientist-animal interactions* (pp. 383–396). New York: Cambridge University Press.

Leslie, A. M. (1987). Pretense and representation: Origins of "theory of mind." *Psychological Review, 94,* 412–426.

Leslie, A. M. (1988). Some implications of pretense for mechanisms underlying the child's theory of mind. In J. Astington, P. Harris, & D. Olson (Eds.), *Developing theories of mind* (pp. 19–46). New York: Cambridge University Press.

Lethmate, J., & Dücker, G. (1973). Untersuchungen zum Selbsterkennen im Spiegel bei Orangutans und einigen Anderen affenarten [Investigation of self-recognition in the mirror by orangutans and some other types of apes]. *Zeitschrift für Tierpsychologie, 33,* 248–269.

Levinson, B. M. (1984). Human/companion animal therapy. *Journal of Contemporary Psychotherapy, 14,* 131–144.

Levvis, G. W. (1992). Why we would not understand a talking lion. *Between the Species, 8,* 156–162.

Levvis, M. A. (1992). The value of judgments regarding the value of animals. *Between the Species, 8,* 150–155.

Lewis, C., & Mitchell, P. (Eds.). (1994). *Children's early understanding of mind.* Hillsdale, NJ: Lawrence Erlbaum.

Lewis, M., & Brooks-Gunn, J. (1979). *Social cognition and the acquisition of self.* New York: Plenum Press.

Libet, B. (1985). Unconscious cerebral initiative and the role of conscious will in voluntary action. *Behavioral and Brain Sciences, 8,* 529–566.

Lieberman, P. (1975). *On the origins of language: An introduction to the evolution of human speech.* New York: Macmillan.

Liebert, A. (1909). Der Anthropomorphismus der Wissenschaft [Anthropomorphism in science]. *Zeitschrift für Philosophie und Philosophische Kritik, 136,* 1–22.

Likona, T. (Ed.). (1976). *Moral development and behavior.* New York: Holt, Rinehart & Winston.

Lin, A. C., Bard, K. A., & Anderson, J. R. (1992). Development of self-recognition in chimpanzees *(Pan troglodytes). Journal of Comparative Psychology, 106,* 120–127.

Linden, E. (1974). *Apes, men, and language.* New York: Penguin.

Linden, E. (1986). *Silent partners: The legacy of the ape language experiments.* New York: Times Books.

Lintz, G. D. (1942). *Animals are my hobby.* New York: McBride.

Lock, A. (1980). *The guided reinvention of language.* London: Academic Press.

Lockwood, R. (1985). Anthropomorphism is not a four letter word. In M. W. Fox & L. D. Mickley (Eds.), *Advances in animal welfare science 1985/86* (pp. 185–199). Washington, DC: Humane Society of America.

Longino, H. E. (1990). *Science as social knowledge.* Princeton, NJ: Princeton University Press.

Lorenz, K. (1937). Über die Bildung des Instinktsbegriffes [On the development of the concept of instinct]. *Die Naturwissenschaften, 25,* 289–300; 307–318; 324–331.

Lorenz, K. (1950). The comparative method in studying innate behavior patterns. *Symposia of the Society of Experimental Biology, 4,* 221–268.

Lorenz, K. (1966). *On aggression.* New York: Harcourt, Brace & World.

Lorenz, K. (1970). Do animals undergo subjective experience? In K. Lorenz, *Studies in animal and human behavior* (Vol. 2, pp. 323–337). Cambridge, MA: Harvard University Press. (Original work published 1963.)

Lovejoy, A. O. (1978). *The great chain of being.* Cambridge, MA: Harvard University Press. (Original work published 1936.)

Loveland, K. A. (1984). Learning about points of view: Spatial perspective and the acquisition of "I/you." *Journal of Child Language, 11,* 535–556.

Lynch, J. J. (1992). *Toward an interspecific psychology.* Unpublished doctoral dissertation, The Claremont Graduate School, Claremont, CA.

MacKinnon, J. (1978). *The ape within us.* St. James Place, London: Collins.

Maddison, W., & Maddison, D. (1994). *MacClade.* Sunderland, MA: Sinaeur Associates.

Mahoney, M. J., & DeMonbreun, B. G. (1978). Consequences of confirmation and disconfirmation in a simulated research environment. *Quarterly Journal of Experimental Psychology, 30,* 395–406.

Maida, A. S., Wainer, J., & Cho, S. (1991). A syntactic approach to introspection and reasoning about the beliefs of other agents. *Fundamenta Informaticae, 15*(3–4), 333–356.

Maier, N. R. F. (1960). Maier's law. *American Psychologist, 15,* 208–212.

Mandler, J. M. (1988). How to build a baby: On the development of an accessible representational system. *Cognitive Development, 3,* 113–136.

Manning, A., & Dawkins, M. S. (1992). *An introduction to animal behaviour* (4th ed.). Cambridge, UK: Cambridge University Press.

Maple, T. L. (1980). *Orang-utan behavior.* New York: Van Nostrand Reinhold.

Martin, J. A. (1987). Structural equation modeling: A guide for the perplexed. *Child Development, 58,* 33–37.

Martin, P. (1990). Psychoimmunology: Relations between brain, behavior, and immune function. In P. P. G. Bateson & P. H. Klopfer (Eds.), *Perspectives in Ethology* (Vol. 8, pp. 173–214). New York: Plenum.

Maser, J. D., & Gallup, G. G., Jr. (1990). Theism as a byproduct of natural selection. *Journal of Religion, 70,* 515–532.

Mason, W. A. (1968). Scope and potential of primate research. *Science and Psychoanalysis, 12,* 101–112.

Mason, W. A. (1976). Windows on other minds. *Science, 194,* 930–931.

Mason, W. A. (1979). Environmental models and mental modes: Representational processes in the great apes. In D. A. Hamburg & E. R. McCown (Eds.), *The great apes* (pp. 277–293). Menlo Park, CA: Benjamin/Cummins.

Mathieu, M., & Bergeron, G. (1983). Piagetian assessment on cognitive development in chimpanzee (*Pan troglodytes*). In A. B. Chiarelli & R. S. Corruccini (Eds.), *Primate behavior and sociobiology* (pp. 143–147). Berlin: Springer-Verlag.

Matsuzawa, T. (1986). Pattern construction by a chimpanzee. *Primate Reports, 14*, 225–226.

Maynard Smith, J. (1965). The evolution of alarm calls. *American Naturalist, 99*, 59–63.

Maynard Smith, J. (1974). The theory of games and evolution of animal conflicts. *Journal of Theoretical Biology, 47*, 209–221.

Maynard Smith, J. (1979). Game theory and the evolution of behavior. *Proceedings of the Royal Society of London, B205*(1161), 475–488.

Maynard Smith, J. (1982). Do animals convey information about their intentions? *Journal of Theoretical Biology, 97*, 1–5.

Maynard Smith, J., & Parker, G. A. (1976). The logic of asymmetric contests. *Animal Behaviour, 24*, 159–175.

McAdams, D. (1993). *The stories we live by: Personal myths and the making of the self.* New York: William Morrow.

McCleery, R. H. (1989). Review of Colgan's *Animal Motivation. Animal Behaviour, 38*, 1091–1092.

McDougall, W. (1923). *Outline of psychology* (2nd ed.). London: Methuen.

McFarland, D. (1989a). *Problems of animal behaviour.* New York: John Wiley & Sons.

McFarland, D. (1989b). Review of Colgan's *Animal Motivation. Ethology, 83*, 170–171.

McGrew, W. C. (1992). *Chimpanzee material culture: Implications for human evolution.* New York: Cambridge University Press.

McIlvane, W. J., & Stoddard, L. T. (1985). Complex stimulus relations and exclusion in severe mental retardation. *Analysis and Intervention in Developmental Disabilities, 5*, 307–321.

McKinney, M., & McNamara, K. (1991). *Heterochrony.* New York: Plenum.

McKinney, W. T., & Bunney, W. E. (1969). Animal model of depression. *Archives of General Psychiatry, 21*, 240–248.

McMahan, P. (1978). Natural history of the coyote. In R. L. Hall & H. S. Sharp (Eds.), *Wolf and man: Evolution in parallel* (pp. 41–54). New York: Academic Press.

McNamee, T. (1982). Breath-holding in Grizzly Country. *Audobon, 84*(6), 69–83.

McPhail, E. M. (1987). The comparative psychology of intelligence. *Behavioral and Brain Sciences, 10*, 645–656.

Mead, G. H. (1974). *Mind, self and society from the standpoint of a social behaviorist.* Chicago: University of Chicago Press. (Original work published 1934.)

Meltzoff, A. N. (1988). The human infant as *Homo imitans.* In T. R. Zentall & B. G. Galef, Jr. (Eds.), *Social learning: Psychological and biological perspectives* (pp. 319–341). Hillsdale, NJ: Lawrence Erlbaum.

Meltzoff, A. N. (1990). Foundations for developing a concept of self: The role of imitation in relating self to other and the value of social mirroring, social modeling, and self practice in infancy. In D. Cicchetti & M. Beeghly (Eds.), *The self in transition: Infancy to childhood* (pp. 139–164). Chicago: University of Chicago Press.

Meltzoff, A. N., & Moore, M. (1977). Imitation of facial and manual gestures by neonates. *Science, 198,* 75–78.

Menzel, E. W., Jr. (1974). A group of young chimpanzees in a one-acre field. In A. M. Schrier & F. Stollnitz (Eds.), *Behaviour of non-human primates* (Vol. 5, pp. 83–153). New York: Academic Press.

Menzel, E. W., Jr. (1986). How can you tell if an animal is intelligent? In R. J. Schusterman, J. A. Thomas, & F. G. Woods (Eds.), *Dolphin cognition and behavior: A comparative approach* (pp. 167–205). Hillsdale, NJ: Lawrence Erlbaum.

Menzel, E. W., Jr., & Everett, J. W. (1978). *Cognitive aspects of foraging behavior.* Paper presented at Animal Behavior Society, Seattle, WA.

Menzel, E. W., Jr., & Johnson, M. (1978). Should mentalist concepts be defended or assumed? *Behavioral and Brain Sciences, 4,* 586–587.

Merleau-Ponty, M. (1962). *Phenomenology of perception.* New York: Humanities Press.

Merleau-Ponty, M. (1963). *The structure of behavior.* Boston: Beacon Press.

Michel, G. F. (1991). Human psychology and the minds of other animals. In C. A. Ristau (Ed.), *Cognitive ethology: The minds of other animals. Essays in honor of Donald R. Griffin* (pp. 253–272). Hillsdale, NJ: Lawrence Erlbaum.

Midgley, M. (1978). *Beast and man: The roots of human nature.* Sussex: Harvester Press.

Midgley, M. (1979). Gene juggling. *Philosophy, 54,* 439–459.

Mignault, C. (1985). Transition between sensorimotor and symbolic activities in nursery-reared chimpanzees (*Pan troglodytes*). *Journal of Human Evolution, 14,* 747–758.

Miles, H. L. (1975). The communicative competence of child and chimpanzee. In D. Premack (Chair), *Linguistic competence of apes.* Conference on *Origins and Evolution of Language and Speech,* New York Academy of Sciences, NY.

Miles, H. L. (1976). The communicative competence of child and chimpanzee. *Annals of the New York Academy of Sciences, 260,* 592–597.

Miles, H. L. (1978a). Language acquisition in apes and children. In F. C. C. Peng (Ed.), *Sign language and language acquisition in man and ape: New dimensions in comparative pedolinguistics* (pp. 103–120). Boulder, CO: Westview Press.

Miles, H. L. (1978b). The use of sign language by two chimpanzees. *Dissertation Abstracts International, 39,* 11A.

Miles, H. L. (1980a). Acquisition of gestural signs by an infant orangutan (*Pongo pygmaeus*). *American Journal of Physical Anthropology, 52,* 256–257.

Miles, H. L. (1980b). Language in primates: Historical overview and philosophical implications. In J. de Luce & H. T. Wilder (Chairs), *Language in Primates: Implications for Linguistics, Anthropology, Psychology, and Philosophy,* Miami University, Oxford, OH.

Miles, H. L. (1983). Apes and language: The search for communicative competence. In J. de Luce & H. T. Wilder (Eds.), *Language in primates: Perspectives and implications* (pp. 43–61). New York: Springer-Verlag.

Miles, H. L. (1986). How can I tell a lie? Apes, language, and the problem of deception. In R. W. Mitchell & N. S. Thompson (Eds.), *Deception: Perspectives on human and nonhuman deceit* (pp. 245–266). Albany: SUNY Press.

Miles, H. L. (1990). The cognitive foundations for reference in a signing orangutan. In S. T. Parker & K. R. Gibson (Eds.), *"Language" and intelligence in monkeys and apes* (pp. 511–538). New York: Cambridge University Press.

Miles, H. L. (1993a). Language and the orangutan: The old "person" of the forest. In P. Cavalieri & P. Singer (Eds.), *The great ape project: Equality beyond humanity* (pp. 42–57). London: Fourth Estate.

Miles, H. L. (1993b). The forgotten ape: "Orangutans are orange." In K. R. Gibson, B. King, & S. T. Parker (Chairs), *Continuities and Discontinuities: Apes and Humans,* 92nd Annual American Anthropological Association Meeting, Washington, DC.

Miles, H. L. (1994). ME CHANTEK: The development of self-awareness in a signing orangutan. In S. T. Parker, R. W. Mitchell, & M. L. Boccia (Eds.), *Self-awareness in animals and humans* (pp. 254–272). New York: Cambridge University Press.

Miles, H. L., & Harper, S. (1994). "Ape language" studies and the study of human language origins. In D. Quiatt & J. Itani (Eds.), *Hominid culture in primate perspective* (pp. 253–278). Denver: University Press of Colorado.

Mill, J. S. (1948). Psychology and ethology. In W. Dennis, *Readings in the history of psychology* (pp. 169–177). New York: Appleton Century Crofts. [Based on Book 6, On the logic of the moral sciences, chaps. 3–5 in Mill's *A System of Logic*]. (Original work published 1843.)

Miller, L. L. (1992). Molecular anthropomorphism. *Journal of Chemical Education, 69,* 141–142.

Millikan, R. G. (1984). *Language, thought, and other biological categories.* Cambridge, MA: Bradford Books/MIT Press.

Millikan, R. G. (1993). *White Queen psychology and other essays for Alice.* Cambridge, MA: Bradford Books/MIT Press.

Millikan, R. G. (1995). Pushmi-pullyu representations. In J. Tomberlin (Ed.), *Philosophical perspectives* (Vol. 9, pp. 185–200). Atascadero, CA: Ridgeview.

Mills, E. (1976). *The grizzly: Our greatest wild animal.* Sausalito: Comstock. (Original work published 1919.)

Mills, W. (1898). *The nature and development of animal intelligence.* London: T. Fisher Unwin.

Mitchell, R. W. (1986). A framework for discussing deception. In R. W. Mitchell & N. S. Thompson (Eds.), *Deception: Perspectives on human and nonhuman deceit* (pp. 3–40). New York: SUNY Press.

Mitchell, R. W. (1987). A comparative developmental approach to understanding imitation. In P. Klopfer & P. Bateson (Eds.), *Perspectives in ethology* (Vol. 7, pp. 183–215). New York: Plenum.

Mitchell, R. W. (1988). Ontogeny, biography, and evidence for tactical deception. *Behavioral and Brain Sciences, 11,* 259–260.

Mitchell, R. W. (1990). The concept of play. In M. Bekoff & D. Jamieson (Eds.), *Interpretation and explanation in the study of animal behavior* (Vol. 1, pp. 197–227). Boulder, CO: Westview Press.

Mitchell, R. W. (1992). Developing concepts in infancy: Animals, self perception, and two theories of mirror-self-recognition. *Psychological Inquiry, 3,* 127–130.

Mitchell, R. W. (1993a). Anthropomorphism, discontinuity, and the adequacy of evidence. In K. R. Gibson, B. King, & S. T. Parker (Chairs), *Continuities and Discontinuities: Apes and Humans,* 92nd Annual American Anthropological Association Meeting, Washington, DC.

Mitchell, R. W. (1993b). Mental models of mirror-self-recognition: Two theories. *New Ideas in Psychology, 11,* 295–325.

Mitchell, R. W. (1993c). Recognizing one's self in a mirror? A reply to Gallup & Povinelli, de Lannoy, Anderson, and Byrne. *New Ideas in Psychology, 11,* 351–377.

Mitchell, R. W. (1993d). Animals as liars: The human face of nonhuman duplicity. In M. Lewis & C. Saarni (Eds.), *Lying and deception in everyday life* (pp. 59–89). New York: Guilford Press.

Mitchell, R. W. (1994a). The evolution of primate cognition: Simulation, self-knowledge, and knowledge of other minds. In D. Quiatt & J. Itani (Eds.), *Hominid culture in primate perspective* (pp. 177–232). Denver: University Press of Colorado.

Mitchell, R. W. (1994b). Multiplicities of self. In S. T. Parker, R. W. Mitchell, & M. L. Boccia (Eds.), *Self-awareness in animals and humans* (pp. 81–107). New York: Cambridge University Press.

Mitchell, R. W. (1995). Review of Marshall Thomas's *The hidden life of dogs. Society and Animals, 4,* 100–103.

Mitchell, R. W. (1996). *Anthropomorphizing animals: A history.* Unpublished manuscript.

Mitchell, R. W., & Hamm, M. (in press). The interpretation of animal behavior: Anthropomorphism or behavior reading? *Behaviour.*

Mitchell, R. W., & Thompson, N. S. (1986a). Deception in play between dogs and people. In R. W. Mitchell & N. S. Thompson (Eds.), *Deception: Perspectives on human and nonhuman deceit* (pp. 193–205). Albany: SUNY Press.

Mitchell, R. W., & Thompson, N. S. (Eds.). (1986b). *Deception: Perspectives on human and nonhuman deceit.* New York: SUNY Press.

Mitchell, R. W., & Thompson, N. S. (1991). Projects, routines and enticements in dog-human play. In P. P. G. Bateson & P. H. Klopfer (Eds.), *Perspectives in ethology* (Vol. 9, pp. 189–216). New York: Plenum.

Mivart, St. G. J. (1889). *Origin of human reason.* London: Kegan Paul, Trench & Co.

Mivart, St. G. J. (1898). *The groundwork of science: A study of epistemology.* New York: G. P. Putnam's Sons.

Montgomery, S. (1991). *Walking with the great apes: Jane Goodall, Dian Fossey, and Biruté Galdikas.* Boston: Houghton Mifflin Co.

Moore, C., Pure, K., & Furrow, D. (1990). Children's understanding of the modal expression of speaker certainty and uncertainty and its relation to the development of a representational theory of mind. *Child Development, 61,* 722–730.

Morgan, C. L. (1894). *An introduction to comparative psychology.* London: Walter Scott.

Morgan, C. L. (1900). *Animal behaviour.* London: Edward Arnold.

Morgan, C. L. (1904). *An introduction to comparative psychology* (2nd ed.). London: Walter Scott.

Morgan, E. (1989). The aquatic ape theory and the origin of speech. In J. Wind, E. G. Pulleyblank, E. de Grolier & B. H. Bichakjian (Eds.), *Studies in language origins* (Vol. 1, pp. 199–207). Amsterdam: John Benjamins.

Morris, M. (1986). Large scale deceit: Deception by captive elephants? In R. W. Mitchell & N. S. Thompson (Eds.), *Deception: Perspectives on human and nonhuman deceit* (pp. 183–191). Albany: SUNY Press.

Morris, R., & Morris, D. (1966). *Men and apes.* New York: McGraw-Hill.

Morton, D. B., Burghardt, G. M., & Smith, J. A. (1990). Critical anthropomorphism, animal suffering, and the ecological context. *Hastings Center Report, 20*(3), 13–19. (Supplement on Animals, Science, and Ethics.)

Morton, D. B., & Griffiths, P. H. M. (1985). Guidelines on the recognition of pain, distress, and discomfort in experimental animals and an hypothesis for assessment. *Veterinary Record, 116,* 431–436.

Morton, E. S. (1977). On the occurrence and significance of motivation/structural rules in some bird and mammal sounds. *American Naturalist, 111,* 855–969.

Moses, L. J., & Flavell, J. H. (1990). Inferring false belief from actions and reactions. *Child Development, 61,* 929–945.

Moss, C. (1975). *Portraits in the wild: Behavior studies of East African mammals.* Boston: Houghton Mifflin.

Moss, C. (1987). *Elephant memories: Thirteen years in the life of an African family.* New York: Ballantine Books.

Mounoud, P., & Vinter, A. (1981). Representation and sensorimotor development. In Butterworth, G. (Ed.), *Infancy and epistemology* (pp. 200–235). Brighton, UK: Harvester Press.

Moynihan, M. (1982). Why is lying about intentions rare during some kinds of contests? *Journal of Theoretical Biology, 97,* 7–12.

Moynihan, M. (1983a). Notes on the behavior of *Euprymna scolopes* (Cephalopoda: *Sepiolidae*). *Behaviour, 85,* 25–41.

Moynihan, M. (1983b). Notes on the behavior of *Idiosepius pygmaeus* (Cephalopoda: *Idiosepiidae*). *Behaviour, 85,* 42–57.

Moynihan, M. (1985). *Communication and non-communication by cephalopods.* Bloomington: Indiana University Press.

Moynihan, M., & Rodaniche, A. F. (1982). *The behavior and natural history of the Caribbean Reef Squid Sepioteuthis sepioidea. Advances in Ethology, 25,* 1–150.

Munn, C. A. (1986). Birds that "cry wolf." *Nature, 319,* 143–145.

Müller, F. M. (1873a, May). Lectures on Mr. Darwin's philosophy of language: First lecture. *Fraser's Magazine, 7,* 525–540.

Müller, F. M. (1873b, June). Lectures on Mr. Darwin's philosophy of language: Second lecture. *Fraser's Magazine, 7,* 659–678.

Müller, F. M. (1873c, July). Lectures on Mr. Darwin's philosophy of language: Third lecture. *Fraser's Magazine, 8,* 1–25.

Müller, F. M. (1887). *The science of thought.* London: Longmans Green & Co.

Myers, G. (1990). *Writing biology: Texts in the social construction of knowledge.* Madison: University of Wisconsin Press.

Mynatt, C. R., Doherty, M. E., & Tweney, R. D. (1978). Consequences of confirmation and disconfirmation in a simulated research environment. *Quarterly Journal of Experimental Psychology, 30,* 395–406.

Nagel, T. (1974). What is it like to be a bat? *Philosophical Review, 83,* 435–450.

Nakajima, S. (1992). Evaluation of animal "intelligence" by undergraduate students. *Japanese Journal of Psychonomic Science, 11,* 27–30.

Napier, J. R., & Napier, P. H. (1967). *A handbook of living primates.* New York: Academic Press.

Neisser, U. (1982). *Memory observed: Remembering in natural contexts.* New York: W. H. Freeman & Co.

Newell, A., & Simon, H. A. (1972). *Human problem solving.* New York: Prentice-Hall.

Newport, E. (1976). Motherese: The speech of mothers to young children. In N. Costellan, D. Pisoni, & G. Potts (Eds.), *Cognitive theory* (Vol. 2, pp. 177–217). Hillsdale, NJ: Lawrence Erlbaum.

Nietzsche, F. (1966). *Werke in Drei Bänder* (Vol. 3). Munich: Carl Hanser. (Original works published after 1871.)

Nisbett, R. E., & Wilson, T. D. (1977). Telling more than we can know: Verbal reports on mental processes. *Psychological Review, 84,* 231–259.

Nissen, H. W., & Riesen, A. H. (1964). The eruption of the permanent dentition of chimpanzee. *American Journal of Physical Anthropology, 22,* 285–294.

Nitecki, M. H., & Kitchell, J. A. (Eds.). (1986). *Evolution of animal behavior: Paleontological and field approaches.* New York: Oxford University Press.

Norbeck, E., & Lock, M. (Eds.). (1987). *Health, illness and medical care in Japan*. Honolulu: University of Hawaii Press.

Noske, B. (1989). *Humans and other animals: Beyond the boundaries of anthropology*. London: Pluto Press.

Novak, M. A., & Suomi, S. (1988). Psychological well-being of primates in captivity. *American Psychologist, 43,* 765–773.

Ohnuki-Tierney, E. (1987). *The monkey as mirror. Symbolic transformations in Japanese history and ritual*. Princeton: Princeton University Press.

Ohnuki-Tierney, E. (1995). Representations of the monkey in Japanese culture. In R. Corbey & B. Theunissen (Eds.), *Ape, man, apeman: Changing views since 1600* (pp. 297–308). Leiden, the Netherlands: Department of Prehistory of Leiden University.

Olton, D. S. (1978). Characteristics of spatial memory. In S. H. Hulse, H. Fowler, & W. K. Honig (Eds.), *Cognitive processes in animal behavior* (pp. 341–373). Hillsdale, NJ: Lawrence Erlbaum.

Olton, D. S. (1979). Mazes, maps and memory. *American Psychologist, 34,* 583–596.

O'Neill, D. K., & Gopnik, A. (1991). Young children's ability to identify the sources of their beliefs. *Developmental Psychology, 27,* 390–397.

Ortony, A. (1975). Why metaphors are necessary and not just nice. *Educational Theory, 25,* 45–53.

O'Sullivan, C., & Yeager, C. P. (1989). Communicative context and linguistic competence: The effects of social setting on a chimpanzee's conversational skill. In R. A. Gardner, B. T. Gardner, & T. E. Van Cantfort (Eds.), *Teaching sign language to chimpanzees* (pp. 269–279). Albany: SUNY Press.

Oxford English Dictionary. (1987). *The compact edition of the Oxford English dictionary* (Vol. 1: A–O). Oxford: Oxford University Press. (Original work published 1971.)

Oxnard, C. (1986). *The order of man*. New Haven, CT: Yale University Press.

Packard, A. (1972). Cephalopods and fish: The limits of convergence. *Biological Review, 47,* 241–307.

Packard, A. (1982). Morphogenesis of chromatophore patterns in cephalopods: Are morphological units the same? *Malacologia, 23,* 193–201.

Packard, A., & Hochberg, F. G. (1977). Skin patterns in Octopus and other genera. In M. Nixon & J. B. Messenger (Eds.), *Symposia of the Zoological Society of London; No. 38. The biology of cephalopods* (pp. 191–231). New York: Academic Press.

Packer, C., & Pusey, A. E. (1983). Adaptations of female lions to infanticide by incoming males. *American Naturalist, 121,* 716–728.

Panel Report. (1987). Colloquium on Recognition and Alleviation of Pain and Distress. *Journal of the American Veterinary Medical Association, 191,* 1186–1192.

Paprotte, W., & Dirven, R. (Eds.). (1985). *The ubiquity of metaphor: Metaphor in language and thought.* Amsterdam: John Benjamins.

Parker, G. A. (1974). Assessment strategy and the evolution of fighting behavior. *Journal of Theoretical Biology, 47,* 223–243.

Parker, S. T. (1977). Piaget's sensorimotor series in an infant macaque: A model for comparing unstereotyped behavior and intelligence in human and nonhuman primates. In S. Chevalier-Skolnikoff & F. Poirier (Eds.), *Primate biosocial development: Biological, social and ecological determinants* (pp. 43–113). New York: Garland.

Parker, S. T. (1987). The origin of symbolic communication: An evolutionary cost/benefit model. In J. Mongangero, A. Tryphon, & S. Dionnet (Eds.), *Symbolism and knowledge,* Cahier No. 8 (pp. 7–27). Geneva: Jean Piaget Archives Foundation.

Parker, S. T. (1990). The origins of comparative developmental evolutionary studies of primate mental abilities. In S. T. Parker & K. R. Gibson (Eds.), *"Language" and intelligence in monkeys and apes* (pp. 3–63). New York: Cambridge University Press.

Parker, S. T. (1991). A developmental model for the origins of self-recognition in great apes. *Human Evolution, 6,* 435–449.

Parker, S. T. (1993). Imitation and circular reactions as evolved mechanisms for cognitive construction. *Human Development, 36,* 309–323.

Parker, S. T. (1994). Incipient mirror self-recognition in zoo chimpanzees and gorillas. In S. T. Parker, R. W. Mitchell, & M. L. Boccia (Eds.), *Self-awareness in animals and humans* (pp. 301–307). New York: Cambridge University Press.

Parker, S. T. (1996). Using cladistic analysis of comparative data to reconstruct the evolution of cognitive development in hominids. In E. Martins (Ed.), *Phylogenies and the comparative method in animal behavior* (pp. 361–398). New York: Oxford University Press.

Parker, S. T., & Gibson, K. R. (1979). A developmental model for the evolution of language and intelligence in early hominids. *Behavioral and Brain Sciences, 2,* 367–407.

Parker, S. T., & Gibson, K. R. (Eds.). (1990). *"Language" and intelligence in monkeys and apes.* New York: Cambridge University Press.

Parker, S. T., Mitchell, R. W. & Boccia, M. L. (Eds.). (1994). *Self-awareness in animals and humans*. New York: Cambridge University Press.

Patterson, F. G. P. (1978). The gestures of a gorilla: Language acquisition in another pongid. *Brain and Language, 5*, 72–97.

Patterson, F. G. P. (1980). Innovative uses of language by a gorilla: A case study. In K. E. Nelson (Ed.), *Children's language* (Vol. 2, pp. 497–561). New York: Gardner Press.

Patterson, F. G. P. (1986). The mind of the gorilla: Conversation and conservation. In K. Benirschke (Ed.), *Primates: The road to self-sustaining populations* (pp. 933–947). New York: Springer-Verlag.

Patterson, F. G. P., & Cohn, R. (1994). Self-recognition and self-awareness in lowland gorillas. In S. T. Parker, R. W. Mitchell, & M. L. Boccia (Eds.), *Self-awareness in animals and humans* (pp. 273–290). New York: Cambridge University Press.

Patterson, F. [G. P.], & Linden, E. (1981). *The education of Koko*. New York: Holt, Rinehart & Winston.

Pavlov, I. P. (1960). *Lectures on conditioned reflexes*. New York: Dover. (Original work published 1927.)

Pellizzer, G., & Georgopoulos, A. P. (1993). Mental rotation of the intended direction of movement. *Current Directions in Psychological Science, 2*, 12–17.

Pepperberg, I. M. (1981). Functional vocalizations by an African grey parrot (*Psittacus erithacus*). *Zeitschrift für Tierpsychologie, 55*, 139–160.

Pepperberg, I. M. (1990a). Cognition in an African grey parrot: Further evidence for comprehension of categories and labels. *Journal of Comparative Psychology, 104*, 41–52.

Pepperberg, I. M. (1990b). Some cognitive capacities of an African grey parrot (*Psittacus erithacus*). *Advances in the Study of Behavior, 19*, 357–409.

Pepys, S. (1970). *The diary of Samuel Pepys* (Vol. 2). Berkeley: University of California Press. (Original work published 1661.)

Perner, J. (1988). Higher order beliefs and intentions in children's understanding of social interaction. In J. W. Astington, P. L. Harris, & D. R. Olson (Eds.), *Developing theories of mind* (pp. 271–294). Cambridge, UK: Cambridge University Press.

Perner, J. (1991). *Understanding the representational mind*. Cambridge, MA: MIT Press.

Perner, J., Frith, U., Leslie, A. M., & Leekman, S. R. (1989). Explorations of the autistic child's theory of mind: Knowledge, belief, and communication. *Child Development, 60*, 689–700.

Peters, C. B. (1991). Apes, humans, and culture: What primatological discourse tells us about ourselves. In J. D. Loy & C. B. Peters (Eds.), *Understanding behavior: What primate studies tell us about human behavior* (pp. 242–255). New York: Oxford University Press.

Peters, R. (1978). Communication, cognitive mapping, and strategy in wolves and hominids. In R. L. Hall & H. S. Sharp (Eds.), *Wolf and man: Evolution in parallel* (pp. 95–107). New York: Academic Press.

Peterson, D., & Goodall, J. (1993). *Visions of Caliban*. Boston: Houghton Mifflin.

Peyrefitte, A. (1989). *L'empire immobile: Le choc des mondes* [The immoble empire: The clash of worlds]. Paris: Fayard.

Phillips, M. T. (1993). Savages, drunks and lab animals: The researcher's perception of pain. *Society and Animals, 1,* 61–81.

Phillips, M. T. (1994). Proper names and the social construction of biography: The negative case of laboratory animals. *Qualitative Sociology, 17,* 119–142.

Piaget, J. (1933). Children's philosophies. In C. Murchison (Ed.), *A handbook of child psychology* (2nd ed., pp. 534–547). Worcester, MA: Clark University.

Piaget, J. (1952). *The origins of intelligence in children*. New York: Norton.

Piaget, J. (1954). *The construction of reality in the child*. New York: Basic Books.

Piaget, J. (1962). *Play, dreams and imitation in childhood*. New York: Norton.

Piaget, J. (1967). *The child's conception of the world*. Totowa, NJ: Littlefield, Adams. (Original work published 1929.)

Piaget, J. (1970). *Genetic epistemology*. New York: Norton.

Piaget, J. (1976). *The grasp of consciousness*. Cambridge, MA: Harvard University Press.

Pickering, A. (Ed.). (1992). *Science as practice and culture*. Chicago: University of Chicago Press.

Plous, S. (1993a). Psychological mechanisms in the human use of animals. *Journal of Social Issues, 49,* 11–52.

Plous, S. (Ed.). (1993b). The role of animals in human society. *Journal of Social Issues, 49*(1).

Polanyi, M. (1967). *The tacit dimension*. New York: Doubleday.

Popper, K. R. (1959). *The logic of scientific discovery*. London: Hutchinson.

Popper, K. R. (1968). *Conjectures and refutations*. London: Routledge and Kegan Paul.

Porter, D. G. (1992). Ethical scores for animal experiments. *Nature, 356*, 101–102.

Povinelli, D. J. (1987). Monkeys, apes, mirrors and minds: The evolution of self-awareness in primates. *Human Evolution, 2*, 493–509.

Povinelli, D. J. (1991). *Social intelligence in monkeys and apes.* Unpublished doctoral dissertation, Yale University, New Haven, CT.

Povinelli, D. J. (1993). Reconstructing the evolution of mind. *American Psychologist, 48*, 493–509.

Povinelli, D. J. (1994). How to create self-recognizing gorillas (but don't try it on macaques). In S. T. Parker, R. W. Mitchell, & M. L. Boccia (Eds.), *Self-awareness in animals and humans* (pp. 291–300). New York: Cambridge University Press.

Povinelli, D. J., & deBlois, S. (1992). Young children's (*Homo sapiens*) understanding of knowledge formation in themselves and others. *Journal of Comparative Psychology, 106*, 228–238.

Povinelli, D. J., & Eddy, T. J. (1996). What young chimpanzees know about seeing. *Monographs of the Society for Research in Child Development, 61*(2, Serial No. 247).

Povinelli, D. J. & Godfrey, L. R. (1993). The chimpanzee's mind: How noble in reason? How absent of ethics? In M. Nitecki (Ed.), *Evolutionary Ethics* (pp. 277–324). Albany: SUNY Press.

Povinelli, D. J., Nelson, K. E., & Boysen, S. T. (1990). Inferences about guessing and knowing by chimpanzees (*Pan troglodytes*). *Journal of Comparative Psychology, 104*, 203–210.

Povinelli, D. J., Nelson, K. E., & Boysen, S. T. (1992). Comprehension of role reversal in chimpanzees: Evidence of empathy? *Animal Behaviour, 43*, 633–640.

Povinelli, D. J., Parks, K. A., & Novak, M. A. (1991). Do rhesus monkeys (*Macaca mulatta*) attribute knowledge and ignorance to others? *Journal of Comparative Psychology, 105*, 318–325.

Povinelli, D. J., Parks, K. A., & Novak, M. A. (1992). Role reversal by rhesus monkeys, but no evidence of empathy. *Animal Behaviour, 44*, 269–281.

Povinelli, D. J., Rulf, A. B., Landau, K., & Bierschwale, D. (1993). Self-recognition in chimpanzees: Distribution, ontogeny, and patterns of emergence. *Journal of Comparative Psychology, 107*, 347–372.

Premack, D. (1971). Language in chimpanzee? *Science, 172*, 808–822.

Premack, D. (1976). *Intelligence in ape and man.* Hillsdale, NJ: Lawrence Erlbaum.

Premack, D. (1986). *Gavagai!*. Cambridge, MA: MIT Press.

Premack, D. (1988a). "Does the chimpanzee have a theory of mind?" revisited. In R. Byrne & A. Whiten (Eds.), *Machiavellian intelligence* (pp. 160–179). New York: Oxford University.

Premack, D. (1988b). Minds with and without language. In L. Weiskrantz (Ed.), *Thought without language* (pp. 44–65). Oxford: Oxford University Press.

Premack, D., & Premack, A. J. (1983). *The mind of an ape*. New York: W. W. Norton & Co.

Premack, D., & Woodruff, G. (1978). Does the chimpanzee have a theory of mind? *Behavioral and Brain Sciences, 4*, 515–526.

Preston, D. J. (1981). Meshie Mungkut. *Natural History, 90*(12), 74–76.

Pryor, K. (1984). *Don't shoot the dog: The new art of teaching and training*. New York: Bantam.

Purton, A. C. (1970). *Philosophical aspects of explanation in ethology*. Unpublished doctoral dissertation, University of London, London, UK.

Putnam, H. (1981). *Reason, truth and history*. Cambridge, UK: Cambridge University Press.

Putnam, H. (1988). *Representation and reality*. Cambridge, MA: Harvard University Press.

Quiatt, D., & Reynolds, V. (1993). *Primate behavior: Information, social knowledge, and the evolution of culture*. Cambridge, UK: Cambridge University Press.

Quine, W. V. O. (1960). *Word and object*. Cambridge, MA: MIT Press.

Quine, W. V. O., & Ullian, J. (1970). *The web of belief* (2nd ed.). New York: Random House.

Rachels, J. (1990). *Created from animals: The moral implications of Darwinism*. New York: Oxford University Press.

Radner, D., & Radner, M. (1989). *Animal consciousness*. Buffalo, NY: Prometheus Books.

Rajecki, D. W. (Ed.). (1983). *Comparing behavior: Studying man studying animals*. Hillsdale, NJ: Lawrence Erlbaum.

Rajecki, D. W., Rasmussen, J. L., & Craft, H. D. (1993). Labels and the treatment of animals: Archival and experimental cases. *Society and Animals, 1*, 45–60.

Rasmussen, J. L., Rajecki, D. W., & Craft, H. D. (1993). People's perceptions of animal mentality: Ascriptions of "thinking." *Journal of Comparative Psychology, 107*, 283–290.

Raven, H. C. (1932). Meshie: The child of a chimpanzee. *Natural History, 32*(2), 607–617.

Real, L. A. (1992). Information processing and the evolutionary ecology of cognitive architecture. *American Naturalist, 140,* S108–S145.

Reber, A. S. (1992). The cognitive unconscious: An evolutionary perspective. *Consciousness and Cognition, 1,* 93–133.

Redshaw, M. (1978). Cognitive development in human and gorilla infants. *Journal of Human Evolution, 7,* 122–141.

Regan, T. (1983). *The case for animal rights.* Berkeley: University of California Press.

Rescorla, R. A. (1967). Pavlovian conditioning and its proper control procedures. *Psychological Review, 74,* 71–80.

Rescorla, R. A., & Wagner, A. R. (1972). A theory of Pavlovian conditioning: Variations in the effectiveness of reinforcement and nonreinforcement. In A. H. Black & N. F. Prokasy (Eds.), *Classical conditioning II: Current research and theory* (pp. 64–99). New York: Appleton Century Crofts.

Reynolds, P. C. (1976). Play, language and human evolution. In J. S. Bruner, A. Jolly, & K. Sylva (Eds.), *Play: Its role in development and evolution* (pp. 621–635). New York: Basic Books.

Ribot, T. (1915). *L'évolution des idées générales* [The evolution of general ideas] (4th ed.). Paris: F. Alcan.

Richards, P. (1995). Local understandings of primates and evolution: Some Mende beliefs concerning chimpanzees. In R. Corbey & B. Theunissen (Eds.), *Ape, man, apeman: Changing views since 1600* (pp. 265–274). Leiden, the Netherlands: Department of Prehistory of Leiden University.

Richards, R. J. (1987). *Darwin and the emergence of evolutionary theories of mind and behavior.* Chicago: University of Chicago Press.

Richards, R. J. (1992). *The meaning of evolution: The morphological construction and ideological reconstruction of Darwin's theory.* Chicago: University of Chicago Press.

Ricoeur, P. (1981). The model of the text: Meaningful action considered as a text. In J. B. Thompson (Ed.), *Paul Ricoeur: Hermeneutics and the human sciences* (pp. 197–222). Cambridge, UK: Cambridge University Pless.

Ricoeur, P. (1984). *Time and narrative* (2 vol.). Chicago: University of Chicago.

Ridley, M. (1984). *Classification and evolution.* Oxford: Oxford University Press.

Ridley, M. (1992). Animist debate. *Nature, 359,* 280.

Ristau, C. A. (1986). Do animals think? In R. J. Hoage & L. Goodman (Eds.), *Animal intelligence*. Washington, DC: Smithsonian.

Ristau, C. (1991a). Aspects of the cognitive ethology of an injury-feigning bird, the piping plovers. In C. A. Ristau (Ed.), *Cognitive ethology: The minds of other animals* (pp. 91–126). Hillsdale, NJ: Lawrence Erlbaum.

Ristau, C. A. (Ed.). (1991b). *Cognitive ethology: The minds of other animals. Essays in honor of Donald R. Griffin*. Hillsdale, NJ: Lawrence Erlbaum.

Ristau, C. A., & Robbins, D. (1982). Language in the great apes: A critical review. *Advances in the Study of Behavior, 12,* 141–253.

Robert, S. (1986). Ontogeny of mirror behavior in two species of great apes. *American Journal of Primatology, 10,* 109–117.

Roberts, W. A. (1979). Spatial memory in the rat on a hierarchical maze. *Learning and Motivation, 10,* 117–140.

Roberts, W. H. (1929). A note on anthropomorphism. *Psychological Review, 36,* 95–96.

Rogers, C. (1965). *Client-centered therapy*. Boston: Houghton Mifflin. (Original work published 1951.)

Roitblat, H. L., Bever, T., & Terrace, H. (Eds.). (1984). *Animal cognition*. Hillsdale, NJ: Lawrence Erlbaum.

Roitblat, H. L., & von Fersen, L. (1992). Comparative cognition: Representations and processes in learning and memory. *Annual Review of Psychology, 43,* 671–710.

Rollin, B. E. (1981). *Animal rights and human morality*. Buffalo: Prometheus Books.

Rollin, B. E. (1987). Laws relevant to animal research in the United States. In A. A. Tuffery (Ed.), *Laboratory Animals* (pp. 323–333). London: John Wiley.

Rollin, B. E. (1989). *The unheeded cry: Animal consciousness, animal pain and science*. Oxford: Oxford University Press.

Rollin, B. E. (1990). How the animals lost their minds: Animal mentation and scientific ideology. In M. Bekoff & D. Jamieson (Eds.), *Interpretation and explanation in the study of animal behavior* (Vol. 1, pp. 375–398). Boulder, CO: Westview Press.

Rollin, B. E. (1992). *Animal rights and human morality* (2nd ed.). Buffalo: Prometheus Books. (Original work published 1982.)

Romanes, E. (1908). *The life and letters of George John Romanes, written and edited by his wife* (6th ed.). London: Longmans Green & Co.

Romanes, G. J. (1882). *Animal intelligence*. London: Kegan Paul.

Romanes, G. J. (1883). *Mental evolution in animals*. London: Kegan Paul, Trench & Co.

Romanes, G. J. (1888). *Mental evolution in man: Origin of human faculty*. London: Kegan Paul, Trench & Co.

Romanes, G. J. (1896). *Darwin and after Darwin* (Vol. 1). Chicago: Open Court.

Romanes, G. J. (1900). *Mental evolution in animals*. New York: D. Appleton & Co. (Original work published 1883.)

Rorty, R. (1980). *Philosophy and the mirror of nature*. Oxford: Blackwell.

Rosenberg, A. (1990). Is there an evolutionary biology of play? In M. Bekoff & D. Jamieson (Eds.), *Interpretation and explanation in the study of animal behavior* (Vol. 1, pp. 180–196). Boulder, CO: Westview Press.

Rosenblueth, A., Wiener, N., & Bigelow, J. (1943). Behavior, purpose and teleology. *Philosophy of Science, 10*, 18–24.

Ross, L. (1977). The intuitive psychologist and his shortcomings: Distortions in the attribution process. In L. Berkowitz (Ed.), *Advances in experimental social psychology* (Vol. 10, pp. 174–187). New York: Academic Press.

Rumbaugh, D. M. (1977). *Language learning by a chimpanzee: The Lana project*. New York: Academic Press.

Rumbaugh, D. M., Gill, T. V., & von Glasersfeld, E. C. (1973). Reading and sentence completion by a chimpanzee (*Pan*). *Science, 182*, 731–733.

Rumbaugh, D. M., Gill, T. V., von Glasersfeld, E. C., Warner, H., & Pisani, P. (1975). Conversations with a chimpanzee in a computer-controlled environment. *Biological Psychiatry, 10*, 627–641.

Rumbaugh, D. M., & Pate, J. L. (1984). The evolution of cognition in primates: A comparative perspective. In H. L. Roitblat, T. G. Bever, & H. S. Terrace (Eds.), *Animal cognition* (pp. 569–587). Hillsdale, NJ: Lawrence Erlbaum.

Russell, B. (1927). *Outline of philosophy*. New York: Median.

Russon, A. (1990). The development of peer social interaction in infant chimpanzees: Comparative, social, Piagetian, and brain perspectives. In S. T. Parker & K. R. Gibson (Eds.), *"Language" and intelligence in monkeys and apes* (pp. 379–418). New York: Cambridge University Press.

Russon, A., & Galdikas, B. (1993). Imitation in free-ranging rehabilitant orangutans (*Pongo pygmaeus*). *Journal of Comparative Psychology, 107*, 147–160.

Russow, L.-M. (1982). It's not like that to be a bat. *Behaviorism, 10*, 55–63.

Ryle, G. (1949). *The concept of mind*. London: Hutchinson.

Sapontzis, S. F. (1987). *Morals, reasons, and animals.* Philadelphia: Temple University Press.

Sappington, A. A. (1990). The independent manipulation of intellectually and emotionally based beliefs. *Journal of Research in Personality, 24,* 487–509.

Sartre, J. P. (1966). *Being and nothingness.* New York: Washington Square Press.

Savage-Rumbaugh, E. S. (1979). Symbolic communication: Its origins and early development in the chimpanzee. *New Directions for Child Development, 3,* 1–15.

Savage-Rumbaugh, E. S. (1984). *Pan paniscus* and *Pan troglodytes*: Contrasts in preverbal competence. In R. Susman (Ed.), *The pygmy chimpanzee: Evolutionary biology and behavior* (pp. 395–414). New York: Plenum.

Savage-Rumbaugh, E. S. (1986). *Ape language: From conditioned response to symbol.* New York: Columbia University Press.

Savage-Rumbaugh, E. S., & Lewin, R. (1994). *Kanzi: The ape at the brink of the human mind.* New York: John Wiley & Sons.

Savage-Rumbaugh, E. S., & McDonald, K. (1988). Deception and social manipulation in symbol-using apes. In R. Byrne & A. Whiten (Eds.), *Machiavellian intelligence: Social expertise and the evolution of intellect in monkeys, apes and humans* (pp. 224–237). Oxford: Clarendon Press.

Savage-Rumbaugh, E. S., Murphy, J., Sevcik, R. A., Brakke, K. E., Williams, S. L., & Rumbaugh, D. M. (1993). *Language comprehension in ape and child. Monographs of the Society for Research in Child Development, 58*(3–4, Serial No. 233).

Savage-Rumbaugh, E. S., Pate, J., Lawson, J., Smith, S., & Rosenbaum, S. (1983). Can a chimpanzee make a statement? *Journal of Experimental Psychology: General, 112,* 457–492.

Savage-Rumbaugh, E. S., & Rumbaugh, D. M. (1978). Symbolization, language and chimpanzees: A theoretical re-evaluation based on initial language acquisition processes in four young *Pan troglodytes. Brain and Language, 6,* 265–300.

Savage-Rumbaugh, E. S., Rumbaugh, D. M., & Boysen, S. (1978). Linguistically mediated tool use and exchange by chimpanzees (*Pan troglodytes*). *Behavioral and Brain Sciences, 4,* 539–554.

Savage-Rumbaugh, E. S., Rumbaugh, D. M., & McDonald, K. (1985). Language learning in two species of apes. *Neuroscience & Biobehavioral Review, 9,* 653–665.

Savage-Rumbaugh, E. S., Rumbaugh, D. M., & Smith, S. T. (1980). Reference: The linguistic essential. *Science, 210,* 922–925.

Scheler, M. (1967). Toward a stratification of the emotional life. In N. Lawrence & D. O'Connor (Eds.), *Readings in existential phenomenology* (pp. 19–31). Englewood Cliffs, NJ: Prentice-Hall.

Schiller, C. H. (Ed.). (1957). *Instinctive behavior.* New York: International Universities Press.

Schusterman, R. J., & Gisiner, R. (1988). Artificial language comprehension in dolphins and sea lions: The essential cognitive skills. *Psychological Record, 38,* 311–348.

Schusterman, R. J., & Gisiner, R. (1989). Please parse the sentence: Animal cognition in the procrustean bed of linguistics. *Psychological Record, 39,* 3–18.

Schusterman, R. J., Gisiner, R., Grimm, B. K., & Hanggi, E. B. (1993). Behavior control by exclusion and attempts at establishing semanticity in marine mammals using matching-to-sample paradigms. In H. Roitblat, L. Herman, & P. Nactigall (Eds.), *Language and communications: A comparative perspective* (pp. 249–275). Hillsdale, NJ: Lawrence Erlbaum.

Schusterman, R. J., & Kastak, D. (1993). A California sea lion (*Zalophus californianus*) is capable of forming equivalence relations. *Psychological Record, 43,* 823–839.

Schusterman, R. J., & Kastak, D. (1995). There is no substitute for an experimental analysis of marine mammal cognition. *Marine Mammal Science, 11,* 263–267.

Schusterman, R. J., & Krieger, K. (1984). California sea lions are capable of semantic comprehension. *Psychological Record, 34,* 3–23.

Schusterman, R. J., & Krieger, K. (1986). Artificial language comprehension and size transposition by a California sea lion (*Zalophus californianus*). *Journal of Comparative Psychology, 100,* 348–355.

Scott, P. S., & Fuller, J. (1965). *Genetics and the social behavior of the dog.* Chicago: University of Chicago Press.

Searle, J. (1992). *The rediscovery of the mind.* Cambridge, MA: Harvard University Press.

Sebeok, T. A. (1981). Apespeak. *The Sciences, 22,* 4.

Sebeok, T. A., & Rosenthal, R. (Eds.). (1981). *The Clever Hans phenomenon: Communication with horses, whales, apes, and people. Annals of the New York Academy of Sciences, 354.*

Sebeok, T. A., & Umiker-Sebeok, J. (Eds.). (1980). *Speaking of apes: A critical anthology of two-way communication with man.* New York: Plenum Press.

Seligman, M. E. P. (1970). On the generality of the laws of learning. *Psychological Review, 77,* 406–418.

Seligman, M. E. P., & Hager, J. L. (1972). *Biological boundaries of learning.* NY: Appleton Century Crofts.

Sellars, W. F. (1956). Empiricism and the philosophy of mind. *Minnesota Studies in the Philosophy of Science, 1,* 253–329.

Sellars, W. F. (1963). *Science, perception and reality.* New York: Humanities Press.

Selman, R. (1980). *The growth of interpersonal understanding.* New York: Academic Press.

Semin, G. R., & Gergen, K. J. (Eds.). (1990). *Everyday understanding: Social and scientific implications.* Newbury Park, CA: Sage.

Serpell, J. (1992). *Effects of pet-acquisition on human health: The search for a mechanism.* Paper presented at Sixth International Conference on Human Animal Interactions, Montreal, Canada.

Seyfarth, R. M., Cheney, D. L., & Marler, P. (1980). Vervet monkey alarm calls: Semantic communication in a free-ranging primate. *Animal Behaviour, 28,* 1070–1094.

Shaffer, P. (1973). *Equus: A play in two acts.* New York: Samuel French.

Shapiro, G. L. (1978). Sign acquisition in a home-reared/free-ranging orangutan: Comparisons with other signing apes. *American Journal of Primatology, 3,* 121–129.

Shapiro, K. (1985). *Bodily reflective modes: A phenomenological method for psychology.* Durham: Duke University Press.

Shapiro, K. (1986). Verification: Validity or understanding. *Journal of Phenomenological Psychology, 17,* 167–179.

Shapiro, K. (1989). The death of the animal: Ontological vulnerability. *Between The Species, 5,* 183–195.

Shapiro, K. (1990a). Animal rights and humanism: The charge of speciesism. *Journal of Humanistic Psychology, 30*(2), 9–37.

Shapiro, K. (1990b). Understanding dogs through kinesthetic empathy, social construction, and history. *Anthrozoös, 3,* 184–195.

Shatz, M., & Gelman, R. (1973). *The development of communication skills: Modifications in the speech of young children as a function of listener. Monographs of the Society for Research in Child Development, 38*(2, Serial No. 152).

Shatz, M., Wellman, H. M., & Silber, S. (1983). The acquisition of mental verbs: A systematic investigation of the first reference to mental state. *Cognition, 14,* 301–321.

Shepard, P. (1978). *Thinking animals: Animals and the development of human intelligence*. New York: Viking Press.

Sherman, P. W. (1988). The levels of analysis. *Animal Behaviour, 36*, 616–619.

Shimp, C. P. (1989). Contemporary behaviorism versus the old behavioral straw man in Gardner's *The mind's new science: A history of the cognitive revolution*. *Journal of the Experimental Analysis of Behavior, 51*, 163–171.

Sidman, M. (1971). Reading and auditory-visual equivalences. *Journal of Speech and Hearing Research, 14*, 5–13.

Sidman, M. (1990). Equivalence relations: Where do they come from? In D. E. Blackman & H. Lejeune (Eds.), *Behavior analysis in theory and practice: Contributions and controversies* (pp. 93–114). Hillsdale, NJ: Lawrence Erlbaum.

Sidman, M., & Tailby, W. (1982). Conditional discrimination vs. matching to sample: An expansion of the testing paradigm. *Journal of the Experimental Analysis of Behavior, 37*, 5–22.

Sidman, M., Wynne, C. K., Maguire, R. W., & Barnes, T. (1989). Functional classes and equivalence relations. *Journal of the Experimental Analysis of Behavior, 52*, 261–274.

Siegler, R. (1986). *Children's thinking*. Englewood Cliffs, NJ: Prentice Hall.

Silverman, P. S. (1983). Attributing mind to animals: The role of intuition. *Journal of Social and Biological Structures, 6*, 231–247.

Silverman, P. S. (1986). Can a pigtail macaque learn to manipulate a thief? In R. W. Mitchell & N. S. Thompson (Eds.), *Deception: Perspectives in human and nonhuman deceit* (pp. 151–167). Albany, NY: SUNY Press.

Simon, H. A. (1979). *Models of thought*. New Haven: Yale University Press.

Singer, P. (1975). *Animal liberation*. New York: Avon.

Singer, P. (1990). *Animal liberation* (2nd ed.). New York: The New York Review of Books.

Singer, P. (1992, April 9). Bandit and friends. *The New York Review of Books, 39*, 9–13.

Skinner, B. F. (1953). *Science and human behavior*. New York: MacMillan.

Skinner, B. F. (1957). *Verbal behavior*. New York: Appleton Century Crofts.

Skinner, B. F. (1961). The operational analysis of psychological terms. In B. F. Skinner (Ed.), *Cumulative record* (pp. 272–286). New York: Appleton Century Crofts.

Skinner, B. F. (1974). *About behaviorism*. New York: Knopf.

Skinner, B. F. (1987). What ever happened to psychology and the science of behavior? *American Psychologist, 42,* 780–786.

Skinner, B. F. (1989). *Recent issues in the analysis of behavior.* Columbus: Merrill.

Skinner, B. F. (1990). Can psychology be a science of mind? *American Psychologist, 45,* 1206–1210.

Slobin, D. I. (1979). *Psycholinguistics* (2nd ed.). Glenview, IL: Scott, Foresman.

Slobodchikoff, C. N., Kiriazis, J., Fisher, C., & Creef, E. (1991). Semantic information distinguishing individual predators in the alarm calls of Gunnison's prairie dogs. *Animal Behaviour, 42,* 713–719.

Small, M. F. (1990). Review of Haraway's *Primate visions. American Journal of Physical Anthropology, 82,* 527–528.

Smith, W. J. (1986). An "informational" perspective on manipulation. In R. W. Mitchell & N. S. Thompson (Eds.), *Deception: Perspectives on human and nonhuman deceit* (pp. 71–97). Albany: SUNY Press.

Smuts, B. B. (1985). *Sex and friendship in baboons.* Hawthorne, NY: Aldine.

Snow, C. E. (1977). The development of conversation between mothers and babies. *Journal of Child Language, 4,* 1–22.

Snowdon, C. T. (1991). Review of Ristau's *Cognitive Ethology. Science, 251,* 813–814.

Sober, E. (in press). Morgan's canon. In D. Cummins & C. Allen (Eds.), *The evolution of mind.* Oxford: Oxford University Press.

Sokolowski, R. (1985). *Moral action: A phenomenological study.* Bloomington: Indiana University Press.

Spinoza, B. (1955). *The chief works of Benedict de Spinoza: On the improvement of the understanding; The ethics; Correspondence.* New York: Dover. (Original works published after 1660.)

Srinivasan, M. V. (1992). Distance perception in insects. *Psychological Science, 1,* 22–26.

Stack, G. J. (1980). Nietzsche and anthropomorphism. *Critica: Revista Hispanoamericana de Filosofia, 12,* 41–71.

Stamps, J. A. (1991). Why evolutionary issues are reviving interest in proximate behavioral mechanisms. *American Zoologist, 31,* 338–348.

Stebbins, S. (1993). Anthropomorphism. *Philosophical Studies, 69,* 113–122.

Stern, W. (1924). *Psychology of early childhood.* New York: Henry Holt & Co.

Sternberg, R. J., Conway, B. E., Ketron, J. L., & Bernstein, M. (1981). People's conceptions of intelligence. *Journal of Personality and Social Psychology, 41*, 37–55.

Stevens, S. S. (1939). Psychology and the science of sciences. *Psychological Bulletin, 36*, 221–263.

Stevenson-Hinde, J., & Zunz, M. (1978). Subjective assessment of individual rhesus monkeys. *Primates, 21*, 473–482.

Stich, L. D. (1983). *From folk psychology to cognitive science.* Cambridge, MA: MIT Press.

Stolti, C. (1990). *The art of crossing cultures.* Yarmouth, ME: Intercultural Press.

Strum, S. C. (1987). *Almost human.* New York: Random House.

Strum, S. C. (1988). Social strategies and primate psychology. *Behavioral and Brain Sciences, 11*, 264–265.

Suarez, S. D., & Gallup, G. G., Jr. (1981). Self-recognition in chimpanzees and orangutans, but not gorillas. *Journal of Human Evolution, 10*, 175–188.

Suls, J., Wan, C. K., & Sanders, G. S. (1988). False consensus and false uniqueness in estimating the prevalence of health-protective behaviors. *Journal of Applied Social Psychology, 18*, 66–79.

Suomi, S. J., & Immelmann, K. (1983). On the process and product of cross-species generalization. In D. W. Rajecki (Ed.), *Comparing behavior: Studying man studying animals* (pp. 203–224). Hillsdale, NJ: Lawrence Erlbaum.

Swartz, K. B. (1990). The concept of mind in comparative psychology. *Annals of the New York Academy of Sciences, 602*, 105–111.

Swartz, K. B., & Evans, S. (1991). Not all chimpanzees (*Pan troglodytes*) show self-recognition. *Primates, 32*, 483–496.

Swartz, K. B., & Evans, S. (1994). Social and cognitive factors in chimpanzee and gorilla mirror behavior and self-recognition. In S. T. Parker, R. W. Mitchell, & M. L. Boccia (Eds.), *Self-awareness in animals and humans* (pp. 189–206). New York: Cambridge University Press.

Tamir, P., & Zohar, A. (1991). Anthropomorphism and teleology in reasoning about biological phenomena. *Science Education, 75*, 57–67.

Taylor, C. (1964). *The explanation of behavior.* London: Routledge & Kegan Paul.

Taylor, S. E. (1982). The availability bias in social perception and interaction. In D. Kahneman, P. Slovic, & A. Tversky (Eds.), *Judgment under uncertainty: Heuristics and biases* (pp. 190–200). New York: Cambridge University Press.

Temerlin, M. (1975). *Lucy: Growing up human*. Palo Alto, CA: Science & Behavior Books.

Terrace, H. S. (1979). *Nim: A chimpanzee who learned sign language*. New York: Knopf.

Terrace, H. S. (1984). Animal cognition. In H. L. Roitblat, T. G. Bever, & H. S. Terrace (Eds.), *Animal cognition* (pp. 7–28). Hillsdale, NJ: Lawrence Erlbaum.

Terrace, H. S., Petitto, L. A., Sanders, R. J., & Bever, T. G. (1979). Can an ape create a sentence? *Science, 206,* 891–902.

Thayer, H. S. (1972). Pragmatism. In P. Edwards (Ed.), *Encyclopedia of philosophy* (Vol. 6, pp. 430–436). New York: Macmillan.

Thines, G. (1977). *Phenomenology and the science of behaviour*. London: George Allen & Unwin.

Thomas, E. M. (1993). *The hidden life of dogs*. Boston: Houghton Mifflin.

Thomas, H. M. (1979). Editorial comment: A lesson from archaeology. *Biology and Human Affairs, 44,* 1–6.

Thomas, K. (1983). *Man and the natural world: A history of modern sensibility*. New York: Pantheon.

Thomas, R. K. (1988). Misdirection and misuse of anecdotes and mental state concepts. *Behavioral and Brain Sciences, 11,* 265–266.

Thomas, R. K., & Noble, L. M. (1988). Visual and olfactory oddity learning in rats: What evidence is necessary to show conceptual behavior? *Animal Learning and Behavior, 16,* 157–163.

Thompson, N. S. (1976). My descent from the monkey. In P. P. G. Bateson & P. K. Klopfer (Eds.), *Perspectives in Ethology* (Vol. 2, pp. 221–230). New York: Plenum Press.

Thompson, N. S. (1987). Natural design and the future of comparative psychology. *Journal of Comparative Psychology, 101,* 282–286.

Thompson, N. S. (1994). The many perils of ejective anthropomorphism. *Behavior and Philosophy, 22,* 59–70.

Thompson, R. L., & Boatright-Horowitz, S. L. (1994). The question of mirror-mediated self-recognition in apes and monkeys: Some new results and reservations. In S. T. Parker, R. W. Mitchell, & M. L. Boccia (Eds.), *Self-awareness in animals and humans* (pp. 330–349). New York: Cambridge University Press.

Thompson, S. (1955). *Motif-index of folk-literature* (rev. ed). Bloomington, IN: Indiana University Press.

Thorpe, W. H. (1972). A comparison of vocal communication in animals and man. In R. A. Hinde (Ed.), *Non-verbal communication* (pp. 27–47). Cambridge, UK: Cambridge University Press.

Thorpe, W. H. (1974a). *Animal nature and human nature*. Garden City, NY: Doubleday.

Thorpe, W. H. (1974b). Reductionism in biology. In T. J. Ayala & T. Dobzhansky (Eds.), *Studies in the philosophy of biology: Reduction and related problems* (pp. 109–138). Berkeley: University of California Press.

Thorpe, W. H. (1979). *The origins and rise of ethology*. London: Heinemann Educational Books.

Timberlake, W. (1990). Natural learning in laboratory paradigms. In D. A. Dewsbury (Ed.), *Contemporary issues in comparative psychology* (pp. 7–54). Sunderland, MA: Sinauer Associates, Inc.

Tinbergen, N. (1951). *The study of instinct*. Oxford: Clarendon Press.

Tinbergen, N. (1960). Discussant for Panel four: The evolution of mind. In S. Tax & C. Callender (Eds.), *Evolution after Darwin* (Vol. 3, pp. 175–206). Chicago: University of Chicago Press.

Tinbergen, N. (1963). On aims and methods of ethology. *Zeitschrift für Tierpsychologie, 20*, 410–433.

Tinbergen, N. (1968). *Curious naturalists*. Garden City, NY: Anchor Books.

Tolman, E. C. (1932). *Purposive behavior in animals and men*. New York: The Century Co.

Tolman, E. C. (1948). Cognitive maps in rats and men. *Psychogical Review, 55*, 189–208.

Toulmin, S. (1961). *Foresight and understanding*. New York: Harper & Row.

Trevarthen, C. (1977). Descriptive analysis of infant communicative behavior. In H. R. Scheffer (Ed.), *Studies in mother-infant interaction* (pp. 227–270). New York: Academic Press.

Trevarthen, C. (1978). Modes of perceiving and modes of acting. In H. J. Pick (Ed.), *Psychological modes of perceiving and processing information* (pp. 100–136). Hillsdale, NJ: Lawrence Erlbaum.

Trevarthen, C. (1979a). Communication and cooperation in early infancy: A description of primary intersubjectivity. In M. Bullowa (Ed), *Before speech: The beginning of interpersonal communication* (pp. 321–347). Cambridge, UK: Cambridge University Press.

Trevarthen, C. (1979b). Instincts for human understanding and for cultural cooperation: Their development in infancy. In M. von Cranach, K. Foppa, W. Lepenies, & D. Ploog (Eds.), *Human ethology* (pp. 530–571). New York: Cambridge University Press.

Trevarthen, C. (1980). The foundations of intersubjectivity: Development of interpersonal and cooperative understanding in infants. In D. Olsen (Ed.), *The social foundations of language and thought: Essays in honor of J. S. Bruner* (pp. 316–342). New York: W. W. Norton.

Trevarthen, C., & Hubley, P. (1978). Secondary intersubjectivity: Confidence, confiding and acts of meaning in the first year. In A. Lock (Ed.), *Action, gesture, and symbol* (pp. 183–229). New York: Academic Press.

Trivers, R. L. (1971). The evolution of reciprocal altruism. *Quarterly Review of Biology, 46,* 35–57.

Trivers, R. L. (1972). Parental investment and sexual selection. In B. Campbell (Ed.), *Sexual selection and the descent of man 1871–1971* (pp. 136–179). Chicago: Aldine.

Trivers, R. L. (1985). *Social evolution.* Menlo Park, CA: Benjamin/Cummings.

Trivers, R. L. (1991). Deceit and self-deception: The relationship between communication and consciousness. In M. H. Robinson & L. Tiger (Eds.), *Man and beast revisited* (pp. 175–192). Washington, DC: Smithsonian Institution Press.

Turiel, E. (1983). Domains and categories in social-cognitive development. In W. F. Overton (Ed.), *The relationship between social and cognitive development* (pp. 53–89). Hillsdale, NJ: Lawrence Erlbaum.

Tuttle, R. H. (1986). *Apes of the world.* Park Ridge, NJ: Noyes.

Urmson, J. O. (1956). *Philosophical analysis: Its development between the wars.* Oxford: Clarendon Press.

U. S. Congress. (1985). Animal Welfare Act (P.L. 99–198). *Federal Register, 54*(49), 10822–10954.

Užgiris, I., & Hunt, J. M. (1975). *Assessment in infancy: Ordinal scales of development.* Urbana: University of Illinois Press.

van Rooijen, J. (1983). Awareness and Griffin's circular reasoning. *Animal Behaviour, 31,* 613–614.

Varela, F. J. (1984). The creative circle: Sketches on the natural history of circularity. In P. Watzlawick (Ed.), *The invented reality* (pp. 309–323). New York: Norton.

Vaughn, W., Jr. (1988). Formation of equivalence sets in pigeons. *Journal of Experimental Psychology: Animal Behavior Processes, 14*, 36–42.

Vedeler, D. (1987). Infant intentionality and the attribution of intentions to infants. *Human Development, 30*, 1–17.

Velmans, M. (1991). Is human information processing conscious? *Behavioral and Brain Sciences, 14*, 651–726.

von Fersen, L., Wynne, C. D. L., Delius, J. D., & Staddon, J. E. R. (1991). Transitive inference formation in pigeons. *Journal of Experimental Psychology: Animal Behavior Processes, 17*, 334–341.

von Glasersfeld, E. (1984). An introduction to radical constructivism. In P. Watzlawick (Ed.), *The invented reality* (pp. 17–40). New York: Norton.

von Uexküll, J. (1909). *Umwelt und Innenwelt der Tiere* [Environment and inner world of animals]. Berlin: Jena.

von Uexküll, J. (1957). A stroll through the worlds of animals and men. In C. H. Schiller (Ed.), *Instinctive behavior* (pp. 5–80). New York: International Universities Press. (Reprinted from von Uexküll, J. [1934]. *Streifzüge durch die Umwelten von Tieren und Menschen*. Berlin: Springer.)

von Uexküll, J. (1985). Environment (Umwelt) and the inner world of animals (C. J. Mellor & D. Gove, Trans.). In G. M. Burghardt (Ed.), *The foundations of comparative ethology* (pp. 222–245). New York: Van Nostrand Reinhold. (Reprinted from von Uexküll, J. [1909]. *Umwelt and Innenwelt der Tiere*. Berlin: Jena.)

Vygotsky, L. S. (1978). Interaction between learning and development. In L. S. Vygotsky, *Mind and society* (pp. 79–91). Cambridge, MA: Harvard University Press. (Original work published 1930.)

de Waal, F. B. M. (1982). *Chimpanzee politics*. London: Jonathan Cape.

de Waal, F. B. M. (1986). Deception in the natural communication of chimpanzees. In R. W. Mitchell & N. S. Thompson (Eds.), *Deception: Perspectives on human and nonhunan deceit* (pp. 221–244). Albany: SUNY Press.

de Waal, F. B. M. (1989). *Peacemaking among primates*. Cambridge, MA: Harvard University Press.

de Waal, F. B. M. (1991). Complementary methods and convergent evidence in the study of primate social cognition. *Behaviour, 118*, 297–320.

de Waal, F. B. M. (1996). *Good natured: The origins of right and wrong in humans and other animals*. Cambridge, MA: Harvard University Press.

de Waal, F. B. M., & van Roosmalen, A. (1979). Reconciliation and consolation among chimpanzees. *Behavioral Ecology and Sociology, 5*, 55–66.

Waddington, C. H. (1960). *The ethical animal.* London: Allen & Unwin.

Wagner, H. (Ed.). (1970). *Alfred Schutz on phenomenology and social relations: Selected writings.* Chicago: University of Chicago Press.

Waldron, T. P. (1985). *Principles of language and mind.* London: Routledge & Kegan Paul.

Walker, S. (1983). *Animal thought.* London: Routledge & Kegan Paul.

Wallace, A. F. C. (1952). The modal personality structure of the Tuscarora Indians as revealed by the Rorschach Test. *Bureau of American Ethnology Bulletin, 150.* Washington, DC: Smithsonian Institution.

Wallman, J. (1992). *Aping language.* New York: Cambridge University Press.

Walls, S. C. (1991). Ontogenetic shifts in the recognition of siblings and neighbours by juvenile salamanders. *Animal Behaviour, 42,* 423–434.

Warden, C. J. (1927). *A short outline of comparative psychology.* New York: W. W. Norton.

Warden, C. J., & Warner, L. H. (1928). The sensory capacities and intelligence of dogs, with a report on the ability of the noted dog "Fellow" to respond to verbal stimuli. *Quarterly Review of Biology, 3,* 1–28.

Washburn, M. F. (1908). *The animal mind.* New York: Macmillan.

Wasserman, E. A., Kiedinger, R. E., & Bhatt, R. S. (1988). Conceptual behavior in pigeons: Categories, subcategories, and pseudocategories. *Journal of Experimental Psychology: Animal Behavior Processes, 14,* 235–246.

Watson, J. B., & Rayner, B. (1920). Conditioned emotional reactions. *Journal of Experimental Psychology, 3,* 1–14.

Webster's New Collegiate Dictionary. (1977). H. B. Woolf (Ed. in chief). Springfield, MA: G. C. Merriam Co.

Weiskrantz, L. (Ed.). (1988). *Thought without language.* Oxford: Oxford University Press.

Weiss, M. (1987). Nucleic acid evidence bearing on hominoid relationships. *Yearbook of Physical Anthropology, 30,* 41–74.

Weller, B. (1980). Gorilla stories. *Gorilla, 3*(2), 2–4.

Wells, M. J. (1978). *Octopus: Physiology and behavior of an advanced invertebrate.* London: Chapman & Hall.

Werner, H. (1948). *Comparative psychology of mental development.* New York: Follett.

Werth, F., & Flaherty, J. (1986). A phenomenological approach to human deception. In R. W. Mitchell & N. S. Thompson (Eds.), *Deception: Perspectives on human and nonhuman deceit* (pp. 293–311). Albany: SUNY Press.

Wertz, F. (1986). The question of reliability. *Journal of Phenomenological Psychology, 17*, 181–205.

Wheeler, O. (1916). *Anthropomorphism and science: A study of the development of ejective cognition in the individual and the race.* London: Allen & Unwin.

Wheeler, W. M. (1902). "Natural history," "ecology" or "ethology"? *Science, 15*, 971–976.

Wheeler, W. M. (1905). Ethology and the mutation theory. *Science, 21*, 535–540.

Wheeler, W. M. (1908). Comparative ethology of the European and North American ants. *Journal für Psychologie und Neurologie, 13*, 404–435.

Whiten, A. (Ed.). (1990). *Natural theories of mind: Evolution, development and simulation of everyday mindreading.* Oxford: Blackwell.

Whiten, A. (1992). Review of Griffin's *Animal Minds. Nature, 360*, 118–119.

Whiten, A. (1993). Evolving a theory of mind: The nature of non-verbal mentalism in other primates. In S. Baron-Cohen, H. Tager-Flusberg, & D. J. Cohen (Eds.), *Understanding other minds: Perspectives from autism* (pp. 367–396). New York: Oxford University Press.

Whiten, A., & Byrne, R. W. (1986). The St. Andrews catalogue of tactical deception in primates. *St. Andrews Psychological Reports, No. 10.*

Whiten, A., & Byrne, R. W. (1988). Tactical deception in primates. *Behavioral and Brain Sciences, 11*, 233–244.

Whiten, A., Byrne, R. W., & Henzi, S. P. (1987). The behavioral ecology of mountain baboons. *International Journal of Primatology, 8*, 367–388.

Whiten, A., & Ham, R. (1992). On the nature and evolution of imitation in the animal kingdom: Reappraisal of a century of research. In P. J. B. Slater, J. S. Rosenblatt, C. Beer, & M. Milinski (Eds.), *Advances in the study of behavior* (Vol. 21, pp. 239–283). New York: Academic Press.

Whiten, A., & Perner, J. (1991). Fundamental issues in the multidisciplinary study of mindreading. In A. Whiten (Ed.), *Natural theories of mind* (pp. 1–18). Oxford: Basil Blackwell.

Whitman, C. O. (1986). Animal behavior. In J. Maienschein (Ed.), *Defining biology: Lectures from the 1890s* (pp. 219–272). Cambridge, MA: Harvard University Press. (Original work published 1898.)

Whitman, C. O. (1899). Myths in animal psychology. *The Monist, 9*, 524–537.

Wickler, W. (1965). Über den taxonomischen Wert homologer VeIhaltensmerk-male [The taxonomic usefulness of homologous characteristic behaviors]. *Naturwissenschaften, 52,* 441–444.

Wiley, E. O. (1981). *Phylogenetics.* New York: Wiley.

Wilkinson, D. (1984). A new job for Koko. *Gorilla, 7*(2), 5.

Williams, G. C. (1966). *Adaptation and natural selection.* Princeton, NJ: Princeton University Press.

Williams, G. C. (1992). *Natural selection: Domains, levels, and challenges.* New York: Oxford University Press.

Wilson, E. O. (1975). *Sociobiology.* Cambridge, MA: Belknap Press.

Wimmer, H., Hogrefe, G.-J., & Perner, J. (1988). Children's understanding of informational access as a source of knowledge. *Child Development, 59,* 386–396.

Wimmer, H., & Perner, J. (1983). Beliefs about beliefs: Representation and con-straining function of wrong beliefs in young children's understanding of deception. *Cognition, 13,* 103–128.

Wind, J. (1989). The evolutionary history of the human speech organs. In J. Wind, E. G. Pulleyblank, E. de Grolier, & B. H. Bichakjian (Eds.), *Studies in language origins* (Vol. 1, pp. 173–197). Amsterdam: John Benjamins.

Wiser, M., & Carey, S. (1983). When heat and temperature were one. In D. Gen-tner & A. L. Stevens (Eds.), *Mental models* (pp. 267–297). Hillsdale, NJ: Lawrence Erlbaum.

Witmer, L. (1909). A monkey with a mind. *Psychological Clinic, 3,* 179–205.

Wittgenstein, L. (1953). *Philosophical investigations.* New York: The Macmillan Co.

Wuensch, K. L., George, P. S., Poteat, G. M., Castellow, W. A., & Pryor, W. H., Jr. (1992). *Ethical ideology and support for animal research.* Paper presented at Animal Behavior Society Meeting, Kingston, Canada.

Wuensch, K. L., Poteat, G. M., & Jernigan, L. M. (1991). *Support for animal rights and perceived similarity between humans and other animals.* Paper presented at Animal Behavior Society Meeting, Wilmington, NC.

Wynne-Edwards, V. C. (1962). *Animal dispersion in relation to social behaviour.* Edinburgh: Oliver & Boyd.

Yerkes, R. M. (1925). *Almost human.* New York: Century Co.

Yerkes, R. M. (1927). A program for anthropoid research. *American Journal of Psychology, 39,* 181–199.

Yoerg, S. I. (1991). Ecological frames of mind: The role of cognition in behavioral ecology. *Quarterly Review of Biology, 66*, 287–301.

Yoerg, S. I. (1992). Mentalist imputations. [Review of D. R. Griffin's *Animal minds*]. *Science, 258*, 830–831.

Yoerg, S. I., & Kamil, A C. (1991). Integrating cognitive ethology with cognitive psychology. In C. A. Ristau (Ed.), *Cognitive ethology* (pp. 273–289). Hillsdale, NJ: Lawrence Erlbaum.

Young, J. Z. (1977). Brain, behavior and evolution of cephalopods. In M. Nixon & J. B. Messenger (Eds.), *Symposia of the Zoological Society of London; No. 38. The biology of cephalopods* (pp. 377–434). London: Academic Press.

Young, R. M. (1985). *Darwin's metaphor: Nature's place in Victorian culture*. Cambridge, UK: Cambridge University Press.

Zabel, C. J., Glickman, S. E., Frank, L. G., Woodmansee, K. B., & Keppel, G. (1992). Coalition formation in a colony of prepubertal hyenas. In A. H. Harcourt & F. B. M. de Waal (Eds.), *Coalitions and alliances in humans and other animals* (pp. 114–135). New York: Oxford University Press.

Zajonc, R. B. (1980). Feeling and thinking: Preferences need no inferences. *American Psychologist, 35*, 151–175.

Zak, S. (1989). Ethics and animals. *Atlantic Monthly, 263*(3), 69–74.

Zihlman, A. L., Cronin, J. E., Cramer, D. L., & Sarich, V. M. (1978). Pygmy chimpanzee as a possible prototype for the common ancestor of humans, chimpanzees, and gorillas. *Nature, 275*, 744–746.

Zuckerman, S. (1991, May 30). Apes R not us. *New York Review of Books, 38*, 43–49.

Zuriff, G. E. (1985). Disbelieving cognitive psychology: A review of Stich's *From folk psychology to cognitive science*. *Journal of the Experimental Analysis of Behavior, 44*, 391–396.

Author Index

Agassi, J., 3, 52
Akins, K., 281
Akmajian, A., 365
Albert, M., 343
Alcock, J., 324–25
Alexander, R. D., 62
Allen, C., 10, 260, 313, 322, 327, 329, 332, 409, 411, 412, 417, 421, 423, 425, 429
Allen, G., 323, 330
Altmann, J., 134, 137
Altmann, S. A., 142, 224, 366
Andersen, S. M., 68
Anderson, J. G. T., 367
Anderson, J. R., 300, 301, 305, 306, 349
Anonymous, 213, 254
Anscombe, G. E., 288
Antinucci, F., 349, 353
Aquinas, T., 386
Archer, J., 329
Aristotle, 51
Asquith, P., xix, 4, 22, 27, 29, 39–40, 49, 92, 98, 100, 317, 409, 410, 411, 412, 413, 415–16, 417, 418, 429
Astington, J., 350, 352
August, P. V., 367
Austin, J. L., 199, 419
Avis, J., 98
Ayer, A. J., 199

Bacon, F., 51, 54
Baldwin, J. M., 348, 351
Balfour, D., 263, 272, 332, 425
Barber, T. X., 410
Bard, K. A., 305, 306, 349
Barlow, G. W., 256
Barnes, T., 379
Baron, R. M., 65, 69
Bartlett, F. C., 299
Basen, J. A., 300, 301
Bates, E., 351, 404
Bateson, G., 396
Bateson, M. C., 396
Bateson, P., 263, 350
Bazar, J., 390
Beaver, B. V., 179
Beck, A. T., 182
Beck, B. B., 327, 331
Beckstead, J. W., 247, 251
Beer, C. G., 7, 171, 193, 198, 207–8, 238, 326–27, 409, 412, 415, 416, 429
Beer, T., 25
Bekoff, M., 10, 151, 237, 238, 239, 254, 260, 263, 264, 313, 314–17, 322, 323, 325, 326, 327–28, 329–30, 331, 332–33, 409, 411, 412, 417, 421, 423, 425, 429
Bell, A. G., 19
Bellugi, U., 392
Beloff, J., 25

Benedict, H., 82
Bennett, J., 167, 173, 326, 331
Bennett, W. L., 158–62
Bentham, J., 239
Berger, P. L., 282
Bergeron, G., 349, 353
Bernstein, M., 238
Berry, D. S., 65
Berry, J. K., 27, 247
Betchart, N. S., 242, 247, 248, 251
Bethe, A., 25
Bever, T., 349, 384, 397
Bhatt, R. S., 260
Bickerton, D., 367
Biddle, B. J., 176
Bierens de Haan, J. A., 25, 261, 267, 268
Bierschwale, D., 101, 299, 305–7, 309, 426
Bigelow, J., 172
Binet, M., 336, 346
Bishop, N., 349
Bitterman, M. E., 94
Black, M., 30, 206
Blough, D. S., 343
Boakes, R., 173, 336
Boatright-Horowitz, S. L., 306
Boccia, M. L., 233, 310, 349
Boersman, D., 82
Booth, W. C., 23, 30
Boren, J. L., 304
Bornstein, M. H., 181
Botkin, P., 350, 352
Boucher, D. H., 70, 71
Bouissac, P., 71, 72
Boyd, R., 27, 30
Boysen, S. T., 87, 96, 349, 390, 397
Bradford, S. A., 343
Bradshaw, J. L., 300
Brainerd, C. J., 262
Brakke, K. E., 391, 400
Bretherton, I., 351, 353
Breuggeman, J. A., v, 127, 166
Bridgeman, P. W., 199
Broadhurst, P. L., 336
Brockmann, J., 327

Brooks, D. R., 272, 359
Brooks-Gunn, J., 350
Brown, R., 351
Bruner, J., 351
Budd, R. L., 400
de Buffon, G. L. L., 95
Bundy, R. A., 157, 300, 303, 306
Bunney, W. E., 182
Burghardt, G. M., 9, 10, 27, 41, 46, 67, 71, 73, 84, 156, 168, 237, 238, 251, 254, 255, 257–63, 265, 267–73, 313, 317, 327, 329, 401, 409, 412, 421, 425, 430
Burke, K., 120–21
Butterworth, G. E., 420
Buytendijk, F. J. J., 129, 261
Byne, W., 152
Byrne, R. W., xix, 7, 27, 48, 62, 63, 127, 134, 135–36, 138–50, 151, 156, 157, 162–63, 168, 172, 173, 175, 177, 230, 234, 272, 288, 329, 349, 351, 354, 409, 422–25, 430

Caine, N. G., 175
Calhoun, S., 306
Callicott, J. B., 27, 28
Campbell, C. B. G., 83
Candland, D. K., 45, 401
Capaldi, E. J., 349
Caporael, L. R., 4, 6, 10, 43, 59, 61–62, 63, 69, 151, 153, 296, 409, 410, 411, 412, 413, 417, 418, 420, 421, 430
Carey, S., 64
Carpenter, C. R., 27
Carruthers, P., 323–24
Cartmill, M., 167
Case, R., 351
Catania, A. C., 371
Cavalieri, P., 73, 417
Cenami Spada, E., xvii, 4–5, 6, 37, 47, 409, 410, 412, 415, 416, 427, 430
Chandler, M., 99
Cheney, D. L., xvi, 48, 127, 175–77, 182, 203, 230, 234, 269, 319, 327, 332, 349, 350, 367, 384, 391, 392
Chevalier-Skolnikoff, S., 349, 353, 355

Chihara, C. S., 201, 202
Cho, S., 93
Chomsky, N., 365
Church, R. M., 343
Churchland, P. M., 201, 209
Churchland, P. S., 209
Cirillo, L., 23, 30, 31
Clark, S. R. L., 323
Clements, K. C., 314
Cohn, R., 301, 398
Colgan, P., 203, 205, 208, 317, 318–19
Collins, H., 7, 153, 169, 401
Conway, B. E., 238
Cooley, C. H., 157, 300
Cooper, R. W., 296, 308
Cosmides, L., 340
Couch, J. B., 389, 400
Coussi-Korbel, S., 144
Craft, H. D., 238, 240, 251
Craik, K. J. W., 147
Cramer, D. L., 400
Creef, E., 367–68
Crider, C., 23, 30, 31
Crocker, D. R., 125, 252
Cronin, H., 322–23, 330
Cronin, J. E., 400
Crook, J. H., 365, 368
Cuvier, G. L., 95

Darwin, C., 4, 12–18, 25, 39, 95,
    101–2, 154, 206, 256, 259, 339, 344,
    348, 359, 408–9, 413, 427
Dasser, V., 79, 173, 326, 328
Davidson, K., 399, 402
Davies, N. B., 255
Davis, H., 10, 153, 263, 272, 332, 335,
    343, 344–45, 409, 412, 413, 414,
    415, 417, 420, 425, 430
Davis, K. E., 78
Davis, W., 154, 155
Dawes, R. M., 61–62, 69, 145
Dawkins, M. S., 256
Dawkins, R., 205–7, 208, 368
deBlois, S., 87, 97, 99
Delius, J. D., 343
Demers, R., 365

DeMonbreun, B. G., 173
Dennett, D. C., 60, 83, 127, 128, 146,
    150, 172, 176, 207, 238, 288, 289,
    291, 295, 328, 329, 331, 424
Descartes, R., 38, 45, 199, 268, 365,
    386, 387
Dettwyler, K. A., 153
DeVore, I., 222–23, 224, 229
Dewsbury, D. A., 254, 255, 303
Dickens, C., 18
Dirven, R., 30
Doherty, M. E., 173
Dolhinow, P., 349
Donceel, S. J., 291
Doré, F. Y., 182, 349
Dostoyevsky, F., 262
Dretske, F., 203, 204, 208
Driscoll, J. W., 247, 251
Dube, W. V., 376
Dücker, G., 300, 303–4, 305
Dumas, C., 182, 349
Dunbar, R. I. M., 60, 62, 230, 331
Dunlop, J., 251

Earle, H., 175
Eddy, T. J., 5–6, 9, 77, 79, 82, 83–84,
    88, 93, 96, 100, 101, 151, 164, 167,
    240, 244, 335, 346, 408, 409, 411,
    413, 415, 420, 424, 430
Edwards, D., 268
Eglash, A. R., 300
Eibl-Eibesfeldt, I., 181
Einstein, A., 410
Ekman, P., 102
Eliot, G., 203
Elstein, A. S., 180
Empson, W., 206
Epstein, R., 315
Erickson, H. H., 127
Essock-Vitale, S., 234
Estep, D. Q., 317
Etienne, B., 219
Evans, S., 7, 10, 101, 158, 296, 299,
    301–7, 409, 411, 413, 420, 422, 423,
    424, 425, 430
Everett, J. W., 173

Faraday, M., 17
Faust, D., 61
Fegley, S., 305
Feigl, H., 199
Feinberg, R. R., 83
Feldman, M. S., 158–62
Fentress, J. C., 329
Feyerabend, P. K., 129, 202
Fields, R., 82
Fisher, C., 367–68
Fisher, J. A., 22, 23, 33, 34, 60, 273, 317, 410
Fisler, G. F., 221
Flaherty, J., 163
Flavell, J. H., 97, 350, 352
Fodor, J. A., 201, 202, 203, 204, 320
Forguson, L., 128
Fossey, D., 160, 234
Fouts, D. H., 367
Fouts, R. S., 46, 367, 400, 401
Fox, M. W., 177, 285
Frank, L. G., 326
Fraser, A. F., 114
Frayer, D. W., 154
Freeman, D., 386
Freeman, K., 38, 51
Freud, S., 54–55
von Frisch, K., 258
Frith, U., 103
Fritz, A. S., 99
From, F., 291
Fry, C., 350, 352
Fuller, J., 285
Furnam, A., 238
Furness, W. H., 388
Furrow, D., 97

Gagliardi, G. J., 304
Galdikas, B., 349, 386
Galdikas-Brindamour, B.; *see* Galdikas, B.
Galef, B. G., Jr., 314
Gallistel, C. R., 344, 350
Gallup, G. G., Jr., 4–6, 9, 48, 64, 77, 78, 79, 80, 82, 83–84, 85, 88–89, 90, 93, 151, 157–58, 164, 167, 172, 173,

176, 181, 182, 233, 240, 244, 247, 297–310, 345, 346, 356, 408, 409, 411, 413, 415, 420, 424, 426, 431
Galvin, S. L., 8, 9, 83, 93, 151, 164, 167, 237, 251, 261, 346, 409, 424, 431
Garcia, J., 345
Gardner, B. T., 46, 349, 354, 367, 384, 388–89, 398, 400
Gardner, R. A., 46, 349, 354, 367, 388–89, 398, 400
Gargett, R. H., 153
Garner, R. L., 226, 387, 388
Gavan, J. A., 309
Gelman, R., 82, 344, 350
Gendlin, E. T., 254, 281
Gentner, D., 414
Georgopoulos, A. P., 257
Gergen, K. J., 238
Gibbon, J., 343
Gibson, K. R., 259, 349, 366, 368
Gill, T. V., 46, 388, 391
Gisiner, R., 11, 370, 371–73, 376, 379–82, 409, 414, 431
Gittleman, J. L., 271, 334
Gladwin, T., 102
von Glasersfeld, E. C., 42, 46, 388, 391
Glass, A. J., 182
Glickman, S. E., 326
Godfrey, L. R., 98
Goethe, J. W. von, 51
Goldiamond, I., 378
Goldman-Rakic, P. S., 94
Goodall, J. van Lawick-, 57, 89, 135, 144, 148, 174, 224, 233–34, 262, 283, 300, 303, 332, 349
Goodman, N., 119–20
Gopnik, A., 97
Gordon, W. J. J., 64
Gould, S. J., 283, 359
Graf, P., 97
Graham, P., 182
Gray, P. H., 25
Green, G., 376
Greenfield, P., 349, 351

Griffin, D. R., xix, 9, 10, 25, 26–27, 48, 60, 95, 127, 160, 168, 172, 173, 179, 216, 256, 258, 259, 260, 262, 271, 296, 303, 314, 316–18, 320, 322–24, 326–27, 329–31, 332, 336, 338–39, 341–43, 345, 349, 366, 415
Griffiths, P. H. M., 128, 129, 178
Grimm, B. K., 371, 373, 376, 382
de Grolier, E., 368
Gruber, H. E., 13
Gruen, L., 239, 263, 332
Guralnick, M. J., 82
Gustafson, D., 329
Guthrie, S. E., 3, 4, 50, 53, 55, 90, 167, 393, 408, 409, 410, 412, 420, 431
Gyger, M., 207, 367

Haeckel, E., 359
Hafez, E. S. E., 115
Hager, J. L., 339
Hailman, J. P., 25, 26, 314
Hala, S., 99
Hall, G. S., 348
Ham, R., 349
Hamilton, W. D., 228, 229
Hamm, M., 3, 151, 164–65, 167, 416, 424
Handler, P., 106
Hanggi, E. B., 371, 373, 376, 382
Hanlon, R. T., 214
Hansen, B., 61
Hanson, N. R., 202
Haraway, D., 159, 167, 395
Harcourt, A. H., 27, 33
Hardy, A., 214
Harlow, H., 349
Harnish, R., 365
Harper, S., xix, 383, 391
Harris, B., 299 (1979)
Harris, P., 98, 350, 352, 420
Harris, R., 23
Harrison, B., 387
Harrison, R., 106, 115
Hart, D., 305
Harvey, P. H., 272
Hauser, M. D., 322, 327, 329, 332

Hausfater, G., 138
Hawking, S. W., 213
Hayes, C., 46, 348, 388
Hayes, K., 388
Hearne, V., 410
Hebb, D. O., xvi, 40, 128, 168, 170–71, 179, 257, 261, 263, 267, 268, 273, 417, 423
Hediger, H., xiii
Hegel, G. W. F., 365
Heidegger, M., 286
Heider, F., 65, 78, 79
Heinroth, K., 25
Heinroth, O., 25
Hempelmann, F., 254
Hennig, W., 359
Henzi, S. P., 139–40
Herder, J. G., 365
Herman, L. M., 46, 370–71, 372, 375, 376
Herrnstein, R. J., 203, 343
Herzog, H. A., Jr., 8, 9, 83, 93, 151, 164, 167, 237, 238, 242, 247, 248, 251, 254, 261, 273, 346, 409, 424, 431
Hess, E., 254
Hess, L., 387
Hesse, M., 31
Hetts, S., 317
Heyes, C., 4, 6, 10, 59, 148, 153, 296, 316–17, 318, 319, 322, 325, 409, 410, 411, 412, 413, 417, 418, 420, 421, 431
Hill, S. D., 157, 300, 303, 306
Hinde, R. A., 258–59, 349
Hirsch, E. D., Jr., 161
Hobbes, T., 365
Hobhouse, L. T., 422
Hochberg, F. G., 214
Hockett, C. F., 366
Hodges, B. H., 69
Hodos, W., 83
Hoffman, M. L., 351
Hoffman, R. R., 24, 30
Hoffmann, R., 267
Hogrefe, G.-J., 97
Holyoak, K. J., 182

Honig, W. K., 343
Hopson, J., 399, 402
Houghton, W. E., 340
Houston, A. I., 319
Howes, D., 413
Hoyningen-Huene, P., 173, 326, 328
Hoyt, A. M., 387
Hrdy, S. B., 134
Hubley, P., 116, 118
Hull, C. L., 408
Humboldt, W. von, 365
Hume, D., 50, 51, 54, 128
Humphrey, N. K., 62–63, 73, 78, 146, 147, 160, 173, 232–35, 317, 318
Hunt, J. M., 351, 353
Huntingford, F., 317, 318, 330
Hurlbert, E. M., 331
Huxley, J. S., 26–27

Immelmann, K., 172
Inagaki, K., 60
Ingold, T., 318
Izard, C. E., 259

Jacobsen, C., 387
Jacobsen, M. M., 387
Jacobson, J. L., 82
Jager, J., 286
James, W., 199, 209, 340
Jamieson, D., 238, 264, 313, 314–15, 323, 325, 327–28, 331, 332
Jamison, W. V., 70–71
Jarvis, P., 350, 352
Jasper, J. M., 238
Jennings, H. S., 269
Jernigan, L. M., 240
Johnson, C. N., 97
Johnson, E., 323
Johnson, M., 23, 31–32
Johnson, M. H., 349
Johnson-Laird, P. N., 340
Jolly, A., 135, 232, 349, 387
Jones, E. E., 68, 78

Kahneman, D., 56, 61
Kamil, A. C., 95, 314, 324, 343

Kamin, A., 343
Kant, E., 41, 199
Karakashian, S. J., 207
Kastak, D., 376, 378, 382
Katz, D., 285
Kean, K. J., 65
Kearton, C., 387
Kellert, S. R., 27–28, 82–83, 247, 251, 423
Kelley, H. H., 78
Kellogg, L. A., 46, 348, 387, 388, 395–96
Kellogg, W. N., 46, 348, 387, 388, 395–96
Kennedy, J. S., xvi, 4, 24, 27, 39, 40–41, 42, 44, 46, 47, 49, 52, 60, 317–18, 321, 331, 410, 412
Kepler, J., 64
Keppel, G., 326
Ketron, J. L., 238
Keyan, R., 365
Kiedinger, R. E., 260
Kim, J., 320
Kinji, I., 429
Kiriazis, J., 11, 365, 367–68, 391, 404, 409, 411, 431
Kitchell, J. A., 171
Kitchell, R. L., 127
Kitchener, R. F., 127
Klima, E., 392
Klopfer, P. H., 25, 26
Knoll, E., 4, 12, 25, 408–9, 411, 412, 413, 431
Köhler, W., 173, 269, 348
Kohts, N., 348, 387
Konner, M., 362
van de Kragt, A. J. C., 61–62, 69
Krebs, J. R., 206–7, 208, 255, 368
Krieger, K., 46, 371–72, 382
Kuhn, T. S., 202, 331
Kummer, H., 173, 326, 328

Laidler, K., 388
Lakoff, G., 23, 31–32
Lamarck, J. B., 95
Landau, K., 101, 299, 305–7, 309, 426

Landau, M., 401
Langer, J., 350, 352
Lapore, E., 204
Latour, B., 61
Lauder, G. V., 255
Lawrence, E. A., 156
Leahy, M. P. T., 323
Leakey, L. S. B., 224
Leakey, R., 153
Ledbetter, D. H., 300, 301
Lee, D. N., 342
Leekman, S. R., 103
Lehman, H., 6, 104, 317, 332, 335, 409, 412, 414, 416, 417, 418, 421, 424, 431
Lennon, E., 145
Leslie, A. M., 97, 103, 351, 420
Lethmate, J., 300, 303–4, 305
Levinson, B. M., 181
Levvis, G. W., 332
Levvis, M. A., 332
Lewin, R., 392
Lewis, C., 420
Lewis, M., 350
Libet, B., 340
Lieberman, P., 366, 368, 388
Liebert, A., 51
Likona, T., 351
Lin, A. C., 305, 306, 349
Linden, E., 174, 236, 386, 398, 402
Linnaeus, C., 95
Lintz, G. D., 387
Lloyd Morgan, C.; *see* Morgan, C. L.
Lock, A., 401
Lock, M., 29
Lockwood, R., 238, 345
London, J., 262
Longino, H. E., 71
Lorenz, K., 26, 126, 199, 208, 252, 263
Lovejoy, A. O., 386
Loveland, D. H., 203, 343
Loveland, K. A., 419
Luckens, D. R., Jr., 309
Luckmann, T., 282
Luh, H.-K., 334
Lunch, W. M., 70–71
Lynch, J. J., 332

MacKinnon, J., 387
Maddison, D., 358
Maddison, W., 358
Maguire, R. W., 379
Mahoney, M. J., 173
Maida, A. S., 93
Maier, N. R. F., 303, 310
Mandler, J. M., 352
Manning, A., 257
Maple, T. L., 386
Marino, L., 5–6, 9, 77, 151, 167, 431–32
Marler, P., 207, 332, 367
Marlin, M. M., 176
Marshall Thomas, E.; *see* Thomas, E. M.
Martin, J. A., 176
Martin, P., 257
Maser, J. D., 90
Mason, W. A., 227–28, 229, 260, 317, 327
Mathieu, M., 349, 353
Matsuzawa, T., 389
Maynard Smith, J., 215, 229
McAdams, D., 401
McCleery, R. H., 319
McClure, M. K., 157, 300, 303, 306
McDonald, K., 46, 143
McDougall, W., 259
McFarland, D., 319, 320, 322
McGrew, W. C., 325
McIlvane, W. J., 376
McKinney, M., 359, 362
McKinney, W. T., 182
McLennan, D. A., 272, 359
McMahan, P., 422
McNamara, K., 359, 362
McNamee, T., 156, 422
McPhail, E. M., 94, 95
Mead, G. H., 351
Mead, M., 396
Meltzoff, A. N., 280, 352, 353, 362, 419
Menzel, E., Jr., 89, 127, 135, 136, 148, 172, 173, 370
Merleau-Ponty, M., 9, 261, 279, 294

Messenger, J. B., 214
Michel, G. F., xiv, 324
Midgley, M., 205
Mignault, C., 349, 353
Miles, H. L., 3, 11, 46, 85, 143, 151,
  254, 304, 349, 383, 389–94,
  399–401, 409, 411, 414, 419, 423,
  425, 432
Mill, J. S., 24
Miller, F. G., 83
Miller, L. L., 24
Millikan, R. G., 6–7, 68, 154, 189, 190,
  193–96, 318, 408, 409, 416
Mills, E., 156, 422
Mills, W., 143, 425
Misovich, S. J., 65
Mitchell, P., 420
Mitchell, R. W., xvii, 3, 4, 7, 9, 22, 37,
  48, 61, 67, 70, 127, 135–36, 138,
  143, 148, 151–56, 158, 160, 163,
  164–65, 167, 172, 173, 224, 233,
  237, 240, 254, 259, 262, 272, 291,
  296, 304, 310, 314, 349, 351, 354,
  383, 384, 393, 401, 407, 409, 410,
  411, 413, 416, 419, 420, 421,
  423–25, 432
Mivart, St. G. J., 18–19, 156, 296
Montet-White, A., 154
Montgomery, S., 332
Moore, C., 97
Moore, M., 352, 353
Morgan, C. L., 20, 25, 62, 139, 143,
  172–73, 257, 269, 273, 326, 346,
  413, 422
Morgan, E., 368
Morrel-Samuels, P., 370
Morris, D., 386
Morris, M., 7, 160
Morris, R., 386
Morton, D. B., 71, 128, 129, 178, 179,
  268
Morton, E. S., 367
Morton, J., 349
Moses, L. J., 97
Moss, C., 132, 220–21
Mounoud, P., 352

Moynihan, M., 8, 213, 214, 215, 217,
  409, 412, 416, 423, 432
Munn, C. A., 140–41
Murphy, J., 391, 400
Müller, F. M., 18
Myers, G., 331, 332
Mynatt, C. R., 173

Nagel, T., 264, 281, 293, 414, 425
Nakajima, S., 9, 28, 167
Napier, J. R., 308
Napier, P. H., 308
Neisser, U., 299
Nelkin, D., 238
Nelson, K. E., 87, 96
Newell, A., 146
Newport, E., 82
Nietzsche, F., 51, 53, 55
Nisbett, R. E., 421
Nissen, H. W., 308
Nitecki, M. H., 171
Noble, L. M., 343
Norbeck, E., 29
Noske, B., 293
Novak, M. A., 87, 96, 179, 263, 272

O'Neill, D. K., 97
O'Sullivan, C., 398
Ohnuki-Tierney, O., 28, 29
Olsen, D., 351, 352
Olson, K., 82
Olton, D. S., 216, 343
Orbell, J. M., 61–62, 69
Ortony, A., 32
Oxnard, C., 95

Packard, A., 218
Packer, C., 138
Pagel, M. D., 272
Paprotte, W., 30
Parker, G. A., 215, 229
Parker, S. T., xix, 10, 158, 168, 171,
  233, 259, 301, 310, 348, 349, 352,
  353, 355, 356, 359, 362, 366, 368,
  409, 412, 413, 414, 422, 425, 432
Parks, K. A., 87, 96

Parsons, B., 152
Pascal, B., 56
Pate, J. L., 94, 356
Patterson, F. G. P., 46, 174, 233, 301,
    349, 389, 398–99, 401–2
Paul-Brown, D., 82
Pavlov, I. P., 59
Pellizzer, G., 257
Pepperberg, I. M., 46, 314, 343, 390,
    419
Pepys, S., 387
Perner, J., 97, 103, 146, 420
Peters, C. B., 262, 332
Peters, R., 422
Peterson, D., 300
Petitto, L. A., 384, 397
Peyrefitte, A., 219
Pérusse, R., 344–45
Phillips, M. T., 252, 332
Phythyon, S., 309
Piaget, J., 57, 63, 162, 172, 269, 349,
    350, 351, 352–53, 355–56, 359, 362
Pickering, A., 61
Pickert, R., 367
Pinch, T., 7, 153, 169, 401
Pisani, P., 391
Pittman, R. B., 242, 247, 248, 251
Plooij, F., 147–48, 162
Plous, S., 164
Poirier, F., 349
Polanyi, M., 279–80
Popper, K. R., 173, 202
Porter, D. G., 263
Poteat, G. M., 240
Potter, J., 268
Povinelli, D. J., 6, 48, 60, 64, 77, 79, 82,
    83–84, 85, 87, 88, 92, 93, 96, 97, 98,
    99, 100, 101, 147, 164, 167, 240,
    244, 299, 300, 304, 305–7, 309–10,
    346, 349, 408, 409, 411, 413, 415,
    420, 421, 424, 426, 432–33
Premack, A. J., 396
Premack, D., 46, 48, 79, 86, 87, 96, 99,
    102, 172, 173, 325, 328, 339, 349,
    351, 388, 389–90, 396
Preston, D. J., 395

Preuss, T. M., 94
Pryor, K., 67
Pure, K., 97
Purton, A. C., 30
Pusey, A. E., 134
Putnam, H., 37, 45

Quiatt, D., 8, 220, 234, 409, 411, 416,
    424, 425, 433
Quine, W. V. O., 107, 202, 418

Rachels, J., 330
Radner, D., 365
Radner, M., 365
Rajecki, D. W., 237, 238, 240, 251
Rasmussen, J. L., 238, 240, 251
Raven, H. C., 387, 395
Rayner, B., 299
Real, L. A., 330, 332
Reber, A. S., 262
Reddish, P. E., 342
Redshaw, M., 349
Regan, T., 239
Reite, M., 175
Rescorla, R. A., 343
Reyna, V. F., 262
Reynolds, P. C.,
Reynolds, V., 234
Ribot, T., 50
Richards, D. G., 46, 371, 372, 375,
    376
Richards, P., 417
Richards, R. J., 62, 350
Ricoeur, P., 292
Ridley, M., 52, 317, 359
Riesen, A. H., 308
Ristau, C. A., 207, 265, 327–29, 332,
    349
Robert, S., 303
Roberts, W. A., 343
Roberts, W. H., 296
Rodaniche, A. F., 217
Rodriguez, M., 307
Rogers, C., 278–79
Rogers, L. J., 300
Roitblat, H. L., 314, 349

Rollin, B. E., 6, 7, 52, 72, 125, 126, 127, 128, 130, 168, 172, 238, 239, 252, 257, 263, 409, 411, 412, 414, 415, 421, 423–25, 433
Romanes, E., 16
Romanes, G. J., 4, 9, 12–13, 16–20, 25, 41, 62, 130, 131, 171, 172, 257, 259, 262, 269, 335, 348, 408–9, 413, 427
van Roosmalen, A., 27
Rorty, R., 199
Rosenberg, A., 320
Rosenblueth, A., 172
Rosenthal, R., 389
Ross, G. A., 83
Ross, L., 68
Rulf, A. B., 101, 299, 305–7, 309, 426
Rumbaugh, D. M., 46, 94, 356, 388, 391, 396–97, 400
Russell, B., 21
Russell, R., 6, 116, 385, 409, 411, 413, 417, 419, 422, 433
Russon, A., 349
Russow, L.-M., 264
Rutter, M., 182
Ryle, G., 120, 200, 201, 365

Sanders, G. S., 84
Sanders, R. J., 297, 384
Sapontzis, S. F., 238, 239
Sappington, A. A., 159
Sarason, S. B., 102
Sarich, V. M., 400
Sartre, J. P., 289
Savage-Rumbaugh, E. S., 46, 143, 349, 389, 391, 392, 396–97, 400–401, 404
Scheler, M., 295
Schiller, C. H., 208
Schusterman, R. J., 11, 46, 370–73, 375–76, 378–82, 409, 414, 433
Scott, P. S., 285
Searle, J., 40
Sebeok, T. A., 383–84, 389
Seligman, M. E. P., 339
Sellars, W. F., 64, 202
Selman, R., 351
Semin, G. R., 238

Serpell, J., 251
Seton, E. T., v, 262
Sevcik, R. A., 391, 400
Seyfarth, R. M., xvi, 48, 127, 175–77, 182, 203, 230, 234, 269, 319, 327, 332, 349, 350, 367, 384, 391, 392
Shaffer, P., 237
Shakespeare, W., 17, 55, 204
Shapiro, G. L., 389
Shapiro, K., 9, 238, 261, 277, 283, 292, 294–95, 409, 411, 413, 415, 421, 425, 433
Shatz, M., 82, 97
Shepard, P., 282
Sherman, P. W., 259
Shimp, C. P., 343
Shulman, L. S., 180
Sidman, M., 370, 371, 378–79, 382
Siegler, R., 352
Silber, S., 97
Silverman, P. S., 8, 170, 172, 177, 179, 182, 272, 409, 412, 419, 421, 423, 434
Simmel, M., 65, 79
Simon, H. A., 146, 180
Singer, P., 73, 129, 178, 239, 244, 323, 417
Skinner, B. F., 47, 94, 201, 257, 261, 262, 264, 265, 268, 292, 338, 339
Slobin, D. I., 351
Slobodchikoff, C. N., 11, 365, 367–68, 391, 404, 409, 411, 434
Slovic, P., 61
Smith, J., 351
Smith, J. A., 71, 268
Smith, S. T., 396
Smuts, B., 27, 48, 176–77, 222–24, 226, 230, 234, 416, 423
Snow, C. E., 174
Snowdon, C. T., 300, 325–26
Sober, E., 326
Sokolowski, R., 279
Southcombe, A., 399
Spada, E. Cenami; *see* Cenami Spada, E.
Spencer, H., 200, 209

Spinoza, B., 51, 54
Spinozzi, G., 353
Sprafka, S. A., 180
Srinivasan, M. V., 270–71
Stack, G. J., 51
Staddon, J. E. R., 343
Stamps, J. A., 255
Stebbins, S., 30
Stern, W., 166
Sternberg, R. J., 238
Stevens, S. S., 199
Stevenson-Hinde, J., 175
Stich, L. D., 203, 209, 238, 339
Stoddard, L. T., 376
Stolti, C., 168
Strum, S. C., 27, 143, 220–26, 230, 234, 387
Suarez, S. D., 82, 158, 173, 181, 299, 300, 301, 303, 304
Sugiyama, K., 60
Suls, J., 84
Suomi, S. J., 172, 179, 263, 272
Swartz, K. B., 7, 10, 101, 158, 296, 299, 301–7, 409, 411, 413, 420, 422, 423, 424, 425, 434

Tailby, W., 379
Tamir, P., 24
Tansey, E. M., 214
Taylor, C., 204–5
Taylor, S. E., 68
Temerlin, M., 387
Terrace, H. S., 46, 339, 349, 383–84, 389, 393, 397–98, 401, 403
Thayer, H. S., 178
Thinés, G., 261
Thomas, E. M., xvi, 410
Thomas, H. M., 26
Thomas, K., 13
Thomas, R. K., 296, 343, 345
Thompson, N. S., 3, 48, 67, 127, 135, 136, 155, 157, 237, 254, 268, 272, 291, 349, 351, 354, 414, 416, 420, 434
Thompson, R. K. R., 296
Thompson, R. L., 306

Thompson, S., 53
Thompson Seton, E.; *see* Seton, E. T.
Thorpe, W. H., 25, 216, 366
Timberlake, W., 314
Tinbergen, N., 26, 107–8, 126, 199, 254, 255, 263, 265, 275, 429
Tolman, E. C., 216, 269, 292
Toth, N., 402
Toulmin, S., 202
Townsend, S. E., 239, 263, 313, 325, 327, 331
Trevarthen, C., 116–22
Trivers, R. L., 135, 229, 270–71
Turiel, E., 351
Tuttle, R. H., 303
Tversky, A., 56, 61
Tweney, R. D., 173

U. S. Congress, 179
von Uexküll, J., xvi, 9, 25, 252, 260–61, 264, 265, 266, 269–70
Ulbaek, I., 79
Ullian, J., 107
Umiker-Sebeok, J., 389
Urmson, J. O., 200
Uzgiris, I., 351, 353

Van Cantfort, T. E., 367, 389, 398
van Rooijen, J., 338
Varela, F. J., 45
Vaughn, W., Jr., 378
Vedeler, D., 117
Velmans, M., 338, 340, 341, 347
Verplanck, W. S., 254
Vinter, A., 352
Visalberghi, E., 353
Volterra, V., 353
von Fersen, L., 314, 343
Vygotsky, L. S., 351, 401

de Waal, F. B. M., xiv, 8, 27, 33, 34, 40, 48, 89, 135, 136, 144, 155, 164, 173, 230, 262, 263, 272–73, 325, 326, 327, 328, 329, 354, 391, 409, 410, 413, 416, 417, 422, 424, 434
Waddington, C. H., 235

Wagner, A. R., 343
Wagner, H., 282
Wainer, J., 93
Waldron, T. P., 365
Walker, S., 172
Wallace, A. F. C., 102
Wallman, J., 47, 384, 390
Wallnau, L. B., 300, 304
Walls, S. C., 343
Wan, C. K., 84
Warden, C. J., 94, 390
Warner, H., 391
Warner, L. H., 390
Washburn, M. F., 270
Wasserman, E. A., 260
Watson, J. B., 25, 268, 299
Weiskrantz, L., 338
Weiss, M., 357
Weller, B., 399
Wellman, H. M., 97, 420
Wells, M. J., 216
Werner, H., 269
Werth, F., 163
Wertz, F., 292
Wheeler, O., 25
Wheeler, W. M., 25
Whiten, A., xix, 27, 48, 62, 63, 127,
    138–50, 156, 162, 168, 172, 173,
    175, 177, 230, 234, 272, 313, 317,
    320, 324, 328, 329, 331, 349, 351,
    354, 422, 423
Whitman, C. O., 25
Whorf, B. L., 339
Wickler, W., 171
Wiener, N., 172
Wiley, E. O., 359
Williams, G. C., 228, 318, 320–21

Williams, S. L., 391, 400
Wilson, E. O., 423
Wilson, T. D., 421
Wimmer, H., 97
Wind, J., 368
Wiser, M., 64
Witmer, L., 346, 388
Wittgenstein, L., 200–201, 293, 416,
    418
Wolz, J. P., 46, 371, 372, 375, 376
Woodmansee, K. B., 326
Woodruff, G., 86, 96, 172, 173, 349,
    351
Wright, J., 350, 352
Wuensch, K. L., 240
Wynne, C. D. L., 343
Wynne, C. K., 379
Wynne-Edwards, V. C., 228

Xenophanes, 37–38, 45, 51

Yeager, C. P., 398
Yerkes, R. M., 27, 349, 387
Yoerg, S. I., 260, 262, 317, 324, 330–31,
    343
Yoshioka, J. G., 387
Young, J. Z., 216
Young, R. M., 409

Zabel, C. J., 326
Zajonc, R. B., 69
Zak, S., 178
Zihlman, A. L., 400
Zohar, A., 24
Zuckerman, S., 318–19
Zunz, M., 175
Zuriff, G. E., 339

Subject Index

abstraction, 17, 19–20
adaptation, 95, 255, 257
adaptationist, 271–72
affection, 84, 91, 240–43, 245–51, 263, 425
affirming the consequent, 336–38, 347, 417
alarm calls, 139–42, 157, 367–68
amorphism, 4–5, 40, 42–43, 49, 411
amphibian, 83, 196–97, 241, 248–49, 413
analogy, 47, 52–53, 113–15, 151, 171, 174, 183, 194, 269, 275, 325, 336, 345, 365, 381–82, 414–15, 420. *See also* mental state attribution
anecdotes, xiii, 3, 7–8, 10, 11, 14–15, 17–20, 25, 125–26, 129–33, 134–50, 151–52, 155–56, 158–67, 172, 174–75, 177, 178–79, 209, 213–14, 258, 269–73, 296–97, 299–300, 304, 309, 314–15, 319, 324, 329, 332, 354, 384, 387, 397, 401–2, 404, 407, 411, 421–27; and rarity, 134–35, 145, 150, 174–75, 272, 423
Animal Language Research, 11, 46–48, 370–82. *See also* language
animal rights/welfare, 70–71, 174, 178–79, 183–5, 238–42, 246–48, 250–53, 266, 294, 323–24, 330, 332–33, 345, 395, 426–27

animism, 5, 57, 63, 71, 79, 393
anthropic principle, 213, 218–19
anthropocentrism, 11, 350, 365, 368, 384–85, 389, 390–93, 397, 403–4, 411
anthropomorphism, xiv–xvi, 3–11, 12–21, 22–34, 37–49, 50–58, 59–73, 77–91, 92–103, 104–15, 118, 120, 125–33, 151–69, 179, 189, 193, 213, 222, 230–33, 238–40, 258, 265, 273, 275, 276, 293–94, 296–97, 299, 309, 314–16, 317, 319, 322, 324, 329, 331, 335–39, 342–47, 350, 361, 370, 384–85, 387, 390–93, 394–95, 407–8, 427; and affectional bonds, 84, 91, 242, 245, 248–51, 425; and assumptions/presuppositions, 5, 45–49, 152, 156–58, 160–63, 168–69, 296, 336–39, 410, 411, 420; categorical, 33, 410; as category mistake, 38, 49; and cognition, 59, 62–64, 72, 83–84, 242, 248, 250–51, 410; as comforting, 51, 54–55, 56, 79, 408; as coordination system between/within species, 59, 64–73; critical, 10, 27, 67, 268–69, 271, 273, 275, 329, 401; crude, 271, 409–10; and cultural stereotype, 165–66; as default, 59, 62–64, 72, 410; and definition, 51, 92–93,

anthropomorphism (*continued*)
104–6, 115, 151, 293–94, 335, 417;
development of, 96–99; and
familiarity, 51, 54–55, 56, 79, 84,
88, 128–29, 165–66, 168, 251, 423,
425; as frame of reference, 6,
118–22, 159, 411; and game-
theory, 55–56; generic, 27; with
geometric shapes, 65, 418; global,
407, 408–9, 420, 426; and
hindsight, 46, 58, 410; and
hypothesis, xv–xvi, 33, 48, 167,
345, 412, 414; inaccurate, 409–16,
426; as innate, 45, 49, 318, 321, 346,
412; and legal system, 151, 158–62;
as misattribution, xv, 52, 105, 181;
mock, 27; and narrative, 151,
155–56, 158–69; as null
hypothesis, 167, 348, 354, 361, 409;
and pattern-discovery, 52; and
perception, 56–58, 412; and
philosophy, 50–51; and
phylogenetic scale/status, 83, 167,
179, 248–51; and phylogeny, 156,
165–66, 296; physical, 38;
pragmatic, 3, 385–86, 402–4; as
projection, 21, 51, 62, 174, 181, 193,
297, 424; reflective, 401, 403; and
science, 3, 7–10, 22–23, 50–52,
60–61, 64, 72–73, 125–33; and
scientists, 252–53; and sentience,
242, 248–51; and similarity to
humans, 79, 83, 91, 155–56,
164–69; situational, 33, 410;
subjective, 408, 411, 416–21, 426;
uncritical, xvi, 409–410; and
undergraduates, 9, 164–67,
240–51, 339. *See also* mental state
attribution
ape, xiii, 4, 14, 18, 73, 85, 102–3, 137,
149, 227–28, 233, 296–97, 333, 353,
354, 357, 360, 361, 383–404, 413–14,
418, 422; bonobo (pygmy
chimpanzee), 301, 358, 360, 389,
392, 400, 402, 404; chimpanzee,
xiii–xiv, 33–34, 46, 56–57, 83,

85–89, 91, 96, 100–102, 144, 147–48,
156, 157–58, 160, 162–63, 164–65,
173, 224, 226, 230–31, 232–34, 241,
243–44, 245, 250, 296–301, 323, 325,
332, 337–38, 346, 348–49, 358, 360,
366–67, 383–84, 388–91, 393,
394–400, 403, 413, 415–16, 422,
425–26; gorilla, 20, 21, 96, 100, 160,
233–34, 301, 358, 360, 383, 389,
398–99, 401–2, 425; gibbon, 358,
360; orangutan, 23, 64, 85, 96, 173,
301, 303–5, 358, 389, 399, 401, 402
art, 53, 154, 261
association, 17, 19–20, 149
attractiveness, 241, 246
attribution: fundamental, error, 68;
theory, 78, 83. *See also* mental state
attribution
autapomorphism, 101
awareness. *See* consciousness

behavior, 190–93, 204–5
behaviorism, xv, 3, 10, 11, 20, 25, 39,
59, 66–67, 129, 148–49, 170–71,
173, 180, 201, 256, 258, 264, 268,
271, 274, 292, 313, 318, 329,
339–41, 343, 348, 396, 412, 413
bicultural, 388, 389
bird, xiv, 9, 15, 26, 80, 83, 100, 106,
107–8, 115, 134, 140, 193, 203,
207–8, 237, 241, 244, 248–49, 324,
328, 333, 341–42, 367, 390, 419
blindfold, 81, 84
bluff, 215
boredom, 106, 115

category mistake, 38, 49, 200
causality, 109–10, 111–12, 120,
255–56, 260, 353, 355, 359, 422
cladistics, 356–61
Clever Hans, 383
cognition, xv, 5, 10, 34, 48, 60–62, 73,
83–84, 94–95, 100–101, 173, 176,
182, 189, 216, 242, 245–48, 250–51,
257, 258–60, 264–65, 313–14,
317–21, 328–31, 335, 339, 342–45,

384, 412, 414; comparative, 259, 343–45; meta-, 99–101, 102–3; unconscious, 262. *See also* ethology, intentionality

color, 154, 204, 273; in coleoid cephalopod, 214, 216–17

common sense, 7, 72–73, 125–33, 238, 239–40, 250, 252, 263, 336, 421, 426–27

communication, 11, 18, 19, 118, 206–8, 214–17, 365–69, 391, 394–404. *See also* deception, language

compassion, 153

conceptual thought. *See* abstraction

conditioning, xiv, 142–49, 157, 175, 232, 349, 352, 371–82, 400, 422

confirmation bias, 173–74

conscience, 15

consciousness, xv–xvi, 9–10, 25, 34, 146, 172, 193, 195, 219, 239, 240–41, 243–46, 250, 252, 256–58, 260–71, 273–76, 294, 316, 317, 319, 322–27, 335, 338–42, 347; and introspective modeling, 78, 82, 96, 101, 340–41, 420–21; self-, 10, 14–15, 20, 48, 77, 79–80, 82, 85–86, 96, 101, 157–58, 172, 173, 219, 230, 232–34, 239, 240–41, 243, 246, 291, 296–310, 346, 349, 350, 357, 360, 362, 411, 413, 420, 421, 425–26; and language, 268, 340, 346, 418–21

construction, social, 277, 282–83, 292, 293

continuity. *See* evolution, homology

control, 179–81, 183, 255–57, 269, 423, 426

convergence, 181–82, 214

counterworld, 261, 264, 269–70

counting, 344–45, 346, 349, 350

cry, 28, 114

cultural hybrid, 385

culture, 4, 21, 22–23, 53, 165–66, 251, 386, 387–90, 391, 403, 411; American, 21; of communication with apes, 394–403; differences in,

27–29, 131, 168; Dyak, 386; Egyptian, ancient, 154, 386; English, 13–17; German, 21; Greek, ancient, 37–38, 386; Iroquois, 102; Japanese, 23, 28–29; laboratory, 252, 396; Mende, 417; North Carolina cockfighter, 240; Ojibwa, 28; similarities in, 98; Truk, 102; Western, 23, 27–29, 199–202, 286, 386–87, 417

daily interaction, 178–81, 235, 426–27

death, 28–29, 90; burial after, 153–54

deception, xiii, 18, 27, 48, 84, 86, 89, 130, 131, 135–50, 154–57, 160–61, 162–63, 164–65, 172, 175, 180–81, 207, 230, 240–41, 243, 272, 295, 304, 321, 326, 328, 329, 349, 351, 354; counter-, 147–48, 162–63

deliberation, 232–34, 340

depression, 182–83, 412

description, 118–22; re-, 43. *See also* observation report

development, 255, 257, 260, 385, 404, 424, 425; of anthropomorphism, 96–99; evolution of, 359, 362; language, 174, 177, 349, 351, 361, 385–86, 392, 396, 403–4; of mental state attribution, 60, 96–99; of mind, 10, 17, 182, 348–57, 359, 361–62

ecology, 8, 11, 70–71, 229, 255, 368

ecological validity, 174–77

eject, 17–18, 25, 62

ejectivism, 269

embodiment, 279–81, 294

emotion, 26, 29, 66–67, 84, 98, 102, 105–6, 165–67, 170, 181, 238, 239, 240–41, 243, 244, 246, 250, 251, 257, 259, 260, 276, 340, 367

empathy, 9, 26, 84, 86, 167, 177–78, 238, 265, 277–82, 285–86, 289–92, 294, 351, 400, 421

empiricism, 199, 202

enculturation, 387–90, 395, 400–401

environmentalism, 70–71
epigenesis, 10, 350–53, 356, 359, 361–62
epiphenomenalism, 111, 115, 347
equivalence relations, 11, 378–82
ethics. *See* animal rights/welfare
ethology, 252–53, 255–57, 264–65, 271, 275–76, 349, 427; cognitive, 26, 46, 158–63, 198, 205, 240, 259, 260, 269, 313–32, 349–50, 359; history of, 24–27, 198–99, 205–9
ethos, 25–26
evidence, 106–12, 115, 155, 174–81, 419, 424
evolution, xiv–xv, 16, 129, 181, 218, 228, 251, 255, 257, 260, 285, 334, 348–50, 353, 359, 362, 422; of language, 19, 368–69, 371, 382; of mind, xiv–xv, 4, 10, 12–20, 25, 39, 84, 93–102, 136, 168, 262, 263, 313–14, 356–61, 387; of social intelligence, 62–64, 65, 78
experience, 43, 45, 257, 265–69, 273–76, 278–80, 294. *See also* consciousness
experiment, 8, 176–77, 179, 272, 424
explanation, 111–12
extraterrestrials, 91, 402, 416

fantasy, 48
fascism, 323
feeling. *See* emotion, perception
fish, 17, 18, 83, 91, 167, 197, 232, 241, 248–49
focal animal approach, 220–36
foundationalism, 199, 201
friendship, 27, 223–24, 234, 251, 263, 351, 416–17
function, 175, 190–96, 214, 255, 257, 260, 269, 285, 336
functionalism, 238, 320, 412

game theory, 55–56, 215, 222
Gegenwelt, 261
gender, 220–21, 247, 248, 250, 251, 332–33

God, 52, 56, 89–90, 102, 226, 336
gods, 37–38, 51, 89–90
grief, 283, 416–17

helping, 132
heterochrony, 362
history, 4, 23, 199–202, 277, 281, 283, 292, 293, 295, 424, 425
holism, 202–4
hominoid, 85, 96
homology, 52, 171, 172, 387, 391, 413–14, 425
homosexuality, 152
human, 3, 15, 83, 85, 96, 121–22, 153–54, 163, 164–65, 218, 227–28, 232, 272, 357, 360, 361, 383, 385–86, 391–92, 412–13, 414, 422, 427; -animal interaction, 272, 386, 393, 394–402; autistic, 102–3; child, 16, 19, 64–65, 67, 79, 82, 96–99, 165–67, 177, 262, 272, 336, 348, 352, 354, 362, 378, 384–86, 393, 395–96, 398–99, 404, 400–401, 404, 414; evolution, 56–57; hominid, 153, 220, 402; *Homo erectus*, 90; *Homo habilis*, 90; hunter-gatherer, 362; infant, 18, 81–82, 116–22, 272, 280, 305, 352–5, 385, 418; mother-infant interaction, 116–22; Neanderthal, 90, 153
hypothesis, xv–xvi, 33, 48, 107, 115, 167, 345, 412, 414; null, 167, 173, 348, 354, 355, 361, 409; testing, 172–75, 180, 181–83, 222, 268, 354

identification, 221–22, 223
imaginary entities, 78, 89–91
imitation, 19, 65, 117, 278, 280, 290, 349, 352–53, 355, 357, 359, 360, 362, 392, 396, 419
infanticide, 134
instinct, 12, 13, 19, 208, 255, 256
intelligence, 62–63, 84, 94–95, 172, 176–77, 232–36, 240–41, 243, 246, 355, 360, 361–62

intentionality, xiv, 6–7, 19, 65–66, 69,
89, 98, 107, 116–17, 136, 146–49,
154, 172, 175, 189–97, 204–9, 215,
229–34, 240, 257, 259, 279–80,
288–92, 319, 321–22, 326, 329, 393,
416
International Ethological Congress,
253, 261
intersubjectivity, 116–19, 122
introspection, 262, 273, 275, 418
intuition, 170–71, 172–73, 179, 181,
239, 265, 266, 267, 421, 425
invertebrate, 241, 250, 251; ant, xv,
241, 243, 244, 248–49, 254, 336;
arthropod, 219; caterpillar, 17;
cockroach, 79, 83, 167;
Coelenterata, 18; coleoid
cephalopod, 8, 213–18, 239;
Crustacea, 17, 18; cuttlefish, 213;
earwig, 17; Echinodermata, 18;
firefly, 160–61; fly, 196–97, 336;
honeybee, 194–96, 203, 258;
Hymenoptera, 18; insect, 9, 69, 91,
270–71, 342; limpet, 17; mollusk,
18, 244; octopus, 213, 216; oyster,
239; protozoa, 336; shrimp, 239;
slug, 346; spider, 17, 65, 241, 243,
248–49, 336; squid, 213, 214–15,
217–18; wasp, 336; worm, 241,
244, 248–49, 270

joking, 399, 400
jury, 130, 158–63, 168, 262

language, 11, 22–23, 57, 99, 118, 339,
340, 365–67, 368–69, 371, 382, 384;
as abbreviation, 121–22;
acquisition, 174, 177, 349, 351, 361,
385–86, 392, 396, 403–4; animal,
11, 19, 46–48, 57, 367–69, 370–82,
390, 419; ape, 11, 236, 354, 366–67,
383–404, 414, 418, 425; and
consciousness, 268, 340, 346,
418–21; and entitling, 121–22;
everyday/ordinary, 39–45, 202;
evolution of, 19, 368–69, 371, 382;

and experience, 43, 45; and games,
200; indicative/referential, 118,
120; and mind, 202–3, 209, 418–21;
and ostensive learning/definition,
109, 117, 120, 417; sign-, 11, 354,
366–67, 388–89, 392, 398–99, 401,
425; sociocultural context of,
394–402; use, 200, 418–19
learning. See conditioning

machines, xv, 5, 38–39, 44, 79, 93–94,
365, 387, 416
Maier's Law, 303, 310
mammal, 79, 244, 250, 251, 424;
anteater, xvi; bat, 241, 244, 245,
248–49, 251, 258, 338, 414; bear, 9,
155–56, 164–65, 259, 422; canid,
285, 333, 422; cat, 9, 18, 84, 91,
105–6, 191–92, 197, 203–5, 237,
241, 243, 245, 248–49, 251, 324,
401; cow, 15, 324, 415; coyote, 422;
dog, xiv, xvi, 4, 9–10, 14–16, 18, 19,
21, 60, 65–67, 77, 78, 79, 83, 84,
105, 111–12, 114, 131, 144, 145,
154–55, 157, 164–67, 184, 237, 241,
243, 244–45, 248–49, 250, 251, 272,
277, 278, 282, 283–92, 324, 333–4,
354, 390, 422; dolphin, 9, 11, 65,
241, 243, 245, 248–49, 333, 370–71,
375, 376, 382, 414, 424; elephant, 9,
83, 114, 132, 153, 164–65, 220;
guinea pig, 53; horse, 114, 237,
384; hyena, 326; lion, 134; mouse,
53, 191–92, 203–5, 241, 243, 244,
245, 248–49, 251, 252; otter, 9,
164–65; pig, 9, 53, 83, 241, 244–45,
248–49, 251, 324; pinniped, 370;
porpoise, 83, 370; prairie dog, 367;
rat, xiv, 53, 241, 245, 248–49, 251,
339, 422; rodent, 70, 333; sea lion,
11, 372–82; sheep, 232, 324;
squirrel, 288, 289; ursid, 422; wolf,
xiv, 14, 232, 285, 422. See also
primate
manipulation, 206–8, 230, 232–33, 368
materialism, 111–12, 115, 412

meaning, 199–200, 202, 279, 293, 295,
    367–68, 379; literal, 31–32, 39
mechanism, 257, 260
mechanomorphism, 5, 43, 44
me-morphism, 130
memory, 299
mentalism, 3, 268, 274, 320, 343
mental state. *See* mind
mental state attribution, 5–6, 17–18,
    25–26, 41–42, 45, 59–63, 65–73,
    77–91, 92–103, 104–15, 116–22,
    126–33, 165–67, 170–85, 193,
    202–3, 208, 239–53, 254–76, 304,
    320–22, 325, 349, 350–51, 353, 409,
    416–21, 426
metaphor, 4, 7, 22–24, 29–34, 38–45,
    64, 205–7, 225, 226, 232, 409, 415–16
mimicry, 215–18
mind, 59–60, 256; in animals, 25–27,
    29, 32–34, 44–45, 48, 104–15, 126,
    128–33, 176–77, 184–85, 198,
    206–9, 237–53, 262, 318, 319,
    328–29, 336, 350, 355;
    development of, 10, 17, 182,
    348–57, 359, 361–62; evolution of,
    xiv–xv, 4, 10, 12–20, 25, 39, 84,
    93–102, 136, 168, 262, 263, 313–14,
    356–61, 387; language of, 202–3,
    209, 418–21; and privacy, 26–27,
    98, 108–9, 115, 320–21; theory of,
    68, 78, 83, 100, 103, 304, 349,
    350–51, 353. *See also* mental state
    attribution
mindless conversation, 81
model: animal, of humans, 127–29,
    181–83, 227–28, 271; introspective,
    78, 82, 96, 101, 147
monogamy, 134
morality, 70–73, 127–28, 351; in
    animals, 18, 34, 147–48
Morgan's Canon, 20, 326, 346
motherese, 81–82
motivation. *See* intentionality

naming, 332–33
narrative. *See* story

negation, 196
neuroscience, 256–58

object permanence, 343, 353, 355, 357,
    359, 360
objectification, 291
objectivism, 25
objectivity, 11, 39–46, 263–64, 297,
    309, 411
observation, 120, 126–27, 132, 154–55,
    175–76, 220–36, 272; reports,
    106–12, 115, 155
ontogeny. *See* development
operationism, 199, 201, 202, 209
ordinary person, 160–63, 424
ostensive: definition, 117, 120, 417;
    learning, 107, 109, 110

pain, 18, 48, 108, 109, 110–11, 114,
    115, 127–30, 178–79, 184, 239–47,
    250, 252, 263, 264, 272–73, 295,
    324, 412
*Pan*morphism, 85, 87, 92, 100, 102
parsimony, xiv, 147, 172–74, 179–80,
    184, 315–16, 325–26, 331, 370, 376,
    378, 413, 415, 422
Pascal's Wager, 56
perception, 5, 17, 18, 56–58, 69–70, 97,
    109–11, 278–79, 295, 326, 342, 412,
    419–21
personal identity, 219
personality, 175, 393, 400
personal space, 285
personhood, 291
personification, 53
perspective, 286, 293, 411, 418;
    another's, 8, 116–22, 162, 221–32,
    253, 260–62, 275, 279, 350, 418,
    424–25; observer, 225–32
pet, 70, 83, 88, 181, 183–84, 237, 244,
    250, 282, 284, 286, 336, 387–88,
    393, 424
phenomenology, 9–10, 261, 265,
    277–95, 421
phylogenetic scale/status, 83–84,
    167, 179, 248–51

phylogeny, xiv–xv, 65, 94, 99, 156,
165–66, 255–56, 271, 296, 356–61,
387, 422
plan. *See* intention
plausibility, 7–8, 127, 130–32, 143,
149, 159–62, 424
play, 64–65, 154–55, 269, 272, 288,
291, 321–22, 333, 349, 353
pleasure, 18, 111–12, 147
point of view. *See* perspective
*Pongo*morphism, 100
positivism, 198, 199–200, 292–94
pragmatism, 170, 177–81
prediction, 179–81, 183, 206–7, 273,
423
pretense, 18, 155–56, 165, 226, 304,
351, 353, 359, 381–82, 392, 413
primate, 8, 14–15, 27, 33, 46, 70, 84,
95, 96, 135–49, 156, 227, 220–36,
272–73, 300–301, 327, 336, 350,
355, 357–61, 386, 424; baboon, 14,
16, 139–42, 143–45, 157, 220–25,
231, 234; cebus, 357–58, 360, 422;
gelada, 230; langur, 134; lemur,
232; mangabey, 144; monkey,
14–16, 21, 78, 83, 137, 149, 227–28,
233–34, 333, 353, 354, 357, 384,
386, 391; prosimian, 232;
macaque, 88, 166–67, 179, 180–81,
182, 224, 357–58, 360, 361;
strepsirhine, 137; vervet monkey,
175–76, 203, 230–31, 367; tarsier,
358. *See also* ape, hominoid,
human
primatology, 220–36, 424
privacy, 26–27, 98, 108–9, 115, 257,
265–69, 273–75, 320–21, 324, 421
private experience, 257, 265–69,
273–76, 320–21, 324, 421. *See also*
consciousness
protective objectification, 177–78
psychological diversity, 94–99, 413
psychology: and behavior, 167,
418–21; comparative, 12–21,
22–23, 94–95, 176, 254, 256,
343–44, 427; comparative

developmental evolutionary,
349–61; ejective, 62; folk, 59–61,
63, 66–73, 111, 202–3, 205, 209,
237, 268, 316, 324, 329, 339–40,
342, 409, 412, 418–21; functional,
95; history of comparative, 24–27
purpose, 6–7, 189–97

questions, 43, 47–49, 392

rationality. *See* reason
realism, 109–10
reason, 13, 17, 29, 54, 56, 61–62,
240–41, 243, 244, 246, 250, 262,
340, 414
recapitulationism, 10, 17, 348, 350,
351, 356–61, 362
recept, 19–20
reflex, 59, 208, 412
religion, 15, 53, 54, 57, 153, 251, 408
representation, 154, 194–97, 209, 257,
352
reptile, 83, 232, 241, 248–49
robot, 93–94

sample size, 333–34, 397, 425
sampling methods, 221–24
scanning, 42–43, 270–71
scheme, 352–53
science, 3, 7–10, 22–23, 50–52, 60–61,
64, 72–73, 152–53, 158, 168–69,
266–67, 292, 421
searching, 42–43
selection, 12, 56, 61–64, 91, 206,
228–29, 232, 313, 409
self, 286–87
self-awareness. *See* consciousness
self-concept, 300
self-recognition. *See* consciousness
self/other, 219, 279, 300
sentience, 245–47, 252–53
simulation, 393
skepticism, 126, 128, 314–28, 421
social roles in ape-human
interaction, 386, 394–402
sociobiology, 205–8, 255, 323, 416

soul, 29, 34

space, 285–87, 343, 353, 355, 359

story, 7–9, 132, 151, 155–56, 158–69, 214, 261, 271–72, 384, 399, 401–2, 404, 421, 423–25

subjective analogical inference. *See* mental state attribution

subjectivity, 26, 38–46, 116–20, 122, 162, 178, 182, 239, 251–52, 256, 257–60, 265, 276, 295, 329, 330, 404, 408, 411, 416–21, 426

symbolism, 57, 156, 172, 390, 392–93. *See also* language

sympathy, 18, 48, 84, 86, 235, 248, 265, 267

talk, 69–70

teleology, 51

text, 292–93

theism, 89–90, 386

theory, 177–78, 181–85, 202; of mind, 68, 78, 83, 100, 103, 304, 349, 350–51, 353. *See also* mental state attribution

thought, 193, 195, 335, 338–42, 347

time, 343, 353, 355

tool, 57, 232, 357, 360, 368, 396, 400, 402

training, 269, 331, 411; dog, 66–67; sea lion, 371–81

transaction, 231, 234–36

translocation, 224–27, 231–32

translation, 42–44

Umwelt, xvi, 261, 283

validity, 174–77, 292

value, 69–73, 121–22, 126, 170, 172–73, 417; survival, 190, 192, 255, 257

Venus fly traps, 19

verificationism, 126–27, 199–200, 202, 236

well-being. *See* pain

working definitions, 48–49

yawning, 67

zoomorphism, 5, 57, 84–89, 420

68595679R00299

Made in the USA
Lexington, KY
15 October 2017